Matrices with Applications in Statistics

The Wadsworth Statistics/Probability Series

Series Editors

Bickel, P.; Doksum, K.; and Hodges, J. L., Jr. *Festschrift for Erich L. Lehmann.*

Graybill, F. *Matrices with Applications in Statistics,* Second Edition.

Matrices with Applications in Statistics

Second Edition

Franklin A. Graybill
Colorado State University,
Fort Collins

Wadsworth International Group
Belmont, California
A Division of Wadsworth, Inc.

Statistics Editor: John Kimmel

Production Editor: Diane Sipes

Copy Editor: Janet Greenblatt

Printed in the United States of America
1 2 3 4 5 6 7 8 9 10—87 86 85 84 83

Library of Congress Cataloging in Publication Data

Graybill, Franklin A.
Matrices with applications in statistics.

(Wadsworth international statistics/probability series)
Rev. ed. of: Introduction to matrices with applications in statistics. 1969.
Includes bibliographical references and index.
1. Matrices. 2. Linear models (Statistics).
I. Title. II. Series.
QA188.G715 1983 512.9'434 82-8485
ISBN 0-534-98038-4 AACR2

Preface

In 1951 Professor Oscar Kempthorne introduced me to the theory and application of the linear statistical model. This subject is one of the most useful and most used in all of applied statistics. My interest in the area has remained constant over the years and, in fact, I have been involved in teaching and doing research in the subject since 1952.

The only mathematics required to enable a student to study the theory of the linear statistical model is calculus and matrix or linear algebra. A number of topics in matrix algebra that are useful in a study of the theory of the linear statistical model are generally not available in a first or second course in matrix or linear algebra. In fact, a number of these topics are not yet available in textbooks. Over the years, as I have taught courses in the theory of the linear statistical model, I have collected a number of results that have proved to be helpful and interesting. These results are the reason for this book and they constitute the main content. I have tried to organize these results so that the proofs of theorems rely on the results of theorems on previous pages. This has been fairly difficult, since many of the results are somewhat unrelated. A large number of them appear as problems.

The prerequisite for reading this book is one undergraduate course in matrix or linear algebra. I cover the material in the book in about one quarter of a year's course in the theory of the linear statistical model, and I have had students who have worked through the material who have had no previous formal course in matrices. The book should prove useful for anyone who takes courses in regression and correlation, analysis of variance, least squares, linear statistical models, multivariate analysis, or econometrics; and it could serve as a resource book for many other subjects.

Originally, Dr. Max Stein, Professor of Mathematics at Colorado State University, and I planned to write a book together. The first part was to be material in a traditional first course in matrices and determinants for college sophomores. The second part was to be material for students in statistics. The resulting manuscript was too large and unwieldy, since it covered too much material. Hence, we decided to publish two books. The part written by Dr. Stein was published in 1967 by Wadsworth Publishing Company, Inc., under the title *Introduction to Matrices and Determinants.* The part of the joint manuscript that I had written is the material in this book.

The book has 12 chapters in addition to an introduction. It contains more than 450 problems and more than 80 worked examples. The first three chapters review material that would normally be covered in a first course in matrix algebra. In fact, much of the material in the first three chapters is taken directly from Dr. Stein's book. I take this opportunity to thank him for giving me permission to do this.

The book is written at an elementary level with the hope that students with a minimum of mathematics can read it if they need to become acquainted with some of the material that it contains.

A number of people gave me a considerable amount of help in writing this book. I am especially indebted to the following graduate students at Colorado State University: Robert Crovelli, Patrick Eicker, Al Kingman, Carl Meyer, and Sing-chou Wu, who read the manuscript as I wrote it and worked the problems. Leo A. Katz of Michigan State University and Ingram Olkin of Stanford University offered many good suggestions when they reviewed the manuscript. I also wish to thank Mrs. Kathy Deason and Mrs. Lilly Steinhorst who typed the manuscript (some of it many times as it changed) and did many other jobs that are necessary for a poorly handwritten manuscript to become a finished product.

Preface to Second Edition

In the second edition most of the original material has been preserved, but several new topics have been added. The most important additions are (1) material on least squares with constraints, (2) vector of a matrix, (3) permutation of a direct product of matrices, (4) circulants, (5) dominant diagonal matrices, (6) Vandermonde matrices, (7) permutation matrices, (8) Hadamard matrices, (9) Toeplitz matrices, (10) M and Z matrices, (11) inverse positive matrices. Chapter 11 of the first edition, Computing Techniques, has been replaced with M matrices, Z matrices, and inverse positive matrices.

I thank various faculty in the Department of Statistics and the Department of Mathematics at Colorado State University for help during the preparation of this text. I also want to thank Ms. Waydene Casey, who helped me in several ways in preparing the manuscript, and Roger Horn, Andre Khuri, Carl Meyer, and Juliet Shaffer for the helpful suggestions received in preparing this revision.

Contents

Contents

This book is affectionately dedicated to the following people:

Mrs. Hope Carr and Mrs. Laura Good for helping me in many ways when I was young;

Dan and Kathy for the many years of enjoyment they have given me and continue to give me;

Tonia, Jules, and Emily, who are delights and the hope of the future;

Finally to my wife, Jeanne, who has been my constant inspiration for so many years.

Introduction

Because matrices are used so extensively in the theory and application of statistics, it is impossible in this one book to discuss all of the subjects in statistics in which they play a significant role. Therefore, this book is primarily concerned with the areas of *multivariate analysis* and *the linear model.* The theory of the linear model is actually a part of multivariate analysis, but they are often discussed separately. These two topics include important areas in statistics, such as design of experiments; analysis of variance; correlation; regression; least squares; components of variance—areas that comprise a large segment of the theory and application of statistics. Since we make no systematic development of these topics, we shall discuss each briefly here and refer to this discussion throughout the book as we point out statistical examples where matrices are used. It is assumed that the reader interested in the statistical applications is acquainted with the basic theory of statistics involved.

Multivariate Analysis. In the theory of multivariate analysis, one considers the joint distribution of n random variables $y_1, y_2, \ldots, y_n$, which we generally write as the elements of a column vector $\mathbf{y}$, where

$$\mathbf{y} = \begin{bmatrix} y_1 \\ y_2 \\ \vdots \\ y_n \end{bmatrix};$$

we call this a random $n \times 1$ vector. We are generally interested in the (arithmetic) mean, called the expected value, of each component, in the variance of each component, and the covariance of each pair of elements of $\mathbf{y}$.

The mean of $\mathbf{y}$, which we denote by $\boldsymbol{\mu}$, is an $n \times 1$ vector whose i-th component is the mean of y_i; that is,

$$\boldsymbol{\mu} = \begin{bmatrix} \mu_1 \\ \vdots \\ \mu_n \end{bmatrix} = \begin{bmatrix} \mathscr{E}(y_1) \\ \vdots \\ \mathscr{E}(y_n) \end{bmatrix} = \mathscr{E}(\mathbf{y}),$$

where $\mathscr{E}(y_i)$ stands for the expected value of the random variable y_i and $\mathscr{E}(\mathbf{y})$ is defined in terms of these elements.

To systematize the variances and covariances of the elements of $\mathbf{y}$, we define an $n \times n$ matrix $\mathbf{V}$, called the covariance of $\mathbf{y}$. The ij-th element v_{ij} of $\mathbf{V}$, when $i \neq j$, is the covariance of y_i and y_j; the i-th diagonal element v_{ii} of $\mathbf{V}$ is the variance of y_i. Thus the matrix $\mathbf{V}$ is a symmetric matrix and is non-negative. Often nothing more is known about $\mathbf{V}$; on the other hand, it is sometimes known that $\mathbf{V}$ has a certain form or pattern (see Chapter 8). The theory of multivariate analysis often centers around an analysis of a covariance matrix **V**. When this is the case, it may be necessary to find the determinant of **V**, the characteristic roots of **V**, the inverse of **V** if it exists, and perhaps to determine these and other quantities for certain submatrices of **V**. Among other things, it is often necessary to find marginal and conditional distributions of a subset of **y**; sometimes it is required to find the moments of **y** or the moment generating (or characteristic) function of **y**. Also it may be necessary to transform from the vector **y** to new vector **x**, and this transformation may require the evaluation of a Jacobian; it may be necessary to find maximum likelihood or least squares estimators of parameters in the distribution of **y**. Many of these problems can be solved by manipulating vectors and matrices, as we shall point out from time to time (see Chapter 10).

Most of the theory and applications of multivariate analysis involve the normal (or Gaussian) distribution. When this is the distribution under study, the theory of matrices and vectors is particularly helpful. Discussions of multivariate analysis can be found in several good texts.

Linear Model. As stated above, the theory of the linear model (sometimes referred to as the general linear hypothesis) can be considered as a part of *multivariate analysis*. However, it is often considered as a separate subject.

The model can be written as

$$\mathbf{y} = \mathbf{X}\boldsymbol{\beta} + \mathbf{e},$$

where **y** is an $n \times 1$ random vector of observations, **X** is an $n \times p$ known matrix of constants, $\boldsymbol{\beta}$ is a $p \times 1$ vector of unknown parameters, and **e** is a vector of unknown errors. The e_i are generally assumed to have a mean of zero and to have variance σ^2 (unknown), and each pair $e_i, e_j, i \neq j$, is assumed to be uncorrelated. For a discussion of the details of how this model is derived, see [G–7].

We can write this model as

$$\mathbf{y} = \boldsymbol{\mu} + \mathbf{e},$$

where of course $\boldsymbol{\mu} = \mathbf{X}\boldsymbol{\beta} = \mathscr{E}(\mathbf{y})$ and one of the objectives is to estimate $\boldsymbol{\beta}$ and σ^2. The method of estimation is usually *least squares* or *maximum likelihood*. If we denote the estimators by $\hat{\boldsymbol{\beta}}$ and $\hat{\sigma}^2$, respectively, then $\hat{\boldsymbol{\mu}} = \mathbf{X}\hat{\boldsymbol{\beta}}$ is a formula for predicting the mean of **y** for various values of the matrix **X**. Clearly, the system of equations

$$\mathbf{y} = \mathbf{X}\boldsymbol{\beta}$$

will, in general, not have a solution $\boldsymbol{\beta}$ for an observed vector **y** and matrix **X**. If no solution exists, it may be desirable to find some kind of approximate (say least squares) solution. This is discussed in Chapters 6 and 7.

Often one wants to test certain hypotheses about the parameters β_i. This is generally done by the technique called analysis of variance. The procedure is to partition $\mathbf{y}'\mathbf{y}$ into a set of quadratic forms such that the following equation obtains:

$$\mathbf{y}'\mathbf{y} = \mathbf{y}'\mathbf{A}_1\mathbf{y} + \mathbf{y}'\mathbf{A}_2\mathbf{y} + \cdots + \mathbf{y}'\mathbf{A}_k\mathbf{y}.$$

The procedures available to test certain hypotheses require that each quadratic form $\mathbf{y}'\mathbf{A}_i\mathbf{y}$ be distributed as a noncentral chi-square variable and that the set of quadratic forms be pairwise independent. The matrices $\mathbf{A}_i$ depend on the elements of $\mathbf{X}$, and sometimes these matrices have a very special structure. The important theorems in determining whether or not $\mathbf{y}'\mathbf{A}_i\mathbf{y}$ are pairwise independent and are distributed as a chi-square random variable are discussed in [G–7]. In general, it is required that the $\mathbf{A}_i$ be idempotent and that $\mathbf{A}_i\mathbf{A}_j = \mathbf{0}$ for all $i \neq j$. These ideas are discussed in Chapters 7, 9, and 12.

Prerequisite Matrix Theory

1

1.1 Introduction

Since this book assumes that the reader has had a course that includes a number of theorems on matrices and vectors, in the first three chapters we shall state without proof some theorems that are generally proved in a first course. These are stated here for the sake of completeness and so that we can refer to them in later chapters. They do not necessarily appear in the same order that would be normal in a text on matrix algebra.

1.2 Notation and Definitions

In this book, matrices are denoted by boldface uppercase letters—for example, **A**, **B**, **U**, **X**, **Z**. We define a matrix to be a *rectangular array* of elements, called scalars, from a field F. Rather than making specific reference to the field F, we shall assume

Most of the results of the first three chapters of this book are taken (some directly) from *An Introduction to Matrices and Determinants* by F. Max Stein, Wadsworth, 1967. Most of the material in these chapters can be found in any undergraduate text in matrix algebra.

it is the field of real numbers unless explicitly stated otherwise. Thus, a scalar will always be a real number, unless otherwise stated. The set of real numbers is denoted by R.

The matrix $\mathbf{A}$ has elements denoted by a_{ij}, where j refers to the column and i to the row. We sometimes write

$$\mathbf{A} = [a_{ij}].$$

If $\mathbf{A}$ denotes a matrix, then $\mathbf{A}'$ will denote the transpose of $\mathbf{A}$, and if $\mathbf{A}$ has an inverse, it will be denoted by $\mathbf{A}^{-1}$. The determinant of $\mathbf{A}$ will be denoted by either $|\mathbf{A}|$ or det ($\mathbf{A}$). An identity matrix will be denoted by $\mathbf{I}$ (to designate the size of the identity, we shall use $\mathbf{I}_n$ to represent the $n \times n$ identity matrix), and $\mathbf{0}$ will denote a null matrix. The size (or order) of a matrix is the number of its rows by the number of its columns. For example, a matrix $\mathbf{A}$ of size $n \times m$, or an $n \times m$ matrix $\mathbf{A}$, will be a matrix $\mathbf{A}$ with n rows and m columns. If $m = 1$, the matrix will sometimes be called an $n \times 1$ (column) vector. The rank of the matrix $\mathbf{A}$ will sometimes be denoted by $\rho(\mathbf{A})$.

Given the matrices $\mathbf{A} = [a_{ij}]$ and $\mathbf{B} = [b_{ij}]$, the product $\mathbf{AB} = \mathbf{C} = [c_{ij}]$ is defined as the matrix $\mathbf{C}$ with pq-th element equal to

$$\sum_{s=1}^{n} a_{ps} b_{sq}.$$

For $\mathbf{AB}$ to be defined, the number of columns in $\mathbf{A}$ must equal the number of rows in $\mathbf{B}$. For $\mathbf{A} + \mathbf{B}$ to be defined, $\mathbf{A}$ and $\mathbf{B}$ must have the same size; $\mathbf{A} + \mathbf{B} = \mathbf{C}$ gives $c_{ij} = a_{ij} + b_{ij}$. If k is a scalar and $\mathbf{A}$ is a matrix, then $k\mathbf{A}$ (and $\mathbf{A}k$) is defined to be a matrix $\mathbf{B}$ such that each element of $\mathbf{B}$ is the corresponding element of $\mathbf{A}$ multiplied by k.

A diagonal matrix $\mathbf{D}$ is defined as a square matrix whose off-diagonal elements are zero; that is, if $\mathbf{D} = [d_{ij}]$, then $d_{ij} = 0$ if $i \neq j$.

1.3 Inverse

Let $\mathbf{A}$ be a square matrix. If there exists a matrix $\mathbf{B}$ such that $\mathbf{AB} = \mathbf{I}$, then $\mathbf{B}$ is called the inverse of $\mathbf{A}$, denoted by $\mathbf{A}^{-1}$. Also if $\mathbf{AB} = \mathbf{I}$, then it can be shown that $\mathbf{BA} = \mathbf{I}$. When there exists a matrix $\mathbf{B}$ such that $\mathbf{AB} = \mathbf{BA} = \mathbf{I}$, the matrix $\mathbf{A}$ is said to be nonsingular; in the contrary case, $\mathbf{A}$ is said to be singular.

Theorem 1.3.1

If a matrix has an inverse, the inverse is unique.

Theorem 1.3.2

If $\mathbf{A}$ *has an inverse, then* $\mathbf{A}^{-1}$ *has an inverse and* $(\mathbf{A}^{-1})^{-1} = \mathbf{A}$.

Theorem 1.3.3

If $\mathbf{A}$ *and* $\mathbf{B}$ *are nonsingular matrices, then* $\mathbf{AB}$ *has an inverse and* $(\mathbf{AB})^{-1} = \mathbf{B}^{-1}\mathbf{A}^{-1}$. *This can be extended to any finite number of matrices.*

Theorem 1.3.4

If $\mathbf{A}$ *is a nonsingular matrix and* k *is a nonzero scalar, then*

$$(k\mathbf{A})^{-1} = (\mathbf{A}k)^{-1} = \frac{1}{k}\mathbf{A}^{-1}.$$

1.4 Transpose of a Matrix

If the rows and columns of a matrix $\mathbf{A}$ are interchanged, the resulting matrix is called the *transpose* of $\mathbf{A}$ and denoted by $\mathbf{A}'$. If $\mathbf{A}$ has size $m \times n$, then $\mathbf{A}'$ has size $n \times m$.

Theorem 1.4.1

If $\mathbf{A}$ *and* $\mathbf{B}$ *are* $m \times n$ *matrices, and* a *and* b *are scalars, then*

$$(a\mathbf{A})' = (\mathbf{A}a)' = \mathbf{A}'a = a\mathbf{A}'$$

and

$$(a\mathbf{A} + b\mathbf{B})' = a\mathbf{A}' + b\mathbf{B}'.$$

Theorem 1.4.2

If $\mathbf{A}$ *is any matrix, then*

$$(\mathbf{A}')' = \mathbf{A}.$$

Theorem 1.4.3

Let $\mathbf{A}$ *and* $\mathbf{B}$ *be* $m \times n$ *matrices; then*

$$\mathbf{A}' = \mathbf{B}'$$

if and only if

$$\mathbf{A} = \mathbf{B}.$$

Theorem 1.4.4

Let $\mathbf{A}$ *and* $\mathbf{B}$ *be any matrices such that* $\mathbf{AB}$ *is defined; then*

$$(\mathbf{AB})' = \mathbf{B}'\mathbf{A}'.$$

This can be extended to any finite number of matrices.

Theorem 1.4.5

If $\mathbf{D}$ *is a diagonal matrix, then* $\mathbf{D} = \mathbf{D}'$.

If $\mathbf{A} = \mathbf{A}'$, then $\mathbf{A}$ is called a *symmetric* matrix, and if $\mathbf{A} = -\mathbf{A}'$, then $\mathbf{A}$ is called a *skew-symmetric* matrix.

Theorem 1.4.6

If $\mathbf{A}$ *is any matrix, then* $\mathbf{A}'\mathbf{A}$ *and* $\mathbf{A}\mathbf{A}'$ *are symmetric.*

Theorem 1.4.7

If $\mathbf{A}$ *is a nonsingular matrix, then* $\mathbf{A}'$ *and* $\mathbf{A}^{-1}$ *are nonsingular and* $(\mathbf{A}')^{-1} = (\mathbf{A}^{-1})'$.

1.5 Determinants

Assuming that you are acquainted with the definition of the determinant of a square matrix $\mathbf{A}$, we now state some theorems that will be useful in later developments of the topic. The matrices **A**, **B**, **C** discussed in this section are assumed to have size $n \times n$.

Theorem 1.5.1

All rows of a matrix may be interchanged with the corresponding columns of the matrix without changing the (value of the) determinant of the matrix; that is, $|\mathbf{A}| = |\mathbf{A}'|$.

Theorem 1.5.2

Any theorem about det (**A**) *that is true for rows (columns) of a matrix* **A** *is also true for columns (rows).*

Theorem 1.5.3

If two rows (columns) of a matrix are interchanged, the determinant of the matrix changes sign.

Theorem 1.5.4

If each element of the i-th row of an $n \times n$ *matrix* **A** *contains a given factor k, then we may write* $|\mathbf{A}| = k|\mathbf{B}|$, *where the rows of* **B** *are the same as the rows of* **A** *except that the number k has been factored from each element of the i-th row of* **A**.

Theorem 1.5.5

If each element of a row of the matrix **A** *is zero, then* $|\mathbf{A}| = 0$.

Theorem 1.5.6

If two rows of a matrix **A** *are identical, then* $|\mathbf{A}| = 0$.

Theorem 1.5.7

The determinant of a matrix is not changed if the elements of the i-th row are multiplied by a scalar k and the results are added to the corresponding elements of the h-th row, $h \neq i$.

Theorem 1.5.8

If **A** *and* **B** *are* $n \times n$ *matrices, then*

$$\det(\mathbf{AB}) = [\det(\mathbf{A})][\det(\mathbf{B})].$$

This result can be extended to any finite number of matrices.

Let **A** be any $m \times n$ matrix. From this matrix, if one deletes any set of $r < m$ rows and any set of $s < n$ columns, the matrix of the remaining elements is a submatrix of **A**. If the $i_1, i_2, \ldots, i_h$ rows and the $i_1, i_2, \ldots, i_h$ columns of a matrix **A** are deleted, and if **A** is an $n \times n$ matrix with $n > h$, the matrix of the remaining elements is called a *principal matrix* of **A** and the determinant of this matrix is called a *principal minor* of **A**. If **A** is an $n \times n$ matrix and if the last $n - r$ rows and $n - r$ columns of **A** are deleted, the resulting matrix is called the *leading principal matrix of order r* and the determinant of this matrix is called the *leading principal minor of order r of A*.

If **A** is an $n \times n$ matrix and if the i-th row and j-th column are deleted, the determinant of the remaining matrix, denoted by m_{ij}, is called the *minor* of a_{ij}. We call A_{ij} the cofactor of the element a_{ij}, where

$$A_{ij} = (-1)^{i+j} m_{ij}.$$

Theorem 1.5.9

Let A_{ij} *be the cofactor of* a_{ij}*; then*

$$\det(\mathbf{A}) = \sum_{j=1}^{n} a_{ij} A_{ij}$$

for any i.

Theorem 1.5.10

Let A_{ij} *be the cofactor of* a_{ij}*; then*

$$\det(\mathbf{A}) = \sum_{i=1}^{n} a_{ij} A_{ij}$$

for any j.

Theorem 1.5.11

Let A_{ij} *be the cofactor of* a_{ij}*; then*

$$\sum_{j=1}^{n} a_{ij} A_{kj} = 0$$

for any $k \neq i$.

Theorem 1.5.12

If $\mathbf{A} = [a_{ij}]$ *and* $\mathbf{B} = [b_{ij}]$ *are* $n \times n$ *matrices that are identical for all elements except for corresponding elements in the k-th row and if* $\mathbf{C} = [c_{ij}]$ *is an* $n \times n$ *matrix, then* $\det(\mathbf{A}) + \det(\mathbf{B}) = \det(\mathbf{C})$, *where* $c_{ij} = a_{ij}$, *except in the k-th row, in which* $c_{kj} = a_{kj} + b_{kj}$; $j = 1, 2, \ldots, n$.

Theorem 1.5.13

Let $\mathbf{A}$ *be an* $n \times n$ *matrix;* $|\mathbf{A}| = 0$ *if and only if* $\mathbf{A}$ *is a singular matrix.*

1.6 Rank of Matrices

An $n \times m$ matrix $\mathbf{A}$ is said to be of rank r if the size of the largest nonsingular square submatrix of $\mathbf{A}$ is r.

It is generally difficult to find the rank of a matrix by using this definition, since a matrix contains many submatrices. We shall give some theorems on the rank of a matrix after first defining elementary transformations.

Each of the following operations is called an elementary transformation of a matrix $\mathbf{A}$:

(1) The interchange of two rows (or two columns) of $\mathbf{A}$.
(2) The multiplication of the elements of a row (or a column) of $\mathbf{A}$ by the same nonzero scalar k.
(3) The addition of the elements of a row of $\mathbf{A}$, after they have been multiplied by the scalar k, to the corresponding elements of another row of $\mathbf{A}$. A corresponding statement can be made regarding columns.

We define *the inverse of an elementary transformation* as the transformation that restores the resulting matrix to the original form of $\mathbf{A}$.

We observe that the inverse of each elementary transformation is a transformation of the same type. For (1), if the same two rows (or columns) are again interchanged, the resulting matrix is again $\mathbf{A}$. For (2), if the elements of the altered row (or column) are multiplied by $1/k$, the matrix $\mathbf{A}$ is again obtained. Finally, for (3), if the same row (or column) of $\mathbf{A}$ is multiplied by $-k$ and the result added to the corresponding elements of the other row (or column) referred to in (3), then $\mathbf{A}$ is again the original matrix. These results can be stated in the form of the theorem that follows.

Theorem 1.6.1

The inverse of an elementary transformation of a matrix is an elementary transformation of the same type.

The following theorem is useful in finding the rank of a matrix.

Theorem 1.6.2

The size and rank of a matrix are not altered by an elementary transformation of the matrix.

Two matrices that have the *same size and the same rank* are said to be *equivalent.*

Theorem 1.6.3

Two matrices that are equivalent can be transformed from one to the other by a succession of elementary transformations.

We now show how the various elementary transformations can be accomplished through multiplication by certain matrices called *elementary transformation matrices,* or simply *elementary matrices.* These matrices will be denoted by $\mathbf{E}_1$, $\mathbf{E}_2$, and $\mathbf{E}_3$; the subscripts correspond to the three types of transformations discussed above.

(1) $\mathbf{E}_1$ is an elementary matrix that interchanges two rows (or columns) of $\mathbf{A}$.
(2) $\mathbf{E}_2$ is an elementary matrix that multiplies a row (or column) of $\mathbf{A}$ by the nonzero scalar k.
(3) $\mathbf{E}_3$ is an elementary matrix that adds the scalar k times each element in a row (column) to the corresponding element in another row (column).

Since the verification that the elementary matrices $\mathbf{E}_1$, $\mathbf{E}_2$, and $\mathbf{E}_3$ perform the transformations as given is rather lengthy, we content ourselves with illustrations using 3×3 matrices; note, however, that the elementary transformation matrices are all square, but the matrix upon which they operate need not be square.

The interchange of two *rows* of $\mathbf{A}$ can be accomplished by interchanging the corresponding *rows* of $\mathbf{I}$ to get the *elementary row matrix* $\mathbf{E}_1$ and then *premultiplying* $\mathbf{A}$ by $\mathbf{E}_1$. The first and third rows of $\mathbf{A}$ are interchanged in the following illustration.

$$\mathbf{E}_1\mathbf{A} = \begin{bmatrix} 0 & 0 & 1 \\ 0 & 1 & 0 \\ 1 & 0 & 0 \end{bmatrix} \begin{bmatrix} 2 & 1 & 3 \\ 1 & 0 & 2 \\ 4 & 1 & 2 \end{bmatrix} = \begin{bmatrix} 4 & 1 & 2 \\ 1 & 0 & 2 \\ 2 & 1 & 3 \end{bmatrix}.$$

If two *columns* of $\mathbf{I}$ are interchanged to get an *elementary column matrix* $\mathbf{E}_1$ and if $\mathbf{A}$ is then *postmultiplied* by $\mathbf{E}_1$, the corresponding *columns* of $\mathbf{A}$ are interchanged.

$$\mathbf{AE}_1 = \begin{bmatrix} 2 & 1 & 3 \\ 1 & 0 & 2 \\ 4 & 1 & 2 \end{bmatrix} \begin{bmatrix} 1 & 0 & 0 \\ 0 & 0 & 1 \\ 0 & 1 & 0 \end{bmatrix} = \begin{bmatrix} 2 & 3 & 1 \\ 1 & 2 & 0 \\ 4 & 2 & 1 \end{bmatrix}.$$

If the elements of a *row* of $\mathbf{I}$ are multiplied by the nonzero scalar k to get the elementary row matrix $\mathbf{E}_2$, *premultiplying* $\mathbf{A}$ by $\mathbf{E}_2$ multiplies the elements of the corresponding row of $\mathbf{A}$ by k. If the elements of a *column* of $\mathbf{I}$ are multiplied by the nonzero scalar k to get $\mathbf{E}_2$, *postmultiplying* $\mathbf{A}$ by $\mathbf{E}_2$ multiplies the elements of the corresponding column of $\mathbf{A}$ by k. These two situations are illustrated below.

$$\mathbf{E}_2\,\mathbf{A} = \begin{bmatrix} 1 & 0 & 0 \\ 0 & k & 0 \\ 0 & 0 & 1 \end{bmatrix} \begin{bmatrix} 2 & 1 & 3 \\ 1 & 0 & 2 \\ 4 & 1 & 2 \end{bmatrix} = \begin{bmatrix} 2 & 1 & 3 \\ k & 0 & 2k \\ 4 & 1 & 2 \end{bmatrix}.$$

$$\mathbf{AE}_2 = \begin{bmatrix} 2 & 1 & 3 \\ 1 & 0 & 2 \\ 4 & 1 & 2 \end{bmatrix} \begin{bmatrix} 1 & 0 & 0 \\ 0 & 1 & 0 \\ 0 & 0 & k \end{bmatrix} = \begin{bmatrix} 2 & 1 & 3k \\ 1 & 0 & 2k \\ 4 & 1 & 2k \end{bmatrix}.$$

If the elements of the i-th *row* of $\mathbf{I}$ are multiplied by the scalar k and then the products are added to the corresponding elements of the j-th *row* to get the elementary row matrix $\mathbf{E}_3$, *premultiplying* $\mathbf{A}$ by $\mathbf{E}_3$ multiplies the elements of the i-th *row* by k and adds the products to the corresponding elements of the j-th *row*. If the elements of the i-th *column* of $\mathbf{I}$ are multiplied by the scalar k and the products are added to the corresponding elements of the j-th *column* to get $\mathbf{E}_3$, the elementary column matrix, then *postmultiplying* $\mathbf{A}$ by $\mathbf{E}_3$ multiplies the elements of the i-th *column* of $\mathbf{A}$ by k and adds the products to the corresponding elements of the j-th *column*. For example,

$$\mathbf{E}_3\,\mathbf{A} = \begin{bmatrix} 1 & 0 & 0 \\ k & 1 & 0 \\ 0 & 0 & 1 \end{bmatrix} \begin{bmatrix} 2 & 1 & 3 \\ 1 & 0 & 2 \\ 4 & 1 & 2 \end{bmatrix} = \begin{bmatrix} 2 & 1 & 3 \\ 2k+1 & k+0 & 3k+2 \\ 4 & 1 & 2 \end{bmatrix}.$$

$$\mathbf{AE}_3 = \begin{bmatrix} 2 & 1 & 3 \\ 1 & 0 & 2 \\ 4 & 1 & 2 \end{bmatrix} \begin{bmatrix} 1 & 0 & 0 \\ 0 & 1 & 0 \\ k & 0 & 1 \end{bmatrix} = \begin{bmatrix} 2+3k & 1 & 3 \\ 1+2k & 0 & 2 \\ 4+2k & 1 & 2 \end{bmatrix}.$$

Note that the elementary matrices are nonsingular, since they are equivalent to $\mathbf{I}$.

An immediate consequence of the definitions of elementary matrices is the following theorem.

Theorem 1.6.4

Every elementary matrix has an inverse of the same type.

We now consider a few of the important properties and consequences of the use of elementary matrices, giving these properties in the form of theorems.

Theorem 1.6.5

Any nonsingular matrix can be written as the product of elementary matrices.

Theorem 1.6.6

The size and rank of a matrix **A** *are not altered by premultiplying or postmultiplying* **A** *by elementary matrices.*

Theorem 1.6.7

If matrices **A** *and* **B** *are nonsingular, then for any matrix* **C**, *the matrices* **C**, **AC**, **CB**, *and* **ACB** *all have the same rank (assuming all multiplications are defined).*

Theorem 1.6.8

If **A** *is an* $m \times n$ *matrix of rank* r, *then there exist nonsingular matrices* **P** *and* **Q** *such that* **PAQ** *is equal to*

$$\mathbf{I}, \quad [\mathbf{I}, \mathbf{0}], \quad \begin{bmatrix} \mathbf{I} \\ \mathbf{0} \end{bmatrix}, \quad \text{or} \quad \begin{bmatrix} \mathbf{I} & \mathbf{0} \\ \mathbf{0} & \mathbf{0} \end{bmatrix},$$

where $m = n = r$; $m = r < n$; $m > r = n$; $m > r$, $n > r$, *respectively, and where* **I** *is the* $r \times r$ *identity matrix.*

Theorem 1.6.9

Two matrices, **A** *and* **B**, *of the same size are equivalent if and only if* **B** *can be obtained by premultiplying and postmultiplying* **A** *by a finite number of elementary matrices.*

Theorem 1.6.10

The rank of the product of matrices **A** *and* **B** *cannot exceed the rank of either* **A** *or* **B**.

Theorem 1.6.11

A nonsingular matrix $\mathbf{A}$ *can always be reduced to* $\mathbf{I}$ *by elementary row (or column) transformation matrices only.*

Theorem 1.6.12

If $\mathbf{A}$ *is an* $m \times n$ *matrix of rank* r *and* $\mathbf{X}$ *is a matrix of size* $n \times p$ *and the product of these matrices is the zero matrix, that is,*

$$\mathbf{AX} = \mathbf{0}, \tag{1.6.1}$$

then

(i) *there exists a matrix* $\mathbf{X}$ *of rank* $n - r$ *such that* Eq. 1.6.1 *is satisfied,*
(ii) *the rank of* $\mathbf{X}$ *cannot exceed* $n - r$.

Theorem 1.6.13

If $\mathbf{A}$ *is a square matrix of size* n *that has rank* r, *then there exists a nonzero matrix* $\mathbf{X}$ *such that* $\mathbf{AX} = \mathbf{0}$ *if and only if* $r < n$.

Theorem 1.6.14

If $\mathbf{A}$ *is an* $m \times n$ *matrix and if* $m < n$, *then there exists a nonzero matrix* $\mathbf{X}$ *such that* $\mathbf{AX} = \mathbf{0}$.

Theorem 1.6.15

The rank of $\mathbf{A} + \mathbf{B}$ *is less than or equal to the rank of* $\mathbf{A}$ *plus the rank of* $\mathbf{B}$.

Theorem 1.6.16

If $\mathbf{A}$ *is an* $n \times n$ *matrix, then* $\det(\mathbf{A}) = 0$ *if and only if rank* $(\mathbf{A}) < n$.

1.7 Quadratic Forms

We shall define a function f of n variables $x_1, x_2, \ldots, x_n$ by

$$f = \left\{(x_1, x_2, \ldots, x_n, y) : y = \sum_{j=1}^{n} \sum_{i=1}^{n} a_{ij} x_i x_j, \right.$$

$$\text{where } a_{ij} \text{ is a given set of numbers, } \left. -\infty < x_i < \infty, i = 1, 2, \ldots, n \right\}. \tag{1.7.1}$$

In other words, the value of f at the point $\mathbf{x}$ is $f(\mathbf{x})$, where

$$f(\mathbf{x}) = \sum_{j=1}^{n} \sum_{i=1}^{n} a_{ij} x_i x_j. \tag{1.7.2}$$

This can also be written as

$$f(\mathbf{x}) = \mathbf{x}'\mathbf{A}\mathbf{x} \tag{1.7.3}$$

where $\mathbf{A}$ is an $n \times n$ matrix with ij-th element equal to a_{ij} and $\mathbf{x}$ is an $n \times 1$ vector with i-th element equal to x_i.

Definition 1.7.1

Quadratic Form. *The function f defined by* Eq. (1.7.1) *is defined to be a quadratic form (in the n variables x_i).*

For brevity we shall refer to $\mathbf{x}'\mathbf{A}\mathbf{x}$ as a quadratic form and to $\mathbf{A}$ as the matrix of the quadratic form.

Theorem 1.7.1

The matrix of a quadratic form can always be chosen to be symmetric.

Because of this theorem, every quadratic form in this book will be considered to have a symmetric matrix unless explicitly stated otherwise. Since quadratic forms play such an important role in statistics, all of Chapter 12 is devoted to this topic. The few theorems stated in this section are usually proved in a first course in linear algebra.

In a quadratic form, $\mathbf{x}'\mathbf{A}\mathbf{x}$, it is often desirable to change from the variables x_i to the variables y_i by the set of linear equations $\mathbf{y} = \mathbf{C}^{-1}\mathbf{x}$, where $\mathbf{C}$ is an $n \times n$ nonsingular matrix. When this is done, the quadratic form $\mathbf{x}'\mathbf{A}\mathbf{x}$ becomes

$$\mathbf{x}'\mathbf{A}\mathbf{x} = \mathbf{y}'\mathbf{C}'\mathbf{A}\mathbf{C}\mathbf{y} = \mathbf{y}'(\mathbf{C}'\mathbf{A}\mathbf{C})\mathbf{y} = \mathbf{y}'\mathbf{B}\mathbf{y},$$

where $\mathbf{B}$ replaces $\mathbf{C}'\mathbf{A}\mathbf{C}$. In this case, we say that $\mathbf{A}$ and $\mathbf{B}$ are congruent.

Definition 1.7.2

Congruent Matrices. *Two matrices* **A** *and* **B** *are defined to be congruent if and only if there exists a nonsingular matrix* **C** *such that* $\mathbf{B} = \mathbf{C}'\mathbf{A}\mathbf{C}$, *and we refer to* **C** *as a congruent transformation of the matrix* **A**.

In this book we are generally interested in a congruent transformation of a symmetric matrix **A**.

Theorem 1.7.2

The matrix **B** *resulting from a congruent transformation on a symmetric matrix* **A** *is symmetric.*

Theorem 1.7.3.

Let **A** *be an* $n \times n$ *(real) symmetric matrix of rank r; then there exists a nonsingular (real) matrix* **C** such *that* $\mathbf{C}'\mathbf{A}\mathbf{C} = \mathbf{D}$, *where* **D** *is a diagonal (real) matrix with exactly r nonzero diagonal elements.*

This is equivalent to saying that **A** is congruent to a diagonal matrix **D** with exactly r nonzero diagonal elements.

Theorem 1.7.4

If **A** *and* **B** *are congruent matrices, they have the same rank.*

Theorem 1.7.5

Let **A** *be an* $n \times n$ *symmetric (real) matrix of rank r. There exists a nonsingular (real) matrix* **C** *such that* $\mathbf{C}'\mathbf{A}\mathbf{C} = \mathbf{D}$, *where*

$$\mathbf{D} = \begin{bmatrix} \mathbf{I}_p & \mathbf{0} & \mathbf{0} \\ \mathbf{0} & -\mathbf{I}_{r-p} & \mathbf{0} \\ \mathbf{0} & \mathbf{0} & \mathbf{0} \end{bmatrix}.$$

In other words, **A** is congruent to a diagonal matrix **D**, with $p \geq 0$ diagonal elements equal to $+1$ and $r - p \geq 0$ diagonal elements equal to -1. For a given symmetric matrix **A**, there may be many nonsingular matrices **C**, such that $\mathbf{C}'\mathbf{A}\mathbf{C}$ is a diagonal matrix with only $+1$, -1, and 0 on the diagonal, but for any such matrix, the integers r and p remain the same. The integer p is called the *index* of the symmetric matrix **A**.

If the above ideas are applied to a quadratic form $\mathbf{x}'\mathbf{A}\mathbf{x}$ through the change of variables from x_i to y_i by $\mathbf{x} = \mathbf{Cy}$, we get

$$\mathbf{x}'\mathbf{Ax} = \mathbf{y}'(\mathbf{C}'\mathbf{AC})\mathbf{y} = \mathbf{y}'\mathbf{Dy} = y_1^2 + \cdots + y_p^2 - y_{p+1}^2 - \cdots - y_r^2,$$

and p is called the index and r the rank of the quadratic form $\mathbf{x}'\mathbf{Ax}$. When $r = p = n$, the quadratic form (and also the matrix $\mathbf{A}$) is called *positive definite*. When $r = p < n$, the quadratic form (and also the matrix $\mathbf{A}$) is called *positive semidefinite*. These ideas, which are extremely important in statistics, are elaborated upon in other chapters.

Reminder: Recall that according to the convention we are using all matrices are real unless stated otherwise. In particular, $\mathbf{C}$ is a *real* matrix in Theorems 1.7.3 through 1.7.5. Of course, $\mathbf{A}$, $\mathbf{B}$, and $\mathbf{D}$ in those theorems are also real. We have emphasized this by including the word "real" in parentheses in Theorems 1.7.3 and 1.7.5. It is obvious that if $\mathbf{A}$ and $\mathbf{C}$ are real matrices, then $\mathbf{C}'\mathbf{AC}$ is also a real matrix, but it is not obvious that there exists a real matrix $\mathbf{C}$ such that $\mathbf{C}'\mathbf{AC}$ is a *diagonal real* matrix even if $\mathbf{A}$ is a real matrix. For example, if $\mathbf{A}$ is a *real* $n \times n$ matrix of rank r, there may not exist a *real* nonsingular matrix $\mathbf{C}$ such that

$$\mathbf{C}'\mathbf{AC} = \begin{bmatrix} \mathbf{I}_r & \mathbf{0} \\ \mathbf{0} & \mathbf{0} \end{bmatrix}.$$

That is, $\mathbf{A}$ may not be congruent to a diagonal matrix with only r values of plus one and $n - r$ zeroes on the diagonal. Contrast this with Theorem 1.7.5. For example, consider the 2×2 matrix

$$\mathbf{A} = \begin{bmatrix} -1 & 0 \\ 0 & 0 \end{bmatrix}.$$

There is no *real* 2×2 nonsingular matrix $\mathbf{C}$ such that

$$\mathbf{C}'\mathbf{AC} = \begin{bmatrix} 1 & 0 \\ 0 & 0 \end{bmatrix}.$$

There is, however, a complex matrix $\mathbf{C}$ such that

$$\mathbf{C}'\mathbf{AC} = \begin{bmatrix} 1 & 0 \\ 0 & 0 \end{bmatrix},$$

and, in general, there always exists a complex matrix $\mathbf{C}$ such that

$$\mathbf{C}'\mathbf{AC} = \begin{bmatrix} \mathbf{I}_r & \mathbf{0} \\ \mathbf{0} & \mathbf{0} \end{bmatrix},$$

when $\mathbf{A}$ has rank r.

In most first courses in matrix algebra, the technique for finding a matrix **C** in Theorem 1.7.5 is discussed.

Theorem 1.7.6

Let **C** *be an* $m \times n$ *matrix of rank* r; *then the rank of* **CC**′ *is* r, *and the rank of* **C**′**C** *is* r.

Theorem 1.7.7

Let **C** *be an* $m \times n$ *matrix of rank* r; *then* **C**′**C** *and* **CC**′ *are either positive definite or positive semidefinite. If the rank of* **C**′**C** *or* **CC**′ *is equal to its size, then the matrix is positive definite; otherwise it is positive semidefinite. If* **A** *is* $n \times n$ *positive definite or positive semidefinite, it can be written as* $\mathbf{A} = \mathbf{K}'\mathbf{K}$, *where* **K** *is* $r \times n$ *and rank of* **K** *equals rank of* **A**.

1.8 Orthogonal Matrices

Computing the inverse of a nonsingular matrix is a tedious and time consuming task, but computing the transpose of a matrix is very easy. With some matrices, called orthogonal matrices, the inverse is equal to the transpose and, hence, for these matrices the inverse is easy to compute. However, the fact that computation of the inverse is easy is not the primary reason why orthogonal matrices are important. Later we shall examine various applications that show why they are important in statistics.

Definition 1.8.1

Orthogonal Matrices. *Let* **P** *be an* $n \times n$ *matrix.* **P** *is defined to be an orthogonal matrix if and only if* $\mathbf{P}^{-1} = \mathbf{P}'$.

Note. Recall that in this book we use the term *orthogonal matrix* to mean *real orthogonal matrix.*

Theorem 1.8.1

An orthogonal matrix is nonsingular.

Theorem 1.8.2

Let the $n \times n$ *matrix* **P** *be partitioned as* $[\mathbf{p}_1, \ldots, \mathbf{p}_n]$, *where* $\mathbf{p}_i$ *is an* $n \times 1$

matrix (vector) consisting of the elements in the i-th *column of* $\mathbf{P}$. *A necessary and sufficient condition that* $\mathbf{P}$ *is an orthogonal matrix is*

(1) $\mathbf{p}_i'\mathbf{p}_i = 1$ for $i = 1, 2, \ldots, n$,

(2) $\mathbf{p}_i'\mathbf{p}_j = 0$ for $i = 1, 2, \ldots, n; j = 1, 2, \ldots, n; i \neq j$.

Theorem 1.8.3

A necessary and sufficient condition that an $n \times n$ *matrix* $\mathbf{P}$ *is an orthogonal matrix is* $\mathbf{P}'\mathbf{P} = \mathbf{I}$.

Theorem 1.8.4

The determinant of an orthogonal matrix is equal to either $+1$ *or* -1.

Theorem 1.8.5

The product of a finite number of $n \times n$ *orthogonal matrices is an orthogonal matrix.*

Theorem 1.8.6

The inverse (and hence the transpose) of an orthogonal matrix is an orthogonal matrix.

Theorem 1.8.7

Let $\mathbf{A}$ *be an* $n \times n$ *matrix and let* $\mathbf{P}$ *be an* $n \times n$ *orthogonal matrix; then* $\det(\mathbf{A}) = \det(\mathbf{P}'\mathbf{A}\mathbf{P})$.

The final theorem of this section will be used many times in the following pages of this book.

Theorem 1.8.8

Let $\mathbf{A}$ *be any (real)* $n \times n$ *matrix. There exists a (real) orthogonal matrix* $\mathbf{P}$ *such that* $\mathbf{P}'\mathbf{A}\mathbf{P} = \mathbf{D}$, *where* $\mathbf{D}$ *is a (real) diagonal matrix, if and only if* $\mathbf{A}$ *is symmetric.*

Reminder: The word "real" has been inserted as a reminder, since this is such an important theorem for later work.

Problems

1. Find the determinant of the matrix **A** where

$$\mathbf{A} = \begin{bmatrix} 1 & -1 & 0 \\ 2 & 1 & 1 \\ 1 & 0 & 3 \end{bmatrix}.$$

2. In Prob. 1, find $\mathbf{A}^{-1}$.
3. In Prob. 1, find $(\mathbf{A}')^{-1}$ and $(\mathbf{A}^{-1})'$, and hence demonstrate Theorem 1.4.7.
4. In Prob. 1, find m_{11}, m_{12}, m_{13}, where m_{ij} is the minor of a_{ij}.
5. In Prob. 4, find A_{11}, A_{12}, A_{13}, where A_{ij} is the cofactor of a_{ij}.
6. Use the results of Prob. 5 and Theorem 1.5.9 to evaluate det (**A**).
7. Use Prob. 5 to demonstrate Theorem 1.5.11.
8. Consider the matrices **A**, **B**, and **C**, where **A** is defined as in Prob. 1 and

$$\mathbf{B} = \begin{bmatrix} 2 & 3 & 1 \\ 2 & 1 & 1 \\ 1 & 0 & 3 \end{bmatrix}, \qquad \mathbf{C} = \begin{bmatrix} 3 & 2 & 1 \\ 2 & 1 & 1 \\ 1 & 0 & 3 \end{bmatrix}.$$

Note that **A**, **B**, and **C** are identical except for the first row, and for that row the relationship $c_{ij} = a_{ij} + b_{ij}$; $j = 1, 2, 3$; holds. Show that det (**A**) + det (**B**) = det (**C**), and hence demonstrate Theorem 1.5.12.

9. Suppose you want to perform the following operations on a 3×3 matrix, **A**:
(1) Interchange the first and third rows.
(2) Interchange the first and third columns.
(3) Multiply the first row by -2 and add the result to the third row.
(4) Multiply the first column by -2 and add the result to the third column.
(5) Multiply the second row by -2 and add the result to the third row.
(6) Multiply the second column by -2 and add the result to the third column.
(7) Multiply the second row by $1/2$.
(8) Multiply the third row by $-1/12$.
Find the eight elementary matrices that perform the operations, and, for each, state whether it involves postmultiplying or premultiplying **A**.
10. If **A** in Prob. 9 is defined by

$$\mathbf{A} = \begin{bmatrix} 0 & 4 & 2 \\ 4 & 2 & 0 \\ 2 & 0 & 1 \end{bmatrix},$$

find the resulting matrix after performing the eight elementary transformations.

11. In Prob. 9, find the inverse of each of the eight elementary transformation matrices.
12. Use the results of Prob. 9 and 10 to find $\mathbf{A}^{-1}$.
13. Find a set of elementary matrices such that **A** in Prob. 10 is equal to the product of these elementary matrices, thus demonstrating Theorem 1.6.5.
14. For the matrix **A** where

$$\mathbf{A} = \begin{bmatrix} 1 & 2 & -1 & 1 \\ 2 & 1 & 0 & 3 \\ 0 & -3 & 2 & 1 \\ -3 & 0 & -1 & -5 \end{bmatrix},$$

find matrices **P** and **Q** such that

$$\mathbf{PAQ} = \begin{bmatrix} \mathbf{I}_r & \mathbf{0} \\ \mathbf{0} & \mathbf{0} \end{bmatrix},$$

where r is the rank of **A**. This demonstrates Theorem 1.6.8.

15. For the matrix **A** where

$$\mathbf{A} = \begin{bmatrix} 1 & -1 & 0 \\ 1 & 1 & 2 \\ 2 & 0 & 2 \end{bmatrix},$$

find a 3×3 matrix **X** of rank $3 - r$ such that $\mathbf{AX} = \mathbf{0}$ (r is the rank of **A**).

16. Find a nonsingular matrix **C** such that

$$\mathbf{C}'\mathbf{AC} = \begin{bmatrix} 1 & 0 & 0 \\ 0 & -1 & 0 \\ 0 & 0 & 0 \end{bmatrix},$$

where

$$\mathbf{A} = \begin{bmatrix} 0 & -1 & -2 \\ -1 & -1 & -1 \\ -2 & -1 & 0 \end{bmatrix}.$$

17. Use the matrix **A** in Prob. 16 and the result of Prob. 16 to transform the quadratic form $\mathbf{x}'\mathbf{A}\mathbf{x}$ to $y_1^2 - y_2^2$; that is, find **C** such that $\mathbf{y} = \mathbf{C}\mathbf{x}$.

18. Find the entries x_i in the matrix $\mathbf{P}$ so that $\mathbf{P}$ is an orthogonal matrix where

$$\mathbf{P} = \frac{1}{2}\begin{bmatrix} 1 & 1 & 1 & 1 \\ 1 & 1 & -1 & -1 \\ 1 & -1 & 1 & -1 \\ x_1 & x_2 & x_3 & x_4 \end{bmatrix}.$$

19. Find det ($\mathbf{P}$) in Prob. 18 and hence demonstrate Theorem 1.8.4.
20. Find a 3×3 matrix $\mathbf{A}$ such that det ($\mathbf{A}$) = 1 but such that $\mathbf{A}$ is not an orthogonal matrix.
21. If $\mathbf{A}$ is defined by

$$\mathbf{A} = \begin{bmatrix} 1 & 1 & 0 & -1 \\ 2 & 1 & 0 & 1 \\ 1 & 1 & 0 & 2 \\ 1 & -1 & 1 & 1 \end{bmatrix},$$

find det ($\mathbf{A}$) and det ($\mathbf{P}'\mathbf{A}\mathbf{P}$), where $\mathbf{P}$ is defined in Prob. 18, and show that the two are equal. This result demonstrates Theorem 1.8.7.

Prerequisite Vector Theory

2

2.1 Introduction and Definitions

Vectors play a very important role in many branches of mathematics and especially in statistics. In this chapter, we define vectors, vector spaces, and subspaces and state without proof some theorems that are generally discussed in a first course in linear algebra.

We shall use the following definition of a vector in this book, although it certainly is not the most general definition.

Definition 2.1.1

(n-component) Vector. *Let n be a positive integer and let $a_1, \ldots, a_n$ be any elements from a field F. The ordered n-tuple*

$$\mathbf{a} = \begin{bmatrix} a_1 \\ \vdots \\ a_n \end{bmatrix}$$

is defined as a (n-component) vector, sometimes called an $n \times 1$ vector.

We shall denote vectors by lowercase boldface letters. *Unless explicitly stated otherwise, the field F will be the field of real numbers.* Strictly speaking, **a** is a column vector, but we shall omit the word column. The symbol $\mathbf{a}'$ (transpose of **a**) is used for a row vector. Note that an $n \times 1$ vector is a special case of a matrix, and all the rules for addition, subtraction, multiplication, and transposition for matrices also hold for vectors. Multiplication of a matrix by a scalar also holds for multiplication of a vector by a scalar. (Recall that in this book a scalar is a real number unless explicitly stated otherwise.) That is, for any $n \times 1$ vectors **a** and **b** and any two scalars a and b, we get

$$a\mathbf{a} + b\mathbf{b} = \begin{bmatrix} aa_1 + bb_1 \\ aa_2 + bb_2 \\ \vdots \\ aa_n + bb_n \end{bmatrix}.$$

In fact, any notation or theorem that is valid for $n \times m$ matrices when $m = 1$ is valid for $n \times 1$ vectors.

2.2 Vector Space

We are interested not only in a single vector but also in a certain collection or set of vectors which we call a *vector space*.

Definition 2.2.1

Vector Space. *Let V_n be a set of n-component vectors such that for every two vectors in V_n, the sum of the two vectors is also in V_n, and for each vector in V_n and each scalar, the product is in V_n. This set V_n is called a vector space.*

Note: This definition states that if a collection of $n \times 1$ vectors is "closed" with respect to addition and with respect to multiplication by a scalar, then the set is a vector space V_n.

Theorem 2.2.1

Let R_n be the set of all $n \times 1$ vectors for a fixed positive integer n—that is,

$$R_n = \{\mathbf{a} : \mathbf{a}' = [a_1, \ldots, a_n]; \; -\infty < a_i < \infty, i = 1, 2, \ldots, n\};$$

then R_n is a vector space.

Thus, the collection of *all* ordered n-tuples of real numbers is a vector space. Geometrically the $n \times 1$ vector $\mathbf{a}$ can be viewed as a point in n-dimensional space or as the directed line segment from the origin, which is the point $\mathbf{0}' = [0, 0, \ldots, 0]$, to the point $\mathbf{a}$. If $n = 3$, R_3 is the space we generally think of in 3-dimensional geometry. In statistics our interest is generally centered around certain subsets of the vectors in R_n. However, we want this subset to satisfy certain conditions. For example, suppose we are interested in the line in R_3 that goes through the points $\mathbf{0}$ and $\mathbf{a}$ where $\mathbf{a}' = [1, -1, 0]$. It is intuitively clear that for every real number λ, the point $\lambda\mathbf{a}$ is a point on the line, and every point on the line is represented by $\lambda\mathbf{a}$ for some real number λ; that is, the line is defined as the set $\mathscr{L}$, where $\mathscr{L} = \{(x_1, x_2, x_3): \mathbf{x} = \lambda\mathbf{a}; \lambda \in R\}$.

Also the plane through the three points $\mathbf{a}' = [0, 1, 1]$, $\mathbf{b}' = [1, 2, 1]$, $\mathbf{0}' = [0, 0, 0]$ is defined to be the set of points $\mathscr{P}$ where

$$\mathscr{P} = \{(x_1, x_2, x_3): \mathbf{x} = \lambda_1\mathbf{a} + \lambda_2\mathbf{b}; \lambda_1 \in R, \lambda_2 \in R\}.$$

Also for any two real numbers λ_1^0 and λ_2^0 the point $\mathbf{x}^0$, where $\mathbf{x}^0 = \lambda_1^0\mathbf{a} + \lambda_2^0\mathbf{b}$, is a point on the plane $\mathscr{P}$ that goes through $\mathbf{a}$, $\mathbf{b}$, and the origin $\mathbf{0}$. Thus the line and the plane illustrated above have one thing in common; both involve *linear combinations* of certain vectors in R_3. There are many other situations in which linear combinations of vectors are important, so we shall devote the next few sections to this problem.

Note: For, say, $n = 3$, the vector space that can be derived from the two vectors

$$\mathbf{a} = \begin{bmatrix} 0 \\ 1 \\ 1 \end{bmatrix}; \qquad \mathbf{b} = \begin{bmatrix} 0 \\ 1 \\ -1 \end{bmatrix}$$

by the process discussed in Def. 2.2.1 does not include the vector

$$\mathbf{c} = \begin{bmatrix} 1 \\ 0 \\ 0 \end{bmatrix}.$$

On the other hand, the vector space R_3 includes every possible 3×1 vector. Therefore, R_3 is certainly not the only vector space for $n = 3$.

2.3 Vector Subspaces

In statistics, and particularly in this book, we generally take R_n, for a given positive integer n, to be the basic vector space, and every n-component vector will be a member of this set. By R_n, we shall hereafter always mean the vector space defined in Theorem 2.2.1. However, when we want the basic vector space under discussion to consist of a set of n-component vectors that is not the vector space R_n, we shall use some other symbol, such as V_n, S_n, and so forth.

We are generally interested in a subset of a vector space, V_n, if this subset is itself a vector space. These are called subspaces of V_n.

Definition 2.3.1

Subspace. *Let S_n be a subset of vectors in the vector space V_n. If the set S_n is itself a vector space, then S_n is called a (vector) subspace of the (vector) space V_n.*

To determine whether or not S_n, a subset of vectors in the vector space V_n, is itself a vector space, the following theorem is useful.

Theorem 2.3.1

If S_n is a subset of the vectors in the vector space V_n such that, for each and every two vectors $\mathbf{s}_1$ and $\mathbf{s}_2$ in S_n, the vector $a_1\mathbf{s}_1 + a_2\mathbf{s}_2$ is in S_n for all real numbers a_1 and a_2, then S_n is a subspace of V_n.

Example 2.3.1. The set of vectors defined by $a\mathbf{a}'$ for $\mathbf{a}' = [1, -1, 1]$ and for each and every real number a is a subspace of R_3. Also the set of vectors $\mathbf{a}' = [a_1, a_2, 0]$ for all real numbers a_1 and a_2 is a subspace of R_3. But the set of vectors defined by $b\mathbf{b}'$ for $\mathbf{b}' = [1, 2]$ and all real b is not a subspace of R_3, because $\mathbf{b}'$ is not a member of R_3; it is however a subspace of R_2.

Note: The set of vectors

$$V_3 = \{\mathbf{v} : \mathbf{v}' = [a_1, a_2, 0];\ a_i \in R\}$$

is a vector space, and $V_3 \subset R_3$; that is to say, V_3 is a subspace of R_3. Also the set of vectors

$$S_3 = \{\mathbf{s} : \mathbf{s}' = [0, a, 0];\ a \in R\}$$

is a vector space, and $S_3 \subset V_3$; that is, S_3 is a subspace of V_3. Also the set $\{\mathbf{0}\}$ consisting of the single vector $\mathbf{0}$ is a vector space. Hence we have $\{\mathbf{0}\} \subset S_3 \subset V_3 \subset R_3$. Also

$$S_3^* = \{\mathbf{s}^* : \mathbf{s}^{*\prime} = [a, 0, 0];\ a \in R\}$$

is a vector space, and it is a subspace of V_3 and of R_3; but S_3^* is not a subspace of S_3. The set of two vectors $U = \{\mathbf{u}_1, \mathbf{u}_2\}$, where $\mathbf{u}_1' = [0, 1, 0]$ and $\mathbf{u}_2' = [0, 2, 0]$, is a *subset* of V_3, of S_3, and of R_3, but U is *not* a *subspace.*

Theorem 2.3.2

The set $\{\mathbf{0}\}$, *where* $\mathbf{0}$ *is the* $n \times 1$ *null vector, is a subspace of every vector space* V_n. *Every vector space* V_n *is a subspace of itself.*

2.4 Linear Dependence and Independence

When a set of n-component vectors is under study, it is important to be able to determine whether some of the vectors can be obtained as linear combinations of other vectors. To be able to do this, we need a definition and some theorems on linear dependence and independence of a set of vectors.

Definition 2.4.1

Linear Dependence and Independence. *Let* $\{\mathbf{v}_1, \mathbf{v}_2, \ldots, \mathbf{v}_m\}$ *be a set of m vectors each with n components,* so that $\mathbf{v}_i \in R_n$; $i = 1, 2, \ldots, m$. *This set of m vectors is defined to be linearly dependent* (*or a linearly dependent set*) *if and only if there exists a set of scalars* $\{c_1, c_2, \ldots, c_m\}$, *at least one of which is not equal to zero, such that*

$$\sum_{i=1}^{m} c_i \mathbf{v}_i = \mathbf{0}.$$

If the only set of scalars $\{c_1, c_2, \ldots, c_m\}$ *such that* $\sum_{i=1}^{m} c_i \mathbf{v}_i = \mathbf{0}$ *is the set* $\{0, 0, \ldots, 0\}$, *then the set of vectors is defined to be linearly independent.*

Example 2.4.1. Consider the two vectors $\mathbf{v}_1' = [1, -1, 3]$, $\mathbf{v}_2' = [1, 1, 1]$. To determine two scalars c_1, c_2 such that $c_1\mathbf{v}_1 + c_2\mathbf{v}_2 = \mathbf{0}$, we obtain

$$c_1 \begin{bmatrix} 1 \\ -1 \\ 3 \end{bmatrix} + c_2 \begin{bmatrix} 1 \\ 1 \\ 1 \end{bmatrix} = \begin{bmatrix} 0 \\ 0 \\ 0 \end{bmatrix},$$

from which we get

$$c_1 + c_2 = 0$$
$$-c_1 + c_2 = 0$$
$$3c_1 + c_2 = 0,$$

and the only solution is $c_1 = c_2 = 0$. Hence, $\mathbf{v}_1$ and $\mathbf{v}_2$ are linearly independent. However, the two vectors $\mathbf{v}_1' = [1, 1, 3]$, $\mathbf{v}_2' = [4, 4, 12]$ are linearly dependent, since $-4\mathbf{v}_1 + \mathbf{v}_2 = \mathbf{0}$.

In the following theorems we assume that each vector has n-components.

Theorem 2.4.1

If the vector $\mathbf{0}$ *is included in a set of n-component vectors, the set is linearly dependent.*

Theorem 2.4.2

If $m > 1$ *vectors are linearly dependent, it is always possible to express at least one of them as a linear combination of the others.*

Theorem 2.4.3

In the set of m vectors $\{\mathbf{v}_1, \mathbf{v}_2, \ldots, \mathbf{v}_m\}$, *if there are s vectors,* $s \leq m$, *that are linearly dependent, then the entire set of m vectors is linearly dependent.*

Theorem 2.4.4

If the set of m vectors $\{\mathbf{v}_1, \mathbf{v}_2, \ldots, \mathbf{v}_m\}$ *is a linearly independent set, while the set of* $m + 1$ *vectors* $\{\mathbf{v}_1, \mathbf{v}_2, \ldots, \mathbf{v}_m, \mathbf{v}_{m+1}\}$ *is a linearly dependent set, then* $\mathbf{v}_{m+1}$ *is expressible as a linear combination of* $\mathbf{v}_1, \mathbf{v}_2, \ldots, \mathbf{v}_m$.

We shall have occasion to write the m vectors $\mathbf{v}_1, \mathbf{v}_2, \ldots, \mathbf{v}_m$ as follows:

$$\mathbf{V} = [\mathbf{v}_1, \mathbf{v}_2, \ldots, \mathbf{v}_m] = \left[\begin{bmatrix} v_{11} \\ v_{21} \\ \vdots \\ v_{n1} \end{bmatrix} \begin{bmatrix} v_{12} \\ v_{22} \\ \vdots \\ v_{n2} \end{bmatrix} \cdots \begin{bmatrix} v_{1m} \\ v_{2m} \\ \vdots \\ v_{nm} \end{bmatrix} \right],$$

which we shall write as

$$\mathbf{V} = \begin{bmatrix} v_{11} & v_{12} & & v_{1m} \\ v_{21} & v_{22} & \cdots & v_{2m} \\ \vdots & \vdots & & \vdots \\ v_{n1} & v_{n2} & \cdots & v_{nm} \end{bmatrix}.$$

Thus, we have an $n \times m$ matrix whose columns consist of m vectors, each with n components. On the other hand, we could view $\mathbf{V}'$ as a matrix whose columns consist of n vectors, each with m elements. We shall call $\mathbf{V}$ a matrix of m column vectors, each with n components, or simply a matrix of vectors.

Note: In this book when we refer to a matrix of vectors, we mean the set of vectors formed by the *columns* of the matrix, unless explicitly stated otherwise.

Theorem 2.4.5

A necessary and sufficient condition for the set of $n \times 1$ vectors $\{\mathbf{v}_1, \mathbf{v}_2, \ldots, \mathbf{v}_m\}$ to be a linearly dependent set is that the rank r of the matrix of the vectors be less than the number of vectors m; that is, $r < m$.

Theorem 2.4.6

If the rank of the matrix of the set of $n \times 1$ vectors $\{\mathbf{v}_1, \mathbf{v}_2, \ldots, \mathbf{v}_m\}$ is r, then r must be less than or equal to m, and if $r > 0$, there exist exactly r of these vectors that are linearly independent, while each of the remaining $m - r$ (if $m - r > 0$) vectors is expressible as a linear combination of these r vectors.

Theorem 2.4.7

The set of $n \times 1$ vectors $\{\mathbf{v}_1, \mathbf{v}_2, \ldots, \mathbf{v}_m\}$ is always a linearly dependent set if $m > n$.

Example 2.4.2. Determine whether or not the vectors $\mathbf{v}_1$, $\mathbf{v}_2$, $\mathbf{v}_3$ below are linearly independent.

$$\mathbf{v}_1 = \begin{bmatrix} 1 \\ 1 \\ 0 \\ -1 \end{bmatrix}; \quad \mathbf{v}_2 = \begin{bmatrix} 2 \\ 0 \\ 1 \\ -1 \end{bmatrix}; \quad \mathbf{v}_3 = \begin{bmatrix} 0 \\ -2 \\ 1 \\ 1 \end{bmatrix}.$$

Using Theorem 2.4.5, we get

$$\mathbf{V} = \begin{bmatrix} 1 & 2 & 0 \\ 1 & 0 & -2 \\ 0 & 1 & 1 \\ -1 & -1 & 1 \end{bmatrix},$$

and clearly the rank of this matrix is equal to two so the three vectors are linearly dependent; from Theorem 2.4.6, there are exactly two linearly independent vectors. Any two of these three vectors are linearly independent.

2.5 Basis of a Vector Space

It is useful to be able to determine a subset of vectors in a vector space V_n such that each vector in V_n can be derived as a linear combination of the vectors in this subset. Such a set is said to generate, or span, the vector space V_n.

Theorem 2.5.1

Let $\{\mathbf{v}_1, \ldots, \mathbf{v}_m\}$ be a set of vectors in V_n and let the set of vectors V be defined by

$$V = \left\{\mathbf{v} : \mathbf{v} = \sum_{i=1}^{m} c_i \mathbf{v}_i \,;\, c_i \in R\right\};$$

then V is a subspace of V_n.

This theorem asserts that if we start with any set of vectors in V_n, then the set V, obtained by every possible linear combination of these vectors, is itself a vector space and is a subspace of V_n.

Definition 2.5.1

Generating (or Spanning) Vectors. *Let V_n be a vector space. If each vector in V_n can be obtained by a linear combination of the vectors in the set $\{\mathbf{v}_1, \ldots, \mathbf{v}_m\}$, then the set of vectors $\{\mathbf{v}_1, \ldots, \mathbf{v}_m\}$ is said to generate (or span) V_n.*

We say that the space V_n is generated, or spanned, by the vectors $\mathbf{v}_1, \ldots, \mathbf{v}_m$

(or by the set of vectors $\{\mathbf{v}_1, \ldots, \mathbf{v}_m\}$). We note that the set $\{\mathbf{0}\}$—the set consisting of the zero vector only—is a vector space. We also note that every vector space must include a zero vector and, for any vector space V except the space consisting only of $\mathbf{0}$, there are many sets that span the space V. Nothing was stated as to whether the vectors that span a space are linearly dependent or linearly independent. However, when the set is linearly independent, we give it a special name: a *basis* (set).

Definition 2.5.2

Basis. *Let* $\{\mathbf{v}_1, \ldots, \mathbf{v}_m\}$ *be a set of linearly independent vectors in* V_n *that span* V_n. *Then the set is called a basis for* V_n. *For the special vector space* $\{\mathbf{0}\}$ *we shall say that* $\mathbf{0}$ *is a basis (even though it is not linearly independent).*

In general, a basis for a vector space V_n is not unique, and hence there are many different bases for V_n. However the *number* of vectors in any basis for V_n is unique.

Theorem 2.5.2

If $\{\mathbf{v}_1, \ldots, \mathbf{v}_m\}$, $\{\mathbf{u}_1, \ldots, \mathbf{u}_q\}$ *are two bases for* V_n, *then* $m = q$; *that is, any two bases for a given vector space contain the same number of vectors.*

It might be noted that no basis can contain the zero vector unless it is the only vector in the basis, in which case the vector space consists of only the zero vector.

A special name is applied to the number of vectors in a basis: *dimension.*

Definition 2.5.3

Dimension. *Let* V_n *be any vector space except* $\{\mathbf{0}\}$. *Let the number of vectors in a basis of* V_n *be* m. *Then* m *is defined to be the dimension of* V_n. *The dimension of the vector space* $\{\mathbf{0}\}$ *is defined to be zero.*

Note: It is not necessary that $m = n$, but it can be so. However, m cannot be greater than n, because of Theorem 2.4.7.

Example 2.5.1. Let V_3 be the vector space spanned by the vectors

$$\mathbf{v}_1 = \begin{bmatrix} 1 \\ 1 \\ 0 \end{bmatrix}; \quad \mathbf{v}_2 = \begin{bmatrix} 1 \\ -1 \\ 0 \end{bmatrix}; \quad \mathbf{v}_3 = \begin{bmatrix} 1 \\ 0 \\ 0 \end{bmatrix}; \quad \mathbf{v}_4 = \begin{bmatrix} 2 \\ 0 \\ 0 \end{bmatrix}.$$

This space has dimension two, since the rank of $\mathbf{V} = [\mathbf{v}_1, \mathbf{v}_2, \mathbf{v}_3, \mathbf{v}_4]$ is two and, by Theorem 2.4.6, at least one set of two of these vectors is linearly independent, and each of the other two vectors is a linear combination of these two. Hence, these two vectors form a basis for the vector space V_3. However, not every two vectors are linearly independent, and hence not every two vectors span V_3. For example, $\mathbf{v}_3$ and $\mathbf{v}_4$ do not span V_3, but $\mathbf{v}_1$ and $\mathbf{v}_3$ span V_3; also $\mathbf{v}_1$ and $\mathbf{v}_2$ span V_3. Each of the sets $\{\mathbf{v}_1, \mathbf{v}_2\}$; $\{\mathbf{v}_1, \mathbf{v}_3\}$; $\{\mathbf{v}_1, \mathbf{v}_4\}$; $\{\mathbf{v}_2, \mathbf{v}_3\}$; $\{\mathbf{v}_2, \mathbf{v}_4\}$ is a basis for V_3. Note also that the three vectors $\mathbf{v}_1, \mathbf{v}_2, \mathbf{v}_3$ span V_3, but they do not form a basis, since they are not linearly independent.

Since every vector in a vector space V_n can be expressed as a linear combination of the vectors in any basis set, it is important to know whether or not a particular vector in V_n can be *uniquely* expressed. This is the context of the next theorem.

Theorem 2.5.3

Let the set of vectors $\{\mathbf{v}_1, \ldots, \mathbf{v}_m\}$ be a basis for the vector space V_n ($V_n \neq \{\mathbf{0}\}$). Let $\mathbf{v}$ be any vector in V_n. There is one and only one ordered set of scalars $\{c_1, c_2, \ldots, c_m\}$ such that

$$\mathbf{v} = \sum_{i=1}^{m} c_i \mathbf{v}_i .$$

In other words, $\mathbf{v}$ is a unique linear combination of a given basis.

Theorem 2.5.4

If $r > 0$ is the rank of the matrix of the vectors $\mathbf{v}_1, \mathbf{v}_2, \ldots, \mathbf{v}_m$ that span the vector space V_n, then there are exactly r linearly independent vectors in the set and every vector in V_n can be expressed uniquely as a linear combination of these r vectors.

Theorem 2.5.5

If the vector space V_n is spanned by a set of m vectors, and if the matrix of these vectors has rank r, then any set of $r + 1$ vectors in V_n is linearly dependent.

As stated earlier, there is generally more than one basis for a given vector space V_n. Of course if B_1 and B_2 are two bases for V_n, then each vector in B_1 must be a linear combination of the vectors in B_2, and vice versa. We now state a theorem about the relationship of bases in V_n.

Theorem 2.5.6

Let $\mathbf{V} = [\mathbf{v}_1, \mathbf{v}_2, \ldots, \mathbf{v}_m]$ *be a matrix consisting of a set of vectors that is a basis for* V_n *and let* $\mathbf{U} = [\mathbf{u}_1, \mathbf{u}_2, \ldots, \mathbf{u}_q]$ *be a matrix that is any set of vectors in* V_n. *The set of vectors in* $\mathbf{U}$ *is a basis set for* V_n *if and only if* $m = q$ *and there exists a nonsingular* $m \times m$ *matrix* $\mathbf{A}$ *such that* $\mathbf{U} = \mathbf{VA}$.

Therefore, if we have a basis set for V_n consisting of $\mathbf{v}_1, \mathbf{v}_2, \ldots, \mathbf{v}_m$, we can change to a new basis set $\{\mathbf{u}_1, \mathbf{u}_2, \ldots, \mathbf{u}_m\}$ by the formula $\mathbf{U} = \mathbf{VA}$, where $\mathbf{A}$ is any nonsingular $m \times m$ matrix.

Theorem 2.5.7

Let $\{\mathbf{v}_1, \mathbf{v}_2, \ldots, \mathbf{v}_m\}$, $m > 1$, *be a basis for the vector space* V_n *and let* $\mathbf{v}$ *be any vector in* V_n *such that* $\mathbf{v} = \sum_{i=1}^{m} c_i \mathbf{v}_i$. *If* $c_t \neq 0$ *for some* t, *then the set* $\{\mathbf{v}_1, \mathbf{v}_2, \ldots, \mathbf{v}_{t-1}, \mathbf{v}, \mathbf{v}_{t+1}, \ldots, \mathbf{v}_m\}$ *is a basis for* V_n. *However if* $c_t = 0$, *then the set* $\{\mathbf{v}_1, \mathbf{v}_2, \ldots, \mathbf{v}_{t-1}, \mathbf{v}, \mathbf{v}_{t+1}, \ldots, \mathbf{v}_m\}$ *is a linearly dependent set and hence is not a basis for* V_n.

Note: By using this theorem, it is possible to replace a vector $\mathbf{v}_t$ in a basis by another vector $\mathbf{v}$, but it is not possible to replace it by just any vector.

Theorem 2.5.8

Let $\{\mathbf{v}_1, \ldots, \mathbf{v}_q\}$ *be a set of linearly independent vectors in* V_n. *Then this set is a subset of a basis for* V_n.

Note: If the dimension of V_n is q, then this set is a basis for V_n. This theorem states that any linearly independent set of vectors that is not a basis can be extended to a basis by including additional vectors in the set.

2.6 Inner Product and Orthogonality of Vectors

Two concepts that play an important role in a discussion of vector spaces and especially in the application to statistics are the *inner product* of two vectors and *orthogonality* of two vectors.

Definition 2.6.1

Inner Product. *Let* $\mathbf{x}$ *and* $\mathbf{y}$ *be two vectors in* V_n. *The inner product of* $\mathbf{x}$ *and* $\mathbf{y}$, *which we shall denote by* $\mathbf{x} \cdot \mathbf{y}$, *is defined to be the scalar* $\sum_{i=1}^{n} x_i y_i$.

Note: The inner product is such that $\mathbf{x} \cdot \mathbf{y} = \mathbf{y} \cdot \mathbf{x}$. Actually, if we view the vectors $\mathbf{x}$ and $\mathbf{y}$ as $n \times 1$ matrices, then $\mathbf{x}'\mathbf{y}$ is a 1×1 matrix that is equal to

$$\left[\sum_{i=1}^{n} x_i y_i\right] = [\mathbf{x} \cdot \mathbf{y}].$$

Thus the inner product is defined to be the scalar that is the element of the 1×1 matrix $\mathbf{x}'\mathbf{y}$.

Definition 2.6.2

Orthogonal Vectors. *Let* $\mathbf{x}$ *and* $\mathbf{y}$ *be two vectors in* V_n. $\mathbf{x}$ *and* $\mathbf{y}$ *are defined to be orthogonal if and only if the inner product is equal to zero (or if and only if* $\mathbf{x}'\mathbf{y}$ *is equal to the* 1×1 *zero matrix).*

Note: The zero vector in V_n is orthogonal to each and every vector in V_n.

Definition 2.6.3

Normal Vectors. *A vector* $\mathbf{x}$ *in* V_n *is defined to be a normal vector if and only if the inner product of* $\mathbf{x}$ *with itself is equal to plus one (or if and only if* $\mathbf{x}'\mathbf{x}$ *is equal to the* 1×1 *identity matrix;* that is, $\mathbf{x}'\mathbf{x} = [1]$).

Note: When there is no chance for ambiguity, we shall not distinguish between $\mathbf{x} \cdot \mathbf{y}$ and $\mathbf{x}'\mathbf{y}$, that is, between the 1×1 matrix $[a]$ and the scalar a, since for our work the two quantities have equivalent algebraic properties. In fact, we shall generally use $\mathbf{x}'\mathbf{y} = a$ rather than $\mathbf{x}'\mathbf{y} = \mathbf{a}$ or $\mathbf{x}'\mathbf{y} = [a]$.

Since, unless explicitly stated otherwise, we assume that the elements of the vectors in this book are real numbers, it follows that $\mathbf{x}'\mathbf{x}$ is the zero vector if and only if $\mathbf{x} = \mathbf{0}$.

It may be desirable to have a basis for V_n such that the vectors in the basis are pairwise orthogonal. This can always be done and the basis is referred to as an orthogonal basis. If, in addition, the vectors in the basis are normal vectors, the basis is called an orthonormal basis.

Definition 2.6.4

Orthogonal and Orthonormal Bases. *If* $\{\mathbf{v}_1, \mathbf{v}_2, \ldots, \mathbf{v}_m\}$ *is a basis for* V_n *such that* $\mathbf{v}_i'\mathbf{v}_j = 0$ *for all* $i \neq j = 1, 2, \ldots, m$, *then the basis is defined to be an orthogonal basis for* V_n. *If in addition* $\mathbf{v}_i'\ \mathbf{v}_i = 1$, *for* $i = 1, 2, \ldots, m$, *the basis is defined to be an orthonormal basis.*

Theorem 2.6.1

Every vector space has an orthogonal basis.

Note: If a basis for a vector space consists of a single vector, then we shall call it an orthogonal basis.

Theorem 2.6.2

Every vector space V_n except $\{\mathbf{0}\}$ has an orthonormal basis.

Theorem 2.6.3

Let $\{\mathbf{v}_1, \mathbf{v}_2, \ldots, \mathbf{v}_m\}$ be a set of vectors in V_n such that each and every distinct pair of vectors is orthogonal; that is, $\mathbf{v}_i' \mathbf{v}_j = 0$ *for all* $i \neq j$. *If none of the vectors is the zero vector, then the set of vectors is a linearly independent set.*

Theorem 2.6.4

Any set of q nonzero pairwise orthogonal vectors in V_n is a subset of a basis for V_n.

The next theorem presents a method for constructing a set of q orthonormal vectors that spans the same vector space as that spanned by a given basis $\{\mathbf{v}_1, \mathbf{v}_2, \ldots, \mathbf{v}_q\}$.

Theorem 2.6.5

Let $\{\mathbf{v}_1, \mathbf{v}_2, \ldots, \mathbf{v}_q\}$ be a basis of the vector space $V_n \neq \{\mathbf{0}\}$. Then the set of q vectors $\{\mathbf{z}_1, \mathbf{z}_2, \ldots, \mathbf{z}_q\}$ is also a basis for V_n and they are an orthonormal set. The $\mathbf{z}_i$ are defined by

$$\mathbf{y}_1 = \mathbf{v}_1; \qquad \mathbf{z}_1 = \frac{\mathbf{y}_1}{\sqrt{\mathbf{y}_1'\mathbf{y}_1}}$$

$$\mathbf{y}_2 = \mathbf{v}_2 - \frac{\mathbf{y}_1'\mathbf{v}_2}{\mathbf{y}_1'\mathbf{y}_1}\mathbf{y}_1; \qquad \mathbf{z}_2 = \frac{\mathbf{y}_2}{\sqrt{\mathbf{y}_2'\mathbf{y}_2}}$$

$$\vdots \qquad \vdots \qquad \vdots$$

$$\mathbf{y}_q = \mathbf{v}_q - \frac{\mathbf{y}_1'\mathbf{v}_q}{\mathbf{y}_1'\mathbf{y}_1}\mathbf{y}_1 - \frac{\mathbf{y}_2'\mathbf{v}_q}{\mathbf{y}_2'\mathbf{y}_2}\mathbf{y}_2 - \cdots - \frac{\mathbf{y}_{q-1}'\mathbf{v}_q}{\mathbf{y}_{q-1}'\mathbf{y}_{q-1}}\mathbf{y}_{q-1}; \qquad \mathbf{z}_q = \frac{\mathbf{y}_q}{\sqrt{\mathbf{y}_q'\mathbf{y}_q}}$$

Example 2.6.1. Find an orthonormal basis for the vector space spanned by the two vectors $\mathbf{v}_1$ and $\mathbf{v}_2$ where

$$\mathbf{v}_1 = \begin{bmatrix} 1 \\ 0 \\ 2 \end{bmatrix}; \qquad \mathbf{v}_2 = \begin{bmatrix} 1 \\ -1 \\ 1 \end{bmatrix}.$$

Since the rank of the matrix

$$\mathbf{V} = [\mathbf{v}_1, \mathbf{v}_2] = \begin{bmatrix} 1 & 1 \\ 0 & -1 \\ 2 & 1 \end{bmatrix}$$

is equal to two, the two vectors are linearly independent and, hence, are a basis. To find an orthonormal basis, we use Theorem 2.6.5.

$$\mathbf{y}_1 = \begin{bmatrix} 1 \\ 0 \\ 2 \end{bmatrix}; \qquad \mathbf{z}_1 = \frac{1}{\sqrt{5}} \begin{bmatrix} 1 \\ 0 \\ 2 \end{bmatrix};$$

$$\mathbf{y}_2 = \begin{bmatrix} 1 \\ -1 \\ 1 \end{bmatrix} - \frac{3}{5} \begin{bmatrix} 1 \\ 0 \\ 2 \end{bmatrix} = \frac{1}{5} \begin{bmatrix} 2 \\ -5 \\ -1 \end{bmatrix}; \qquad \mathbf{z}_2 = \frac{1}{\sqrt{30}} \begin{bmatrix} 2 \\ -5 \\ -1 \end{bmatrix}.$$

It is easily demonstrated that $\mathbf{z}_1'\mathbf{z}_2 = 0$, $\mathbf{z}_1'\mathbf{z}_1 = 1$, $\mathbf{z}_2'\mathbf{z}_2 = 1$, and hence $\{\mathbf{z}_1, \mathbf{z}_2\}$ is a set of two orthonormal vectors. To demonstrate that they span the same space as $\mathbf{v}_1$ and $\mathbf{v}_2$, we must show that $\mathbf{v}_1$ and $\mathbf{v}_2$ are each linear combinations of $\mathbf{z}_1$ and $\mathbf{z}_2$. Clearly,

$$\mathbf{v}_1 = \sqrt{5}\,\mathbf{z}_1; \; \mathbf{v}_2 = \frac{3}{\sqrt{5}} \mathbf{z}_1 + \frac{\sqrt{30}}{5} \mathbf{z}_2 .$$

Since $\mathbf{z}_1$ and $\mathbf{z}_2$ span the same space as $\mathbf{v}_1$ and $\mathbf{v}_2$ and since $\mathbf{z}_1$ and $\mathbf{z}_2$ are orthonormal and hence linearly independent, they are a basis.

Problems

1. Show that the three vectors

$$\mathbf{e}_1 = \begin{bmatrix} 1 \\ 0 \\ 0 \end{bmatrix}; \qquad \mathbf{e}_2 = \begin{bmatrix} 0 \\ 1 \\ 0 \end{bmatrix}; \qquad \mathbf{e}_3 = \begin{bmatrix} 0 \\ 0 \\ 1 \end{bmatrix}$$

are a basis for R_3.

2. In Prob. 1, generalize to R_n.
3. For any vector $\mathbf{x}' = [x_1, x_2, x_3]$ in R_3, find scalars c_1, c_2, c_3 such that

$$\mathbf{x} = \sum_{i=1}^{3} c_i \mathbf{e}_i .$$

4. In Prob. 3, generalize to R_n.
5. Show that the vector $\mathbf{v}$ is in the vector space spanned by $\mathbf{v}_1$ and $\mathbf{v}_2$ where

$$\mathbf{v}_1 = \begin{bmatrix} 1 \\ 1 \\ 0 \\ 1 \end{bmatrix}; \qquad \mathbf{v}_2 = \begin{bmatrix} -1 \\ 1 \\ 1 \\ 0 \end{bmatrix}; \qquad \mathbf{v} = \begin{bmatrix} 5 \\ 1 \\ -2 \\ 3 \end{bmatrix}.$$

6. Show that the four vectors below are linearly dependent.

$$\mathbf{v}_1 = \begin{bmatrix} 1 \\ -1 \\ 1 \\ 0 \end{bmatrix}; \qquad \mathbf{v}_2 = \begin{bmatrix} 2 \\ 1 \\ 1 \\ 0 \end{bmatrix}; \qquad \mathbf{v}_3 = \begin{bmatrix} 0 \\ -6 \\ 2 \\ 0 \end{bmatrix}; \qquad \mathbf{v}_4 = \begin{bmatrix} 0 \\ -3 \\ 1 \\ 0 \end{bmatrix}.$$

7. In Prob. 6, find a set of two linearly independent vectors.
8. Is the vector $\mathbf{v}$ in the vector space spanned by the four vectors in Prob. 6, where $\mathbf{v}' = [1\ 1\ 0\ 1]$?
9. In Prob. 6, show that $\mathbf{v}_2$ can be expressed as a linear combination of the other three vectors.
10. In Prob. 6, find two different bases for the vector space spanned by the four vectors.
11. In Prob. 10, find a nonsingular 2×2 matrix $\mathbf{A}$ that relates the two bases (see Theorem 2.5.6).

12. Prove Theorem 2.5.3 by using the definition of linear independence of vectors.
13. Find an orthonormal basis for the vector space spanned by

$$\mathbf{v}_1 = \begin{bmatrix} 1 \\ 1 \\ -1 \end{bmatrix}; \quad \mathbf{v}_2 = \begin{bmatrix} 2 \\ 1 \\ 0 \end{bmatrix}; \quad \mathbf{v}_3 = \begin{bmatrix} 3 \\ 2 \\ 0 \end{bmatrix}.$$

Linear Transformations and Characteristic Roots

3

3.1 Linear Transformations

The theorems discussed in this chapter—first a few on *linear transformations* then some about *characteristic vectors* and *characteristic roots*—are stated without proofs. Like those in the first two chapters, these theorems are usually proved in a first course in linear algebra.

Let $\mathbf{A}$ be an $m \times n$ matrix, let $\mathbf{x}$ be any vector in R_n, and define the $m \times 1$ vector $\mathbf{y}$ by the equation

$$\mathbf{y} = \mathbf{A}\mathbf{x}. \tag{3.1.1}$$

$\mathbf{y}$ is a vector in R_m, and we say "the vector $\mathbf{x}$ is transformed to the vector $\mathbf{y}$ by the transformation $\mathbf{A}$." We can view Eq. (3.1.1) as moving the vector $\mathbf{x}$ in R_n to the vector $\mathbf{y}$ in R_m (or as moving the point $\mathbf{x}$ to the point $\mathbf{y}$). Let $\mathbf{x}_1$ and $\mathbf{x}_2$ be any two vectors in R_n; then $\mathbf{x}_1$ is transformed to $\mathbf{y}_1$ and $\mathbf{x}_2$ is transformed to $\mathbf{y}_2$ by the transformation $\mathbf{A}$ if $\mathbf{y}_1$ and $\mathbf{y}_2$ are defined by

$$\mathbf{y}_1 = \mathbf{A}\mathbf{x}_1, \qquad \mathbf{y}_2 = \mathbf{A}\mathbf{x}_2 .$$

Also, if we define the vector $\mathbf{x}_3$ to be equal to $c_1\mathbf{x}_1 + c_2\mathbf{x}_2$, where c_1 and c_2 are any two real numbers, then by the transformation **A** the vector $\mathbf{x}_3$ is transformed to the vector $\mathbf{y}_3$, where

$$\begin{aligned}\mathbf{y}_3 = \mathbf{A}\mathbf{x}_3 &= \mathbf{A}(c_1\mathbf{x}_1 + c_2\mathbf{x}_2) = \mathbf{A}c_1\mathbf{x}_1 + \mathbf{A}c_2\mathbf{x}_2 \\ &= c_1\mathbf{A}\mathbf{x}_1 + c_2\mathbf{A}\mathbf{x}_2 = c_1\mathbf{y}_1 + c_2\mathbf{y}_2,\end{aligned}$$

that is,

$$\mathbf{y}_3 = c_1\mathbf{y}_1 + c_2\mathbf{y}_2.$$

Thus if we know that, by the transformation **A**, the vector $\mathbf{x}_1$ is transformed to $\mathbf{y}_1$ and $\mathbf{x}_2$ is transformed to $\mathbf{y}_2$, then we know that the vector $c_1\mathbf{x}_1 + c_2\mathbf{x}_2$ is transformed to $c_1\mathbf{y}_1 + c_2\mathbf{y}_2$.

The transformation of vectors in R_n defined by Eq. (3.1.1) is called a *linear homogeneous* transformation and generally is referred to as simply a *linear* transformation. Note that $\mathbf{0} = \mathbf{A0}$; that is, a zero vector is transformed to a zero vector, and for this reason the transformation is called *homogeneous*. It is called *linear* because, for any two vectors $\mathbf{x}_1$ and $\mathbf{x}_2$ in R_n and any two scalars c_1 and c_2, we obtain

$$\mathbf{A}(c_1\mathbf{x}_1 + c_2\mathbf{x}_2) = c_1(\mathbf{A}\mathbf{x}_1) + c_2(\mathbf{A}\mathbf{x}_2);$$

in other words, the transformation **A** of a linear combination of two vectors is obtained by taking the same linear combination of the two transformed vectors.

Suppose that

$$\mathbf{y} = \mathbf{A}\mathbf{x}$$

is a transformation of **x** to **y** and

$$\mathbf{z} = \mathbf{B}\mathbf{y}$$

a transformation of **y** to **z**. Then by substitution we obtain

$$\mathbf{z} = \mathbf{B}\mathbf{y} = \mathbf{B}(\mathbf{A}\mathbf{x}) = (\mathbf{BA})\mathbf{x},$$

and **BA** is the transformation that "moves **x** directly to **z**." This can be extended to any finite number of transformations.

We are interested in questions such as the following: If each vector **x** in the vector space R_n is transformed by the equation $\mathbf{y} = \mathbf{A}\mathbf{x}$, does the resulting set of transformed vectors form a vector space? Or, more specifically, if each vector **x** in any vector space

V_n is transformed by the equation $\mathbf{y} = \mathbf{A}\mathbf{x}$, does the resulting set of vectors form a vector space? The answer lies in the next theorem.

Theorem 3.1.1

Let S be a set of vectors that results from transforming each vector in a vector space V_n by the transformation $\mathbf{A}$, so that

$$S = \{\mathbf{y} : \mathbf{y} = \mathbf{A}\mathbf{x};\ \mathbf{x} \in V_n\};$$

then S is a vector space.

Sometimes it is useful to be able to relate basis vectors in one space to basis vectors in a space resulting from a linear transformation. In this connection the next theorem can be helpful.

Theorem 3.1.2

Let S be the vector space that results from transforming the vector space V_n by $\mathbf{y} = \mathbf{A}\mathbf{x}$; that is $S = \{\mathbf{y} : \mathbf{y} = \mathbf{A}\mathbf{x};\ \mathbf{x} \in V_n\}$. Let $\{\mathbf{x}_1, \ldots, \mathbf{x}_q\}$ be a set of vectors that span the vector space V_n, then the set of vectors $\{\mathbf{y}_1, \ldots, \mathbf{y}_q\}$ span the vector space S where $\mathbf{y}_i = \mathbf{A}\mathbf{x}_i$, $i = 1, 2, \ldots, q$.

Example 3.1.1. Consider the transformation $\mathbf{A}$ that transforms a vector space V_2 to a vector space S_2 where

$$\mathbf{A} = \begin{bmatrix} 1 & 2 \\ 3 & 6 \end{bmatrix}.$$

Suppose that the vector space V_2 has the two vectors $\mathbf{x}_1$ and $\mathbf{x}_2$ as a basis, where

$$\mathbf{x}_1 = \begin{bmatrix} 1 \\ 2 \end{bmatrix}; \qquad \mathbf{x}_2 = \begin{bmatrix} 0 \\ -1 \end{bmatrix}.$$

Find the dimension of S_2.

The set of vectors $\mathbf{y}_1$ and $\mathbf{y}_2$, where

$$\mathbf{y}_1 = \mathbf{A}\mathbf{x}_1 = \begin{bmatrix} 5 \\ 15 \end{bmatrix}; \qquad \mathbf{y}_2 = \mathbf{A}\mathbf{x}_2 = \begin{bmatrix} -2 \\ -6 \end{bmatrix},$$

spans S_2, by Theorem 3.1.2. The dimension of S_2 is one, since the rank of $[\mathbf{y}_1, \mathbf{y}_2]$ is one.

Example 3.1.2. In Example 3.1.1 find a basis for S_2. Since S_2 has dimension one, any nonzero vector in S_2 is a basis. Thus, the vector $\mathbf{y}' = [5, 15]$ is a basis.

3.2 Characteristic Roots and Vectors

Among the questions that are important in considering transformations of vectors is one that concerns the transformation of a vector into a multiple of itself. That is to say, suppose we are discussing the transformation described by the $n \times n$ matrix $\mathbf{A}$. For this transformation, does there exist a vector $\mathbf{x}$ in R_n such that

$$\mathbf{A}\mathbf{x} = \lambda\mathbf{x} \tag{3.2.1}$$

for some real number λ? If such a vector $\mathbf{x}$ and a real number λ do exist, then, since $\lambda\mathbf{x}$ is a multiple of $\mathbf{x}$, the vector $\mathbf{x}$ is transformed to a multiple of itself. Assuming that Eq. (3.2.1) obtains, we can write it as

$$(\mathbf{A} - \lambda\mathbf{I})\mathbf{x} = \mathbf{0}, \tag{3.2.2}$$

and clearly $\mathbf{x} = \mathbf{0}$ satisfies Eq. (3.2.2) for any scalar λ; but this merely states that in a linear homogeneous transformation, the origin (vector $\mathbf{0}$) transforms to itself. Therefore, we pose this question: Is there any *nonzero* vector $\mathbf{x}$ and a scalar λ such that Eq. (3.2.1) is satisfied?

We know from the elementary theory of the solutions of linear equations that a nonzero solution to Eq. (3.2.2) exists if and only if the determinant of the matrix $\mathbf{A} - \lambda\mathbf{I}$ is equal to zero, that is, if and only if $|\mathbf{A} - \lambda\mathbf{I}| = 0$. But the determinant of $\mathbf{A} - \lambda\mathbf{I}$ is an n-th degree polynomial in λ, and in place of $|\mathbf{A} - \lambda\mathbf{I}| = 0$, we can write

$$a_n\lambda^n + a_{n-1}\lambda^{n-1} + \cdots + a_1\lambda + a_0 = 0. \tag{3.2.3}$$

This equation is called the *characteristic equation* of the $n \times n$ matrix $\mathbf{A}$. We know from elementary algebra that this equation has exactly n roots; in other words, there are exactly n values of λ (not necessarily all distinct) such that $|\mathbf{A} - \lambda\mathbf{I}| = 0$. These n roots are called the *characteristic roots* (sometimes called eigenvalues), or simply the *roots* of the $n \times n$ matrix $\mathbf{A}$. However some of these roots may not be real numbers but may be complex numbers even if **A** is a real matrix. If λ_1 is a root of **A**, and if $\mathbf{x}_1$ is a vector corresponding to this root, that is, if

$$\mathbf{A}\mathbf{x}_1 = \lambda_1 \mathbf{x}_1, \quad \mathbf{x}_1 \neq \mathbf{0},$$

then $\mathbf{x}_1$ is defined to be a *characteristic vector* of the $n \times n$ matrix **A** corresponding to the root λ_1. If λ_1 is a complex number, then the elements of $\mathbf{x}_1$ may not be real numbers. The following theorem answers one of the questions presented above.

Theorem 3.2.1

Let **A** *be an* $n \times n$ *(real) matrix. There always exist n complex numbers* $\lambda_1, \ldots, \lambda_n$ *(called characteristic roots of* **A**) *that satisfy the polynomial equation* $|\mathbf{A} - \lambda\mathbf{I}| = 0$. *Some or all of these roots may not be real numbers.*

Characteristic roots, vectors, and polynomials are important in statistics and we shall now state some theorems that are generally proved in a first course in matrix algebra.

Theorem 3.2.2

Let **A** *be an* $n \times n$ *(real) matrix. A necessary and sufficient condition that there exists a nonzero vector* **x** *that satisfies*

$$\mathbf{A}\mathbf{x} = \lambda\mathbf{x}$$

is that λ *is a characteristic root of A. The characteristic vector* **x** *has elements which are complex numbers that may not be real numbers.*

Theorem 3.2.3

The $n \times n$ *matrix* **A** *has at least one characteristic root equal to zero if and only if* **A** *is singular.*

Theorem 3.2.4

Let **A** *be an* $n \times n$ *matrix, and let* **C** *be any* $n \times n$ *nonsingular matrix. The three matrices* **A**, $\mathbf{C}^{-1}\mathbf{A}\mathbf{C}$, *and* $\mathbf{C}\mathbf{A}\mathbf{C}^{-1}$ *have the same set of characteristic roots.*

Theorem 3.2.5

Let **A** *be an* $n \times n$ *matrix. The two matrices* **A** *and* $\mathbf{A}'$ *have the same set of* *characteristic roots, but a characteristic vector of* **A** *need not be a characteristic vector of* **A**′.

Theorem 3.2.6

If λ is a characteristic root of the $n \times n$ matrix **A** *and* **x** *is a characteristic vector of* **A** *corresponding to the root λ, then λ^k is a characteristic root of $\mathbf{A}^k$ and* **x** *is a characteristic vector of $\mathbf{A}^k$ corresponding to the root λ^k (k is any positive integer).*

Theorem 3.2.7

Let **A** *be an $n \times n$ nonsingular matrix and let λ be a characteristic root of* **A**, *then $1/\lambda$ is a characteristic root of $\mathbf{A}^{-1}$.*

Theorem 3.2.8

If **x** *is a characteristic vector of the $n \times n$ matrix* **A** *corresponding to the root λ of* **A**, *then for any nonzero complex number c, the vector $c\mathbf{x}$ is also a characteristic vector of* **A** *corresponding to the root λ of* **A**.

Reminder: Even though we consider only real matrices **A**, the characteristic roots of **A** and the elements of the characteristic vectors of **A** may *not* be real numbers; they are always complex numbers. It may be important to know conditions under which the characteristic vectors of a matrix are *real* numbers. This is the context of the next theorem.

Theorem 3.2.9

Let **A** *be an $n \times n$ (real) matrix that has a real characteristic root λ; then* **A** *has a real characteristic vector corresponding to the root λ.*

Example 3.2.1. Consider the 2×2 matrix **A** defined by

$$\mathbf{A} = \begin{bmatrix} 1 & 2 \\ 3 & 5 \end{bmatrix}.$$

The roots are given by the solution to $|\mathbf{A} - \lambda\mathbf{I}| = 0$. This gives us

$$|\mathbf{A} - \lambda\mathbf{I}| = \begin{vmatrix} 1-\lambda & 2 \\ 3 & 5-\lambda \end{vmatrix} = (1-\lambda)(5-\lambda) - 6 = 0$$

or

$$\lambda^2 - 6\lambda - 1 = 0.$$

The two roots are $\lambda_1 = 3 + \sqrt{10}$; $\lambda_2 = 3 - \sqrt{10}$. To find a characteristic vector corresponding to the root λ_1, we solve

$$\mathbf{Ax} = \lambda_1 \mathbf{x},$$

which results in two equations and two unknowns:

$$x_1 + 2x_2 = (3 + \sqrt{10})x_1,$$
$$3x_1 + 5x_2 = (3 + \sqrt{10})x_2.$$

A solution is

$$x_1 = 1, \qquad x_2 = 1 + \frac{1}{2}\sqrt{10},$$

which clearly satisfies $\mathbf{Ax} = \lambda_1 \mathbf{x}$. Also, for any complex number c, it is clear that

$$x_1 = c, \qquad x_2 = c + \frac{c\sqrt{10}}{2}$$

are also the elements of a characteristic vector of **A**. Note that since the roots of **A** are real, there exist real characteristic vectors of **A**. However, there *always* exist characteristic vectors of a real matrix that are complex and not real; for instance; in this example, let $c = \sqrt{-1}$.

3.3 Similar Matrices

When considering various transformations that are useful in statistical theory, we sometimes find it helpful to study the relationship between certain matrices. For example, it is useful to know some relationships between two $n \times n$ matrices **A** and **B** such that a nonsingular matrix **Q** exists, so that $\mathbf{B} = \mathbf{Q}^{-1}\mathbf{AQ}$. This may be particularly important when **B** is a diagonal (or a triangular) matrix, since, in this case, we have transformed a matrix **A** into a diagonal (or triangular) matrix by a certain kind of transformation and, for many uses, a diagonal (or triangular) matrix is quite easy to work with. Matrices **A** and **B** above are said to be *similar*, and since this concept is important in statistical work we shall define it and state some theorems that are generally proved in a first course in matrix algebra.

Definition 3.3.1

Similarity Transformation. *Two* $n \times n$ *matrices* $\mathbf{A}$ *and* $\mathbf{B}$ *are defined to be similar if and only if there exists a nonsingular matrix* $\mathbf{Q}$ *such that* $\mathbf{B} = \mathbf{Q}^{-1}\mathbf{AQ}$. *The transformation* $\mathbf{Q}^{-1}\mathbf{AQ}$ *such that* $\mathbf{B} = \mathbf{Q}^{-1}\mathbf{AQ}$ *is called a similarity transformation from* $\mathbf{A}$ *to* $\mathbf{B}$.

Theorem 3.3.1

The determinants of similar matrices are equal.

Theorem 3.3.2

Let $\mathbf{A}$ *be an* $n \times n$ *real matrix. If the characteristic roots of* $\mathbf{A}$ *are distinct, there exists a nonsingular complex matrix* $\mathbf{Q}$ *such that* $\mathbf{Q}^{-1}\mathbf{AQ} = \mathbf{D}$, *where* $\mathbf{D}$ *is a complex diagonal matrix.*

Note: The matrices $\mathbf{Q}$ and $\mathbf{D}$ may not be real even if $\mathbf{A}$ is a real matrix. However, if $\mathbf{A}$ and the characteristic roots of $\mathbf{A}$ are real and if the conditions of the theorem are met, that is, if the roots of $\mathbf{A}$ are distinct, then there exists a *real* nonsingular matrix $\mathbf{Q}$ such that $\mathbf{Q}^{-1}\mathbf{AQ} = \mathbf{D}$, where $\mathbf{D}$ is a *real* diagonal matrix.

The following theorem is widely applicable in statistics.

Theorem 3.3.3

Let $\mathbf{A}$ *be an* $n \times n$ *real matrix. There exists a nonsingular, complex matrix* $\mathbf{Q}$ *(not necessarily real) such that* $\mathbf{Q}^{-1}\mathbf{AQ} = \mathbf{T}$, *where* $\mathbf{T}$ *is an upper, complex triangular matrix (not necessarily real), and the characteristic roots of* $\mathbf{A}$ *are the diagonal elements of* $\mathbf{T}$.

Note: In Theorem 3.3.3, if $\mathbf{A}$ and the roots of $\mathbf{A}$ are real, then there exists a real $\mathbf{Q}$ such that $\mathbf{Q}^{-1}\mathbf{AQ}$ is an upper triangular real matrix $\mathbf{T}$ with the characteristic roots of $\mathbf{A}$ on the diagonal of $\mathbf{T}$.

Theorem 3.3.4

Similar matrices have the same set of characteristic roots.

3.4 Symmetric Matrices

Since quadratic forms play such an important role in statistics, we are led to a discussion of real symmetric matrices.

Definition 3.4.1

Real Symmetric matrix. *An $n \times n$ matrix $\mathbf{A}$ is defined to be a real symmetric matrix if and only if* (1) *the elements of* $\mathbf{A}$ *are real;* (2) $\mathbf{A} = \mathbf{A}'$.

Reminder: We shall generally omit the word "real" and say "$\mathbf{A}$ is an $n \times n$ symmetric matrix" to mean that $\mathbf{A}$ is an $n \times n$ real symmetric matrix.

Theorem 3.4.1

Let $\mathbf{A}$ be an $n \times n$ symmetric matrix. The characteristic roots of $\mathbf{A}$ are real.

Theorem 3.4.2

For each characteristic root of an $n \times n$ symmetric matrix there exists a real characteristic vector.

Note: This theorem states the following: let λ_1 be a characteristic root of the $n \times n$ symmetric matrix $\mathbf{A}$; then there exists a real nonzero vector $\mathbf{x}_1$ such that $\mathbf{A}\mathbf{x}_1 = \lambda_1\mathbf{x}_1$. Of course, since $\mathbf{x}_1$ is a characteristic vector of $\mathbf{A}$ corresponding to the root λ_1, then $a\mathbf{x}_1$ is also a characteristic vector corresponding to λ_1, where a is any nonzero complex number. Thus if a is a complex number that is not real, then $\mathbf{x}_1$ is a complex vector that is not real, so this demonstrates that a complex characteristic vector that is not real also exists. However it is very useful to know that a *real* characteristic vector exists. Therefore, in the remainder of this book, when we say "characteristic vector of a symmetric matrix" we shall *always* mean *real* characteristic vector.

Theorem 3.4.3

Let $\mathbf{A}$ be an $n \times n$ symmetric matrix, let λ_1 and λ_2 be two characteristic roots of $\mathbf{A}$, and let $\mathbf{x}_1$ and $\mathbf{x}_2$ be two characteristic vectors of $\mathbf{A}$ corresponding to λ_1 and λ_2 respectively. If λ_1 and λ_2 are distinct, then $\mathbf{x}_1'\mathbf{x}_2 = 0$; that is, $\mathbf{x}_1$ and $\mathbf{x}_2$ are orthogonal vectors.

The next theorem is used extensively in transforming a real quadratic form to a sum of squares.

Theorem 3.4.4

Let **A** *be an* $n \times n$ *symmetric matrix. There exists an orthogonal matrix* **P** *such that* $\mathbf{P}'\mathbf{A}\mathbf{P} = \mathbf{D}$, *where* **D** *is a diagonal matrix with the characteristic roots of* **A** *displayed on the diagonal of* **D**.

Note: Since $\mathbf{P}' = \mathbf{P}^{-1}$, this theorem states that a symmetric matrix is similar to a diagonal matrix.

Theorem 3.4.5

Let **A** *be an* $n \times n$ *symmetric matrix; then* **A** *has* n *linearly independent (real) characteristic vectors.*

Note: A direct consequence of this last theorem is that a set of characteristic vectors of an $n \times n$ symmetric matrix **A** is a basis for R_n.

Theorem 3.4.6

Let **A** *be an* $n \times n$ *symmetric matrix and let* λ_1 *be a characteristic root of* **A** *with multiplicity* k; *then the matrix* $\mathbf{A} - \lambda_1\mathbf{I}$ *has rank* $n - k$.

In succeeding chapters we state and prove a large number of theorems that deal with characteristic roots and characteristic vectors.

Problems

1. Find the roots of the matrix **A** where

$$\mathbf{A} = \begin{bmatrix} 4 & 1 \\ 3 & 2 \end{bmatrix}.$$

2. Find a characteristic vector associated with each root of the matrix **A** in Prob. 1.
3. For the transformation **A** in Prob. 1, find the vector **y** to which the vector $\mathbf{x}' = [1, -1]$ is transformed.
4. Consider the 2×2 matrices **A** and **B** where

$$\mathbf{A} = \begin{bmatrix} 1 & 1 \\ 2 & -1 \end{bmatrix}; \qquad \mathbf{B} = \begin{bmatrix} 1 & 1 \\ -2 & 3 \end{bmatrix}.$$

If the vector $\mathbf{x}' = [2, -1]$ is transformed to **y** by $\mathbf{y} = \mathbf{A}\mathbf{x}$ and if **y** is transformed to **z** by $\mathbf{z} = \mathbf{B}\mathbf{y}$, find the vectors **y** and **z**.

5. In Prob. 4, suppose that the transformation $\mathbf{C} = \mathbf{BA}$ is considered and the vector $\mathbf{x}$ is transformed to $\mathbf{w}$ by $\mathbf{w} = \mathbf{Cx}$. Find $\mathbf{w}$ and show that it is equal to $\mathbf{z}$.
6. In Prob. 4, suppose $\mathbf{x}$ is first transformed to $\mathbf{u}$ by $\mathbf{u} = \mathbf{Bx}$ and then $\mathbf{u}$ is transformed to $\mathbf{v}$ by $\mathbf{v} = \mathbf{Au}$. Find $\mathbf{u}$ and $\mathbf{v}$ and demonstrate that $\mathbf{v} \neq \mathbf{z}$.
7. In Prob. 6, what are conditions on matrices $\mathbf{A}$ and $\mathbf{B}$ such that, in general, $\mathbf{v} = \mathbf{z}$? In other words, what are conditions such that any $n \times 1$ vector which is transformed by an $n \times n$ matrix $\mathbf{A}$ followed by $\mathbf{B}$ results in the same vector as when it is transformed by $\mathbf{B}$ followed by $\mathbf{A}$?
8. Consider the vector space V_3 spanned by the four vectors

$$\mathbf{v}_1 = \begin{bmatrix} 1 \\ 2 \\ -1 \end{bmatrix}; \quad \mathbf{v}_2 = \begin{bmatrix} 1 \\ 0 \\ -1 \end{bmatrix}; \quad \mathbf{v}_3 = \begin{bmatrix} 1 \\ 4 \\ -1 \end{bmatrix}; \quad \mathbf{v}_4 = \begin{bmatrix} -1 \\ -2 \\ 1 \end{bmatrix};$$

and consider the transformation $\mathbf{A}$ where

$$\mathbf{A} = \begin{bmatrix} 1 & 0 & 2 \\ 1 & -1 & 3 \\ -1 & 2 & -4 \end{bmatrix}.$$

If each vector in V_3 is transformed by $\mathbf{A}$, then by Theorem 3.1.1, the resulting set of vectors is a vector space which we shall denote by S_3. Is the vector $\mathbf{x}' = [1, 1, -1]$ in S_3?
9. In Prob. 8, show that the dimension of S_3 is two.
10. Find the roots of the matrix $\mathbf{A}$ where

$$\mathbf{A} = \begin{bmatrix} 1 & 1 \\ -2 & -1 \end{bmatrix}.$$

11. Show that $\mathbf{A}$ in Prob. 10 has no real characteristic vectors.
12. Find the roots of the symmetric matrix $\mathbf{A}$ where

$$\mathbf{A} = \begin{bmatrix} 1 & 2 \\ 2 & 4 \end{bmatrix}.$$

13. Find real characteristic vectors $\mathbf{x}$ and $\mathbf{y}$ associated with each root in Prob. 12.
14. In Prob. 13, verify that $\mathbf{x}'\mathbf{y} = 0$, thus demonstrating Theorem 3.4.3.
15. Verify that the two vectors $\mathbf{x}$ and $\mathbf{y}$ in Prob. 13 are linearly independent, thus demonstrating Theorem 3.4.5.

16. Demonstrate Theorem 3.4.6 by showing that the rank of the matrix $\mathbf{A} - \lambda_1\mathbf{I}$ in Prob. 12 is one, where λ_1 is either of the roots.
17. Normalize each vector in Prob. 13 and use the two resulting vectors to form the columns of a matrix which we shall denote by **P**. Verify that **P** is a 2×2 orthogonal matrix.
18. In Prob. 17, verify that $\mathbf{P}'\mathbf{A}\mathbf{P} = \mathbf{D}$, where **D** is a diagonal matrix with the roots of **A** as diagonal elements. This is an example of Theorem 3.4.4.
19. Consider the two matrices **A**, **Q**, where

$$\mathbf{A} = \begin{bmatrix} 1 & 2 \\ 3 & -1 \end{bmatrix}; \qquad \mathbf{Q} = \begin{bmatrix} 2 & 1 \\ 4 & 1 \end{bmatrix}.$$

Show that $|\mathbf{A}| = |\mathbf{Q}^{-1}\mathbf{A}\mathbf{Q}|$ for these matrices and thus demonstrate Theorem 3.3.1

20. Show that **A** and **A**′ have the same set of roots, where

$$\mathbf{A} = \begin{bmatrix} 1 & 2 \\ 3 & -1 \end{bmatrix},$$

and thus demonstrate Theorem 3.2.5.

21. In Prob. 20, show that **A** and **A**′ do not have the same characteristic vectors.
22. In Prob. 12, find $\mathbf{A}^3$ and the roots of $\mathbf{A}^3$ and demonstrate Theorem 3.2.6.
23. In Prob. 22, find a set of characteristic vectors for $\mathbf{A}^3$.
24. If $\mathbf{x}_1$ is a characteristic vector corresponding to a root λ_1 of the matrix **A**, show that for any positive integer k the vector $\mathbf{y}_1$ is also a characteristic vector corresponding to the root λ_1 where

$$\mathbf{y}_1 = \mathbf{A}^k\mathbf{x}_1.$$

Geometric Interpretations

4

4.1 Introduction

This chapter, which assumes that the reader is acquainted with the elementary theory of analytic geometry, briefly discusses some geometric interpretations of vectors that may be helpful in statistics. When we interpret vectors geometrically we shall sometimes consider a vector in R_n as a point in n-coordinate space and sometimes as a directed line segment from the origin to the point represented by the vector. One advantage of a "geometrical" interpretation of vectors is that it often allows one to "picture" a situation in two- or three-space and, by intuition or analogy, extend this to n-space.

In two- or three-space analytical geometry, the concepts of points, lines, planes, and so forth, are generally referred to a fixed set of coordinate axes—that is, the x_1 and x_2 axes in two-space or the x_1, x_2, and x_3 axes in three-space. These axes are usually at right angles (orthogonal) to each other. A point, denoted by, say,

$$\mathbf{a} = \begin{bmatrix} a_1 \\ a_2 \end{bmatrix}$$

means: start at the origin and mark off a_1 units on the x_1 axis and at that point draw a line that is perpendicular to the x_1 axis; mark off a_2 units on the constructed line, and this is the point $\mathbf{a}$. Any point in the plane can be located by this procedure. Note that the vectors

$$\mathbf{e}_1 = \begin{bmatrix} 1 \\ 0 \end{bmatrix}, \qquad \mathbf{e}_2 = \begin{bmatrix} 0 \\ 1 \end{bmatrix}$$

represent points one unit along the x_1 and x_2 axes, respectively, and we can write

$$\mathbf{a} = \begin{bmatrix} a_1 \\ a_2 \end{bmatrix} = a_1\mathbf{e}_1 + a_2\mathbf{e}_2 .$$

We, therefore, obtain the relationship between coordinate axes and a basis set of vectors. For n-space geometry, we extend the ideas considered in two- and three-space and define n-coordinate axes $x_1, x_2, \ldots, x_n$, which are mutually orthogonal. Each point in this n-space has n components, so a point $\mathbf{a}$ is defined by

$$\mathbf{a} = \begin{bmatrix} a_1 \\ a_2 \\ \vdots \\ a_n \end{bmatrix}.$$

To locate this point we start at the origin and mark off a_1 units on the x_1 axis; from that point mark off a_2 units parallel to the x_2 axis; at that point mark off a_3 units parallel to the x_3 axis, and so on, until finally, we mark off a_n units parallel to the x_n axis and this locates the point $\mathbf{a}$. On the other hand, we can define n mutually orthogonal n-component vectors,

$$\mathbf{e}_1 = \begin{bmatrix} 1 \\ 0 \\ 0 \\ \vdots \\ 0 \\ 0 \end{bmatrix}; \quad \mathbf{e}_2 = \begin{bmatrix} 0 \\ 1 \\ 0 \\ \vdots \\ 0 \\ 0 \end{bmatrix}; \quad \ldots; \quad \mathbf{e}_n = \begin{bmatrix} 0 \\ 0 \\ 0 \\ \vdots \\ 0 \\ 1 \end{bmatrix}; \tag{4.1.1}$$

and we say that $\mathbf{e}_i$ represents a point that is one unit on the x_i axis. Now these n vectors are a basis in R_n, and any vector $\mathbf{a}$ is a linear combination of these vectors; that is,

$$\mathbf{a} = \sum_{i=1}^{n} a_i \mathbf{e}_i . \tag{4.1.2}$$

Of course, just as there are many basis sets in R_n, there are also many ways to define coordinate axes. However, we shall use axes such that $\mathbf{e}_i$ is a point one unit in the positive direction on the x_i axis.

We are interested in geometrical interpretations only as a means to getting an intuitive, or pictorial, insight into certain concepts in vector spaces. In succeeding chapters we often intermix words that are used in vector spaces and n-coordinate geometry. For example, we interchange the words point and vector.

Now we shall develop a few concepts in the terminology of n-coordinate geometry.

Definition 4.1.1

Euclidean n-space. *The n-component vector space R_n is defined to be a Euclidean space, denoted by E_n, if and only if the distance between any two points (vectors)* $\mathbf{a}$ *and* $\mathbf{b}$ *in R_n is defined to be*

$$d = \left[\sum_{i=1}^{n} (a_i - b_i)^2\right]^{1/2}.$$

Note: This definition of distance can also be written as

$$d = [(\mathbf{a} - \mathbf{b})'(\mathbf{a} - \mathbf{b})]^{1/2}.$$

When there is no chance for ambiguity, we shall use the symbol E_n to denote both a Euclidean n-space and the vector space R_n. As a fundamental reference system in E_n, we shall use rectangular (orthogonal) coordinate axes; and the basis set, when viewed as a vector space, will be the one given by the set of vectors $\mathbf{e}_1, \ldots, \mathbf{e}_n$ in Eq. (4.1.1).

Definitions of lines, planes, and angles in Euclidean n-space will be generalizations of the concepts in two- and three-space of analytic geometry.

4.2 Lines in E_n

From two-space analytic geometry, the line through any two points

$$\mathbf{a} = \begin{bmatrix} a_1 \\ a_2 \end{bmatrix} \quad \text{and} \quad \mathbf{b} = \begin{bmatrix} b_1 \\ b_2 \end{bmatrix}, \quad \mathbf{a} \neq \mathbf{b},$$

is defined to be the set of points $\mathscr{L} = \{(x, y)\}$, where (x, y) satisfies

$$y - a_2 = \frac{b_2 - a_2}{b_1 - a_1}(x - a_1), \quad \text{if } a_1 \neq b_1, \tag{4.2.1}$$

or

$$x = a_1, \quad \text{if } a_1 = b_1.$$

We could, however, just as well define the line that passes through the two points **a** and **b** to be the set of points $\mathscr{L}$ such that

$$\mathscr{L} = \left\{\begin{bmatrix} x \\ y \end{bmatrix} : \begin{bmatrix} x \\ y \end{bmatrix} = \lambda\mathbf{b} + (1 - \lambda)\mathbf{a}\,;\, \lambda \in R\right\}. \tag{4.2.2}$$

If we examine this, we obtain

$$\begin{bmatrix} x \\ y \end{bmatrix} = \begin{bmatrix} \lambda b_1 + (1 - \lambda)a_1 \\ \lambda b_2 + (1 - \lambda)a_2 \end{bmatrix},$$

and finally

$$x = \lambda b_1 + (1 - \lambda)a_1,$$

$$y = \lambda b_2 + (1 - \lambda)a_2.$$

If we solve each equation for λ, we obtain (assume $a_1 \neq a_2$, $b_1 \neq b_2$),

$$\lambda = \frac{x - a_1}{b_1 - a_1} \quad \text{and} \quad \lambda = \frac{y - a_2}{b_2 - a_2}.$$

By equating these, we obtain

$$\frac{y - a_2}{b_2 - a_2} = \frac{x - a_1}{b_1 - a_1},$$

which is equivalent to the line defined by Eq. (4.2.1). If $a_1 = b_1$ or $a_2 = b_2$, then a similar result is obtained. To define a line in E_n, we shall generalize Eq. (4.2.2).

Definition 4.2.1

Line in E_n. *The set $\mathscr{L}$ of points defined below is a line through the two points* $\mathbf{a}$ *and* $\mathbf{b}$ $(\mathbf{a} \neq \mathbf{b})$ *in* E_n.

$$\mathscr{L} = \{\mathbf{x} : \mathbf{x} = \lambda\mathbf{b} + (1 - \lambda)\mathbf{a};\ \lambda \in R\}.$$

The line segment that connects the two points $\mathbf{a}$ *and* $\mathbf{b}$ *in* E_n *is the set of points* ℓ *defined by*

$$\ell = \{\mathbf{x} : \mathbf{x} = \lambda\mathbf{b} + (1 - \lambda)\mathbf{a};\ 0 \leq \lambda \leq 1\}.$$

We shall now discuss the "direction" of lines. We often picture the direction of a line segment with an arrow to indicate the direction, as in Fig. 4.2.1. We assume that the reader is acquainted with the parallelogram law of addition and subtraction of two vectors.

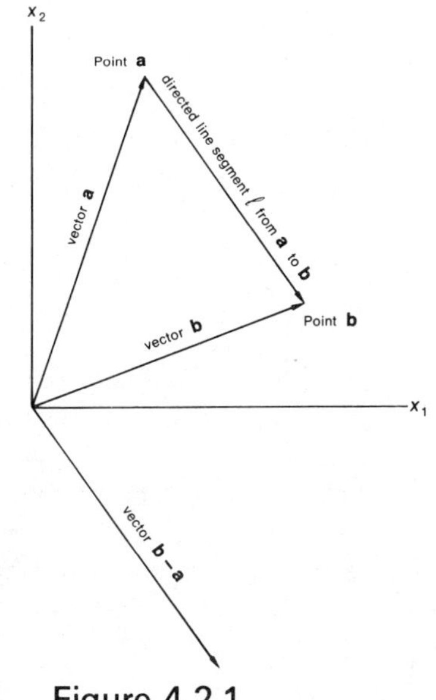

Figure 4.2.1

Note: A line segment—and a directed line segment—determines a line uniquely, since it is a portion of a line. However a line contains many line segments.

In two-space and three-space analytic geometry, two lines intersect if they have at least one point in common. In two-space, if two distinct lines have no points in common, they are defined to be parallel; however, in three-space, this is generally not the definition of parallel lines. In two-space analytic geometry, two lines are defined to be parallel if they have the same slope (direction). If we write the equation of one line in two-space as $y = mx + b$ and that of another as $y' = mx + a$, then these two lines are parallel. We notice that for any value x we obtain $y - y' = b - a$ or $y = y' + c$, where $c = b - a$; in other words, the difference between two ordinate values is a constant. This is the concept that we shall use to define parallel lines in E_n.

Definition 4.2.2

Parallel Lines in $\mathbf{E_n}$. *Let $\mathscr{L}_1$ and $\mathscr{L}_2$ be lines in E_n where*

$$\mathscr{L}_1 = \{\mathbf{y}_1 : \mathbf{y}_1 = \lambda \mathbf{b}_1 + (1 - \lambda)\mathbf{a}_1 ; \lambda \in R\},$$

$$\mathscr{L}_2 = \{\mathbf{y}_2 : \mathbf{y}_2 = \alpha \mathbf{b}_2 + (1 - \alpha)\mathbf{a}_2 ; \alpha \in R\}.$$

The two lines are defined to be parallel if and only if there exists a constant vector $\mathbf{c}$ such that for each point $\mathbf{y}_2$ on $\mathscr{L}_2$ there is a point $\mathbf{y}_1$ on $\mathscr{L}_1$ such that $\mathbf{y}_1 = \mathbf{y}_2 + \mathbf{c}$.

This definition states that if the equation for the line $\mathscr{L}_1$ can be written as $\mathbf{y}_1 = \lambda\mathbf{b} + (1 - \lambda)\mathbf{a}$, then a line $\mathscr{L}_2$ is parallel to $\mathscr{L}_1$ if and only if the equation for $\mathscr{L}_2$ can be written as $\mathbf{y}_2 = \lambda\mathbf{b} + (1 - \lambda)\mathbf{a} + \mathbf{c}$.

Definition 4.2.3

Intersection of Two Lines in $\mathbf{E_n}$. *Two lines (or line segments) $\mathscr{L}_1$ and $\mathscr{L}_2$ in E_n are said to intersect if and only if they have at least one point in common.*

Note: If two lines do *not* intersect they may not be parallel in E_n if $n > 2$. Also, if two lines have two or more points in common, the lines have all points in common and hence are the same line. This is the context of the next theorem.

Theorem 4.2.1

Let $\mathscr{L}_1$ and $\mathscr{L}_2$ be lines in E_n. If they intersect in more than one distinct point the lines are identical ($\mathscr{L}_1 = \mathscr{L}_2$).

Proof: Let $\mathbf{p}_1$ and $\mathbf{p}_2$ be two distinct points of intersection. Then $\mathbf{p}_1$ and $\mathbf{p}_2$ are on $\mathscr{L}_1$, and $\mathscr{L}_1$ can be defined by these two points;

$$\mathscr{L}_1 = \{\mathbf{y}_1 : \mathbf{y}_1 = \lambda\mathbf{p}_1 + (1 - \lambda)\mathbf{p}_2 ; \lambda \in R\}.$$

The same is true for $\mathscr{L}_2$ since, by the hypothesis of the theorem, $\mathbf{p}_1$ and $\mathbf{p}_2$ are on $\mathscr{L}_2$. Therefore,

$$\mathscr{L}_2 = \{\mathbf{y}_2 : \mathbf{y}_2 = \alpha\mathbf{p}_1 + (1-\alpha)\mathbf{p}_2;\ \alpha \in R\},$$

and, clearly, every point on $\mathscr{L}_1$ is also on $\mathscr{L}_2$, and vice versa. Therefore, $\mathscr{L}_1 = \mathscr{L}_2$. ▮

Theorem 4.2.2

If two lines $\mathscr{L}_1$ and $\mathscr{L}_2$ in E_n are parallel and intersect in at least one point, then the two lines are identical.

The proof of this theorem is left for the reader.

Note: Theorems 4.2.1 and 4.2.2 are not necessarily true if lines $\mathscr{L}_1$ and $\mathscr{L}_2$ are replaced by line segments ℓ_1 and ℓ_2.

Example 4.2.1. Consider two lines in E_3. Let the four points **a, b, c, d** be defined by

$$\mathbf{a} = \begin{bmatrix}1\\0\\1\end{bmatrix}; \quad \mathbf{b} = \begin{bmatrix}1\\1\\0\end{bmatrix}; \quad \mathbf{c} = \begin{bmatrix}1\\0\\0\end{bmatrix}; \quad \mathbf{d} = \begin{bmatrix}0\\1\\0\end{bmatrix}.$$

Let the equation for $\mathscr{L}_1$ through the points **a** and **b** be $\mathbf{y} = \lambda\mathbf{b} + (1-\lambda)\mathbf{a}$, which is

$$\begin{bmatrix}y_1\\y_2\\y_3\end{bmatrix} = \begin{bmatrix}1\\\lambda\\1-\lambda\end{bmatrix},$$

and the equation for $\mathscr{L}_2$ through the points **c** and **d** be $\mathbf{x} = \alpha\mathbf{d} + (1-\alpha)\mathbf{c}$, which is

$$\begin{bmatrix}x_1\\x_2\\x_3\end{bmatrix} = \begin{bmatrix}1-\alpha\\\alpha\\0\end{bmatrix}.$$

If these lines intersect, there must be at least one value of λ and α such that $\mathbf{y} = \mathbf{x}$. We would have

$$\begin{bmatrix}1\\\lambda\\1-\lambda\end{bmatrix} = \begin{bmatrix}1-\alpha\\\alpha\\0\end{bmatrix}. \qquad (4.2.3)$$

Clearly there are no α and λ such that Eq. (4.2.3) is satisfied, and hence the lines do not intersect. Next, we examine them to see if they are parallel. We can write

$$\mathbf{y} = (\mathbf{b} - \mathbf{a})\lambda + \mathbf{a}; \qquad \mathbf{x} = (\mathbf{d} - \mathbf{c})\alpha + \mathbf{c},$$

and for the lines to be parallel, there must exist for each value of λ a value of α such that $\mathbf{y} - \mathbf{x} = \mathbf{k}$, where $\mathbf{k}$ is a constant vector (does not depend on α and λ). *This relationship implies that the vector* $\mathbf{a} - \mathbf{b}$ *must be proportional to the vector* $\mathbf{c} - \mathbf{d}$; that is, $\mathbf{a} - \mathbf{b}$ is a scalar constant times $\mathbf{c} - \mathbf{d}$. Clearly this is not true, since

$$\mathbf{a} - \mathbf{b} = \begin{bmatrix} 0 \\ -1 \\ 1 \end{bmatrix} \quad \text{and} \quad \mathbf{c} - \mathbf{d} = \begin{bmatrix} 1 \\ -1 \\ 0 \end{bmatrix},$$

so the lines are not parallel.

Let ℓ_1 be a directed line segment from $\mathbf{0}$ to $\mathbf{a}$ in E_2 and let ℓ_2 be a directed line segment from $\mathbf{0}$ to $\mathbf{b}$ in E_2 (see Fig. 4.2.2). These two intersecting line segments do not

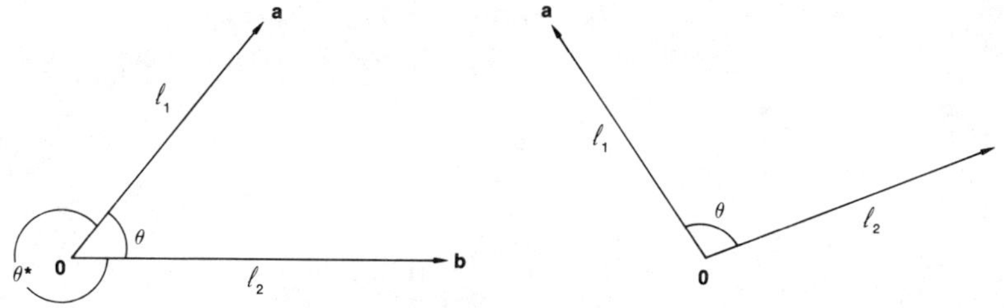

Figure 4.2.2

determine a unique angle, since in Fig. 4.2.2 (a), θ and θ^* are both angles of intersection. However, if we limit the angle to the interval from zero to π, then they determine a unique angle. It can be acute, as in Fig. 4.2.2 (a), or obtuse, as in Fig. 4.2.2 (b). In Fig. 4.2.3, you will note that the angle θ_1 is the unique angle $0 \leq \theta_1 \leq \pi$ between the

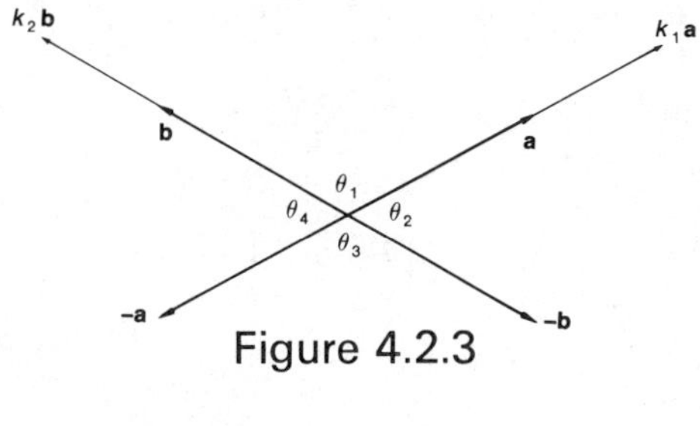

Figure 4.2.3

directed line segments determined by the points **0**, **a** and the points **0**, **b**. However, if k_1 and k_2 are any positive scalars, then θ_1 is also the angle between the directed line segments determined by the points **0**, $k_1\mathbf{a}$ and **0**, $k_2\mathbf{b}$. If k_1 is negative and k_2 is negative or if either is negative, then the angle between the directed line segments may be different.

We can use the law of cosines to compute the angle θ between two directed line segments. Let d_1 be the distance from **0** to **a**; d_2 the distance from **0** to **b**; d_3 the distance between **a** and **b**. By the law of cosines, we obtain (see Fig. 4.2.4)

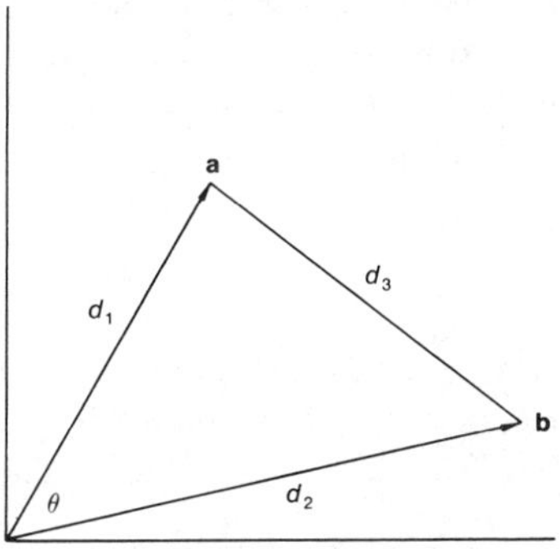

Figure 4.2.4

$$d_3^2 = d_1^2 + d_2^2 - 2d_1 d_2 \cos\theta.$$

Substituting $d_3^2 = (\mathbf{a} - \mathbf{b})'(\mathbf{a} - \mathbf{b})$; $d_1^2 = \mathbf{a}'\mathbf{a}$; $d_2^2 = \mathbf{b}'\mathbf{b}$, we get

$$\cos\theta = \frac{\mathbf{a}'\mathbf{b}}{\sqrt{\mathbf{a}'\mathbf{a}}\sqrt{\mathbf{b}'\mathbf{b}}}, \qquad 0 \le \theta \le \pi. \tag{4.2.4}$$

The formula in Eq. (4.2.4) will allow us to generalize from E_2 to E_n.

Definition 4.2.4

Angle between Two Directed Line Segments in E_n That Intersect at the Origin. *Let ℓ_1 be the directed line segment in E_n from* **0** *to* **a**; *let ℓ_2 be the directed line segment in E_n from* **0** *to* **b**, *where* $\mathbf{a} \ne \mathbf{0}$, $\mathbf{b} \ne \mathbf{0}$. *The angle θ such that $0 \le \theta \le \pi$, formed by ℓ_1 and ℓ_2, is defined by the formula*

$$\cos\theta = \frac{\mathbf{a}'\mathbf{b}}{\sqrt{\mathbf{a}'\mathbf{a}}\sqrt{\mathbf{b}'\mathbf{b}}}, \qquad 0 \le \theta \le \pi. \tag{4.2.5}$$

Note: Remember that even though there are two angles determined by two directed line segments, we always pick the one in the interval $0 \leq \theta \leq \pi$. This formula only defines the angle between two directed line *segments* from the origin to **a** and **b** respectively, which is equivalent to the angle between the two vectors **a** and **b**.

We note that $\cos \theta = 0$ in Eq. (4.2.5), if and only if $\mathbf{a}'\mathbf{b} = 0$, and when this is the case we define the two lines to be perpendicular. If we consider **a** and **b** as vectors, then $\cos \theta = 0$ if and only if they are orthogonal (Sec. 2.6).

We want some way of defining the direction of a line. This direction will be relative to the n-coordinate axes. We note that the point $\mathbf{0}$ and the point $\mathbf{a}$ where $\mathbf{a} \neq \mathbf{0}$ determine a line uniquely. Also, the points $\mathbf{0}$ and $k\mathbf{a}$ for any scalar $k \neq 0$ determine this same line uniquely. We make use of these facts to define direction of a line.

Definition 4.2.5

Direction Vector of a Line in E_n through the Origin. *Let $\mathscr{L}$ be a line through the origin and the point* $\mathbf{a}$ *where* $\mathbf{a} \neq \mathbf{0}$. *The vector* $k\mathbf{a}$ *for any scalar $k \neq 0$ is defined to be a direction vector of $\mathscr{L}$.*

For a directed line segment ℓ from $\mathbf{0}$ to $\mathbf{a}$ where $\mathbf{a} \neq \mathbf{0}$, we shall define a direction cosine vector. The i-th coordinate of this vector is the cosine of the angle γ_i between ℓ and the positive x_i axis (actually the directed line segment from $\mathbf{0}$ to $\mathbf{e}_i$). From Def. 4.2.4, we obtain

$$\cos \gamma_i = \frac{\mathbf{e}_i'\mathbf{a}}{\sqrt{\mathbf{e}_i'\mathbf{e}_i}\sqrt{\mathbf{a}'\mathbf{a}}} = \frac{a_i}{\sqrt{\mathbf{a}'\mathbf{a}}}, \qquad 0 \leq \gamma_i \leq \pi. \tag{4.2.6}$$

Definition 4.2.6

Direction Angles of Directed Line Segment in E_n through the Origin. *Let ℓ be a directed line segment from* $\mathbf{0}$ *to* $\mathbf{a}$; $\mathbf{a} \neq \mathbf{0}$. *Let γ_i be the angle between ℓ and the directed line segment from* $\mathbf{0}$ *to* $\mathbf{e}_i$. *Then γ_i is the angle between ℓ and the* i-th *coordinate axis, and the set of angles $\{\gamma_1, \gamma_2, \ldots, \gamma_n\}$ is defined as the direction angles of ℓ. The formula for computing the γ_i is*

$$\cos \gamma_i = \frac{a_i}{\sqrt{\mathbf{a}'\mathbf{a}}}, \qquad 0 \leq \gamma_i \leq \pi; \quad i = 1, 2, \ldots, n.$$

We call the set of $\cos \gamma_i$; $i = 1, 2, \ldots, n$, the direction cosines of ℓ. We call the set of a_i (the elements of **a**) direction numbers of ℓ. The set of elements in $k\mathbf{a}$, for any $k > 0$, is also called a set of direction numbers for ℓ.

Example 4.2.2. Consider the directed line segment ℓ in E_4 from the origin to the point **a** where

$$\mathbf{a} = \begin{bmatrix} 1 \\ 2 \\ -2 \\ 0 \end{bmatrix}.$$

The direction cosines of ℓ are elements of a vector **c**, where

$$\mathbf{c} = \begin{bmatrix} \cos\gamma_1 \\ \cos\gamma_2 \\ \cos\gamma_3 \\ \cos\gamma_4 \end{bmatrix} = \frac{1}{\sqrt{\mathbf{a}'\mathbf{a}}}\mathbf{a} = \frac{1}{3}\mathbf{a} = \begin{bmatrix} 1/3 \\ 2/3 \\ -2/3 \\ 0 \end{bmatrix}. \tag{4.2.7}$$

One set of direction numbers of ℓ is

$$\mathbf{a} = \begin{bmatrix} 1 \\ 2 \\ -2 \\ 0 \end{bmatrix}.$$

Another is

$$3\mathbf{a} = \begin{bmatrix} 3 \\ 6 \\ -6 \\ 0 \end{bmatrix},$$

and so on. Clearly, the direction cosines of the directed line segment from **0** to 3**a** (and in fact from **0** to $k\mathbf{a}$ for $k > 0$) are the same as those in Eq. (4.2.7). The direction cosines of the directed line segment from **0** to $-\mathbf{a}$ are

$$\mathbf{c}^* = \begin{bmatrix} \cos\gamma_1^* \\ \cos\gamma_2^* \\ \cos\gamma_3^* \\ \cos\gamma_4^* \end{bmatrix} = \frac{-\mathbf{a}}{\sqrt{(-\mathbf{a})'(-\mathbf{a})}} = \begin{bmatrix} -1/3 \\ -2/3 \\ 2/3 \\ 0 \end{bmatrix},$$

and clearly $\mathbf{c}^* = -\mathbf{c}$. However, both of these sets of direction cosines determine the same line $\mathscr{L}$ that passes through **0** and **a** (or **0** and $-\mathbf{a}$). A direction vector of the line $\mathscr{L}$ is the vector **a**. We note that $\mathscr{L}$ is defined by

$$\mathscr{L} = \{\mathbf{y} : \mathbf{y} = \lambda\mathbf{a} + (1 - \lambda)\mathbf{0};\ \lambda \in R\},$$

which reduces to

$$\mathscr{L} = \{\mathbf{y} : \mathbf{y} = \lambda\mathbf{a};\ \lambda \in R\}.$$

Next we shall consider a line through any two distinct points **a** and **b**. The line $\mathscr{L}$ is defined by

$$\mathscr{L} = \{\mathbf{y} : \mathbf{y} = \lambda\mathbf{b} + (1 - \lambda)\mathbf{a};\ \lambda \in R\}.$$

The equation of the line can be written as

$$\mathbf{y} = (\mathbf{b} - \mathbf{a})\lambda + \mathbf{a}; \qquad \lambda \in R,$$

and, hence, the line can be defined by

$$\mathscr{L} = \{\mathbf{y} : \mathbf{y} = (\mathbf{b} - \mathbf{a})\lambda + \mathbf{a};\ \lambda \in R\}.$$

But, by Def. 4.2.2, this line is parallel to the line $\mathscr{L}^*$ that passes through the origin and the point $\mathbf{b} - \mathbf{a}$, where $\mathscr{L}^*$ is defined by

$$\mathscr{L}^* = \{\mathbf{x} : \mathbf{x} = (\mathbf{b} - \mathbf{a})\lambda + (1 - \lambda)\mathbf{0};\ \lambda \in R\},$$

which reduces to

$$\mathscr{L}^* = \{\mathbf{x} : \mathbf{x} = (\mathbf{b} - \mathbf{a})\lambda;\ \lambda \in R\}.$$

But a direction vector of the line $\mathscr{L}^*$ is $\mathbf{b} - \mathbf{a}$ and this uniquely determines the direction of $\mathscr{L}^*$, and since $\mathscr{L}^*$ passes through the origin, $\mathscr{L}^*$ is uniquely determined. Since $\mathscr{L}$ and $\mathscr{L}^*$ are parallel, it seems reasonable to define the direction of $\mathscr{L}$ to be the same as the direction of $\mathscr{L}^*$. This is, in fact, what we shall use to define the direction of a line $\mathscr{L}$ that does not necessarily pass through the origin **0**. A similar result is true for two directed line segments that do not intersect.

Definition 4.2.7

Direction Angles of a Directed Line Segment in E_n. *Let ℓ be a directed line segment in E_n from the point* $\mathbf{a}$ *to the point* $\mathbf{b}$, *where* $\mathbf{a} \neq \mathbf{b}$. *The direction angles γ_i of ℓ; $i = 1, 2, \ldots, n$, are defined by*

$$\cos \gamma_i = \frac{b_i - a_i}{\sqrt{(\mathbf{b} - \mathbf{a})'(\mathbf{b} - \mathbf{a})}}; \qquad i = 1, 2, \ldots, n. \tag{4.2.8}$$

Definition 4.2.8

Direction Vector of a Line in E_n. *A direction vector of the line $\mathscr{L}$ in E_n through the points* $\mathbf{a}$ *and* $\mathbf{b}$, $\mathbf{a} \neq \mathbf{b}$ *is defined to be a vector* $k(\mathbf{b} - \mathbf{a})$, *where k is any nonzero scalar.*

Note: If $\mathscr{L}$ is a line with direction vector $\mathbf{d}$ and if $\mathscr{L}$ passes through the point $\mathbf{c}$, then the equation for $\mathscr{L}$ is $\mathbf{y} = \mathbf{d}\alpha + \mathbf{c}$; $\alpha \in R$. Also, if $\mathscr{L}^*$ is a line through the two points $\mathbf{a}$, $\mathbf{b}$ with $\mathbf{a} \neq \mathbf{b}$, then a direction vector of $\mathscr{L}^*$ is $\mathbf{d}^* = \mathbf{b} - \mathbf{a}$, and the equation for $\mathscr{L}^*$ can be written as $\mathbf{y}^* = \mathbf{d}^*\alpha + \mathbf{b}$; $\alpha \in R$, or as $\mathbf{y}^* = \mathbf{d}^*\alpha + \mathbf{a}$; $\alpha \in R$.

Theorem 4.2.3

Two lines in E_n are parallel if and only if they have proportional direction vectors.

Proof: This follows from Defs. 4.2.2 and 4.2.8. ▮

We also define two lines in E_n to be perpendicular (or orthogonal) if and only if they have orthogonal direction vectors.

Definition 4.2.9

Length of Line Segment in E_n. *Let ℓ be a line segment in E_n between the points* $\mathbf{a}$ *and* $\mathbf{b}$ *with* $\mathbf{a} \neq \mathbf{b}$. *The length of ℓ is defined to be the distance between* $\mathbf{a}$ *and* $\mathbf{b}$.

It is also important to define the distance between a point $\mathbf{p}$ and a line $\mathscr{L}$ through the points $\mathbf{a}$ and $\mathbf{b}$, where $\mathbf{a} \neq \mathbf{b}$. It is clear that, from the point $\mathbf{p}$ to any point $\mathbf{x}$ on $\mathscr{L}$, the distance is

$$D_{\mathbf{x}} = [(\mathbf{p} - \mathbf{x})'(\mathbf{p} - \mathbf{x})]^{1/2}; \tag{4.2.9}$$

and by the distance from a point to a line we shall mean the shortest distance (see Fig. 4.2.5).

Definition 4.2.10

Distance from a Point to a Line. *The distance from a point* **p** *to a line* $\mathscr{L}$ *in* E_n *through the points* **a** *and* **b** *is* D, *where*

$$D = \min_{\mathbf{x} \in \mathscr{L}} [(\mathbf{p} - \mathbf{x})'(\mathbf{p} - \mathbf{x})]^{1/2}. \tag{4.2.10}$$

Since **x** is on $\mathscr{L}$, as λ varies over the real numbers, **x** takes on each value of $\mathscr{L}$; that is, $\mathbf{x} = \lambda(\mathbf{b} - \mathbf{a}) + \mathbf{a}$. Therefore, the distance from **p** to $\mathscr{L}$ can also be defined by

$$D = \min_{\lambda \in R} d(\lambda) = \min_{\lambda \in R} \{[\mathbf{p} - \lambda(\mathbf{b} - \mathbf{a}) - \mathbf{a}]'[\mathbf{p} - \lambda(\mathbf{b} - \mathbf{a}) - \mathbf{a}]\}^{1/2}. \tag{4.2.11}$$

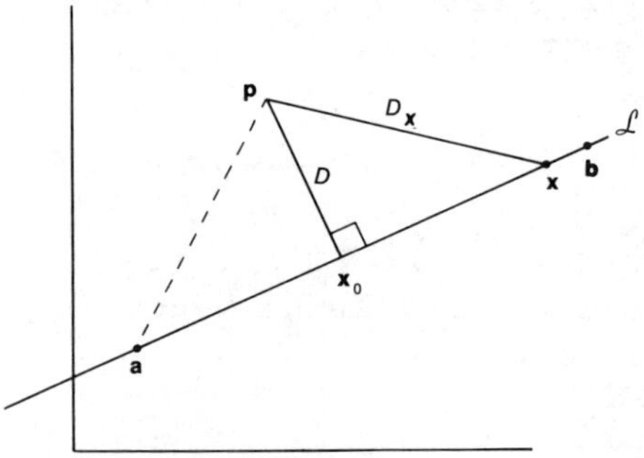

Figure 4.2.5

If we set to zero the derivative of $d(\lambda)$ with respect to λ, we obtain the fact that

$$D = \left\{\frac{(\mathbf{p} - \mathbf{a})'(\mathbf{p} - \mathbf{a})(\mathbf{b} - \mathbf{a})'(\mathbf{b} - \mathbf{a}) - [(\mathbf{p} - \mathbf{a})'(\mathbf{b} - \mathbf{a})]^2}{(\mathbf{b} - \mathbf{a})'(\mathbf{b} - \mathbf{a})}\right\}^{1/2}, \tag{4.2.12}$$

and we have proved the following theorem.

Theorem 4.2.4

The distance D from the point **p** *to the line* $\mathscr{L}$ *in* E_n *through the points* **a** *and* **b** *where* $\mathbf{a} \neq \mathbf{b}$ *is given by* Eq. (4.2.12).

4.3 Planes in E_n

In three-space analytic geometry with coordinate axes $x_1 x_2 x_3$, the equation of a plane can be written as

$$a_1 x_1 + a_2 x_2 + a_3 x_3 = c, \tag{4.3.1}$$

where a_1, a_2, a_3, and c are constants. The plane is defined to be the set of points $\{x_1, x_2, x_3\}$ that satisfy Eq. (4.3.1). If b is any nonzero scalar, and if every term in Eq. (4.3.1) is multiplied by b, we obtain

$$ba_1x_1 + ba_2x_2 + ba_3x_3 = bc,$$

and this is an equation of the same plane as the one defined by Eq. (4.3.1). There are two distinct situations: $c = 0$ and $c \neq 0$; and the equation of a plane can be written with $c = 0$ in Eq. (4.3.1) if and only if $\mathbf{x} = \mathbf{0}$ satisfies the equation; otherwise the equation for a plane can always be written with $c = 1$. We can write Eq. (4.3.1) as

$$\mathbf{a}'\mathbf{x} = c,$$

where $\mathbf{a}$ is the fixed point, $\mathbf{a}' = (a_1, a_2, a_3) \neq \mathbf{0}$, and we could define the plane as the set of points $\mathscr{P}$ such that

$$\mathscr{P} = \{\mathbf{x} : \mathbf{a}'\mathbf{x} = c;\ \mathbf{x} \in R_3\}.$$

Let $\mathbf{x}_1$ and $\mathbf{x}_2$ be two points on the plane $\mathscr{P}$. Then $\mathbf{a}'\mathbf{x}_1 = \mathbf{a}'\mathbf{x}_2 = c$ and $\mathbf{a}'(\mathbf{x}_1 - \mathbf{x}_2) = 0$; hence the line through the points $\mathbf{0}$ and $\mathbf{a}$ is orthogonal to the line through $\mathbf{0}$ and $\mathbf{x}_1 - \mathbf{x}_2$. In fact, let $\mathscr{L}_1$ be the line through two distinct points $\mathbf{x}_1$ and $\mathbf{x}_2$ on $\mathscr{P}$. Then the equation for $\mathscr{L}_1$ is

$$\mathbf{y}_1 = \lambda(\mathbf{x}_1 - \mathbf{x}_2) + \mathbf{x}_2; \qquad \lambda \in R.$$

Let $\mathscr{L}_2$ be the line through the points $\mathbf{0}$ and $\mathbf{a}$. The equation for $\mathscr{L}_2$ is

$$\mathbf{y}_2 = \alpha\mathbf{a}; \qquad \alpha \in R.$$

These two lines are perpendicular for each and every choice of $\mathbf{x}_1$, $\mathbf{x}_2$ on $\mathscr{P}$; that is to say, the line $\mathscr{L}_2$ is orthogonal to each line through every two distinct points on $\mathscr{P}$. On the other hand, each and every point on $\mathscr{P}$ is on a line that is entirely in $\mathscr{P}$; hence, we see that the line $\mathscr{L}_2$ from $\mathbf{0}$ to $\mathbf{a}$ is perpendicular to every line in $\mathscr{P}$, and we say that the line $\mathscr{L}_2$ is normal to the plane. We note that the point $\mathbf{a}$ and the scalar c determine the plane uniquely.

We have shown that two distinct points in E_3 determine a line uniquely, and we shall now show that three linearly independent points in E_3 determine a unique plane. Let $\mathbf{x}_1$, $\mathbf{x}_2$, $\mathbf{x}_3$ be three linearly independent points. (This means that the three vectors are linearly independent.)

The equation of a plane in E_3 is $\mathbf{a}'\mathbf{x} = c$. Assume that $\mathbf{a}$ and c are unknown but we know that the three points $\mathbf{x}_1$, $\mathbf{x}_2$, $\mathbf{x}_3$ are on the plane. We shall show that, from a knowledge of these three points, we can compute $\mathbf{a}$ and the equation of the plane. Since $\mathbf{x}_1$, $\mathbf{x}_2$, $\mathbf{x}_3$ are on the plane, we obtain

$$\mathbf{a}'\mathbf{x}_1 = c, \qquad \mathbf{a}'\mathbf{x}_2 = c, \qquad \mathbf{a}'\mathbf{x}_3 = c;$$

if we combine these into one equation we obtain

$$\mathbf{a}'[\mathbf{x}_1, \mathbf{x}_2, \mathbf{x}_3] = [c, c, c]$$

or

$$\mathbf{a}'\mathbf{X} = c\mathbf{1}',$$

and, since $\mathbf{X}$ is nonsingular, we obtain $\mathbf{a}' = c\mathbf{1}'\mathbf{X}^{-1}$; since $\mathbf{a}'$ cannot be zero, this implies that $c \neq 0$, so let $c = 1$ and $\mathbf{a}' = \mathbf{1}'\mathbf{X}^{-1}$. Thus, from the knowledge of the matrix $\mathbf{X}$ (from the 3 points $\mathbf{x}_1$, $\mathbf{x}_2$, $\mathbf{x}_3$), we can compute $\mathbf{a}'$, and from this we can write the equation of the plane as

$$\mathbf{a}'\mathbf{x} = 1.$$

We could generalize the equation $\mathbf{a}'\mathbf{x} = c$ to define a plane in E_n as follows: let $\mathbf{a}$ be any nonzero vector in E_n and let c be any scalar, then the set of points $\mathscr{P}$ is defined to be a plane in E_n, where

$$\mathscr{P} = \{\mathbf{x} : \mathbf{a}'\mathbf{x} = c; \mathbf{x} \in E_n\}.$$

We note that the plane goes through the origin (that is, $\mathbf{x} = \mathbf{0}$ satisfies the equation of the plane) if and only if $c = 0$.

We shall use a different definition than the one described above, since it will be more suitable for our purposes. We shall then show that the above equation of a plane can be obtained from this definition.

For the definition of a plane in E_n we generalize Def. 4.2.1 of a line in E_n, first discussing a plane that does *not* go through the origin

Definition 4.3.1

Plane in E_n Not through the Origin. *Let $\mathbf{b}_1, \mathbf{b}_2, \ldots, \mathbf{b}_n$ be n points in E_n such that $\mathbf{B} = [\mathbf{b}_1, \mathbf{b}_2, \ldots, \mathbf{b}_n]$ has rank n. The plane through these n points is defined to be the set of points $\mathscr{P}$ where*

$$\mathscr{P} = \{\mathbf{y} : \mathbf{y} = \sum_{i=1}^{n} \lambda_i \mathbf{b}_i; \Sigma\lambda_i = 1; \lambda_i \in R\}. \tag{4.3.2}$$

If we let $\boldsymbol{\lambda}' = [\lambda_1, \ldots, \lambda_n]$, then $\mathbf{1}'\boldsymbol{\lambda} = 1$ and $\mathbf{y} = \mathbf{B}\boldsymbol{\lambda}$. Since $\mathbf{B}$ is nonsingular, we get $\boldsymbol{\lambda} = \mathbf{B}^{-1}\mathbf{y}$ and $\mathbf{1}'\boldsymbol{\lambda} = 1 = \mathbf{1}'\mathbf{B}^{-1}\mathbf{y}$. If we set $\mathbf{1}'\mathbf{B}^{-1} = \mathbf{a}'$, then $\mathbf{a}'\mathbf{y} = 1$ is also the equation of the plane $\mathscr{P}$. If we start with $\mathbf{a}'\mathbf{y} = 1$ and if $\mathbf{b}_1, \ldots, \mathbf{b}_n$ are n points on the plane, we get

$$\mathbf{a}'\mathbf{b}_1 = 1, \ \mathbf{a}'\mathbf{b}_2 = 1, \ \ldots, \ \mathbf{a}'\mathbf{b}_n = 1,$$

or if $\boldsymbol{\lambda}' = [\lambda_1, \lambda_2, \ldots, \lambda_n]$ such that $\Sigma\lambda_i = 1$, then

$$\mathbf{a}'\mathbf{b}_1\lambda_1 = \lambda_1; \quad \mathbf{a}'\mathbf{b}_2\lambda_2 = \lambda_2; \quad \ldots; \mathbf{a}'\mathbf{b}_n\lambda_n = \lambda_n \quad \text{and} \quad \mathbf{a}'\Sigma\mathbf{b}_i\lambda_i = 1$$

and hence the point $\mathbf{w} = \Sigma\lambda_i\mathbf{b}_i$ is on the plane. Thus, we have proved the following theorem.

Theorem 4.3.1

Let $\mathbf{a}$ be any nonzero vector in E_n and let c be any nonzero scalar. The set of points $\mathscr{P}$ defined by

$$\mathscr{P} = \{\mathbf{y} : \mathbf{a}'\mathbf{y} = c; \ \mathbf{y} \in R_n\} \tag{4.3.3}$$

is a plane in E_n that does not go through the origin.

The next two theorems concern lines that lie in planes that do not go through the origin. The first states that if $\mathbf{y}_1$ and $\mathbf{y}_2$ are two distinct points on $\mathscr{P}$, then every point on the line through $\mathbf{y}_1$ and $\mathbf{y}_2$ is on the plane $\mathscr{P}$.

Theorem 4.3.2

Let $\mathscr{L}$ be a line through the points $\mathbf{y}_1$ and $\mathbf{y}_2$ where $\mathbf{y}_1$ and $\mathbf{y}_2 (\mathbf{y}_1 \neq \mathbf{y}_2)$ are both on a plane $\mathscr{P}$ in E_n. Then $\mathscr{L}$ is on $\mathscr{P}$.

Proof: Each and every point $\mathbf{y}$ on $\mathscr{L}$ can be written as $\mathbf{y} = \lambda\mathbf{y}_1 + (1 - \lambda)\mathbf{y}_2$. But

$$\mathbf{a}'\mathbf{y} = \mathbf{a}'\mathbf{y}_1\lambda + \mathbf{a}'\mathbf{y}_2(1 - \lambda) = c\lambda + (1 - \lambda)c = c$$

and hence $\mathbf{y}$ is on $\mathscr{P}$. ▮

Theorem 4.3.3

Let the equation of a plane $\mathscr{P}$ in E_n be $\mathbf{a}'\mathbf{y} = c$. The line through the points $\mathbf{0}$ and $\mathbf{a}$ is orthogonal to every line in $\mathscr{P}$.

Proof: A line $\mathscr{L}_1$ in $\mathscr{P}$ must pass through at least two distinct points, say $\mathbf{y}_1$ and $\mathbf{y}_2$ in $\mathscr{P}$. This implies that $\mathbf{a}'\mathbf{y}_1 = c$, $\mathbf{a}'\mathbf{y}_2 = c$, and $\mathbf{a}'(\mathbf{y}_1 - \mathbf{y}_2) = 0$. The equation of the line $\mathscr{L}_1$ in $\mathscr{P}$ is

$$\mathbf{y} = \lambda(\mathbf{y}_1 - \mathbf{y}_2) + \mathbf{y}_2,$$

and the equation of the line $\mathscr{L}_2$ through the points $\mathbf{0}$ and $\mathbf{a}$ is $\mathbf{x} = \alpha\mathbf{a}$. These two lines are orthogonal, since $\mathbf{a}'(\mathbf{y}_1 - \mathbf{y}_2) = 0$. ▮

If we view the points $\mathbf{y}$ on the plane in Def. 4.3.1 as vectors, we notice that $\mathscr{P}$ is not a vector space, since it does not include the zero vector. However, there are planes such that if the points on the plane are considered as vectors, these planes (sets of vectors) form a vector space. These are of prime importance in statistics, so the remainder of our discussion about planes will concern those that go through the origin.

In E_3 there are three linear surfaces: planes, which have dimension two; lines, which have dimension one; and points, which have dimension zero. They are defined by

plane: $\{\mathbf{y} : \mathbf{y} = \lambda_1\mathbf{a}_1 + \lambda_2\mathbf{a}_2 + (1 - \lambda_1 - \lambda_2)\mathbf{a}_3;\ \lambda_1, \lambda_2 \in R\}$,

line: $\{\mathbf{y} : \mathbf{y} = \lambda_1\mathbf{a}_1 + (1 - \lambda_1)\mathbf{a}_3;\ \lambda_1 \in R\}$,

point: $\{\mathbf{y} : \mathbf{y} = \mathbf{a}_3\}$.

Notice that the planes are determined by three points, $\mathbf{a}_1$, $\mathbf{a}_2$, $\mathbf{a}_3$, say; the lines by two points, $\mathbf{a}_1$, $\mathbf{a}_2$, say. We can get the equation of a plane by using two scalar parameters, λ_1, λ_2 and the equation of a line by using one scalar parameter, λ_1. Now in E_n, a similar situation obtains; a plane defined in Def. 4.3.1 has dimension $n - 1$. However there are planes of dimension $n - 2, n - 3, \ldots, 2, 1, 0$. We shall now give these names.

Definition 4.3.2

Plane of Dimension k through the Origin in E_n. *Let* $\mathbf{a}_1, \ldots, \mathbf{a}_k, \mathbf{0}$ *be* $k + 1$ *points in* E_n $(1 \le k < n)$ *such that, when considered as vectors, the set* $\{\mathbf{a}_1, \mathbf{a}_2, \ldots, \mathbf{a}_k\}$ *is linearly independent. Then a plane in* E_n *of dimension k through the points* $\mathbf{a}_1, \ldots, \mathbf{a}_k$, $\mathbf{0}$ *is defined to be the set of points* $\mathscr{P}_k^n$, *where*

$$\mathscr{P}_k^n = \left\{\mathbf{y} : \mathbf{y} = \sum_{i=1}^{k} \lambda_i \mathbf{a}_i;\ \lambda_i \in R;\ i = 1, 2, \ldots, k\right\}.$$

In vector terminology, the equation for $\mathscr{P}_k^n$ can be written

$$\mathbf{y} = \mathbf{A}\boldsymbol{\lambda}; \qquad \boldsymbol{\lambda} \in R_k; \quad \mathbf{A} = [\mathbf{a}_1, \mathbf{a}_2, \ldots, \mathbf{a}_k].$$

Note: The plane $\mathscr{P}_k^n$ is a vector subspace of E_n of dimension k. We could define a plane of dimension k that does not pass through the origin, but this definition will suffice for our purposes. If $k = 1$, we have defined $\mathscr{P}_1^n$ as a line through $\mathbf{0}$ and $\mathbf{a}_1$.

We are interested next in the intersection of planes of various dimensions in E_n and the "projections" of lines onto planes in E_n.

4.4 Projections

In our discussion of the projections of directed line segments onto lines and onto planes, we first consider the situation for E_2; refer to Fig. 4.4.1.

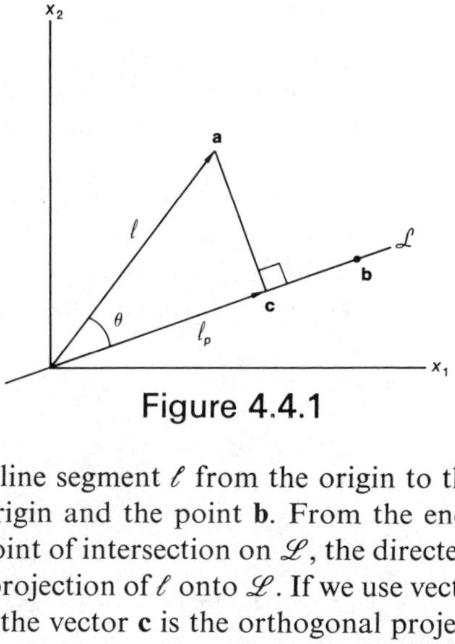

Figure 4.4.1

Consider the directed line segment ℓ from the origin to the point **a** and the line $\mathscr{L}$ that goes through the origin and the point **b**. From the end of ℓ, a perpendicular is drawn to $\mathscr{L}$. If **c** is the point of intersection on $\mathscr{L}$, the directed line segment from $\mathbf{0}$ to **c** is called the orthogonal projection of ℓ onto $\mathscr{L}$. If we use vector, rather than geometric, terminology we say that the vector **c** is the orthogonal projection of the vector **a** onto the vector **b**. Hereafter we shall simply use the word projection to mean "orthogonal projection."

The projection ℓ_p is defined by

$$\ell_p = \{\mathbf{y} : \mathbf{y} = \lambda\mathbf{c}; \quad 0 \le \lambda \le 1\}.$$

Therefore, to determine ℓ_p, we need only determine the point **c**, and we notice that **c** is simply a scalar multiple of **b**; that is, $\mathbf{c} = k\mathbf{b}$. So we want to determine k from a knowledge of **a** and **b**. If we let d denote the directed length of the line segment from **0** to **c**, then

$$d = \sqrt{\mathbf{a}'\mathbf{a}} \cos\theta,$$

where θ is the angle between ℓ and $\mathscr{L}$. If $\mathbf{b}^*$ denotes the unit vector in the direction from **0** to **b**, then

$$\mathbf{b}^* = \frac{\mathbf{b}}{\sqrt{\mathbf{b}'\mathbf{b}}} \quad \text{and} \quad \mathbf{c} = d\mathbf{b}^* = \frac{d\mathbf{b}}{\sqrt{\mathbf{b}'\mathbf{b}}} = \left(\frac{\sqrt{\mathbf{a}'\mathbf{a}}}{\sqrt{\mathbf{b}'\mathbf{b}}} \cos\theta\right)\mathbf{b}.$$

If we substitute for $\cos\theta$, we get

$$\mathbf{c} = \left(\frac{\mathbf{a}'\mathbf{b}}{\mathbf{b}'\mathbf{b}}\right)\mathbf{b},$$

and from this we can obtain the projection ℓ_p. We shall generalize this to E_n. In the special case in E_2 that we have described, the line $\mathscr{L}$ and the directed line segment ℓ start at the origin. We consider the case where a line segment and a line in E_n may not intersect, but we may still want to consider projecting ℓ onto $\mathscr{L}$. What we do is move ℓ parallel to itself and consider a directed line segment from **0** to $\mathbf{a}_2 - \mathbf{a}_1$, instead of from $\mathbf{a}_1$ to $\mathbf{a}_2$. We move $\mathscr{L}$ parallel to itself, so that it also goes through the origin. We then project the transformed directed line segment onto the transformed line. For example, if ℓ is the directed line segment from $\mathbf{a}_1$ to $\mathbf{a}_2$, the equation for ℓ is

$$\mathbf{y} = \mathbf{a}_2 \lambda + (1 - \lambda)\mathbf{a}_1; \qquad 0 \le \lambda \le 1.$$

which can also be written

$$\mathbf{y} = (\mathbf{a}_2 - \mathbf{a}_1)\lambda + \mathbf{a}_1; \qquad 0 \le \lambda \le 1,$$

and a direction vector for ℓ is $\mathbf{a}_2 - \mathbf{a}_1$; hence, if ℓ^* denotes the directed line segment that begins at the origin and has the same length and direction as ℓ, the equation for ℓ^* is

$$\mathbf{y}^* = (\mathbf{a}_2 - \mathbf{a}_1)\lambda^*; \qquad 0 \le \lambda^* \le 1,$$

which is the line segment from **0** to the point $\mathbf{a}_2 - \mathbf{a}_1$.

Now if $\mathscr{L}$ is the line through the two points $\mathbf{b}_1$ and $\mathbf{b}_2$, the equation for $\mathscr{L}$ is

$$\mathbf{x} = \alpha\mathbf{b}_2 + (1 - \alpha)\mathbf{b}_1; \qquad -\infty < \alpha < \infty,$$

which can be written as

$$\mathbf{x} = \alpha(\mathbf{b}_2 - \mathbf{b}_1) + \mathbf{b}_1; \qquad -\infty < \alpha < \infty,$$

and a direction vector for $\mathscr{L}$ is $\mathbf{b}_2 - \mathbf{b}_1$. If $\mathscr{L}^*$ denotes a line through $\mathbf{0}$ and parallel to $\mathscr{L}$, the equation for $\mathscr{L}^*$ can be written as

$$\mathbf{x}^* = \alpha^*(\mathbf{b}_2 - \mathbf{b}_1); \qquad -\infty < \alpha^* < \infty.$$

Therefore, the directed line segment ℓ^* and the line $\mathscr{L}^*$ intersect at $\mathbf{0}$ and we draw a perpendicular from the point $\mathbf{a}_2 - \mathbf{a}_1$ to the line $\mathscr{L}^*$, and the projection of ℓ^* onto $\mathscr{L}^*$ is the directed line segment from $\mathbf{0}$ to $\mathbf{c}$. (See Fig. 4.4.2.) We shall also define this to

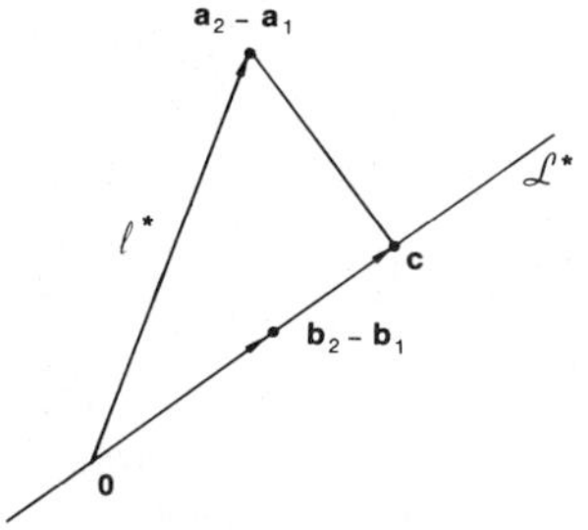

Figure 4.4.2

be the projection of ℓ onto $\mathscr{L}$. Another way to describe this projection is as follows: from $\mathbf{a}_1$, draw a perpendicular to $\mathscr{L}$ and denote by $\mathbf{c}_1$ the point on $\mathscr{L}$ at which the perpendicular intersects $\mathscr{L}$; from $\mathbf{a}_2$, draw a perpendicular to $\mathscr{L}$ and denote by $\mathbf{c}_2$ the point on $\mathscr{L}$ at which the perpendicular intersects $\mathscr{L}$. The directed line segment from $\mathbf{c}_1$ to $\mathbf{c}_2$ is the projection of ℓ onto $\mathscr{L}$. Clearly, these two procedures result in two line segments, $\mathbf{c}_1$ to $\mathbf{c}_2$ and $\mathbf{0}$ to $\mathbf{c}$, that have the same direction and the same length but perhaps a different starting place. The projection is always considered to start from $\mathbf{0}$.

Definition 4.4.1

Projection of a Directed Line Segment onto a Line in E_n. *Let ℓ be a directed line segment from $\mathbf{a}_1$ to $\mathbf{a}_2$ ($\mathbf{a}_1 \neq \mathbf{a}_2$) in E_n and let $\mathscr{L}$ be a line through $\mathbf{b}_1$ and $\mathbf{b}_2$ ($\mathbf{b}_1 \neq \mathbf{b}_2$) in E_n. The projection of ℓ onto $\mathscr{L}$ is defined to be the directed line segment ℓ_p from $\mathbf{0}$ to $\mathbf{c}$, where $\mathbf{c}$ is defined by*

$$\mathbf{c} = \left[\frac{(\mathbf{a}_2 - \mathbf{a}_1)'(\mathbf{b}_2 - \mathbf{b}_1)}{(\mathbf{b}_2 - \mathbf{b}_1)'(\mathbf{b}_2 - \mathbf{b}_1)}\right](\mathbf{b}_2 - \mathbf{b}_1). \tag{4.4.1}$$

Note: If ℓ is a directed line segment from $\mathbf{0}$ to $\mathbf{a}_2$ (that is, if $\mathbf{a}_1 = \mathbf{0}$) and if $\mathscr{L}$ is a line through $\mathbf{0}$ and $\mathbf{b}_2$ (that is, if $\mathbf{b}_1 = \mathbf{0}$), then the projection is the directed line segment from $\mathbf{0}$ to $\mathbf{c}$, where $\mathbf{c}$ is defined by letting $\mathbf{a}_1 = \mathbf{b}_1 = \mathbf{0}$ in Eq. (4.4.1). In this situation, if we use vector terminology, we say that the vector $\mathbf{c}$ is the projection of the vector $\mathbf{a}_2$ onto the vector $\mathbf{b}_2$. We shall discuss this in more detail in Chapter 12.

We are now ready to discuss the projection of a directed line segment onto a plane. In E_3, consider a plane $\mathscr{P}$ through the origin and the points $\mathbf{b}_1$ and $\mathbf{b}_2$, where $[\mathbf{b}_1, \mathbf{b}_2, \mathbf{0}] = \mathbf{B}$ has rank 2 (see Def. 4.3.2). Also consider a directed line segment ℓ through the origin and a point $\mathbf{a} \neq \mathbf{0}$ (see Fig. 4.4.3). From the end of ℓ (the point $\mathbf{a}$) draw a line that is perpendicular to the plane $\mathscr{P}$. Denote by $\mathbf{c}$ the point where the perpendicular intersects $\mathscr{P}$. Then the directed line segment ℓ_p from $\mathbf{0}$ to $\mathbf{c}$ is called the (orthogonal) projection of ℓ onto $\mathscr{P}$.

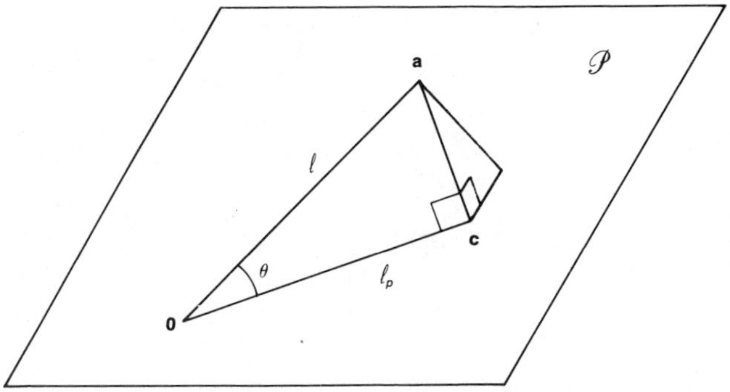

Figure 4.4.3

To find ℓ_p, we must only determine the point $\mathbf{c}$ from a knowledge of $\mathbf{b}_1$, $\mathbf{b}_2$, and $\mathbf{a}$ or, in other words, from a knowledge of $\mathscr{P}$ and ℓ. If the directed line segment from $\mathbf{c}$ to $\mathbf{a}$ is to be perpendicular to $\mathscr{P}$, it must be perpendicular to every line in $\mathscr{P}$. If $\mathscr{P}$ is defined by two points, $\mathbf{b}_1$ and $\mathbf{b}_2$ (see Def. 4.3.3), then the directed line segment from $\mathbf{c}$ to $\mathbf{a}$, which has direction vector $\mathbf{a} - \mathbf{c}$, must be perpendicular to the line through $\mathbf{0}$, $\mathbf{b}_1$ and also perpendicular to the line through $\mathbf{0}$, $\mathbf{b}_2$. This means that

$$(\mathbf{a} - \mathbf{c})'\mathbf{b}_1 = 0 \quad \text{and} \quad (\mathbf{a} - \mathbf{c})'\mathbf{b}_2 = 0. \tag{4.4.2}$$

But, since every point on $\mathscr{P}$ can be written as $\mathbf{y} = \lambda_1 \mathbf{b}_1 + \lambda_2 \mathbf{b}_2$ for some values of λ_1 and λ_2 and since $\mathbf{c}$ is on $\mathscr{P}$, there is a λ_1 and a λ_2 (denote them by λ_1^* and λ_2^*) such that $\mathbf{c} = \lambda_1^*\mathbf{b}_1 + \lambda_2^*\mathbf{b}_2$. By substituting this into Eq. (4.4.2), we get

$$(\lambda_1^*\mathbf{b}_1 + \lambda_2^*\mathbf{b}_2)'\mathbf{b}_1 = \mathbf{a}'\mathbf{b}_1,$$

$$(\lambda_1^*\mathbf{b}_1 + \lambda_2^*\mathbf{b}_2)'\mathbf{b}_2 = \mathbf{a}'\mathbf{b}_2,$$

and these equations have a unique solution for λ_1^* and λ_2^*. We can thus determine $\mathbf{c}$ and hence ℓ_p, the projection of ℓ onto $\mathscr{P}$.

With these ideas, we can formulate the projection of a line in E_n onto a plane of dimension k in E_n.

Definition 4.4.2

Projection of a Directed Line Segment in E_n onto a Plane through the Origin and of Dimension k in E_n. *Let ℓ be a directed line segment in E_n from the origin to the point* $\mathbf{a}$ *where* $\mathbf{a} \neq \mathbf{0}$. *Let $\mathscr{P}_k^n$ be a plane in E_n through the origin and the points* $\mathbf{b}_1, \mathbf{b}_2, \ldots, \mathbf{b}_k$, *where the matrix* $\mathbf{B} = [\mathbf{b}_1, \mathbf{b}_2, \ldots, \mathbf{b}_k]$ *has rank k. Denote by* $\mathbf{c}$ *the point on $\mathscr{P}_k^n$ where a line from* $\mathbf{a}$ *that is perpendicular to $\mathscr{P}_k^n$ intersects $\mathscr{P}_k^n$. The projection of ℓ onto $\mathscr{P}_k^n$ is defined to be the directed line segment from the origin to the point* $\mathbf{c}$.

Note: In vector terminology, this defines the projection of a vector $\mathbf{a}$ onto a subspace, with basis $\{\mathbf{b}_1, \mathbf{b}_2, \ldots, \mathbf{b}_k\}$; and the projection is the vector $\mathbf{c}$.

The following theorem enables us to compute the point $\mathbf{c}$, and hence the projection ℓ_p, from a knowledge of the point $\mathbf{a}$ and the points $\mathbf{b}_1, \mathbf{b}_2, \ldots, \mathbf{b}_k$.

Theorem 4.4.1

Let $\mathscr{P}_k^n$ be the plane through the origin and the points $\mathbf{b}_1, \mathbf{b}_2, \ldots, \mathbf{b}_k$ *in E_n such that the $n \times k$ matrix* $\mathbf{B}$ *has rank k, where* $\mathbf{B} = [\mathbf{b}_1, \mathbf{b}_2, \ldots, \mathbf{b}_k]$. *Let ℓ be the directed line segment from the origin to the point* $\mathbf{a}$ *where* $\mathbf{a} \neq \mathbf{0}$. *The point* $\mathbf{c}$ *that determines the projection of ℓ onto $\mathscr{P}_k^n$ is determined by*

$$\mathbf{c} = \mathbf{B}(\mathbf{B}'\mathbf{B})^{-1}\mathbf{B}'\mathbf{a}. \tag{4.4.3}$$

Proof: By definition, the line segment (denote it by ℓ^*) from the point $\mathbf{a}$ to the point $\mathbf{c}$ is perpendicular to the plane $\mathscr{P}_k^n$, which means it is perpendicular to each and every line in the plane. The equation for ℓ^* can be written as

$$\mathbf{y} = \lambda(\mathbf{c} - \mathbf{a}) + \mathbf{a}; \qquad 0 \leq \lambda \leq 1,$$

and a direction vector for ℓ^* is $\mathbf{c} - \mathbf{a}$. Since the line through $\mathbf{0}$ to $\mathbf{b}_1$ is on $\mathscr{P}_k^n$, it follows that the vector $\mathbf{c} - \mathbf{a}$ is perpendicular to the vector $\mathbf{b}_1$; the same is true of the line through $\mathbf{0}$ and $\mathbf{b}_2$; $\mathbf{0}$ and $\mathbf{b}_3$; ... ; $\mathbf{0}$ and $\mathbf{b}_k$. This gives us the equations

$$(\mathbf{c} - \mathbf{a})'\mathbf{b}_1 = 0; \quad (\mathbf{c} - \mathbf{a})'\mathbf{b}_2 = 0; \quad \ldots; \quad (\mathbf{c} - \mathbf{a})'\mathbf{b}_k = 0,$$

or, equivalently,

$$\mathbf{c}'\mathbf{b}_i = \mathbf{a}'\mathbf{b}_i, \qquad i = 1, 2, \ldots, k,$$

or

$$\mathbf{c}'\mathbf{B} = \mathbf{a}'\mathbf{B}.$$

But since $\mathbf{c}$ is a point on $\mathscr{P}_k^n$, there are k scalars $\lambda_1, \lambda_2, \ldots, \lambda_k$ such that

$$\mathbf{c} = \sum_{i=1}^{k} \lambda_i \mathbf{b}_i \quad \text{or} \quad \mathbf{c} = \mathbf{B}\boldsymbol{\lambda},$$

where $\boldsymbol{\lambda}' = [\lambda_1, \lambda_2, \ldots, \lambda_k]$. If we substitute this for $\mathbf{c}$ in the matrix equation above, we obtain

$$\boldsymbol{\lambda}'\mathbf{B}'\mathbf{B} = \mathbf{a}'\mathbf{B}.$$

Since $\mathbf{B}$ is an $n \times k$ matrix of rank k, $\mathbf{B}'\mathbf{B}$ is nonsingular, and hence

$$\boldsymbol{\lambda}' = \mathbf{a}'\mathbf{B}(\mathbf{B}'\mathbf{B})^{-1}$$

and

$$\mathbf{c} = \mathbf{B}(\mathbf{B}'\mathbf{B})^{-1}\mathbf{B}'\mathbf{a}. \quad \blacksquare$$

Example 4.4.1. Consider the plane $\mathscr{P}_2^3$ through the origin and the two points $\mathbf{b}_1$ and $\mathbf{b}_2$, where $\mathbf{b}_1' = [1, 1, 0]$, $\mathbf{b}_2' = [1, 0, 1]$; determine the projection of ℓ onto $\mathscr{P}_2^3$ where ℓ is a directed line segment from the origin to the point $\mathbf{a}$ where $\mathbf{a}' = [1, 1, 1]$. The point $\mathbf{c}$ is determined by Eq. (4.4.3) and is

$$\mathbf{c} = \begin{bmatrix} 1 & 1 \\ 1 & 0 \\ 0 & 1 \end{bmatrix} \begin{bmatrix} 2 & 1 \\ 1 & 2 \end{bmatrix}^{-1} \begin{bmatrix} 1 & 1 & 0 \\ 1 & 0 & 1 \end{bmatrix} \begin{bmatrix} 1 \\ 1 \\ 1 \end{bmatrix} = \begin{bmatrix} 4/3 \\ 2/3 \\ 2/3 \end{bmatrix},$$

and the equation for ℓ_p is

$$\mathbf{y} = \begin{bmatrix} (4/3)\lambda \\ (2/3)\lambda \\ (2/3)\lambda \end{bmatrix}; \qquad 0 \le \lambda \le 1.$$

It is intuitively clear that the point **c** in E_3 that determines the projection of ℓ onto $\mathscr{P}_2^3$ is the point on the plane that is the shortest distance from the point **a**. This is actually the case for the projection of any line onto any plane $\mathscr{P}_k^n$, as is stated in the next theorem.

Theorem 4.4.2

Let ℓ be the directed line segment from the origin to the point $\mathbf{a}$ $(\mathbf{a} \neq \mathbf{0})$ *in* E_n, *and let* $\mathscr{P}_k^n$ *be a plane of dimension k through the points* $\mathbf{0}, \mathbf{b}_1, \mathbf{b}_2, \ldots, \mathbf{b}_k$ *in* E_n *such that* $\mathbf{B} = [\mathbf{b}_1, \mathbf{b}_2, \ldots, \mathbf{b}_k]$ *has rank k. If the point* **c** *denotes the end point of the projection of ℓ onto* $\mathscr{P}_k^n$, *then* **c** *is the point on* $\mathscr{P}_k^n$ *whose distance from* **a** *is a minimum.*

Proof: The distance $d(\mathbf{x})$ from **a** to **x**, where **x** is any point on $\mathscr{P}_k^n$, is defined by

$$d(\mathbf{x}) = [(\mathbf{a} - \mathbf{x})'(\mathbf{a} - \mathbf{x})]^{1/2};$$

we must prove that the value of **x** on $\mathscr{P}_k^n$ that minimizes $d(\mathbf{x})$ is

$$\mathbf{x}_0 = \mathbf{B}(\mathbf{B}'\mathbf{B})^{-1}\mathbf{B}'\mathbf{a},$$

since by Theorem 4.4.1 this is the value of **c**. Since **x** is on $\mathscr{P}_k^n$, we substitute $\mathbf{x} = \Sigma\lambda_i \mathbf{b}_i = \mathbf{B}\boldsymbol{\lambda}$ into $d(\mathbf{x})$ and we must determine the set of scalars λ_i such that

$$f(\boldsymbol{\lambda}) = [(\mathbf{a} - \mathbf{B}\boldsymbol{\lambda})'(\mathbf{a} - \mathbf{B}\boldsymbol{\lambda})]^{1/2}$$

is a minimum. If we use the calculus to find the value of $\boldsymbol{\lambda}$ such that $f(\boldsymbol{\lambda})$ is a minimum, we get

$$\begin{aligned}\frac{\partial f(\boldsymbol{\lambda})}{\partial \lambda_i} &= \frac{1}{2}[(\mathbf{a} - \mathbf{B}\boldsymbol{\lambda})'(\mathbf{a} - \mathbf{B}\boldsymbol{\lambda})]^{-1/2}\left[-2\mathbf{a}'\mathbf{b}_i + 2\sum_j \lambda_j \mathbf{b}_j' \mathbf{b}_i\right] \\ &= 0, \qquad i = 1, 2, \ldots, k.\end{aligned}$$

This reduces to

$$\boldsymbol{\lambda}'(\mathbf{B}'\mathbf{B}) = \mathbf{a}'\mathbf{B},$$

and, finally, $\boldsymbol{\lambda} = (\mathbf{B}'\mathbf{B})^{-1}\mathbf{B}'\mathbf{a}$.

Therefore, the point **x** on $\mathscr{P}_k^n$ such that the distance from **a** to $\mathscr{P}_k^n$ is a minimum is

$$\mathbf{x}_0 = \mathbf{B}\boldsymbol{\lambda} = \mathbf{B}(\mathbf{B}'\mathbf{B})^{-1}\mathbf{B}'\mathbf{a};$$

and this is equal to **c** by Theorem 4.4.1, and the theorem is proved. ∎

Note: The distance d from the point $\mathbf{a}$ to the plane $\mathscr{P}_k^n$ is defined as the minimum distance from $\mathbf{a}$ to $\mathbf{x}$ for $\mathbf{x}$ on $\mathscr{P}_k^n$ and is equal to

$$\begin{aligned} d &= [(\mathbf{a} - \mathbf{B}(\mathbf{B}'\mathbf{B})^{-1}\mathbf{B}'\mathbf{a})'(\mathbf{a} - \mathbf{B}(\mathbf{B}'\mathbf{B})^{-1}\mathbf{B}'\mathbf{a})]^{1/2} \\ &= \{\mathbf{a}'[\mathbf{I} - \mathbf{B}(\mathbf{B}'\mathbf{B})^{-1}\mathbf{B}']\mathbf{a}\}^{1/2}. \end{aligned}$$

We note that d^2 is a quadratic form in the a_i, and the matrix of the quadratic form is $\mathbf{I} - \mathbf{B}(\mathbf{B}'\mathbf{B})^{-1}\mathbf{B}' = \mathbf{G}$ (say) and $\mathbf{G} = \mathbf{G}'$; $\mathbf{G} = \mathbf{G}^2$; that is, $\mathbf{G}$ is a symmetric idempotent matrix. This fact is of extreme importance in statistics and we shall discuss it in more detail later.

Another approach to a "projection" is to view it as a transformation of one vector space into another vector space so that the transformation satisfies certain properties. This is discussed in Chapter 12.

Problems

1. Find the equation of the line in E_3 through the point $\mathbf{x}' = [1, 1, 0]$ and parallel to the line through the two points $\mathbf{a}' = [0, 1, -1]$, $\mathbf{b}' = [1, 0, 1]$.
2. Find the equation of a line $\mathscr{L}_1$ that goes through the point $\mathbf{x}' = [1, -1, 1]$, intersects another line $\mathscr{L}_2$, and is perpendicular to $\mathscr{L}_2$. The line $\mathscr{L}_2$ goes through the points $\mathbf{a}' = \mathbf{0}$ and $\mathbf{b}' = [2, 1, 1]$.
3. Use Def. 4.2.2 to show that $\mathscr{L}_1$ and $\mathscr{L}_2$ are parallel where $\mathscr{L}_1$ goes through the two points $\mathbf{a}' = [1, 0, 1]$; $\mathbf{b}' = [0, 1, -1]$ and $\mathscr{L}_2$ goes through the two points $\mathbf{c}' = [1, 2, -1]$; $\mathbf{d}' = [2, 1, 1]$.
4. Do the two lines $\mathscr{L}_1$ and $\mathscr{L}_2$ intersect where $\mathscr{L}_1$ goes through the two points $\mathbf{a}' = [1, 1, -1]$; $\mathbf{b}' = [2, 0, 1]$ and $\mathscr{L}_2$ goes through the two points $\mathbf{c}' = [1, 1, 1]$; $\mathbf{d}' = [2, 0, 2]$?
5. Find the angle between the two directed line segments ℓ_1 and ℓ_2 where ℓ_1 is the directed line segment from $\mathbf{0}$ to $\mathbf{a}' = [1, -1, 1]$ and ℓ_2 is the directed line segment from $\mathbf{0}$ to $\mathbf{b}' = [2, 1, -1]$.
6. Find the direction angles of ℓ_1 in Prob. 5.
7. Find the direction angles of the line through the points $\mathbf{a}' = [2, 0, -1]$ and $\mathbf{b}' = [1, -1, 3]$.
8. Find the distance from the point $\mathbf{x}' = [2, 1, -1]$ to the line $\mathscr{L}$ that goes through the two points $\mathbf{a}' = [1, 1, 0]$ and $\mathbf{b}' = [1, -1, 2]$.

9. Find the equation of the line $\mathscr{L}$ that is parallel to $\mathscr{L}_1$ and $\mathscr{L}_2$ of Prob. 3, equidistant from the two lines, and in the same plane as $\mathscr{L}_1$ and $\mathscr{L}_2$.
10. Find the equation of the plane $\mathscr{P}$ that goes through the three points $\mathbf{b}_1' = [1, 1, 1]$, $\mathbf{b}_2' = [1, -1, 0]$, $\mathbf{b}_3' = [0, 1, -1]$. Write the equation in the form $\mathbf{a}'\mathbf{y} = c$.
11. In Prob. 10, find the equation of a line $\mathscr{L}$ that goes through the origin and is perpendicular to every line in the plane $\mathscr{P}$.
12. In Prob. 11, find the point on $\mathscr{P}$ where $\mathscr{L}$ intersects $\mathscr{P}$.
13. In Prob. 12, verify that the distance between $\mathbf{0}$ and the point of intersection of the line $\mathscr{L}$ and the plane $\mathscr{P}$ is $d = [\mathbf{1}'(\mathbf{B}'\mathbf{B})^{-1}\mathbf{1}]^{-1/2}$.
14. In Prob. 13, show that this value of d is the minimum distance between $\mathbf{0}$ and the plane $\mathscr{P}$.
15. Let $\mathscr{L}$ be a line through $\mathbf{0}$ and $\mathbf{a}$, where $\mathbf{a}' = [1, -1, 1]$. Find the equation of the plane $\mathscr{P}$ that is perpendicular to $\mathscr{L}$ and passes through the point $[2, 1, -1]$.
16. Find the projection of the directed line segment ℓ from $\mathbf{a}_1$ to $\mathbf{a}_2$ onto the line $\mathscr{L}$ that goes through $\mathbf{0}$ and $\mathbf{b}$ where $\mathbf{a}_1' = [1, 2, -1]$, $\mathbf{a}_2' = [1, -1, 2]$, $\mathbf{b}' = [1, 1, 1]$.
17. In Prob. 16, sketch the three lines.
18. In Prob. 16, find the angle between ℓ and $\mathscr{L}$.
19. Find the projection of the directed line segment from $\mathbf{a}_1$ to $\mathbf{a}_2$ onto the plane $\mathscr{P}_2^3$ where $\mathscr{P}_2^3$ goes through $\mathbf{0}$, $\mathbf{x}_1' = [1, 1, 1]$ and $\mathbf{x}_2' = [1, -1, 1]$ and where $\mathbf{a}_1' = [1, 0, -1]$, $\mathbf{a}_2' = [2, -1, 1]$.
20. Consider the plane $\mathscr{P}_2^4$ through the points $\mathbf{a}_1$, $\mathbf{a}_2$, $\mathbf{0}$ where $\mathbf{a}_1' = [1, 1, 0, 1]$, $\mathbf{a}_2' = [1, -1, 1, 0]$. Find the projection of the directed line segment ℓ from $\mathbf{0}$ to $\mathbf{b}$ onto $\mathscr{P}_2^4$ where $\mathbf{b}' = [0, 2, -1, 0]$.

Algebra of Vector Spaces

5

5.1 Introduction

The algebra of vector spaces that concerns us in this chapter involves topics that are a part of linear algebra but often are not discussed in a first course. They are: intersection of vector spaces, sum of vector spaces, orthogonal complement of vector spaces, null and column spaces of matrices.

In the theory of sets, it is generally necessary to discuss certain operations on two or more sets. For example, it is useful to discuss intersection of sets, union of sets, complementation of sets, and so forth. A similar situation arises in vector spaces—which of course are sets of vectors. For example, it is important in geometric interpretations to discuss intersections of planes, lines, and so on.

5.2 Intersection and Sum of Vector Spaces

In the theory of sets, if S_1 and S_2 are two subsets of a set E, then the intersection of these subsets is defined as the set of elements S that belong to both S_1 and S_2. This will be used to define the intersection of two sets of vectors.

Definition 5.2.1

Intersection of Two Vector Spaces. *Let S_1 and S_2 be two vector subspaces of E_n. The intersection of these two subspaces, which we denote by $S = S_1 \cap S_2$, is defined as the set of vectors that belong to both S_1 and S_2; that is,*

$$S = \{\mathbf{y} : \mathbf{y} \in S_1 ; \mathbf{y} \in S_2\}.$$

Note: This definition can be extended to include the intersection of any finite number m of subspaces of E_n: If S_i, $i = 1, 2, \ldots, m$ are vector subspaces of E_n, then the intersection S of these subspaces is defined by

$$S = \{\mathbf{y} : \mathbf{y} \in S_1 ; \mathbf{y} \in S_2 ; \ldots ; \mathbf{y} \in S_m\}.$$

We shall now prove that the intersection of two subspaces of E_n is itself a subspace of E_n.

Theorem 5.2.1

Let S_1 and S_2 be vector subspaces of E_n and let $S = S_1 \cap S_2$; then S is a vector subspace of E_n.

Proof: Let $\mathbf{x}_1$ and $\mathbf{x}_2$ be any two vectors in S. We must show that for each and every two scalars a_1 and a_2, the vector $a_1\mathbf{x}_1 + a_2\mathbf{x}_2$ is in S. Now if $\mathbf{x}_1 \in S$, then, by definition, $\mathbf{x}_1 \in S_1$ and $\mathbf{x}_1 \in S_2$; also since $\mathbf{x}_2 \in S$, then $\mathbf{x}_2 \in S_1$ and $\mathbf{x}_2 \in S_2$; but since S_1 and S_2 are vector spaces, it follows that

$$a_1\mathbf{x}_1 + a_2\mathbf{x}_2 \in S_1 \quad \text{and} \quad a_1\mathbf{x}_1 + a_2\mathbf{x}_2 \in S_2$$

for all scalars a_1 and a_2; and hence $a_1\mathbf{x}_1 + a_2\mathbf{x}_2 \in S$. This completes the proof. ∎

Mathematical induction can be used to show that if $S = S_1 \cap S_2 \cap \cdots \cap S_m$ for any positive integer m, then S is a vector subspace of E_n if each S_i is a vector subspace of E_n.

Theorem 5.2.1 has the following geometric interpretation: Let $\mathscr{P}_k^n$ be a plane through the origin and of dimension k in E_n (the points of $\mathscr{P}_k^n$, when viewed as vectors, form a subspace S_1 of E_n); let $\mathscr{P}{*}_t^n$ be a plane through the origin and of dimension t in E_n (the points of $\mathscr{P}{*}_t^n$ when viewed as vectors form a subspace S_2 of E_n). The intersection of the two planes is the set of points that is on both planes, and it is the set S where $S = S_1 \cap S_2$. Hence, the intersection of these two planes is a plane through the origin in E_n.

Corollary 5.2.1

In Theorem 5.2.1, S is a vector subspace of S_1 and also of S_2.

Definition 5.2.2

Sum of Vector Subspaces. *Let S_1 and S_2 be two subspaces of the vector space E_n. The set of vectors S, denoted by $S = S_1 \oplus S_2$, is called the sum of the vector subspaces S_1 and S_2 and is defined by*

$$S = \{\mathbf{y} : \mathbf{y} = \mathbf{x}_1 + \mathbf{x}_2; \quad \mathbf{x}_1 \in S_1; \quad \mathbf{x}_2 \in S_2\}.$$

Note: The sum of two vector spaces is sometimes called the direct sum. This is not the definition of the union of two vector subspaces. The union of two vector subspaces may not be a vector space.

The sum of a finite number h of subspaces of E_n is denoted by $S = S_1 \oplus S_2 \oplus \ldots \oplus S_h$ where S_i is a subspace of E_n, for $i = 1, 2, \ldots, h$, and S is defined by

$$S = \{\mathbf{y} : \mathbf{y} = \mathbf{x}_1 + \mathbf{x}_2 + \cdots + \mathbf{x}_h; \quad \mathbf{x}_i \in S_i, i = 1, 2, \ldots, h\}.$$

We now prove that S is a vector subspace of E_n.

Theorem 5.2.2

Let S_1 and S_2 be vector subspaces of E_n. The sum S of these two subspaces is a subspace of E_n.

Proof: If $\mathbf{y}_1$ and $\mathbf{y}_2$ belong to S, we must prove that $a_1 \mathbf{y}_1 + a_2 \mathbf{y}_2$ belongs to S for all scalars a_1 and a_2. Now if $\mathbf{y}_1$ belongs to S, then there must exist two vectors $\mathbf{x}_1$ and $\mathbf{x}_2$ such that $\mathbf{y}_1 = \mathbf{x}_1 + \mathbf{x}_2$ where $\mathbf{x}_1 \in S_1$, $\mathbf{x}_2 \in S_2$. Also if $\mathbf{y}_2 \in S$, there must exist vectors $\mathbf{z}_1 \in S_1$, $\mathbf{z}_2 \in S_2$ such that $\mathbf{y}_2 = \mathbf{z}_1 + \mathbf{z}_2$. But

$$a_1\mathbf{x}_1 \in S_1, \; a_1\mathbf{x}_2 \in S_2, \; a_2\mathbf{z}_1 \in S_1, \quad \text{and} \quad a_2\mathbf{z}_2 \in S_2.$$

Hence

$$a_1\mathbf{x}_1 + a_2\mathbf{z}_1 \in S_1 \quad \text{and} \quad a_1\mathbf{x}_2 + a_2\mathbf{z}_2 \in S_2,$$

and finally

$$(a_1\mathbf{x}_1 + a_2\mathbf{z}_1) + (a_1\mathbf{x}_2 + a_2\mathbf{z}_2) \in S,$$

but

$$(a_1\mathbf{x}_1 + a_2\mathbf{z}_1) + (a_1\mathbf{x}_2 + a_2\mathbf{z}_2) = a_1(\mathbf{x}_1 + \mathbf{x}_2) + a_2(\mathbf{z}_1 + \mathbf{z}_2) = a_1\mathbf{y}_1 + a_2\mathbf{y}_2,$$

and hence $a_1\mathbf{y}_1 + a_2\mathbf{y}_2 \in S$ and the theorem is proved. ▮

Theorem 5.2.3

Let S_1 and S_2 be subspaces in E_n, and let $S = S_1 \oplus S_2$; then S_1 is a subspace of S; and S_2 is a subspace of S.

Proof: The proof of this theorem follows immediately from the definition of S. ▮

Theorem 5.2.4

Let V_n be a subspace of E_n, and let S_1 and S_2 be subspaces of V_n; then $S_1 \oplus S_2$ and $S_1 \cap S_2$ are subspaces of V_n.

Proof: The proof of this theorem also follows directly from the definition of the sum and intersection of vector spaces. ▮

Example 5.2.1. Let S_1 be a vector subspace of E_4 spanned by $\mathbf{a}_1$, $\mathbf{a}_2$; let S_2 be a vector subspace of E_4 spanned by $\mathbf{b}_1$, $\mathbf{b}_2$, $\mathbf{b}_3$ where

$$\mathbf{a}_1 = \begin{bmatrix} 0 \\ 2 \\ 1 \\ -1 \end{bmatrix}; \quad \mathbf{a}_2 = \begin{bmatrix} 1 \\ 1 \\ 0 \\ 0 \end{bmatrix}; \quad \mathbf{b}_1 = \begin{bmatrix} 1 \\ 0 \\ -1 \\ 0 \end{bmatrix}; \quad \mathbf{b}_2 = \begin{bmatrix} 0 \\ -2 \\ -1 \\ 1 \end{bmatrix}; \quad \mathbf{b}_3 = \begin{bmatrix} 2 \\ 2 \\ -1 \\ -1 \end{bmatrix}.$$

Find a basis for $S = S_1 \oplus S_2$. It is straightforward to show that $\{\mathbf{a}_1, \mathbf{a}_2\}$ is a basis for S_1 and $\{\mathbf{b}_1, \mathbf{b}_2\}$ is a basis for S_2. Clearly, the sum of the two spaces is spanned by the set $\{\mathbf{a}_1, \mathbf{a}_2, \mathbf{b}_1, \mathbf{b}_2\}$. The vector $\mathbf{b}_3$ is not needed, since $\{\mathbf{b}_1, \mathbf{b}_2\}$ is a basis set for S_2. We must find a linearly independent set from $\{\mathbf{a}_1, \mathbf{a}_2, \mathbf{b}_1, \mathbf{b}_2\}$ and, clearly, one such set is $\{\mathbf{a}_1, \mathbf{a}_2, \mathbf{b}_1\}$, so this is a basis for $S_1 \oplus S_2$.

The proof of the final theorem of this section is left for the reader.

Theorem 5.2.5

Let S_1 and S_2 be two subspaces of E_n and let $R_1 = S_1 \cap S_2$ and $R_2 = S_1 \oplus S_2$. The following relationship obtains:

$$\textit{dimension}\ (S_1) + \textit{dimension}\ (S_2) = \textit{dimension}\ (R_1) + \textit{dimension}\ (R_2).$$

5.3 Orthogonal Complement of a Vector Subspace

If a vector $\mathbf{b}$ is orthogonal to each vector in the set $\{\mathbf{c}_1, \mathbf{c}_2, \ldots, \mathbf{c}_h\}$, then clearly $\mathbf{b}$ is orthogonal to each and every vector in the subspace spanned by the set. If S_1 denotes the subspace of E_n spanned by the set of vectors $\{\mathbf{c}_1, \mathbf{c}_2, \ldots, \mathbf{c}_h\}$, then we say that $\mathbf{b}$ is orthogonal to the subspace S_1. However $\mathbf{b}$ also spans a subspace of E_n (denote it by S_2) and so we say that the two subspaces S_1 and S_2 are orthogonal. This concept can be extended to include the case in which S_2 is spanned by a set of vectors instead of a single vector $\mathbf{b}$. We now formulate the definition of orthogonal subspaces.

Definition 5.3.1

Orthogonal Vector Subspaces in E_n. *Let S_1 and S_2 be two subspaces in E_n. If $\mathbf{x}_1'\mathbf{x}_2 = 0$ for each vector $\mathbf{x}_1$ in S_1 and for each vector $\mathbf{x}_2$ in S_2, then S_1 and S_2 are defined to be orthogonal subspaces in E_n and we denote this by $S_1 \perp S_2$.*

Note: If $S_1 \perp S_2$, then $S_1 \cap S_2 = \{\mathbf{0}\}$, but if $S_1 \cap S_2 = \{\mathbf{0}\}$, this does not necessarily imply that $S_1 \perp S_2$.

Theorem 5.3.1

Let $\{\mathbf{a}_1, \mathbf{a}_2, \ldots, \mathbf{a}_h\}$ and $\{\mathbf{b}_1, \mathbf{b}_2, \ldots, \mathbf{b}_m\}$ be two sets of vectors in E_n such that $\mathbf{a}_i'\mathbf{b}_j = 0$ for all i and j. The vector subspace S_1 spanned by $\{\mathbf{a}_1, \mathbf{a}_2, \ldots, \mathbf{a}_h\}$ is orthogonal to the vector subspace S_2 spanned by $\{\mathbf{b}_1, \mathbf{b}_2, \ldots, \mathbf{b}_m\}$.

Proof: Any vector $\mathbf{x}_1$ in S_1 can be written as

$$\sum_{i=1}^{h} a_i \mathbf{a}_i = \mathbf{x}_1$$

for some set of constants $\{a_i\}$, and any vector $\mathbf{x}_2$ in S_2 can be written

$$\sum_{j=1}^{m} b_j \mathbf{b}_j = \mathbf{x}_2$$

for some set of constants $\{b_j\}$. But

$$\mathbf{x}_1'\mathbf{x}_2 = \sum_j \sum_i a_i b_j \mathbf{a}_i' \mathbf{b}_j = 0,$$

and hence, by Def. 5.3.1, $S_1 \perp S_2$. ∎

In the theory of sets, if a set A is a subset of B and if $\bar{A}$ denotes the complement of A relative to B, then $A \cap \bar{A}$ is the null set and $A \cup \bar{A} = B$. Operations similar to these play an important role in vector spaces.

If S_1 is a vector subspace of E_n, there may be many vector subspaces of E_n that are orthogonal to S_1. One of these orthogonal subspaces is of particular importance, and that is the one (denote it by S_2) such that $S_1 \oplus S_2 = E_n$. That is to say, for a given vector subspace S_1 in E_n, we are interested in a vector subspace S_2 in E_n such that $S_1 \perp S_2$ and $S_1 \oplus S_2 = E_n$. S_2 is called the *orthogonal complement* of S_1, and we now formally state the definition.

Definition 5.3.2

Orthogonal Complement of a Vector Subspace in $\mathbf{E}_n$. *Let S_1 be a vector subspace in E_n. The vector subspace S_2 in E_n is defined as the orthogonal complement of S_1 in E_n if and only if $S_1 \perp S_2$ and $S_1 \oplus S_2 = E_n$.*

We sometimes denote the orthogonal complement of a subspace S_1 by $S_1^{\perp}$. We are now ready to state and prove some theorems on orthogonal complements.

Theorem 5.3.2

For a given vector subspace S_1 in E_n, the orthogonal complement $S_1^{\perp}$ always exists and is unique.

Proof: If $S_1 = E_n$, then $S_2 = \{\mathbf{0}\}$ is clearly the only vector subspace that is orthogonal to S_1 and such that $S_1 \oplus S_2 = E_n$. Next assume $S_1 \neq E_n$, and let $\{\boldsymbol{\alpha}_1, \boldsymbol{\alpha}_2, \ldots, \boldsymbol{\alpha}_h\}$ be an orthogonal basis for S_1. By Theorem 2.6.4, this set is part of an orthogonal basis for E_n, which we shall denote by $\{\boldsymbol{\alpha}_1, \boldsymbol{\alpha}_2, \ldots, \boldsymbol{\alpha}_h, \boldsymbol{\beta}_1, \ldots, \boldsymbol{\beta}_{n-h}\}$. Let S_2 be a vector subspace of E_n that has $\{\boldsymbol{\beta}_1, \boldsymbol{\beta}_2, \ldots, \boldsymbol{\beta}_{n-h}\}$ as a basis. Clearly $S_1 \perp S_2$ and $S_1 \oplus S_2 = E_n$, and this proves that an orthogonal complement always exists for a subspace. The proof of uniqueness is left for the reader. ▌

Note: If S_2 is the orthogonal complement of S_1 in E_n, then S_1 is the orthogonal complement of S_2 in E_n.

The next theorem is used extensively in geometric interpretations of the analysis of variance in statistics.

Theorem 5.3.3.

Let S_1 be a subspace of E_n and let $\mathbf{y}$ be any vector in E_n; then $\mathbf{y}$ can be written as the sum of two vectors, $\mathbf{y} = \mathbf{x}_1 + \mathbf{x}_2$, where $\mathbf{x}_1$ is in S_1 and $\mathbf{x}_2$ is in the orthogonal complement of S_1.

Proof: If $\mathbf{y}$ is in S_1, then let $\mathbf{y} = \mathbf{x}_1$ and let $\mathbf{x}_2 = \mathbf{0}$ and the theorem is proved, since the vector $\mathbf{0}$ is in the orthogonal complement of S_1. Next assume $\mathbf{y}$ is not in S_1. Let S_2 be the orthogonal complement of S_1; the fact that $E_n = S_1 \oplus S_2$, means that $\mathbf{y}$ can be written as a linear combination of basis vectors of S_1 plus a linear combination of basis vectors of S_2; that is,

$$\mathbf{y} = \sum_{i=1}^{h} a_i \boldsymbol{\alpha}_i + \sum_{j=h+1}^{n} b_j \boldsymbol{\beta}_j ;$$

but $\sum_{i=1}^{h} a_i \boldsymbol{\alpha}_i$ is a vector in S_1, which we shall label $\mathbf{x}_1$, and $\sum_{j=h+1}^{n} b_j \boldsymbol{\beta}_j$ is a vector in S_2, which we shall label $\mathbf{x}_2$; hence $\mathbf{y} = \mathbf{x}_1 + \mathbf{x}_2$ and the theorem is proved. The $\boldsymbol{\alpha}_i$ and $\boldsymbol{\beta}_i$ are defined in the proof of Theorem 5.3.2. ▮

Note: A result of this theorem is that any vector $\mathbf{y}$ can be written as $\mathbf{y} = \mathbf{x}_1 + \mathbf{x}_2$ where $\mathbf{x}_1'\mathbf{x}_2 = 0$ and where $\mathbf{x}_1$ can be arbitrary except for a scalar multiplier depending on S_1. Also notice that $\mathbf{x}_1$ is the *orthogonal projection* of $\mathbf{y}$ (line through the origin) on the subspace S_1 (plane through the origin).

Corollary 5.3.3

In Theorem 5.3.3, $\mathbf{x}_1$ *and* $\mathbf{x}_2$ *are unique vectors for a given* $\mathbf{y}$ *and a given* S_1.

Example 5.3.1. In E_4 find a basis for the orthogonal complement $S^\perp$ of the vector space S spanned by $\mathbf{a}_1$, $\mathbf{a}_2$ where

$$\mathbf{a}_1 = \begin{bmatrix} 1 \\ 0 \\ -1 \\ 1 \end{bmatrix}; \quad \mathbf{a}_2 = \begin{bmatrix} 1 \\ 1 \\ 0 \\ 1 \end{bmatrix}.$$

Since $\mathbf{a}_1$ and $\mathbf{a}_2$ are linearly independent, they are a basis for S. Using Theorem 2.6.5 to find an orthogonal basis for S, we get

$$\mathbf{y}_1 = \mathbf{a}_1 = \begin{bmatrix} 1 \\ 0 \\ -1 \\ 1 \end{bmatrix}, \qquad \mathbf{y}_2 = \mathbf{a}_2 - \left(\frac{\mathbf{a}_1'\mathbf{a}_2}{\mathbf{a}_1'\mathbf{a}_1}\right)\mathbf{y}_1 = \frac{1}{3}\begin{bmatrix} 1 \\ 3 \\ 2 \\ 1 \end{bmatrix}.$$

Thus $\mathbf{y}_1$ and $\mathbf{y}_2$ are orthogonal and span S (hence they are a basis for S). If we find vectors $\mathbf{x}_1$ and $\mathbf{x}_2$ such that $\{\mathbf{y}_1, \mathbf{y}_2, \mathbf{x}_1, \mathbf{x}_2\}$ is an orthogonal basis for E_4, then $\{\mathbf{x}_1, \mathbf{x}_2\}$ is a basis for $S^\perp$. By inspection, we note that

$$\mathbf{x}_1 = \begin{bmatrix} 1 \\ 0 \\ 0 \\ -1 \end{bmatrix}; \qquad \mathbf{x}_2 = \begin{bmatrix} 1 \\ -2 \\ 2 \\ 1 \end{bmatrix}$$

will be satisfactory.

Example 5.3.2. In Example 5.3.1, show that $\mathbf{x}$ is in $S^\perp$ and $\mathbf{y}$ is in S where a_1, a_2, b_1, b_2 are any scalars and

$$\mathbf{x} = \begin{bmatrix} a_1 + a_2 \\ -2a_2 \\ 2a_2 \\ -a_1 + a_2 \end{bmatrix}; \qquad \mathbf{y} = \begin{bmatrix} b_1 + b_2 \\ 3b_2 \\ -b_1 + 2b_2 \\ b_1 + b_2 \end{bmatrix}.$$

Clearly, $\mathbf{x} = a_1\mathbf{x}_1 + a_2\mathbf{x}_2$ and $\mathbf{y} = b_1\mathbf{y}_1 + 3b_2\mathbf{y}_2$. Also note that $\mathbf{x}'\mathbf{y} = 0$.

Example 5.3.3. Let $\mathbf{y}' = [1, 2, -1]$ and $\mathbf{x}' = [1, 1, 0]$. Find two vectors $\mathbf{x}_1$ and $\mathbf{z}$ such that $\mathbf{y} = \mathbf{x}_1 + \mathbf{z}$, where $\mathbf{x}_1$ is in the subspace (call it S) spanned by $\mathbf{x}$ and $\mathbf{z}$ is in $S^\perp$. Clearly $\mathbf{x}_1 = \lambda\mathbf{x}$ for some scalar λ. We must find a basis for $S^\perp$. Clearly, the basis contains two vectors, and each must be orthogonal to $\mathbf{x}_1$. By inspection, we note that $\{\mathbf{z}_1, \mathbf{z}_2\}$ is a basis for $S^\perp$, where $\mathbf{z}_1' = [1, -1, 1]$, $\mathbf{z}_2' = [1, -1, -2]$. Thus $\mathbf{z}$ must be a linear combination of $\mathbf{z}_1$ and $\mathbf{z}_2$, so we get

$$\mathbf{y} = \lambda\mathbf{x} + a_1\mathbf{z}_1 + a_2\mathbf{z}_2$$

or

$$\begin{bmatrix} 1 \\ 2 \\ -1 \end{bmatrix} = \begin{bmatrix} \lambda + a_1 + a_2 \\ \lambda - a_1 - a_2 \\ a_1 - 2a_2 \end{bmatrix},$$

and we obtain $\lambda = 3/2$; $a_1 = -2/3$; $a_2 = 1/6$;

$$\mathbf{y} = \mathbf{x}_1 + \mathbf{z} = \begin{bmatrix} 3/2 \\ 3/2 \\ 0 \end{bmatrix} + \begin{bmatrix} -1/2 \\ 1/2 \\ -1 \end{bmatrix}.$$

Example 5.3.4. Show that $\mathbf{x}_1$ in Example 5.3.3 is the projection of $\mathbf{y}$ into S. We shall use Theorem 4.4.1. Since S is spanned by the single vector $\mathbf{x}$, it is a basis for S; so set $\mathbf{B} = \mathbf{x}$ and the projection of $\mathbf{y}$ into S is the vector $\mathbf{w}$ where

$$\mathbf{w} = \mathbf{B}(\mathbf{B}'\mathbf{B})^{-1}\mathbf{B}'\mathbf{y} = \begin{bmatrix} 3/2 \\ 3/2 \\ 0 \end{bmatrix},$$

and we observe that this is $\mathbf{x}_1$.

5.4 Column and Null Spaces of a Matrix

In Section 2.4 we used the fact that a matrix could be viewed as a collection of vectors. More specifically, if $\mathbf{A}$ is an $n \times m$ matrix, then we can view the columns of $\mathbf{A}$ as m vectors in E_n. We now define a vector space associated with a matrix $\mathbf{A}$.

Definition 5.4.1

Column Space of a Matrix. *Let* $\mathbf{A}$ *be an* $n \times m$ *matrix; we denote the* m *columns of* $\mathbf{A}$ *as vectors in* E_n, *so that* $\mathbf{A} = [\mathbf{a}_1, \mathbf{a}_2, \ldots, \mathbf{a}_m]$. *The vector space spanned by these* m *column vectors of* $\mathbf{A}$ *is defined as the column space of* $\mathbf{A}$.

Note: Sometimes the *column space of A* is called the *range space of* $\mathbf{A}$. Clearly, the dimension of the column space of $\mathbf{A}$ is equal to the number of linearly independent columns of $\mathbf{A}$, which is equal to the rank of $\mathbf{A}$.

Another way to define the *column space* of an $n \times m$ matrix $\mathbf{A}$ is: the set S of vectors where

$$S = \left\{\mathbf{y} : \mathbf{y} = \sum_{i=1}^{m} b_i \mathbf{a}_i\,;\ b_i \in R\right\},$$

and S is clearly the vector space spanned by the columns of $\mathbf{A}$.

Still another way to define the column space of $\mathbf{A}$ is the set S of vectors where

$$S = \{\mathbf{y} : \mathbf{y} = \mathbf{A}\mathbf{b};\quad \mathbf{b} \in E_m\},$$

and S is clearly the vector space spanned by the columns of $\mathbf{A}$. By using the last definition, we notice that $\mathbf{y} \in S$ if and only if there exists a vector $\mathbf{b}$ in E_m such that $\mathbf{Ab} = \mathbf{y}$.

Theorem 5.4.1

Let $\mathbf{A}$ be an $n \times n$ nonsingular matrix. The column space of $\mathbf{A}$ is E_n.

Proof: For each and every vector $\mathbf{y}$ in E_n, there exists a vector $\mathbf{b}$ such that $\mathbf{Ab} = \mathbf{y}$ (let $\mathbf{b} = \mathbf{A}^{-1}\mathbf{y}$); hence E_n is the column space of $\mathbf{A}$. The proof of this theorem also follows immediately from the fact that, since $\mathbf{A}$ is nonsingular, it has rank n, so the n column vectors are linearly independent and hence they span E_n. ▮

If we examine the set of homogeneous equations $\mathbf{Ax} = \mathbf{0}$, we note that any vector $\mathbf{x}$ that satisfies the system is orthogonal to the rows of the matrix $\mathbf{A}$ and hence orthogonal to the columns of $\mathbf{A}'$. Since these systems occur frequently in statistics, we shall state some definitions and theorems to aid in studying them.

Definition 5.4.2

Null Space of a Matrix. *Let $\mathbf{A}$ be an $n \times m$ matrix. The null space of the matrix $\mathbf{A}$ is defined to be the set of vectors S where*

$$S = \{\mathbf{y} : \mathbf{Ay} = \mathbf{0}; \quad \mathbf{y} \in E_m\}.$$

Theorem 5.4.2

The null space of the $n \times m$ matrix $\mathbf{A}$ is a vector subspace of E_m.

Proof: Clearly, any vector $\mathbf{y}$ in S must have m elements and hence must be in E_m. If $\mathbf{y}_1 \in S$ and $\mathbf{y}_2 \in S$, we must show that $a_1\mathbf{y}_1 + a_2\mathbf{y}_2 \in S$ for all scalars a_1 and a_2. If $\mathbf{Ay}_1 = \mathbf{0}$, and $\mathbf{Ay}_2 = \mathbf{0}$, clearly

$$\mathbf{A}(a_1\mathbf{y}_1 + a_2\mathbf{y}_2) = a_1\mathbf{Ay}_1 + a_2\mathbf{Ay}_2 = \mathbf{0};$$

therefore, $a_1\mathbf{y}_1 + a_2\mathbf{y}_2 \in S$. ▮

Theorem 5.4.3

Let $\mathbf{A}$ be an $n \times m$ matrix. The null space of $\mathbf{A}'$ and the orthogonal complement of the column space of $\mathbf{A}$ are the same.

Proof: If S is the null space of $\mathbf{A}'$, then $\mathbf{y} \in S$ if and only if $\mathbf{A}'\mathbf{y} = \mathbf{0}$. If we denote the columns of $\mathbf{A}$ by $\mathbf{a}_1, \mathbf{a}_2, \ldots, \mathbf{a}_m$, then $\mathbf{y} \in S$ if and only if $\mathbf{a}_i'\mathbf{y} = 0$ for $i = 1, 2, \ldots, m$; hence, if $\mathbf{y}$ belongs to S, $\mathbf{y}$ is orthogonal to each column of $\mathbf{A}$ and hence orthogonal to the vector space spanned by the columns of $\mathbf{A}$. Thus, if $\mathbf{y} \in S$, $\mathbf{y}$ belongs to the orthogonal complement of the column space of $\mathbf{A}$. Now if $\mathbf{y}$ belongs to the orthogonal complement of the column space of $\mathbf{A}$, then $\mathbf{a}_i'\mathbf{y} = 0$ for $i = 1, 2, \ldots, m$, and hence $\mathbf{A}'\mathbf{y} = \mathbf{0}$, and thus $\mathbf{y}$ belongs to the null space of $\mathbf{A}'$; and the proof is complete. ▮

Example 5.4.1. Determine whether or not the vector $\mathbf{b}$ is in the column space of the 3×3 matrix $\mathbf{A}$ where

$$\mathbf{A} = \begin{bmatrix} 2 & 4 & -8 \\ -1 & 2 & -8 \\ 3 & 1 & 3 \end{bmatrix}; \qquad \mathbf{b} = \begin{bmatrix} 1 \\ 1 \\ 1 \end{bmatrix}.$$

We shall solve this problem by finding the rank of $\mathbf{A}$ and then finding the rank of $\mathbf{B}$ where $\mathbf{B} = [\mathbf{A}, \mathbf{b}]$. Clearly, rank $(\mathbf{A})$ = rank $(\mathbf{B})$ if and only if $\mathbf{b}$ is a linear combination of the columns of $\mathbf{A}$ and, hence, if and only if $\mathbf{b}$ is in the vector space spanned by the columns of $\mathbf{A}$. It is easy to show that rank $(\mathbf{A}) = 2$ and rank $(\mathbf{B}) = 3$, so $\mathbf{b}$ is not in the column space of $\mathbf{A}$.

Example 5.4.2. Find a basis set for the null space of $\mathbf{A}$ where $\mathbf{A}$ is defined by

$$\mathbf{A} = \begin{bmatrix} 2 & -1 & 3 \\ 4 & 2 & 1 \\ -8 & -8 & 3 \end{bmatrix}.$$

We shall use Theorem 5.4.3 to find the orthogonal complement of the column space of $\mathbf{A}'$ which is the same as the null space of $\mathbf{A}$:

$$\mathbf{A}' = \begin{bmatrix} 2 & 4 & -8 \\ -1 & 2 & -8 \\ 3 & 1 & 3 \end{bmatrix} = [\mathbf{a}_1^*, \mathbf{a}_2^*, \mathbf{a}_3^*].$$

Now $\{\mathbf{a}_1^*, \mathbf{a}_2^*\}$ is a basis set for the column space of $\mathbf{A}'$. By Theorem 2.6.5, we compute an orthogonal basis for $\mathbf{A}'$. It is $\{\mathbf{y}_1, \mathbf{y}_2\}$ where

$$\mathbf{y}_1 = \begin{bmatrix} 2 \\ -1 \\ 3 \end{bmatrix}; \qquad \mathbf{y}_2 = \begin{bmatrix} 38 \\ 37 \\ -13 \end{bmatrix}.$$

We must determine a vector $\mathbf{x} \neq \mathbf{0}$ such that $\mathbf{y}_1'\mathbf{x} = 0$ and $\mathbf{y}_2'\mathbf{x} = 0$. We get two equations to solve:

$$2x_1 - x_2 + 3x_3 = 0,$$

$$38x_1 + 37x_2 - 13x_3 = 0.$$

A solution is

$$\mathbf{x} = \begin{bmatrix} -7 \\ 10 \\ 8 \end{bmatrix}.$$

Therefore, $S = \{\mathbf{y} : \mathbf{y} = \lambda\mathbf{x};\ \lambda \in R\}$ is the orthogonal complement of the column space of $\mathbf{A}'$; hence it is the null space of $\mathbf{A}$, and $\{\mathbf{x}\}$ is a basis.

Theorem 5.4.4

Let $\mathbf{A}$ *be an* $n \times m$ *matrix and* $\mathbf{B}$ *an* $m \times k$ *matrix. The column space of* $\mathbf{AB}$ *is a subspace of the column space of* $\mathbf{A}$.

Proof: Let $\mathbf{C} = \mathbf{AB}$; then $\mathbf{C}$ is an $n \times k$ matrix. We can write this as

$$[\mathbf{c}_1, \mathbf{c}_2, \ldots, \mathbf{c}_k] = \left[\sum_i \mathbf{a}_i b_{i1}, \sum_i \mathbf{a}_i b_{i2}, \ldots, \sum_i \mathbf{a}_i b_{ik}\right].$$

From this we obtain $\mathbf{c}_j = \sum_i \mathbf{a}_i b_{ij}$, and hence the columns of $\mathbf{C}$ are a linear combination of the columns of $\mathbf{A}$; so the columns of $\mathbf{C}$ are in the subspace spanned by the columns of $\mathbf{A}$, and hence the column space of $\mathbf{C}$ is a subspace of the column space of $\mathbf{A}$. ∎

Corollary 5.4.4

In Theorem 5.4.4, *if the rank of* $\mathbf{AB}$ *is equal to the rank of* $\mathbf{A}$, *then the column space of* $\mathbf{A}$ *is the same as the column space of* $\mathbf{AB}$ (*in particular* $\mathbf{A}$ *and* $\mathbf{AA}'$ *have the same column space*).

Theorem 5.4.5

Let $\mathbf{A}$ *and* $\mathbf{B}$ *be* $n \times m$ *matrices. There exists a nonsingular* $m \times m$ *matrix* $\mathbf{C}$ *such that* $\mathbf{AC} = \mathbf{B}$ *if and only if* $\mathbf{A}$ *and* $\mathbf{B}$ *have the same column space.*

The proof of this theorem is left for the reader.

Theorem 5.4.6

Let $\mathbf{A}$ *and* $\mathbf{B}$ *be* $n \times m$ *matrices. A necessary and sufficient condition that there exists an* $m \times m$ *matrix* $\mathbf{C}$ *such that* $\mathbf{AC} = \mathbf{B}$ *is that the column space of* $\mathbf{B}$ *is a subspace of the column space of* $\mathbf{A}$.

Proof: Let $\mathbf{A} = [\mathbf{a}_1, \mathbf{a}_2, \ldots, \mathbf{a}_m]$ and $\mathbf{B} = [\mathbf{b}_1, \mathbf{b}_2, \ldots, \mathbf{b}_m]$. If the column space of $\mathbf{B}$ is a subspace of the column space of $\mathbf{A}$, then each $\mathbf{b}_j$ is a linear combination of the $\mathbf{a}_i$; that is,

$$\mathbf{b}_j = \sum_{i=1}^{m} \mathbf{a}_i c_{ij}, \quad j = 1, 2, \ldots, m.$$

and hence we can write this as $\mathbf{AC} = \mathbf{B}$. Next, if there does exist an $m \times m$ matrix $\mathbf{C}$ such that $\mathbf{AC} = \mathbf{B}$, then clearly

$$\sum_{i=1}^{m} \mathbf{a}_i c_{ij} = \mathbf{b}_j, \quad j = 1, 2, \ldots, m,$$

and hence $\mathbf{b}_j$ is in the space spanned by the $\mathbf{a}_i$, and this completes the proof of the theorem. ▌

Note: Instead of discussing the column space of matrices we could just as well have discussed the row space (the vector space spanned by the rows of a matrix $\mathbf{A}$). In fact the row space of $\mathbf{A}'$ is the same as the column space of $\mathbf{A}$; hence for each theorem in this section there is an analogous theorem for the row space of the matrix.

5.5 Statistical Applications

A procedure used a great deal in statistics is *hypothesis testing*. For example, a random sample $y_1, \ldots, y_n$ is obtained from a distribution that contains an unknown vector parameter $\boldsymbol{\theta}$, and on the basis of this sample the investigator wants to determine whether or not $\boldsymbol{\theta}$ is in a certain set. In general, the basic assumptions are that the distribution from which the sample is selected depends on the unknown parameter $\boldsymbol{\theta}$ and it is assumed that $\boldsymbol{\theta}$ is in a set which we call the parameter space denoted by Ω. The investigator may have reason to believe that in his situation the unknown $\boldsymbol{\theta}$ is in a subset ω of the parameter space Ω. In other words, it is known that $\boldsymbol{\theta}$ is in Ω, and the hypothesis to be tested is that $\boldsymbol{\theta}$ is in ω. On the basis of the observations in the sample, it will be decided that $\boldsymbol{\theta}$ is in ω or $\boldsymbol{\theta}$ is in $\bar{\omega}$, which is the complement of ω

with respect to Ω. The hypothesis that $\boldsymbol{\theta}$ is in ω is sometimes called the null hypothesis, and the hypothesis that $\boldsymbol{\theta}$ is in $\bar{\omega} = \Omega - \omega$ is sometimes called the alternative hypothesis.

Consider the *linear model* $\mathbf{y} = \mathbf{X}\boldsymbol{\beta} + \mathbf{e}$ which is defined in the Introduction, page 2. We assume that $\mathbf{y}$ is an $n \times 1$ vector, $\mathbf{X}$ is an $n \times p$ known matrix of rank k. We can also write this as $\mathbf{y} = \boldsymbol{\mu} + \mathbf{e}$ where $\boldsymbol{\mu} = \mathbf{X}\boldsymbol{\beta}$ is the mean of $\mathbf{y}$. If $\boldsymbol{\beta}$, the unknown parameter, can take on any value in E_p, then the parameter space Ω for $\boldsymbol{\mu}$ is the $k \times 1$ vector space determined by

$$\Omega = \{\boldsymbol{\mu} : \boldsymbol{\mu} = \mathbf{X}\boldsymbol{\beta}; \quad \boldsymbol{\beta} \in E_p\}.$$

That is, Ω is the column space of $\mathbf{X}$, which is a vector subspace of E_n.

Generally in this model, it is desired to test whether a set of linear restrictions among the β_i exist. For example, one may want to test whether or not the relationships $\beta_1 = \beta_2 = \beta_3 = 0$ hold. Or one may want to test whether or not $\beta_1 - \beta_2 = 4$ and $\beta_1 - 2\beta_3 = 6$ hold, or $\beta_1 = \beta_2 = \beta_3$ or some other linear relationship holds. The linear relationships can be put into the form $\mathbf{G}\boldsymbol{\beta} = \mathbf{0}$ or $\mathbf{G}\boldsymbol{\beta} = \mathbf{g}$ where $\mathbf{G}$ and $\mathbf{g}$ are known. For example, the relationship $\beta_1 = \beta_2 = \beta_3 = 0$ can be written as $\mathbf{G}\boldsymbol{\beta} = \mathbf{0}$ where $\mathbf{G} = \mathbf{I}$. The relationship $\beta_1 - \beta_2 = 4$, $\beta_1 - 2\beta_3 = 6$ can be written as $\mathbf{G}\boldsymbol{\beta} = \mathbf{g}$ where

$$\mathbf{G} = \begin{bmatrix} 1 & -1 & 0 \\ 1 & 0 & -2 \end{bmatrix} \quad \text{and} \quad \mathbf{g} = \begin{bmatrix} 4 \\ 6 \end{bmatrix}.$$

The relationship $\beta_1 = \beta_2 = \beta_3$ can be written as $\mathbf{G}\boldsymbol{\beta} = \mathbf{0}$ where

$$\mathbf{G} = \begin{bmatrix} 1 & -1 & 0 \\ 1 & 0 & -1 \end{bmatrix}.$$

Suppose we want to test a hypothesis that can be written as $\mathbf{G}\boldsymbol{\beta} = \mathbf{0}$ in the linear model $\mathbf{y} = \mathbf{X}\boldsymbol{\beta} + \mathbf{e}$. The parameter space Ω, given above, is a vector space. The parameter space ω is defined by ($\mathbf{X}$ and $\mathbf{G}$ are known)

$$\omega = \{\boldsymbol{\mu} : \boldsymbol{\mu} = \mathbf{X}\boldsymbol{\beta}; \quad \mathbf{G}\boldsymbol{\beta} = \mathbf{0}; \quad \boldsymbol{\beta} \in E_p\}.$$

ω is also a vector space, since the set of vectors $\boldsymbol{\beta}$ in E_p that satisfy $\mathbf{G}\boldsymbol{\beta} = \mathbf{0}$ is the null space of $\mathbf{G}$ and hence a vector space; hence the vectors $\boldsymbol{\mu}$ that satisfy $\boldsymbol{\mu} = \mathbf{X}\boldsymbol{\beta}$, as $\boldsymbol{\beta}$ ranges over the null space of $\mathbf{G}$, is also a vector space. Since Ω and ω are each vector spaces, this hypothesis is called a general linear hypothesis. The parameter space ω in the hypothesis $\mathbf{G}\boldsymbol{\beta} = \mathbf{g}$ is not a vector space, but by transforming the various vectors involved in the linear model $\mathbf{y} = \mathbf{X}\boldsymbol{\beta} + \mathbf{e}$, we can test a hypothesis equivalent to $\mathbf{G}\boldsymbol{\beta} = \mathbf{g}$ such that the parameter space ω is a vector space.

5.6 Functions of Matrices

In this section we briefly discuss certain functions of matrices that are useful in statistics, and in fact we shall use some of these results in subsequent chapters. We generally relate functions of matrices to functions of scalar variables. For example, if the function f is defined by $f(x) = x^2$ for $-\infty < x < \infty$, then $f(a) = a^2$ for any scalar a. If we use this for a matrix function, then for an $n \times n$ matrix $\mathbf{A}$ we get $f(\mathbf{A}) = \mathbf{A}^2$. Similarly, for the scalar polynomial function $p_3(x) = a_3x^3 + a_2x^2 + a_1x + a_0$, the corresponding polynomial function of $\mathbf{A}$ is $p_3(\mathbf{A}) = a_3\mathbf{A}^3 + a_2\mathbf{A}^2 + a_1\mathbf{A} + a_0\mathbf{I}$, where we define $\mathbf{A}^0 = \mathbf{I}$ for any nonzero square matrix $\mathbf{A}$. We now define a polynomial function of matrices.

Definition 5.6.1

Polynomial Function of Matrices. *Let $p_q(x) = a_qx^q + a_{q-1}x^{q-1} + \ldots + a_1x + a_0$ be a q-th degree polynomial in the scalar variable x. The corresponding q-th degree polynomial in the $n \times n$ matrix $\mathbf{A}$ is called a q-th degree polynomial function of the matrix $\mathbf{A}$ and is defined by $p_q(\mathbf{A}) = a_q\mathbf{A}^q + \ldots + a_1\mathbf{A} + a_0\mathbf{I}$.*

The first theorem we state in this section is known as the Cayley-Hamilton theorem. We will omit the proof, since it is usually found in a first course in matrix algebra.

Theorem 5.6.1

Let $p_n(x)$ be the characteristic polynomial of the $n \times n$ matrix $\mathbf{A}$, where $p_n(x) = det(\mathbf{A} - x\,\mathbf{I}) = \sum_{i=0}^{n} c_ix^i$. Then $p_n(\mathbf{A}) = \mathbf{0}$.

If $p_n(x)$ is the characteristic polynomial of the $n \times n$ matrix $\mathbf{A}$, then $p_n(\lambda) = 0$ if and only if λ is a characteristic root of $\mathbf{A}$. The theorem states that the polynomial function of a matrix $\mathbf{B}$, say $p_n(\mathbf{B})$, is such that $p_n(\mathbf{A}) = \mathbf{0}$; that is, a matrix $\mathbf{A}$ is a "root" of its own characteristic polynomial. There may be polynomial functions of matrices $\mathbf{A}$ of degree $k < n$ (where $\mathbf{A}$ is $n \times n$) such that $p_k(\mathbf{A}) = \mathbf{0}$. The smallest degree (say k) polynomial function of the matrix $\mathbf{A}$ such that $a_k = 1$ and $p_k(\mathbf{A}) = \mathbf{0}$ is called the *minimum polynomial* of $\mathbf{A}$.

There are also other functions (scalars) associated with a matrix $\mathbf{A}$ that are useful for describing $\mathbf{A}$. For example, the following are real-valued functions of $\mathbf{A}$ (of the elements a_{ij} of $\mathbf{A}$) that represent a matrix in some useful way: (1) det($\mathbf{A}$); (2) trace($\mathbf{A}$);

(3) largest characteristic root of $\mathbf{A}'\mathbf{A}$; (4) smallest element in $\mathbf{A}$; and so forth. Certain functions of matrices, called *norms,* play a special role, and we shall now define and discuss them briefly.

Definition 5.6.2

Norm of a Matrix. *Let* $\mathbf{A}$ *be any* $n \times n$ *matrix and let* $\|\mathbf{A}\|$ *denote a real-valued (actually non-negative) function of* $\mathbf{A}$ *(of the elements* a_{ij} *of* $\mathbf{A}$*) that satisfies the following.*

(1) $\|\mathbf{A}\| \geq 0$ *and* $\|\mathbf{A}\| = 0$ *if and only if* $\mathbf{A} = \mathbf{0}$.
(2) $\|c\mathbf{A}\| = |c| \cdot \|\mathbf{A}\|$, *where c is any scalar.*
(3) *If* $\mathbf{B}$ *is also any* $n \times n$ *matrix, then* $\|\mathbf{A} + \mathbf{B}\| \leq \|\mathbf{A}\| + \|\mathbf{B}\|$.
(4) *If* $\mathbf{B}$ *is any* $n \times n$ *matrix, then* $\|\mathbf{AB}\| \leq \|\mathbf{A}\| \cdot \|\mathbf{B}\|$.

The function $\|\mathbf{A}\|$ is called a norm of the matrix $\mathbf{A}$.

Example 5.6.1. Consider the non-negative function of the $n \times n$ matrix $\mathbf{A}$ (of the elements of $\mathbf{A}$) defined by $f(\mathbf{A}) = \Sigma\Sigma|a_{ij}|$. We want to show that this is a norm of $\mathbf{A}$.

(1) Clearly, $f(\mathbf{A}) \geq 0$ and $f(\mathbf{A}) = 0$ if and only if $\mathbf{A} = \mathbf{0}$.
(2) It is also quite obvious that for any scalar c we obtain $f(c\mathbf{A}) = \Sigma\Sigma|ca_{ij}| = |c|\Sigma\Sigma|a_{ij}| = |c|f(\mathbf{A})$.
(3) $f(\mathbf{A} + \mathbf{B}) = \Sigma\Sigma|a_{ij} + b_{ij}| \leq \Sigma\Sigma(|a_{ij}| + |b_{ij}|) = f(\mathbf{A}) + f(\mathbf{B})$.
(4) Let $\mathbf{C} = \mathbf{AB}$. Then $f(\mathbf{AB}) = f(\mathbf{C}) = \sum_i\sum_j|c_{ij}| = \sum_i\sum_j|\sum_t a_{it} b_{tj}| \leq \sum_i\sum_j\sum_t|a_{it} b_{tj}| = \sum_i\sum_j\sum_t|a_{it}|\,|b_{tj}| \leq \sum_i\sum_j(\sum_t|a_{it}|\sum_s|b_{sj}|) = f(\mathbf{A})f(\mathbf{B})$.

So clearly, $f(\mathbf{A})$ satisfies the four conditions of Def. 5.6.1, and hence $\Sigma\Sigma|a_{ij}|$ is a norm of the matrix $\mathbf{A}$.

In the next theorem we define some functions of $\mathbf{A}$ that are norms. We distinguish them by subscripts. These norms are useful in various branches of statistics and probability, and some of them will be used later in this book.

Theorem 5.6.2

Let $\mathbf{A}$ *be an* $n \times n$ *matrix. The functions of* $\mathbf{A}$ *(the elements* a_{ij} *of* $\mathbf{A}$*) defined below are norms of* $\mathbf{A}$.

(1) $\|\mathbf{A}\|_1 = \max_j \left(\sum_{i=1}^{n} |a_{ij}|\right)$ = *maximum of sums of absolute values of column elements.*

(2) $\|\mathbf{A}\|_\infty = \max_i \left(\sum_{j=1}^{n} |a_{ij}| \right)$ = *maximum of sums of absolute values of row elements.*

(3) $\|\mathbf{A}\|_2 = [\textit{maximum characteristic root of } \mathbf{A}'\mathbf{A}]^{1/2}$.
This is called the spectral norm.

(4) $\|\mathbf{A}\|_E = [\sum_i \sum_j (|a_{ij}|)^2]^{1/2}$. *This is called a Euclidean (E) norm.*

Proof: The proof that these functions of **A** satisfy the four conditions of a matrix norm, and hence indeed are matrix norms, will be asked for in the problems. ▮

Note: We remind the reader that *unless otherwise stated,* we are considering real matrices. However, Def. 5.6.2 and Theorem 5.6.2 are valid for complex matrices if $\mathbf{A}'$ is replaced by $\mathbf{A}^*$ (conjugate transpose), if $|a_{ij}|$ stands for the modulus of the complex number a_{ij}, and if $|c|$ is the modulus of the complex number c.

The next theorem follows directly from the definition of a norm of a matrix, and the proof will be left for the reader.

Theorem 5.6.3

Consider any $n \times n$ matrix **A**. *Then*

(1) $\|\mathbf{A}^q\| \le (\|\mathbf{A}\|)^q$ *for any positive integer q where $\|\mathbf{A}\|$ is any norm in Theorem* 5.6.2,

(2) $0 \le |(\|\mathbf{A}\| - \|\mathbf{B}\|)| \le \|\mathbf{A} - \mathbf{B}\|$ *for any norm in Theorem* 5.6.2,

(3) $\|\mathbf{A}\|_1 = \|\mathbf{A}'\|_\infty$,

(4) *For any $n \times n$ matrix* **B** *it follows that* $\|\mathbf{AB}\|_E \le \|\mathbf{A}\|_E \, \|\mathbf{B}\|_2$,

(5) $\|\mathbf{A}\|_E^2 = \text{trace}(\mathbf{A}'\mathbf{A}) = \|\mathbf{A}'\|_E^2$,

(6) $\|\mathbf{A}\|_E = \|\mathbf{A}'\|_E$,

(7) $\|\mathbf{A}\|_2 \le \|\mathbf{A}\|_E \le \sqrt{n}\, \|\mathbf{A}\|_2$,

(8) $\|\mathbf{D}\|_m = \max |d_{ii}|$ *for* $m = 1, 2, \infty$ *and* $\mathbf{D} = [d_{ij}]$ *is a diagonal matrix,*

(9) *If* **P** *is any orthogonal $n \times n$ matrix, then* $\|\mathbf{PA}\|_m = \|\mathbf{AP}\|_m = \|\mathbf{P}'\mathbf{AP}\|_m$, *where $m = E$ or* 2,

(10) $\|\mathbf{A}\|_2^2 = \max_{\mathbf{x} \varepsilon S} [\mathbf{x}'\mathbf{A}'\mathbf{A}\mathbf{x}/\mathbf{x}'\mathbf{x}]$, *where S is the set of all $n \times 1$ real vectors except the* **0** *vector,*

(11) *If* **A** *is non-negative, then* $\|\mathbf{A}\|_2$ = *max root of* **A**.

The next topic we consider is a sequence of matrices. For example, let $\{\mathbf{A}^{(k)}\} = \{\mathbf{A}^{(k)}: k = 1, 2, \ldots\} = \{[a_{ij}^{(k)}]\} = \{\mathbf{A}^{(1)}, \mathbf{A}^{(2)}, \ldots\}$ be symbols that represent a sequence of $m \times n$ matrices. Note that $\mathbf{A}^{(k)}$ is not the k-th power of $\mathbf{A}$ but rather represents the k-th matrix in the sequence. Also $a_{ij}^{(k)}$ represents the (ij)-th element in the matrix $\mathbf{A}^{(k)}$. We are interested in whether the sequence $\{\mathbf{A}^{(k)}\}$ converges, but first we must define what we mean by the convergence of a sequence of matrices.

Definition 5.6.3

Convergence of a Sequence of Matrices. *Let $\{\mathbf{A}^{(k)}\}$ denote a sequence of $m \times n$ matrices where $a_{ij}^{(k)}$ is the (ij)-th element in the k-th matrix. The sequence is said to converge to the $m \times n$ matrix $\mathbf{A} = [a_{ij}]$ if and only if the sequence of scalars $[a_{ij}^{(k)}: k = 1, 2, \ldots]$ converges to a_{ij} for each $i = 1, \ldots, m$ and $j = 1, \ldots, n$.*

Example 5.6.2. Consider the two sequences of matrices $\{\mathbf{A}^{(k)}\}$ and $\{\mathbf{B}^{(k)}\}$, where

$$\mathbf{A}^{(k)} = \begin{bmatrix} \dfrac{k}{k-1} & 6 \\ \dfrac{1}{k} & k \end{bmatrix}; \mathbf{B}^{(k)} = \begin{bmatrix} \dfrac{1}{k} & -2 & \dfrac{k-1}{k} \\ \dfrac{1}{2} & \dfrac{1}{k^2} & 1 \end{bmatrix}.$$

Clearly, $\{\mathbf{A}^{(k)}\}$ does not converge, since the element $a_{22}^{(k)} = k$ diverges. However $\{\mathbf{B}^{(k)}\}$ does converge, since each element converges; we write $\lim_{k\to\infty} \mathbf{B}^{(k)} = \mathbf{B}$, where

$$\mathbf{B} = \begin{bmatrix} 0 & -2 & 1 \\ \dfrac{1}{2} & 0 & 1 \end{bmatrix}.$$

Note: The sequence $\{\|\mathbf{A}^{(k)} - \mathbf{A}\|\}$ is a sequence of real (non-negative) numbers. Thus $\lim_{k\to\infty}[\|\mathbf{A}^{(k)} - \mathbf{A}\|]$ converges if and only if $\lim_{k\to\infty} a_k$ converges, where $a_k = \|\mathbf{A}^{(k)} - \mathbf{A}\|$ for any norm in Theorem 5.6.2.

In the next theorem we state some results that relate the limit of a sequence of matrices to the limit of a sequence of scalars (the sequence of scalars will be the sequence of norms of the matrices). This will demonstrate how the theory of limits in sequences of real numbers will be helpful in determining limits of sequences of matrices.

Theorem 5.6.4

Consider an $n \times n$ matrix $\mathbf{A}$, and the sequence of $n \times n$ matrices $\{\mathbf{A}^{(k)}\}$. Let $\|\mathbf{A}\|$ represent any norm in Theorem 5.6.2. The results below follow.

(1) $\lim_{k\to\infty} \mathbf{A}^{(k)} = \mathbf{0}$ *if and only if* $\lim_{k\to\infty} \|\mathbf{A}^{(k)}\| = 0.$

(2) $\lim_{k\to\infty} \mathbf{A}^{(k)} = \mathbf{A}$ *if and only if* $\lim_{k\to\infty} \|\mathbf{A}^{(k)} - \mathbf{A}\| = 0.$

(3) *If* $\lim_{k\to\infty} \mathbf{A}^{(k)} = \mathbf{A}$, *then* $\lim_{k\to\infty} \|\mathbf{A}^{(k)}\| = \|\mathbf{A}\|.$

The proof will be left for the reader.

Note: In (3) of Theorem 5.6.4 it may not be true that $\lim_{k\to\infty} \|\mathbf{A}^{(k)}\| = \|\mathbf{A}\|$ implies that $\lim_{k\to\infty} \mathbf{A}^{(k)} = \mathbf{A}$. Consider the sequence of 2×2 matrices $\{\mathbf{A}^{(k)}\}$, where

$$\mathbf{A}^{(k)} = \begin{bmatrix} 1 & 0 \\ 0 & 4 \end{bmatrix},$$

that is, a constant matrix for each k, and

$$\mathbf{A} = \begin{bmatrix} 4 & 0 \\ 0 & 1 \end{bmatrix}.$$

Clearly $\lim_{k\to\infty} \|\mathbf{A}^{(k)}\| = \|\mathbf{A}\|$ for each norm in Theorem 5.6.2, but $\lim_{k\to\infty} \mathbf{A}^{(k)} \neq \mathbf{A}$.

We will also be interested in the convergence of an ''infinite sum'' of matrices, that is, the sum $\mathbf{A}^{(0)} + \mathbf{A}^{(1)} + \mathbf{A}^{(2)} + \mathbf{A}^{(3)} + \ldots$. This corresponds to the infinite sum of scalars $a_0 + a_1 + a_2 + \ldots$. First we define what we mean by the convergence of an infinite sum of matrices; then we state some theorems that will be useful.

Definition 5.6.4

Convergence of an Infinite Sum of $m \times n$ Matrices. *Let $\{\mathbf{A}^{(k)}: k = 0, 1, 2, \ldots\}$ be a sequence of $m \times n$ matrices. Consider the infinite sum of these matrices; that is, $\Sigma_{k=0}^{\infty} \mathbf{A}^{(k)} = \Sigma_{k=0}^{\infty} [a_{ij}^{(k)}]$. The sum $\Sigma_{k=0}^{\infty} \mathbf{A}^{(k)}$ is defined as convergent (to $\mathbf{A} = [a_{ij}]$) if and only if $\Sigma_{k=0}^{\infty} a_{ij}^{(k)} = a_{ij}$ for each $i = 1, \ldots, m$; $j = 1, \ldots, n$. Otherwise $\Sigma_{k=0}^{\infty} \mathbf{A}^{(k)}$ is defined to be divergent.*

The following theorem follows immediately from this definition.

Theorem 5.6.5

Consider the sequence of $m \times n$ *matrices* $\{\mathbf{A}^{(k)}: k = 0, 1, \ldots\}$ *and the sequence of partial sums* $\{\mathbf{S}_t: t = 0, 1, 2, \ldots\}$, *where* $\mathbf{S}_t = \Sigma_{k=0}^{t} \mathbf{A}^{(k)}$. *Then the infinite sum* $\Sigma_{k=0}^{\infty} \mathbf{A}^{(k)}$ *converges to* $\mathbf{A}$ *if and only if the limit of the sequence* $\mathbf{S}_t$ *is* $\mathbf{A}$ *(that is,* $\lim_{t\to\infty} \mathbf{S}_t = \mathbf{A}$*).*

Example 5.6.3. Determine whether $\Sigma_{k=0}^{\infty} \mathbf{A}^{(k)}$ converges, where

$$\mathbf{A}^{(k)} = \begin{bmatrix} \frac{1}{2^k} & 0 \\ \frac{(-1)^k}{k+1} & \frac{1}{k!} \end{bmatrix}.$$

This sum clearly converges to $\mathbf{A}$, where

$$\mathbf{A} = \begin{bmatrix} 2 & 0 \\ \ell n 2 & e \end{bmatrix}.$$

If

$$\mathbf{A}^{(k)} = \begin{bmatrix} \frac{1}{k+1} & 0 \\ \frac{1}{k^2} & \frac{1}{k!} \end{bmatrix},$$

the sum does *not* converge, since $\sum_{k=0}^{\infty} \left(\frac{1}{k+1}\right)$ diverges.

A sequence of $n \times n$ matrices that is often of interest is $\{\mathbf{A}^k: k = 0, 1, \ldots\}$; that is, the k-th term of the sequence is the k-th power of $\mathbf{A}$. We state some theorems about this sequence and the sum $\Sigma_{k=0}^{\infty} \mathbf{A}^k$.

Theorem 5.6.6

Consider the sequence of $n \times n$ *matrices* $\{\mathbf{A}^k\}$. *Then* $\lim_{k\to\infty} \mathbf{A}^k = \mathbf{0}$ *if* $\|\mathbf{A}\| < 1$ *for any norm defined in Theorem* 5.6.2.

Proof: Let $\|\mathbf{A}\| = a$. By (1) of Theorem 5.6.3, $\|\mathbf{A}^k\| \leq a^k$ and clearly $\lim a^k = 0$ if $a < 1$. ∎

Note: It may be true that $\lim_{k\to\infty} \mathbf{A}^k = \mathbf{0}$ even if $\|\mathbf{A}\| \geq 1$ for one or more of the norms in Theorem 5.6.2.

The norm $\|\mathbf{A}\|_2^2$ is the largest characteristic root of $\mathbf{A}'\mathbf{A}$ and it may be of interest to determine a relationship between the various norms of $\mathbf{A}$ (when $\mathbf{A}$ is $n \times n$) and the roots of $\mathbf{A}$ rather than the roots of $\mathbf{A}'\mathbf{A}$. Note that the roots of $\mathbf{A}'\mathbf{A}$ are real (since $\mathbf{A}'\mathbf{A}$ is real symmetric), but the roots of $\mathbf{A}$ may *not* be real.

Theorem 5.6.7

Let $\mathbf{A}$ be an $n \times n$ matrix and let λ be any characteristic root of $\mathbf{A}$. Then $|\lambda| \leq \|\mathbf{A}\|$ where $\|\mathbf{A}\|$ is any norm defined in Theorem 5.6.2.

Proof: Let λ be any root of $\mathbf{A}$. Thus $\mathbf{Ax} = \lambda\mathbf{x}$ for some $\mathbf{x} \neq \mathbf{0}$. We write $\mathbf{A}[\mathbf{x}, \mathbf{x}, \ldots, \mathbf{x}] = [\lambda\mathbf{x}, \lambda\mathbf{x}, \ldots, \lambda\mathbf{x}]$ or $\mathbf{AX} = \lambda\mathbf{X}$, where $\mathbf{X}$ is the $n \times n$ matrix $[\mathbf{x}, \mathbf{x}, \ldots, \mathbf{x}]$. We get $\|\lambda\mathbf{X}\| = |\lambda| \cdot \|\mathbf{X}\| = \|\mathbf{AX}\| \leq \|\mathbf{A}\| \cdot \|\mathbf{X}\|$, but $\|\mathbf{X}\| \neq 0$, since $\mathbf{x} \neq \mathbf{0}$ ($\mathbf{x}$ is a characteristic vector), so $|\lambda| \leq \|\mathbf{A}\|$ and this proves the theorem. ∎

Often it is not easy to determine if $\lim_{k\to\infty} \mathbf{A}^k = \mathbf{0}$ merely by inspection. Theorem 5.6.6 can be helpful, since some of the norms are easy to compute. But Theorem 5.6.6 is a sufficient condition and not necessary. The next two theorems will sometimes help to determine whether $\lim_{k\to\infty} \mathbf{A}^k = \mathbf{0}$.

Theorem 5.6.8

Let $\{\mathbf{A}^k\}$ be a sequence of $n \times n$ matrices and let $\mathbf{P}$ be any nonsingular $n \times n$ matrix. Define $\mathbf{B}^k$ by $\mathbf{B}^k = (\mathbf{PAP}^{-1})^k$ for $k = 0, 1, \ldots$. Then $\lim_{k\to\infty} \mathbf{B}^k = \mathbf{0}$ if and only if $\lim_{k\to\infty} \mathbf{A}^k = \mathbf{0}$.

Proof: We sketch the proof. Clearly $\mathbf{B}^k = \mathbf{PA}^k\mathbf{P}^{-1}$ so if $\lim_{k\to\infty} \mathbf{A}^k = \mathbf{0}$, it follows that $\lim_{k\to\infty} \mathbf{PA}^k\mathbf{P}^{-1} = \mathbf{0}$. Also $\mathbf{A}^k = \mathbf{P}^{-1}\mathbf{B}^k\mathbf{P}$, and if $\lim_{k\to\infty} \mathbf{B}^k = \mathbf{0}$, it follows that $\lim_{k\to\infty} \mathbf{P}^{-1}\mathbf{B}^k\mathbf{P} = \mathbf{0}$. ∎

Theorem 5.6.9

Let $\mathbf{A}$ be an $n \times n$ matrix. Then $\lim_{k\to\infty} \mathbf{A}^k = \mathbf{0}$ if and only if $|\lambda_i| < 1$ for $i = 1, \ldots, n$, that is, if the modulus of each characteristic root of $\mathbf{A}$ is less than 1.

Proof: We sketch the proof and ask the reader to fill in the details.

Let λ be a characteristic root of $\mathbf{A}$ and $\mathbf{x}$ be a corresponding characteristic vector. Then $\mathbf{Ax} = \lambda\mathbf{x}$ implies $\mathbf{A}^2\mathbf{x} = \lambda\mathbf{Ax} = \lambda^2\mathbf{x}$ and for any positive integer k we get $\mathbf{A}^k\mathbf{x} = \lambda^k\mathbf{x}$. Since $\mathbf{x}$ is a fixed vector ($\mathbf{x} \neq \mathbf{0}$), and by hypothesis $\lim_{k\to\infty} \mathbf{A}^k = \mathbf{0}$, this implies $\lim_{k\to\infty} \lambda^k = 0$, which in turn implies $|\lambda| < 1$. Next we show that if $|\lambda| < 1$, this implies $\lim_{k\to\infty} \mathbf{A}^k = \mathbf{0}$. By Theorem 3.3.3 there exists a nonsingular matrix $\mathbf{P}$ such that $\mathbf{P}^{-1}\mathbf{AP} = \mathbf{T}$, where $\mathbf{T}$ is an upper triangular matrix and where t_{ii} for $i = 1, 2, \ldots, n$ are the characteristic roots of $\mathbf{A}$. By Theorem 5.6.8 it follows that $\lim_{k\to\infty} \mathbf{A}^k = \mathbf{0}$ if and only if $\lim_{k\to\infty} \mathbf{T}^k = \mathbf{0}$. But the roots of $\mathbf{T}^k$ are clearly t_{ii}^k for $i = 1, \ldots, n$. Also $\lim_{k\to\infty} \mathbf{T}^k = \mathbf{0}$ if $\lim_{k\to\infty} t_{ii}^k = 0$ for each $i = 1, \ldots, n$. But $\lim_{k\to\infty} t_{ii}^k = 0$ if $|t_{ii}| < 1$ for each $i = 1, \ldots, n$. This completes the proof. ∎

Corollary 5.6.9

Let $\mathbf{A}$ *be an* $n \times n$ *matrix. If* $\lim_{k\to\infty} \mathbf{A}^k = \mathbf{0}$, *then* $\mathbf{I} - \mathbf{A}$ *is a nonsingular matrix.*

Example 5.6.4. Consider the 3×3 matrix

$$\mathbf{A} = \begin{bmatrix} 0 & .2 & .8 \\ 0 & .8 & -.2 \\ .3 & 0 & .3 \end{bmatrix}$$

For the various norms in Theorem 5.6.2 we get $\|\mathbf{A}\|_1 = 1.3$; $\|\mathbf{A}\|_\infty = 1.0$; $\|\mathbf{A}\|_E = \sqrt{1.54}$. Clearly $\|\mathbf{A}\|_2$ is generally the most difficult norm to compute. From the results of the three norms calculated above we cannot determine if $\lim_{k\to\infty} \mathbf{A}^k = \mathbf{0}$. We would like to use Theorem 5.6.6. Define $\mathbf{B}$ by $\mathbf{B} = \mathbf{P}^{-1}\mathbf{A}\,\mathbf{P}$, which gives

$$\mathbf{B} = \begin{bmatrix} .5 & 0 & 0 \\ 0 & 1 & 0 \\ 0 & 0 & 1 \end{bmatrix} \begin{bmatrix} 0 & .2 & .8 \\ 0 & .8 & -.2 \\ .3 & 0 & .3 \end{bmatrix} \begin{bmatrix} 2 & 0 & 0 \\ 0 & 1 & 0 \\ 0 & 0 & 1 \end{bmatrix} = \begin{bmatrix} 0 & .1 & .4 \\ 0 & .8 & -.2 \\ .6 & 0 & .3 \end{bmatrix}$$

Clearly $\|\mathbf{B}\|_1 = .9$, so by Theorem 5.6.6 we obtain $\lim_{k\to\infty} \mathbf{B}^k = \mathbf{0}$, and then by using Theorem 5.6.8, we get $\lim_{k\to\infty} \mathbf{A}^k = \mathbf{0}$.

We now examine the matrix sum $\mathbf{I} + \mathbf{A} + \mathbf{A}^2 + \ldots = \Sigma_{k=0}^{\infty} \mathbf{A}^k$ and establish conditions that will help determine if the sum is convergent, and if so, to determine the matrix to which it converges.

Theorem 5.6.10

Let $\mathbf{A}$ *be an* $n \times n$ *matrix. The sum* $\Sigma_{k=0}^{\infty} \mathbf{A}^k$ *is convergent if and only if* $\lim_{k\to\infty} \mathbf{A}^k = \mathbf{0}$.

Proof: The proof of the "only if" part is obvious. To prove the "if" part consider the following identity that is true for all $k = 1, 2, \ldots$:

$$(\mathbf{I} - \mathbf{A})(\mathbf{I} + \mathbf{A} + \mathbf{A}^2 + \ldots + \mathbf{A}^k) = \mathbf{I} - \mathbf{A}^{k+1}.$$

Since $\lim_{k\to\infty} \mathbf{A}^k = \mathbf{0}$, this implies that $|\lambda| < 1$ for all characteristic roots λ of $\mathbf{A}$. By Corollary 5.6.9 this implies that $\mathbf{I} - \mathbf{A}$ is nonsingular, so we get for the partial sum $\mathbf{S}_k$

$$\mathbf{S}_k = \sum_{i=0}^{k} \mathbf{A}^i = (\mathbf{I} - \mathbf{A})^{-1} (\mathbf{I} - \mathbf{A}^{k+1}) = (\mathbf{I} - \mathbf{A})^{-1} - (\mathbf{I} - \mathbf{A})^{-1} \mathbf{A}^{k+1}.$$

But clearly $\lim_{k\to\infty} \mathbf{S}_k = (\mathbf{I} - \mathbf{A})^{-1}$, so $\Sigma_{i=0}^{\infty} \mathbf{A}^i$ converges. ∎

Corollary 5.6.10.1

Let $\mathbf{A}$ *be an* $n \times n$ *matrix. If* $|\lambda| < 1$ *for every characteristic root* λ *of* $\mathbf{A}$*, then the following obtain.*

(1) $\Sigma_{i=1}^{\infty} \mathbf{A}^i$ *converges to* $(\mathbf{I} - \mathbf{A})^{-1}$.

(2) *There exists a positive integer* K *such that for all positive integers* $k \geq K$ *the matrix* $\mathbf{B}^k$ *is nonsingular where* $\mathbf{B}^k = \mathbf{I} + \mathbf{A} + \mathbf{A}^2 + \ldots + \mathbf{A}^k$.

Corollary 5.6.10.2

Conditions (1) and (2) of Corollary 5.6.10.1 *are true if* $\|\mathbf{A}\| < 1$, *where* $\|\mathbf{A}\|$ *is any norm in Theorem* 5.6.2.

There will be situations when we will want to discuss the behavior of $n \times n$ matrices when n is large but perhaps the number of nonzero elements in $\mathbf{A}$ is small (or many of the elements are very small). There may be cases when a matrix is difficult to work with (difficult to find the inverse, the characteristic roots, and so on) but a related matrix can be found that is easy to work with. If n is large, it may be that we can work with the simpler matrix and the results will be approximately the same as if we worked with the more complicated matrix. One such complicated matrix is the Toeplitz matrix discussed in Sec. 8.15. Sometimes this kind of matrix can be approximated by a circulant matrix (discussed in Sec. 8.10), which is relatively simple to work with. If for large n a Toeplitz matrix is "approximately the same" as a circulant matrix, then we prefer to work with the simpler circulant matrix. We now describe what we mean by "approximately the same" and discuss some results.

Definition 5.6.5

Asymptotically Equivalent Sequences of Matrices. *Let* $\{\mathbf{A}^{(k)}: k = 1, 2, \ldots\} = \{\mathbf{A}_k\}$ *be a sequence of matrices where* $\mathbf{A}^{(k)}$ *has size* $k \times k$. *Let* $\{\mathbf{B}^{(k)}: k = 1, 2, 3, \ldots\} = \{\mathbf{B}_k\}$ *be a sequence of matrices, where* $\mathbf{B}_k$ *has size* $k \times k$. *The two sequences are defined to be asymptotically equivalent if and only if they satisfy the following.*

(1) $\|\mathbf{A}_k\|_2 \le c < \infty$, $\|\mathbf{B}_k\|_2 \le c < \infty$ *for* $k = 1, 2, 3, \ldots$, *where* c *is a real number that does not depend on* k.

(2) $\lim_{k\to\infty} k^{-1/2} \|(\mathbf{A}_k - \mathbf{B}_k)\|_E = 0.$

Note: This definition does not imply that two matrices are "equal" element by element as k goes to infinity, but rather that $k^{-1/2}$ times the Euclidean norm of the difference of the two matrices approaches zero.

Note: The size of each matrix in the sequence is different, that is, $\mathbf{A}_1$ is 1×1, $\mathbf{A}_2$ is 2×2; that is, the k-th matrix $\mathbf{A}_k$ in the sequence has size $k \times k$, and so forth.

Note: We use the notation $\mathbf{A}_k$ as the k-th element in a sequence of matrices when we want to indicate $\mathbf{A}_k$ has size $k \times k$. On the other hand, $\mathbf{A}^{(k)}$ implies each matrix in the sequence has the same size. When $\|\mathbf{A}_k\|_2 \le c < \infty$ and c is independent of k, we say the sequence $\{\mathbf{A}_k\}$ is uniformly bounded in 2-norm.

Theorem 5.6.11

Let $\{\mathbf{A}_k\}$ *and* $\{\mathbf{B}_k\}$ *be two asymptotically equivalent sequences of matrices.*

(1) *Then* $\lim_{k\to\infty} k^{-1/2} \|\mathbf{A}_k\|_E = \lim_{k\to\infty} k^{-1/2} \|\mathbf{B}_k\|_E$.

(2) *If* $\mathbf{A}_k^{-1}$ *and* $\mathbf{B}_k^{-1}$ *exist for each* $k = 1, 2, \ldots$, *and if* $\|\mathbf{A}_k^{-1}\|_2 \le c < \infty$, $\|\mathbf{B}_k^{-1}\|_2 \le c < \infty$ *for each* $k = 1, 2, \ldots$, *where* c *is a scalar*

that does not depend on k, then $\{\mathbf{A}_k^{-1}\}$ *and* $\{\mathbf{B}_k^{-1}\}$ *are asymptotically equivalent.*

Proof: The result (1) can be written $\lim_{k\to\infty}[\|k^{-1/2}\mathbf{A}_k\|_E - \|k^{-1/2}\mathbf{B}_k\|_E] = 0$. By the note following Example 5.6.2 and (2) of Def. 5.6.5 the result (1) can be proved. To prove (2) we write

$$k^{-1/2}\|(\mathbf{A}_k^{-1} - \mathbf{B}_k^{-1})\|_E = k^{-1/2}\|\mathbf{B}_k^{-1}\,\mathbf{B}_k\mathbf{A}_k^{-1} - \mathbf{B}_k^{-1}\mathbf{A}_k\mathbf{A}_k^{-1}\|_E$$

$$= k^{-1/2}\,\|\mathbf{B}_k^{-1}\,(\mathbf{B}_k - \mathbf{A}_k)\mathbf{A}_k^{-1}\|_E \le \|\mathbf{B}_k^{-1}\|_2\,k^{-1/2}\|(\mathbf{B}_k - \mathbf{A}_k)\,\mathbf{A}_k^{-1}\|_E$$

$$\le \|\mathbf{B}_k^{-1}\|_2\|\mathbf{A}_k^{-1}\|_2\,k^{-1/2}\|\mathbf{B}_k - \mathbf{A}_k\|_E \le c\cdot ck^{-1/2}\|\mathbf{B}_k - \mathbf{A}_k\|_E$$

and the result follows from Def. 5.6.5, since $\{\mathbf{A}_k\}$ and $\{\mathbf{B}_k\}$ are asymptotically equivalent. ∎

Theorem 5.6.12

Let $\{\mathbf{A}_k\}$, $\{\mathbf{B}_k\}$, $\{\mathbf{F}_k\}$, *and* $\{\mathbf{G}_k\}$ *be sequences of matrices.*

(1) *If* $\{\mathbf{A}_k\}$ *is asymptotically equivalent to* $\{\mathbf{B}_k\}$, *and if* $\{\mathbf{F}_k\}$ *is asymptotically equivalent to* $\{\mathbf{G}_k\}$, *then* $\{\mathbf{A}_k\mathbf{F}_k\}$ *is asymptotically equivalent to* $\{\mathbf{B}_k\mathbf{G}_k\}$.

(2) *If in* (1), $\{\mathbf{B}_k\}$ *is asymptotically equivalent to* $\{\mathbf{C}_k\}$, *then* $\{\mathbf{A}_k\}$ *is asymptotically equivalent to* $\{\mathbf{C}_k\}$.

(3) *If* $\{\mathbf{A}_k\mathbf{B}_k\}$ *is asymptotically equivalent to* $\{\mathbf{D}_k\}$ *and* $\|\mathbf{A}_k^{-1}\|_2 \le c < \infty$, *where c is a constant that does not depend on k, then* $\{\mathbf{B}_k\}$ *is asymptotically equivalent to* $\{\mathbf{A}_k^{-1}\mathbf{D}_k\}$.

Proof: (1) Clearly if any two sequences, say $\{\mathbf{M}_k\}$ and $\{\mathbf{N}_k\}$, are uniformly bounded in 2-norm (that is, $\|\mathbf{M}_k\|_2 \le m < \infty$ and $\|\mathbf{N}_k\|_2 \le n < \infty$), then the product sequence $\{\mathbf{M}_k\mathbf{N}_k\}$ is also uniformly bounded in 2-norm (that is, $\|\mathbf{M}_k\mathbf{N}_k\|_2 \le mn < \infty$). We must show that $\lim_{k\to\infty} k^{-1/2}\,\|\mathbf{A}_k\mathbf{F}_k - \mathbf{B}_k\mathbf{G}_k\|_E = 0$. We can write $k^{-1/2}\,\|\mathbf{A}_k\mathbf{F}_k - \mathbf{B}_k\mathbf{G}_k\|_E = k^{-1/2}\,\|\mathbf{A}_k\mathbf{F}_k - \mathbf{A}_k\mathbf{G}_k + \mathbf{A}_k\mathbf{G}_k - \mathbf{B}_k\mathbf{G}_k\|_E = k^{-1/2}\,\|\mathbf{A}_k(\mathbf{F}_k - \mathbf{G}_k) + (\mathbf{A}_k - \mathbf{B}_k)\mathbf{G}_k\|_E \le k^{-1/2}\,\|\mathbf{A}_k(\mathbf{F}_k - \mathbf{G}_k)\|_E + k^{-1/2}\,\|(\mathbf{A}_k - \mathbf{B}_k)\mathbf{G}_k\|_E \le k^{-1/2}\,\|\mathbf{A}_k\|_2\,\|\mathbf{F}_k - \mathbf{G}_k\|_E + k^{-1/2}\,\|\mathbf{G}_k\|_2\,\|\mathbf{A}_k - \mathbf{B}_k\|_E \le k^{-1/2}\,a\,\|\mathbf{F}_k - \mathbf{G}_k\|_E + k^{-1/2}\,g\,\|\mathbf{A}_k - \mathbf{B}_k\|_E \to 0$ as $k \to \infty$, since by hypothesis $\|\mathbf{A}_k\|_2 \le a$, $\|\mathbf{G}_k\|_2 \le g$ and $\lim_{k\to\infty} k^{-1/2}\,\|\mathbf{F}_k - \mathbf{G}_k\|_E = 0$, $\lim_{k\to\infty} k^{-1/2}\,\|\mathbf{A}_k - \mathbf{B}_k\|_E = 0$. Thus $\lim_{k\to\infty} k^{-1/2}\,\|\mathbf{A}_k\mathbf{F}_k - \mathbf{B}_k\mathbf{G}_k\|_E = 0$ and this proves (1). Parts (2) and (3) are proved by a similar method and will be left for the reader. ∎

Problems

1. If $\mathbf{a}_1$, $\mathbf{a}_2$, $\mathbf{a}_3$ span S_1 and $\mathbf{b}_1$, $\mathbf{b}_2$, $\mathbf{b}_3$ span S_2, find a basis for $S_1 \cap S_2$.

$$\mathbf{a}_1 = \begin{bmatrix} 1 \\ 3 \\ 2 \\ -1 \end{bmatrix}; \quad \mathbf{a}_2 = \begin{bmatrix} 0 \\ -1 \\ 2 \\ 1 \end{bmatrix}; \quad \mathbf{a}_3 = \begin{bmatrix} 2 \\ 7 \\ 2 \\ -3 \end{bmatrix};$$

$$\mathbf{b}_1 = \begin{bmatrix} 1 \\ 8 \\ 0 \\ -4 \end{bmatrix}; \quad \mathbf{b}_2 = \begin{bmatrix} 2 \\ 9 \\ -2 \\ -5 \end{bmatrix}; \quad \mathbf{b}_3 = \begin{bmatrix} 1 \\ 0 \\ -1 \\ 3 \end{bmatrix}.$$

2. In Prob. 1, find a basis for $S_1 \oplus S_2$.
3. If $\{\mathbf{a}_1, \ldots, \mathbf{a}_t\}$ spans S_1 and $\{\mathbf{b}_1, \ldots, \mathbf{b}_s\}$ spans S_2, prove that $\{\mathbf{a}_1, \ldots, \mathbf{a}_t, \mathbf{b}_1, \ldots, \mathbf{b}_s\}$ spans $S_1 \oplus S_2$.
4. In Prob. 1, find the dimension of $S_1 \cap S_2$.
5. In Prob. 1, find the dimension of $S_1 \oplus S_2$.
6. Use the results of Probs. 4 and 5 to demonstrate Theorem 5.2.5.
7. Prove Theorem 5.2.5.
8. Find a basis for the orthogonal complement of the subspace in E_3 spanned by $\mathbf{x}$ where $\mathbf{x}' = [1, 1, -1]$.
9. Find the vector $\mathbf{z}$ and the scalar λ such that

$$\mathbf{y} = \lambda \mathbf{x} + \mathbf{z}$$

where

$$\mathbf{y} = \begin{bmatrix} 1 \\ 1 \\ -1 \end{bmatrix}; \quad \mathbf{x} = \begin{bmatrix} 1 \\ 2 \\ -1 \end{bmatrix}$$

and such that $\mathbf{x}$ and $\mathbf{z}$ are orthogonal.

10. Let S be spanned by $\mathbf{x}_1$ and $\mathbf{x}_2$ where $\mathbf{x}_1' = [1, 1, 0, -1]$; $\mathbf{x}_2' = [0, 1, 1, 1]$; the vector $\mathbf{y}$ is defined by $\mathbf{y}' = [1, 2, -1, 1]$. Find vectors $\mathbf{z}_1$ and $\mathbf{z}_2$ such that $\mathbf{y} = \mathbf{z}_1 + \mathbf{z}_2$ where $\mathbf{z}_1$ is in S and $\mathbf{z}_2$ is in $S^{\perp}$.

11. In Prob. 10, show that $\mathbf{z}_1$ is the projection of $\mathbf{y}$ into S by using the results of Theorem 4.4.1.
12. Find the dimension of the column space of $\mathbf{A}$ where

$$\mathbf{A} = \begin{bmatrix} 1 & 1 & 1 \\ 2 & 2 & 2 \\ -1 & 1 & -3 \\ 1 & 2 & 0 \end{bmatrix}.$$

13. If $\mathbf{A}$ and $\mathbf{B}$ are nonsingular $n \times n$ matrices, prove that they have the same column space.
14. Let $\mathbf{A}$ and $\mathbf{B}$ be $m \times n$ matrices of rank n where $n < m$. Show that $\mathbf{A}$ and $\mathbf{B}$ may not have the same column space.
15. Prove Theorem 5.4.5.
16. Let $\mathbf{A}$ and $\mathbf{B}$ be defined by

$$\mathbf{A} = \begin{bmatrix} 1 & 1 & 1 \\ 2 & 2 & 2 \\ -1 & 1 & -3 \end{bmatrix}; \qquad \mathbf{B} = \begin{bmatrix} 0 & 2 & 1 \\ 0 & 4 & 2 \\ 0 & -2 & -1 \end{bmatrix}.$$

Show that the column space of $\mathbf{B}$ is a subspace of the column space of $\mathbf{A}$.
17. In Prob. 16, find a matrix $\mathbf{C}$ such that $\mathbf{AC} = \mathbf{B}$.
18. Prove Corollary 5.3.3.
19. Prove Theorem 5.6.2.
20. Prove Theorem 5.6.4.
21. Suppose $\mathbf{A}$ is an $n \times n$ matrix and $\mathbf{A}^k = \mathbf{0}$ for a positive integer k. Show that $\mathbf{I} - \mathbf{A}$ is nonsingular.
22. In Prob. 21, show $(\mathbf{I} - \mathbf{A})^{-1} = \Sigma_{i=0}^{n-1} \mathbf{A}^i$.
23. By the method in Prob. 22, find the inverse of $\mathbf{I} - \mathbf{A}$ where

$$\mathbf{A} = \begin{bmatrix} 0 & 1 & -1 \\ 0 & 0 & 2 \\ 0 & 0 & 0 \end{bmatrix}.$$

Generalized Inverse; Conditional Inverse

6

6.1 Introduction

In Chapter 1, the inverse of a matrix was defined and various properties were discussed. It was stated that if a matrix $\mathbf{A}$ has an inverse, the matrix must be square and the determinant must be nonzero. It has been stated in previous chapters that the theory of linear models, which includes a considerable part of theoretical and applied statistics, involves the solutions of a system of linear equations

$$\mathbf{A}\mathbf{x} = \mathbf{g} \tag{6.1.1}$$

and functions of the solutions.

If $\mathbf{A}$ is an $n \times n$ nonsingular matrix, the solution to the system in Eq. (6.1.1) exists, is unique, and is given by $\mathbf{x} = \mathbf{A}^{-1}\mathbf{g}$. However, there are cases where $\mathbf{A}$ is not a square matrix and also situations where $\mathbf{A}$ is a square matrix but is singular. In these situations, there may still be a solution to the system, and a unified theory to treat all situations may be desirable. One such theory involves the use of "generalized" and "conditional" inverses of matrices, which are discussed in this chapter.

6.2 Definition and Basic Theorems of Generalized Inverse

Let $\mathbf{A}$ be an $m \times n$ matrix of rank r. We shall investigate a matrix denoted by $\mathbf{A}^-$, which has many of the properties that the inverse of the matrix $\mathbf{A}$ would have if the inverse existed.

Definition 6.2.1

Generalized Inverse. *Let* $\mathbf{A}$ *be an* $m \times n$ *matrix. If a matrix* $\mathbf{A}^-$ *exists that satisfies the four conditions below, we shall call* $\mathbf{A}^-$ *a generalized inverse of* $\mathbf{A}$.

1. $\mathbf{AA}^-$ *is symmetric.*
2. $\mathbf{A}^-\mathbf{A}$ *is symmetric.*
3. $\mathbf{AA}^-\mathbf{A} = \mathbf{A}$.
4. $\mathbf{A}^-\mathbf{AA}^- = \mathbf{A}^-$.

(6.2.1)

We use the terminology "g-inverse" for generalized inverse.

If $\mathbf{A}$ is nonsingular it is clear that $\mathbf{A}^{-1}$ satisfies the conditions of a g-inverse. However, if $\mathbf{A}$ is a square matrix and singular or if $\mathbf{A}$ is not a square matrix, then the problem remains as to whether a matrix $\mathbf{A}^-$ exists that satisfies Eq. (6.2.1). We shall show that for each matrix $\mathbf{A}$, a g-inverse matrix $\mathbf{A}^-$ exists and is unique. We shall also state and prove various properties of this g-inverse.

Theorem 6.2.1

If a g-inverse of an $m \times n$ *matrix* $\mathbf{A}$ *exists, it has order* $n \times m$.

Proof: The proof follows from the fact that $\mathbf{AA}^-$ is symmetric and hence square. ▮

Theorem 6.2.2

If $\mathbf{A}$ *is the null matrix of order* $m \times n$, *then* $\mathbf{A}^-$ *is the null matrix of size* $n \times m$.

Proof: Clearly, $\mathbf{A}^- = \mathbf{0}$ satisfies the conditions in Eq. (6.2.1) if $\mathbf{A} = \mathbf{0}$. ▮

Theorem 6.2.3

For each matrix $\mathbf{A}$, *there is a matrix* $\mathbf{A}^-$ *satisfying the conditions of Eq.* (6.2.1); *that is, each matrix has a g-inverse.*

Proof: If $\mathbf{A} = \mathbf{0}$, then by Theorem 6.2.2, $\mathbf{A}^- = \mathbf{0}$. Assume $\mathbf{A} \neq \mathbf{0}$. By using Theorem 1.6.8, if $\mathbf{A}$ has rank $r > 0$, it can be factored as

$$\mathbf{A} = \mathbf{BC}, \tag{6.2.2}$$

where $\mathbf{B}$ is $m \times r$ of rank r and $\mathbf{C}$ is $r \times n$ of rank r. Note that $\mathbf{B'B}$ and $\mathbf{CC'}$ are both nonsingular. If we define $\mathbf{A}^-$ as

$$\mathbf{A}^- = \mathbf{C}'(\mathbf{CC}')^{-1}(\mathbf{B}'\mathbf{B})^{-1}\mathbf{B}', \tag{6.2.3}$$

then it is easily shown that it satisfies the conditions of Eq. (6.2.1). Note also that if $\mathbf{A}$ is real, then $\mathbf{A}^-$ is real. ▮

The factorization of $\mathbf{A}$ in Eq. (6.2.2) is not unique; however the g-inverse $\mathbf{A}^-$ is unique. This is the context of the next theorem.

Theorem 6.2.4

For each matrix $\mathbf{A}$ *there exists a unique matrix* $\mathbf{A}^-$ *that satisfies the conditions of Eq.* (6.2.1); *that is, each matrix* $\mathbf{A}$ *has a unique g-inverse.*

Proof: Assume that $\mathbf{A}_1^-$ and $\mathbf{A}_2^-$ are two g-inverses of a matrix $\mathbf{A}$. This means that $\mathbf{A}_1^-$ and $\mathbf{A}_2^-$ each satisfy the four conditions of Eq. (6.2.1). We shall show that when this is the case, it follows that $\mathbf{A}_1^- = \mathbf{A}_2^-$. First we show that $\mathbf{AA}_1^- = \mathbf{AA}_2^-$. Multiply $\mathbf{A} = \mathbf{AA}_1^-\mathbf{A}$ on the right by $\mathbf{A}_2^-$ and obtain

$$\mathbf{AA}_2^- = \mathbf{AA}_1^-\mathbf{AA}_2^-.$$

By Eq. (6.2.1), the left hand side, and hence also the right hand side, is symmetric; that is,

$$\mathbf{AA}_1^-\mathbf{AA}_2^- = [\mathbf{AA}_1^-\mathbf{AA}_2^-]'.$$

From this we get

$$\mathbf{AA}_2^- = \mathbf{AA}_1^-\mathbf{AA}_2^- = [(\mathbf{AA}_1^-)(\mathbf{AA}_2^-)]' = (\mathbf{AA}_2^-)'(\mathbf{AA}_1^-)' = (\mathbf{AA}_2^-)(\mathbf{AA}_1^-) = \mathbf{AA}_1^-, \tag{6.2.4}$$

since, by Eq. (6.2.1), $\mathbf{AA}_1^-$ and $\mathbf{AA}_2^-$ are each symmetric.

By a similar procedure (multiplying $\mathbf{A} = \mathbf{AA}_1^-\mathbf{A}$ by $\mathbf{A}_2^-$ on the left, instead of on the right), we obtain

$$\mathbf{A}_1^-\mathbf{A} = \mathbf{A}_2^-\mathbf{A}. \tag{6.2.5}$$

By using the results of Eq. (6.2.4) and Eq. (6.2.5), we get

$$\mathbf{A}_1^- = \mathbf{A}_1^-\mathbf{AA}_1^- = (\mathbf{A}_1^-\mathbf{A})\mathbf{A}_1^- = (\mathbf{A}_2^-\mathbf{A})\mathbf{A}_1^- = \mathbf{A}_2^-(\mathbf{AA}_1^-) = \mathbf{A}_2^-\mathbf{AA}_2^- = \mathbf{A}_2^-, \tag{6.2.6}$$

and the proof is complete. ∎

In the next several theorems, you will note a resemblance between the properties of the g-inverse and a regular inverse.

Theorem 6.2.5

The g-inverse of the transpose of $\mathbf{A}$ *is the transpose of the g-inverse of* $\mathbf{A}$*; that is,* $(\mathbf{A}')^- = (\mathbf{A}^-)'$.

Proof: The proof consists of showing that $(\mathbf{A}^-)'$ is the g-inverse of $\mathbf{A}'$ and since the g-inverse of $\mathbf{A}'$ is unique, it follows that $(\mathbf{A}')^- = (\mathbf{A}^-)'$.

Write

$$\mathbf{A} = \mathbf{BC},$$

as in Eq. (6.2.2), and get

$$\mathbf{A}^- = \mathbf{C}'(\mathbf{CC}')^{-1}(\mathbf{B}'\mathbf{B})^{-1}\mathbf{B}'. \tag{6.2.7}$$

Also

$$\mathbf{A}' = \mathbf{C}'\mathbf{B}',$$

and

$$(\mathbf{A}')^- = \mathbf{B}(\mathbf{B}'\mathbf{B})^{-1}(\mathbf{CC}')^{-1}\mathbf{C}.$$

Take the transpose of $\mathbf{A}^-$ in Eq. (6.2.7) and get

$$(\mathbf{A}^-)' = \mathbf{B}(\mathbf{B}'\mathbf{B})^{-1}(\mathbf{CC}')^{-1}\mathbf{C};$$

hence

$$(\mathbf{A}^-)' = (\mathbf{A}')^-,$$

and, since the g-inverse of a matrix $\mathbf{A}'$ is unique, the theorem is proved.

Theorem 6.2.6

The g-inverse of $\mathbf{A}^-$ is equal to $\mathbf{A}$; that is, $(\mathbf{A}^-)^- = \mathbf{A}$.

Proof: By Def. 6.2.1, the g-inverse of $\mathbf{A}^-$ satisfies

1. $\mathbf{A}^-(\mathbf{A}^-)^- = [\mathbf{A}^-(\mathbf{A}^-)^-]'$,
2. $(\mathbf{A}^-)^-\mathbf{A}^- = [(\mathbf{A}^-)^-\mathbf{A}^-]'$,
3. $\mathbf{A}^-(\mathbf{A}^-)^-\mathbf{A}^- = \mathbf{A}^-$,
4. $(\mathbf{A}^-)^-\mathbf{A}^-(\mathbf{A}^-)^- = (\mathbf{A}^-)^-$.

But if we substitute $\mathbf{A}$ for $(\mathbf{A}^-)^-$, the four equations above are exactly those in Def. 6.2.1, so $\mathbf{A}$ is the unique g-inverse of $\mathbf{A}^-$; that is, $(\mathbf{A}^-)^- = \mathbf{A}$. ∎

Theorem 6.2.7

The rank of the g-inverse of $\mathbf{A}$ is equal to the rank of $\mathbf{A}$.

Proof: If we apply Theorem 1.6.10 to $\mathbf{AA}^-\mathbf{A} = \mathbf{A}$, we get rank $(\mathbf{A})$ = rank $(\mathbf{AA}^-\mathbf{A}) \leq$ rank $(\mathbf{A}^-)$; but from $\mathbf{A}^-\mathbf{AA} = \mathbf{A}^-$, we get rank $(\mathbf{A}^-)$ = rank $(\mathbf{A}^-\mathbf{AA}^-) \leq$ rank $(\mathbf{A})$; hence rank $(\mathbf{A})$ = rank $(\mathbf{A}^-)$. ∎

An extension of this theorem is given in the following corollary.

Corollary 6.2.7

If the rank of the matrix $\mathbf{A}$ is equal to r, the rank of each of the following matrices is also equal to r: $\mathbf{A}^-$, $\mathbf{AA}^-$, $\mathbf{A}^-\mathbf{A}$, $\mathbf{AA}^-\mathbf{A}$, $\mathbf{A}^-\mathbf{AA}^-$.

Theorem 6.2.8

For any matrix $\mathbf{A}$ we get $(\mathbf{A}'\mathbf{A})^- = \mathbf{A}^-\mathbf{A}'^-$.

Proof: The g-inverse of $\mathbf{A'A}$, which is denoted by $(\mathbf{A'A})^-$, must satisfy

1. $$(\mathbf{A'A})(\mathbf{A'A})^- = [(\mathbf{A'A})(\mathbf{A'A})^-]',$$
2. $$(\mathbf{A'A})^-(\mathbf{A'A}) = [(\mathbf{A'A})^-(\mathbf{A'A})]',$$
3. $$(\mathbf{A'A})(\mathbf{A'A})^-(\mathbf{A'A}) = \mathbf{A'A},$$
4. $$(\mathbf{A'A})^-(\mathbf{A'A})(\mathbf{A'A})^- = (\mathbf{A'A})^-.$$

It can be shown by straightforward multiplication that if $\mathbf{A}^-(\mathbf{A}')^-$ is substituted for $(\mathbf{A'A})^-$, the equations above are those in Def. 6.2.1. For example, if we substitute $\mathbf{A}^-(\mathbf{A}')^-$for $(\mathbf{A'A})^-$ in the third equation above, we get

$$(\mathbf{A'A})\mathbf{A}^-(\mathbf{A}')^-(\mathbf{A'A}) = \mathbf{A}'[\mathbf{AA}^-](\mathbf{A}')^-\mathbf{A'A}.$$

We can replace the quantity in the brackets by $(\mathbf{A}')^-\mathbf{A}'$, since by Def. 6.2.1, the g-inverse of $\mathbf{A}$ is such that $\mathbf{AA}^- = (\mathbf{AA}^-)' = (\mathbf{A}^-)'\mathbf{A}'$, and by Theorem 6.2.5, this is equal to $(\mathbf{A}')^-\mathbf{A}'$. So, if we substitute this value of $\mathbf{AA}^-$ into the above, we get

$$\begin{aligned}(\mathbf{A'A})\mathbf{A}^-(\mathbf{A}')^-\mathbf{A'A} &= \mathbf{A}'[\mathbf{AA}^-](\mathbf{A}')^-\mathbf{A'A}\\ &= \mathbf{A}'(\mathbf{A}')^-\mathbf{A}'(\mathbf{A}')^-\mathbf{A'A} = \mathbf{A}'(\mathbf{A}')^-\mathbf{A'A} = \mathbf{A'A}.\end{aligned}$$

We have shown that if we replace $(\mathbf{A'A})^-$ by $\mathbf{A}^-(\mathbf{A}')^-$, the third equation in Def. 6.2.1 is satisfied. By a similar procedure, we can show that the remaining three equations satisfy those in Def. 6.2.1. Hence $\mathbf{A}^-(\mathbf{A}')^-$ is the unique g-inverse of $\mathbf{A'A}$; that is, $(\mathbf{A'A})^- = \mathbf{A}^-(\mathbf{A}')^-$ and the theorem is proved. ∎

Theorem 6.2.9

For any matrix $\mathbf{A}$, *we get* $(\mathbf{AA}^-)^- = \mathbf{AA}^-$ *and* $(\mathbf{A}^-\mathbf{A})^- = \mathbf{A}^-\mathbf{A}$.

This theorem can be proved by a procedure almost identical with the method used to prove Theorem 6.2.8. The details are left for the reader.

Theorem 6.2.10

Let $\mathbf{P}$ *be an* $m \times m$ *orthogonal matrix,* $\mathbf{Q}$ *be an* $n \times n$ *orthogonal matrix and* $\mathbf{A}$ *any* $m \times n$ *matrix. Then* $(\mathbf{PAQ})^- = \mathbf{Q'A^-P'}$.

Proof: Let $\mathbf{B} = \mathbf{PAQ}$. We must show that $\mathbf{B}^- = \mathbf{Q'A^-P'}$ satisfies the four conditions of Eq. (6.2.1). We get

(1) $\mathbf{BB^-} = \mathbf{PAQQ'A^-P'} = \mathbf{PAA^-P'}$. But $\mathbf{PAA^-P'}$ is symmetric, since $\mathbf{AA^-}$ is symmetric. Hence $\mathbf{BB^-}$ is symmetric.
(2) $\mathbf{B^-B} = (\mathbf{Q'A^-P'})(\mathbf{PAQ}) = \mathbf{Q'A^-AQ}$, which is symmetric, and so $\mathbf{B^-B}$ is symmetric.
(3) $(\mathbf{PAQ})(\mathbf{Q'A^-P'})(\mathbf{PAQ}) = \mathbf{PAA^-AQ} = (\mathbf{PAQ})$ or $\mathbf{BB^-B} = \mathbf{B}$.
(4) $(\mathbf{Q'A^-P'})(\mathbf{PAQ})(\mathbf{Q'A^-P'}) = \mathbf{Q'A^-AA^-P'} = \mathbf{Q'A^-P'}$ or $\mathbf{B^-BB^-} = \mathbf{B^-}$.

Hence $\mathbf{Q'A^-P'}$ satisfies the conditions in Eq. (6.2.1), and so $\mathbf{Q'A^-P'}$ is the g-inverse of $\mathbf{PAQ}$. ▮

Note: If the two matrices $\mathbf{G}$, $\mathbf{H}$ are nonsingular, then, by Theorem 1.3.3, we get $(\mathbf{GH})^{-1} = \mathbf{H}^{-1}\mathbf{G}^{-1}$. Theorems 6.2.8 and 6.2.9 state a similar result; that is, $(\mathbf{GH})^- = \mathbf{H^-G^-}$ if $\mathbf{G}$ and $\mathbf{H}$ are *certain* matrices. Theorem 6.2.10 also states a similar result. However, it is *not* true that $(\mathbf{GH})^- = \mathbf{H^-G^-}$ for all matrices $\mathbf{G}$ and $\mathbf{H}$.

Theorem 6.2.11

If $\mathbf{A}$ *is a symmetric matrix, the g-inverse of* $\mathbf{A}$ *is also symmetric; that is, if* $\mathbf{A} = \mathbf{A}'$, *then* $\mathbf{A}^- = (\mathbf{A}^-)'$.

Proof: By Theorem 6.2.5, we obtain $(\mathbf{A}')^- = (\mathbf{A}^-)'$, but, since $\mathbf{A} = \mathbf{A}'$, we obtain $\mathbf{A}^- = (\mathbf{A}^-)'$. ▮

Theorem 6.2.12

If $\mathbf{A} = \mathbf{A}'$, *then* $\mathbf{AA^-} = \mathbf{A^-A}$.

Proof: By Eq. (6.2.1) we get $\mathbf{AA^-} = (\mathbf{AA^-})'$, but $(\mathbf{AA^-})' = \mathbf{A}'^-\mathbf{A}'$, which is equal to $\mathbf{A^-A}$, by using Theorem 6.2.11. ▮

The next several theorems give the form of the g-inverse of special matrices.

Theorem 6.2.13

If the matrix $\mathbf{A}$ *is nonsingular, then* $\mathbf{A}^{-1} = \mathbf{A}^-$.

Proof: The proof is accomplished by showing that $\mathbf{A}^{-1}$ satisfies Eq. (6.2.1). ▮

Theorem 6.2.14

If $\mathbf{A}$ *is symmetric idempotent, then* $\mathbf{A}^- = \mathbf{A}$; that is, *if* $\mathbf{A} = \mathbf{A}'$ *and* $\mathbf{A} = \mathbf{A}^2$, *then* $\mathbf{A}^- = \mathbf{A}$.

Proof: The proof is accomplished by showing that $\mathbf{A}$ satisfies Eq. (6.2.1). ▮

Theorem 6.2.15

Let $\mathbf{D}$ *be an* $n \times n$ *diagonal matrix with diagonal elements* d_{ii}; $i = 1, 2, \ldots, n$. *The g-inverse* $\mathbf{D}^-$ *of* $\mathbf{D}$ *is a diagonal matrix with i-th diagonal element of* $\mathbf{D}^-$ *equal to* d_{ii}^{-1} *if* $d_{ii} \neq 0$ *and equal to zero if* $d_{ii} = 0$.

Proof: The proof is immediately obtained by showing that $\mathbf{D}^-$ satisfies Eq. (6.2.1). ▮

For example, if $\mathbf{D}$ is defined by

$$\mathbf{D} = \begin{bmatrix} 3 & 0 & 0 \\ 0 & 1 & 0 \\ 0 & 0 & 0 \end{bmatrix}, \quad \text{then} \quad \mathbf{D}^- = \begin{bmatrix} \frac{1}{3} & 0 & 0 \\ 0 & 1 & 0 \\ 0 & 0 & 0 \end{bmatrix}.$$

Theorem 6.2.16

If $\mathbf{A}$ *is an* $m \times n$ *matrix of rank* m, *then* $\mathbf{A}^- = \mathbf{A}'(\mathbf{A}\mathbf{A}')^{-1}$ *and* $\mathbf{A}\mathbf{A}^- = \mathbf{I}$. *If the rank of* $\mathbf{A}$ *is* n, *then* $\mathbf{A}^- = (\mathbf{A}'\mathbf{A})^{-1}\mathbf{A}'$ *and* $\mathbf{A}^-\mathbf{A} = \mathbf{I}$. *Also* $\mathbf{A}^- = \mathbf{A}'(\mathbf{A}\mathbf{A}')^- = (\mathbf{A}'\mathbf{A})^-\mathbf{A}'$ *whatever the rank of* $\mathbf{A}$.

The proofs of this theorem and the next theorem are left for the reader.

Theorem 6.2.17

The matrices $\mathbf{A}\mathbf{A}^-$, $\mathbf{A}^-\mathbf{A}$, $\mathbf{I} - \mathbf{A}\mathbf{A}^-$, *and* $\mathbf{I} - \mathbf{A}^-\mathbf{A}$ *are all symmetric idempotent.*

Note: We have stated that it is not *always* true that $(\mathbf{GH})^- = \mathbf{H}^-\mathbf{G}^-$ for all matrices $\mathbf{G}$ and $\mathbf{H}$. However, for certain matrices, this equation is correct (see Theorems 6.2.8, 6.2.9, and 6.2.10). We now state another theorem related to this problem.

Theorem 6.2.18

Let $\mathbf{B}$ *be an* $m \times r$ *matrix of rank* r $(r > 0)$ *and let* $\mathbf{C}$ *be an* $r \times m$ *matrix of rank* r; *then* $(\mathbf{BC})^- = \mathbf{C}^-\mathbf{B}^-$.

Proof: By Theorem 6.2.16, we obtain

$$\mathbf{C}^- = \mathbf{C}'(\mathbf{C}\mathbf{C}')^{-1}; \quad \mathbf{B}^- = (\mathbf{B}'\mathbf{B})^{-1}\mathbf{B}',$$

and hence $\mathbf{C}^-\mathbf{B}^- = \mathbf{C}'(\mathbf{CC}')^{-1}(\mathbf{B}'\mathbf{B})^{-1}\mathbf{B}'$, but this is $(\mathbf{BC})^-$ (see Eq. (6.2.3), where $\mathbf{A} = \mathbf{BC}$). ∎

6.3 Systems of Linear Equations

Chapter 7 will be devoted to the theory of solving systems of linear equations. However, in this section we state some theorems on linear equations that make use of the g-inverse.

Consider the system

$$\mathbf{Ax} = \mathbf{g}, \tag{6.3.1}$$

where $\mathbf{A}$ is an $m \times n$ matrix of known numbers and $\mathbf{g}$ is an $m \times 1$ vector of known numbers. The problem is to see if there are one or more vectors $\mathbf{x}$ that satisfy the system. If there is at least one vector $\mathbf{x}$ that satisfies the system, the system is said to be consistent. In the contrary case, the system is said to be inconsistent.

Theorem 6.3.1

The system of equations in Eq. (6.3.1) *is consistent if and only if* $\mathbf{AA}^-\mathbf{g} = \mathbf{g}$.

Proof: First assume that the system is consistent and let $\mathbf{x}_1$ be a vector satisfying the system; that is,

$$\mathbf{Ax}_1 = \mathbf{g}.$$

Multiply on the left by $\mathbf{AA}^-$ and get

$$\mathbf{AA}^-\mathbf{Ax}_1 = \mathbf{AA}^-\mathbf{g}.$$

But the left hand side is

$$\mathbf{AA}^-\mathbf{Ax}_1 = \mathbf{Ax}_1 = \mathbf{g},$$

so $\mathbf{AA}^-\mathbf{g} = \mathbf{g}$.

Next, assume that $\mathbf{AA}^-\mathbf{g} = \mathbf{g}$. Let $\mathbf{x} = \mathbf{A}^-\mathbf{g}$. If we substitute this value for $\mathbf{x}$ into the system $\mathbf{Ax} = \mathbf{g}$, we get $\mathbf{AA}^-\mathbf{g} = \mathbf{g}$, and hence $\mathbf{x} = \mathbf{A}^-\mathbf{g}$ is a solution, and the proof is complete. ∎

The next theorem, which gives the form of all solutions to the system $\mathbf{Ax} = \mathbf{g}$, will be restated and proved in Chapter 7 (see Corollary 7.3.1.1).

Theorem 6.3.2

If the linear system of equations $\mathbf{Ax} = \mathbf{g}$ *in Eq.* (6.3.1) *has a solution, then for each and every* $n \times 1$ *vector* $\mathbf{h}$, *the vector* $\mathbf{x}$ *is a solution where*

$$\mathbf{x} = \mathbf{A}^{-}\mathbf{g} + (\mathbf{I} - \mathbf{A}^{-}\mathbf{A})\mathbf{h}. \tag{6.3.2}$$

Also every solution to the system can be written in the form of Eq. (6.3.2) *for some* $n \times 1$ *vector* $\mathbf{h}$.

The last theorem in this section is a generalization of Theorem 6.3.2. It concerns the matrix equation

$$\mathbf{AXB} = \mathbf{C}, \tag{6.3.3}$$

where $\mathbf{A}$ is an $m_1 \times m_2$ known matrix, $\mathbf{B}$ is an $m_3 \times m_4$ known matrix and $\mathbf{C}$ is an $m_1 \times m_4$ known matrix. The problem is to determine whether there exists an $m_2 \times m_3$ matrix $\mathbf{X}$ that satisfies the system in Eq. (6.3.3). If the system has a solution, then the problem is to determine the form of all solutions. The following theorem answers these questions, and the proof is omitted, since it is almost identical with the proof of Theorem 6.3.2.

Theorem 6.3.3

If $\mathbf{A}$ *is an* $m_1 \times m_2$ *matrix,* $\mathbf{X}$ *is an* $m_2 \times m_3$ *matrix,* $\mathbf{B}$ *is an* $m_3 \times m_4$ *matrix, and* $\mathbf{C}$ *is an* $m_1 \times m_4$ *matrix, then a necessary and sufficient condition that a matrix* $\mathbf{X}$ *exists that satisfies the matrix equation* $\mathbf{AXB} = \mathbf{C}$ *is* $\mathbf{AA}^{-}\mathbf{CB}^{-}\mathbf{B} = \mathbf{C}$; *the general solution is*

$$\mathbf{X} = \mathbf{A}^{-}\mathbf{CB}^{-} + \mathbf{H} - \mathbf{A}^{-}\mathbf{AHBB}^{-},$$

where $\mathbf{H}$ *is any* $m_2 \times m_3$ *matrix.*

6.4 Generalized Inverses for Special Matrices

In this section, we shall state some additional theorems on g-inverses of special matrices. Some of the proofs are not given, since they are obtained in each case by simply showing that the definition of a g-inverse is satisfied.

Theorem 6.4.1

If c is a nonzero scalar, then $(c\mathbf{A})^- = (1/c)\mathbf{A}^-$.

In particular, $(-\mathbf{A})^- = -(\mathbf{A}^-)$.

Theorem 6.4.2

If $\mathbf{A} = \mathbf{A}_1 + \mathbf{A}_2 + \cdots + \mathbf{A}_t$ *and* $\mathbf{A}_i\mathbf{A}_j' = \mathbf{0}$ *and* $\mathbf{A}_i'\mathbf{A}_j = \mathbf{0}$ *for all* $i, j = 1, \ldots, t$, $i \neq j$, *then* $\mathbf{A}^- = \mathbf{A}_1^- + \cdots + \mathbf{A}_t^-$.

Theorem 6.4.3

Let $\mathbf{A}$ *be any* $n \times m$ *matrix, let* $\mathbf{K}$ *be any* $m \times m$ *nonsingular matrix, and let* $\mathbf{B} = \mathbf{AK}$. *Then* $\mathbf{BB}^- = \mathbf{AA}^-$.

Proof: From $\mathbf{B} = \mathbf{AK}$ we get the chain of equations $\mathbf{AA}^-\mathbf{B} = \mathbf{AA}^-\mathbf{AK} = \mathbf{AK} = \mathbf{B}$; $\mathbf{AA}^-\mathbf{BB}^- = \mathbf{BB}^-$; $\mathbf{BB}^- = (\mathbf{BB}^-)' = [(\mathbf{AA}^-)(\mathbf{BB}^-)]' = (\mathbf{BB}^-)'(\mathbf{AA}^-)' = \mathbf{BB}^-\mathbf{AA}^-$; and finally

$$\mathbf{BB}^- = \mathbf{BB}^-\mathbf{AA}^-.$$

From $\mathbf{A} = \mathbf{BK}^{-1}$ we get the following chain of equations: $\mathbf{BB}^-\mathbf{A} = \mathbf{BB}^-\mathbf{BK}^{-1} = \mathbf{BK}^{-1} = \mathbf{A}$; $\mathbf{BB}^-\mathbf{AA}^- = \mathbf{AA}^-$. If we combine this result with the result above, we obtain $\mathbf{AA}^- = \mathbf{BB}^-$ and the theorem is proved. ∎

Note: In this theorem it may not be true that $\mathbf{B}^-\mathbf{B} = \mathbf{A}^-\mathbf{A}$.

Theorem 6.4.4

If $\mathbf{A}'\mathbf{A} = \mathbf{AA}'$ *then* $\mathbf{A}^-\mathbf{A} = \mathbf{AA}^-$ *and* $(\mathbf{A}^n)^- = (\mathbf{A}^-)^n$ *for any positive integer n.*

Note: If $\mathbf{A}$ is a symmetric matrix, the hypothesis of Theorem 6.4.4 is satisfied.

Note: In general, $(\mathbf{A}^n)^- \neq (\mathbf{A}^-)^n$ for an $m \times m$ matrix $\mathbf{A}$.

Theorem 6.4.5

If $\mathbf{A} = \begin{bmatrix} \mathbf{B} \\ \mathbf{C} \end{bmatrix}$ *and* $\mathbf{BC}' = \mathbf{0}$, *then*

$$\mathbf{A}^- = [\mathbf{B}^-, \mathbf{C}^-], \qquad \mathbf{A}^-\mathbf{A} = \mathbf{B}^-\mathbf{B} + \mathbf{C}^-\mathbf{C},$$

and

$$\mathbf{AA}^- = \begin{bmatrix} \mathbf{BB}^- & \mathbf{0} \\ \mathbf{0} & \mathbf{CC}^- \end{bmatrix}.$$

Theorem 6.4.6

If $\mathbf{A}$ *is an* $m \times n$ *matrix and every row of* $\mathbf{A}$ *is the same, then every column of* $\mathbf{A}^-$ *is also the same.*

The proof of this theorem will be asked for in the problems.

Theorem 6.4.7

If $\mathbf{A} = \begin{bmatrix} \mathbf{B} & \mathbf{0} \\ \mathbf{0} & \mathbf{C} \end{bmatrix}$, *then*

$$\mathbf{A}^- = \begin{bmatrix} \mathbf{B}^- & \mathbf{0} \\ \mathbf{0} & \mathbf{C}^- \end{bmatrix}, \quad \mathbf{A}^-\mathbf{A} = \begin{bmatrix} \mathbf{B}^-\mathbf{B} & \mathbf{0} \\ \mathbf{0} & \mathbf{C}^-\mathbf{C} \end{bmatrix}, \quad \textit{and} \quad \mathbf{AA}^- = \begin{bmatrix} \mathbf{BB}^- & \mathbf{0} \\ \mathbf{0} & \mathbf{CC}^- \end{bmatrix}.$$

Theorem 6.4.8

If $\mathbf{a}$ *is a nonzero vector, then* $\mathbf{a}^- = (\mathbf{a}'\mathbf{a})^{-1}\mathbf{a}'$.

Theorem 6.4.9

If $\mathbf{A}$ *is an* $m \times n$ *matrix with each element equal to unity, then*

$$\mathbf{A}^- = \frac{1}{nm}\mathbf{A}'.$$

Theorem 6.4.10

$\mathbf{A}^- = \mathbf{A}'$ *if and only if* $\mathbf{A}'\mathbf{A}$ *is idempotent.*

Proof: Assume that $\mathbf{A}'\mathbf{A}$ is idempotent. This implies that

$$\mathbf{A}'\mathbf{A} = (\mathbf{A}'\mathbf{A})(\mathbf{A}'\mathbf{A}).$$

Multiply both sides of this equation on the left by $\mathbf{A}^-\mathbf{A}'^-$ and on the right by $\mathbf{A}^-$. The result is

$$\mathbf{A}^-\mathbf{A}'^-\mathbf{A}'\mathbf{A}\mathbf{A}^- = \mathbf{A}^-\mathbf{A}'^-\mathbf{A}'\mathbf{A}\mathbf{A}'\mathbf{A}\mathbf{A}^-.$$

Simplification gives the result

$$\mathbf{A}^- = \mathbf{A}'.$$

Next assume that $\mathbf{A}^- = \mathbf{A}'$ and multiply both sides of the equation on the right by $\mathbf{A}$ to obtain

$$\mathbf{A}^-\mathbf{A} = \mathbf{A}'\mathbf{A},$$

which states that $\mathbf{A}'\mathbf{A}$ is idempotent, since $\mathbf{A}^-\mathbf{A}$ is idempotent, and the theorem is proved. ▌

Corollary 6.4.10

Let $\mathbf{P} = [\mathbf{p}_1, \mathbf{p}_2, \ldots, \mathbf{p}_n]$ be any n distinct columns of an $m \times m$ orthogonal matrix; then $\mathbf{P}^- = \mathbf{P}'$.

The proof of the following theorem is left for the reader.

Theorem 6.4.11

If $\mathbf{A}$ is any $m \times n$ matrix, the following are true.

(1) *The column spaces of $\mathbf{A}$ and $\mathbf{A}\mathbf{A}^-$ are the same.*
(2) *The column spaces of $\mathbf{A}^-$ and $\mathbf{A}^-\mathbf{A}$ are the same.*
(3) *The column space of $\mathbf{I} - \mathbf{A}\mathbf{A}^-$ is the orthogonal complement of the column space of $\mathbf{A}$.*
(4) *The column space of $\mathbf{I} - \mathbf{A}^-\mathbf{A}$ is the orthogonal complement of the column space of $\mathbf{A}'$.*
(5) *The column space of $\mathbf{I} - \mathbf{A}^-\mathbf{A}$ is the same as the null space of $\mathbf{A}$.*
(6) *The column space of $\mathbf{I} - \mathbf{A}\mathbf{A}^-$ is the same as the null space of $\mathbf{A}^-$.*

Theorem 6.4.12

Let $\mathbf{X}$ be an $n \times p$ matrix and suppose that rows $i_1, i_2, \ldots, i_q$ are identical. Then these same rows in $\mathbf{X}'^-$ (columns in $\mathbf{X}^-$) are also identical. These same rows in $\mathbf{X}\mathbf{X}^c$ are also identical for any c-inverse of $\mathbf{X}$, $\mathbf{X}^c$. See section 6.6 for the definition of a c-inverse.

Proof: Let $\mathbf{A}$ be an $n \times m$ matrix whose $i_1, i_2, \ldots, i_q$ rows are identical and let $\mathbf{B}$ be an $m \times p$ matrix. Then clearly the $i_1, i_2, \ldots, i_q$ rows of $\mathbf{AB}$ are identical. Since $\mathbf{X}'^-$ can be written as $\mathbf{X}(\mathbf{X}'\mathbf{X})^- = \mathbf{X}[(\mathbf{X}'\mathbf{X})^-]$ and $\mathbf{XX}^c$ can be written as $\mathbf{X}[\mathbf{X}^c]$, the result follows. ∎

Corollary 6.4.12.1

If the q rows in the matrix $\mathbf{X}$ in Theorem 6.4.12 are proportional, then the corresponding rows in $\mathbf{X}'^-$ (columns in $\mathbf{X}^-$) and in $\mathbf{XX}^c$ are also proportional and have the same coefficients of proportionality.

Corollary 6.4.12.2

If a matrix $\mathbf{X}$ has any zero rows, then the corresponding rows of $\mathbf{X}'^-$ (columns of $\mathbf{X}^-$) and of $\mathbf{XX}^c$ are also zero.

Corollary 6.4.12.3

If a set of rows in a matrix $\mathbf{X}$ are linearly dependent, then the corresponding set of rows in $\mathbf{X}'^-$ (columns of $\mathbf{X}^-$) and in $\mathbf{XX}^c$ are also linearly dependent.

Results similar to Theorem 6.4.12 are true if certain columns of a matrix are identical, proportional, zero, or dependent.

6.5 Computing Formulas for the g-inverse

In this section we shall prove some theorems that can be used to compute the generalized inverse of a matrix. These methods are presented mainly for their theoretical interest.

Theorem 6.5.1

Let $\mathbf{A}$ be an $m \times t$ matrix, let $\mathbf{A}_{t-1}$ be an $m \times (t - 1)$ matrix that consists of the first $t - 1$ columns of $\mathbf{A}$, and let $\mathbf{a}_t$ be the t-th column of $\mathbf{A}$. Thus $\mathbf{A}$ can be written as

$$\mathbf{A} = [\mathbf{A}_{t-1}, \mathbf{a}_t]. \tag{6.5.1}$$

The g-inverse of $\mathbf{A}$ *is equal to* $\mathbf{B}$ *where*

$$\mathbf{B} = \begin{bmatrix} \mathbf{A}_{t-1}^{-} - \mathbf{A}_{t-1}^{-}\mathbf{a}_t\mathbf{b}_t^{-} \\ \mathbf{b}_t^{-} \end{bmatrix}, \tag{6.5.2}$$

where the $1 \times m$ *vector* $\mathbf{b}_t^{-}$ *is the g-inverse of* $\mathbf{b}_t$, *which is defined by*

$$\mathbf{b}_t = \begin{cases} (\mathbf{I} - \mathbf{A}_{t-1}\mathbf{A}_{t-1}^{-})\mathbf{a}_t & \text{if } \mathbf{a}_t \neq \mathbf{A}_{t-1}\mathbf{A}_{t-1}^{-}\mathbf{a}_t. \text{ (Case I)} \\ \dfrac{[1 + \mathbf{a}_t'(\mathbf{A}_{t-1}\mathbf{A}_{t-1}')^{-}\mathbf{a}_t](\mathbf{A}_{t-1}\mathbf{A}_{t-1}')^{-}\mathbf{a}_t}{\mathbf{a}_t'(\mathbf{A}_{t-1}\mathbf{A}_{t-1}')^{-}(\mathbf{A}_{t-1}\mathbf{A}_{t-1}')^{-}\mathbf{a}_t} & \text{if } \mathbf{a}_t = \mathbf{A}_{t-1}\mathbf{A}_{t-1}^{-}\mathbf{a}_t. \text{ (Case II)} \end{cases} \tag{6.5.3}$$

A proof of this theorem can be found in [G–12].

Theorem 6.5.1 can be used to form an iterative scheme to find the g-inverse of a matrix $\mathbf{A}$. Let $\mathbf{A}$ be an $m \times t$ matrix; let $\mathbf{C}_k$ be an $m \times k$ matrix that is composed of the first k columns of $\mathbf{A}$; let $\mathbf{A}_{k-1}$ be a matrix that is composed of the first $k - 1$ columns of $\mathbf{C}_k$, and let $\mathbf{a}_k$ be the k-th column of $\mathbf{C}_k$. Then

$$\mathbf{C}_2 = [\mathbf{A}_1, \mathbf{a}_2] = [\mathbf{a}_1, \mathbf{a}_2],$$

and we compute $\mathbf{C}_2^{-}$ by using Theorem 6.5.1. We can use Theorem 6.4.8 to compute $\mathbf{a}_1^{-}$. Next we define $\mathbf{C}_3$ by

$$\mathbf{C}_3 = [\mathbf{A}_2, \mathbf{a}_3] = [\mathbf{C}_2, \mathbf{a}_3],$$

and we can compute $\mathbf{C}_3^{-}$ by using Theorem 6.5.1, since $\mathbf{C}_2^{-}$ is now known. Next we define $\mathbf{C}_4$ by

$$\mathbf{C}_4 = [\mathbf{A}_3, \mathbf{a}_4] = [\mathbf{C}_3, \mathbf{a}_4],$$

and we can compute $\mathbf{C}_4^{-}$ by Theorem 6.5.1, since $\mathbf{C}_3^{-}$ is known. We continue until we define $\mathbf{C}_t$ by

$$\mathbf{C}_t = [\mathbf{A}_{t-1}, \mathbf{a}_t] = [\mathbf{C}_{t-1}, \mathbf{a}_t],$$

and we can use Theorem 6.5.1 to compute $\mathbf{C}_t^{-}$; but $\mathbf{C}_t = \mathbf{A}$ hence $\mathbf{C}_t^{-} = \mathbf{A}^{-}$. We illustrate this result with a simple example.

Example 6.5.1. We shall use the results of Theorem 6.5.1 to find the g-inverse of $\mathbf{A}$ where

$$\mathbf{A} = \begin{bmatrix} 1 & 0 & -2 \\ 0 & 1 & -1 \\ -1 & 1 & 1 \\ 2 & -1 & 2 \end{bmatrix}.$$

We compute $\mathbf{C}_2^-$ in the following steps:

(1) $\mathbf{C}_2 = [\mathbf{A}_1, \mathbf{a}_2] = \begin{bmatrix} 1 & 0 \\ 0 & 1 \\ -1 & 1 \\ 2 & -1 \end{bmatrix}.$

(2) $\mathbf{A}_1^- = \frac{1}{6}[1 \quad 0 \quad -1 \quad 2].$

(3) $\mathbf{A}_1^- \mathbf{a}_2 = -\frac{3}{6}.$

(4) $\mathbf{A}_1 \mathbf{A}_1^- \mathbf{a}_2 = \begin{bmatrix} -\frac{3}{6} \\ 0 \\ \frac{3}{6} \\ -\frac{6}{6} \end{bmatrix}.$

(5) $\mathbf{a}_2 \neq \mathbf{A}_1 \mathbf{A}_1^- \mathbf{a}_2$, so Case I applies and we get

$$\mathbf{b}_2^- = \frac{2}{3}\left[\frac{3}{6}, \ \frac{6}{6}, \ \frac{3}{6}, \ 0\right] = \left[\frac{1}{3}, \ \frac{2}{3}, \ \frac{1}{3}, \ 0\right].$$

(6) $\mathbf{C}_2^- = \mathbf{A}_2^- = \begin{bmatrix} \mathbf{A}_1^- - \mathbf{A}_1^- \mathbf{a}_2 \mathbf{b}_2^- \\ \mathbf{b}_2^- \end{bmatrix} = \begin{bmatrix} \frac{2}{6} & \frac{2}{6} & 0 & \frac{2}{6} \\ \frac{2}{6} & \frac{4}{6} & \frac{2}{6} & 0 \end{bmatrix}.$

Now we have

$$\mathbf{C}_3^- = \mathbf{A}^- = \begin{bmatrix} \mathbf{C}_2^- - \mathbf{C}_2^- \mathbf{a}_3 \mathbf{b}_3^- \\ \mathbf{b}_3^- \end{bmatrix}$$

where

$$\mathbf{C}_3 = [\mathbf{C}_2, \mathbf{a}_3] = \begin{bmatrix} 1 & 0 & -2 \\ 0 & 1 & -1 \\ -1 & 1 & 1 \\ 2 & -1 & 2 \end{bmatrix}.$$

(7) $\mathbf{C}_2^- \mathbf{a}_3 = \mathbf{A}_2^- \mathbf{a}_3 = \begin{bmatrix} -\frac{2}{6} \\ -\frac{6}{6} \end{bmatrix}.$

(8) $\mathbf{A}_2 \mathbf{A}_2^- \mathbf{a}_3 = \frac{1}{6}\begin{bmatrix} -2 \\ -6 \\ -4 \\ 2 \end{bmatrix}.$

(9) $\mathbf{a}_3 \neq \mathbf{A}_2 \mathbf{A}_2^- \mathbf{a}_3$, so Case I applies and we get

$$\mathbf{b}_3^- = \left[-\frac{1}{5},\ 0,\ \frac{1}{5},\ \frac{1}{5}\right].$$

(10) $\mathbf{C}_3^- = \mathbf{A}^- = \frac{1}{15}\begin{bmatrix} 4 & 5 & 1 & 6 \\ 2 & 10 & 8 & 3 \\ -3 & 0 & 3 & 3 \end{bmatrix}.$

You can verify that the matrix in (10) is actually $\mathbf{A}^-$ by showing that it satisfies Def. 6.2.1.

If $\mathbf{A}$ is an $m \times n$ matrix $(m \geq n)$ of rank n, then $\mathbf{A}'\mathbf{A}$ is an $n \times n$ nonsingular matrix, and in this case, Theorem 6.2.16 can be used to evaluate $\mathbf{A}^-$ by first evaluating the inverse of the nonsingular matrix $\mathbf{A}'\mathbf{A}$. However if the rank of $\mathbf{A}$ is $k < n$, we can use the fact that

$$(\mathbf{A}'\mathbf{A})^-\mathbf{A}' = \mathbf{A}^-(\mathbf{A}')^-\mathbf{A}' = \mathbf{A}^-\mathbf{A}\mathbf{A}^- = \mathbf{A}^-,$$

and we can evaluate $\mathbf{A}^-$ by first evaluating $(\mathbf{A}'\mathbf{A})^-$. We illustrate with an example.

Example 6.5.2. Find the g-inverse of $\mathbf{A}$ where

$$\mathbf{A} = \begin{bmatrix} 1 & 1 \\ 3 & 0 \\ -2 & 1 \\ 0 & 2 \\ -1 & 2 \end{bmatrix}.$$

We obtain

$$\mathbf{A}'\mathbf{A} = \begin{bmatrix} 15 & -3 \\ -3 & 10 \end{bmatrix};$$

since $\mathbf{A}'\mathbf{A}$ has rank 2, it is nonsingular and

$$(\mathbf{A}'\mathbf{A})^{-1} = \frac{1}{141}\begin{bmatrix} 10 & 3 \\ 3 & 15 \end{bmatrix} \quad \text{and} \quad \mathbf{A}^- = \frac{1}{141}\begin{bmatrix} 13 & 30 & -17 & 6 & -4 \\ 18 & 9 & 9 & 30 & 27 \end{bmatrix}.$$

Next, we illustrate the procedure when $\mathbf{A}'\mathbf{A}$ is singular.

Example 6.5.3. Find the g-inverse of the matrix $\mathbf{A}$ where

$$\mathbf{A} = \begin{bmatrix} -1 & -2 \\ 0 & 0 \\ 2 & 4 \\ 1 & 2 \\ 3 & 6 \end{bmatrix}.$$

Now

$$\mathbf{A}'\mathbf{A} = \begin{bmatrix} 15 & 30 \\ 30 & 60 \end{bmatrix} = 15\begin{bmatrix} 1 & 2 \\ 2 & 4 \end{bmatrix},$$

and clearly $\mathbf{A}'\mathbf{A}$ is singular. We can use Theorem 6.5.1 to compute $(\mathbf{A}'\mathbf{A})^-$. We get

$$(\mathbf{A}'\mathbf{A})^- = \frac{1}{375}\begin{bmatrix} 1 & 2 \\ 2 & 4 \end{bmatrix}.$$

So

$$\mathbf{A}^- = (\mathbf{A}'\mathbf{A})^-\mathbf{A}' = \frac{1}{75}\begin{bmatrix} -1 & 0 & 2 & 1 & 3 \\ -2 & 0 & 4 & 2 & 6 \end{bmatrix}.$$

Notice that in this example $\mathbf{A}^- = c\mathbf{A}'$ where $c = 1/75$.

The next theorem and its corollary may be useful for computing the g-inverse of small or medium sized well-conditioned matrices. In addition they have theoretical interest.

Theorem 6.5.2

Let $\mathbf{A}$ *be an* $m \times n$ *matrix with* $n \leq m$ *and let* $\mathbf{B}$ *be any* $n \times n$ *matrix that satisfies*

$$(\mathbf{A}'\mathbf{A})^2\mathbf{B} = \mathbf{A}'\mathbf{A}. \tag{6.5.4}$$

Then the g-inverse of $\mathbf{A}$ *is given by*

$$\mathbf{A}^- = \mathbf{B}'\mathbf{A}'.$$

Proof: Clearly $\mathbf{B} = (\mathbf{A}'\mathbf{A})^-$ is a solution to the equation $\mathbf{A}'\mathbf{A} = (\mathbf{A}'\mathbf{A})^2\mathbf{B}$, and by Theorem 6.3.3, the general solution is

$$\mathbf{B} = [(\mathbf{A}'\mathbf{A})^2]^-(\mathbf{A}'\mathbf{A}) + \mathbf{H} - [\{(\mathbf{A}'\mathbf{A})^2\}^-(\mathbf{A}'\mathbf{A})^2]\mathbf{H},$$

where $\mathbf{H}$ is any $n \times n$ matrix. If we multiply $\mathbf{B}'$ on the right by $\mathbf{A}'$, we get

$$\mathbf{B}'\mathbf{A}' = \mathbf{A}^-,$$

and the theorem is proved. ▮

Corollary 6.5.2

Let $\mathbf{A}$ *be any* $n \times n$ *symmetric matrix; then the g-inverse of* $\mathbf{A}$ *is given by*

$$\mathbf{B}'\mathbf{A}\mathbf{B},$$

where $\mathbf{B}$ *is any solution to the system* $\mathbf{A}^2\mathbf{B} = \mathbf{A}$.

Some additional theorems that are of theoretical interest in evaluating the g-inverse of a matrix follow. In some cases, the proofs are not given but are asked for in the problems.

Theorem 6.5.3

If $\mathbf{B}$ *is an* $m \times m$ *symmetric matrix, the g-inverse of* $\mathbf{B}$ *is given by*

$$\mathbf{B}^- = (\mathbf{B}\mathbf{K})^2\mathbf{B}, \tag{6.5.5}$$

where $\mathbf{K}$ *is any solution to the equation*

$$\mathbf{B}^2\mathbf{K}\mathbf{B}^2 = \mathbf{B}^2. \tag{6.5.6}$$

Proof: Clearly $\mathbf{K} = (\mathbf{B}^2)^-$ is one solution to Eq. (6.5.6), and by Theorem 6.3.3, the general solution is given by

$$\mathbf{K} = (\mathbf{B}^2)^-(\mathbf{B}^2)(\mathbf{B}^2)^- + \mathbf{H} - (\mathbf{B}^2)^-(\mathbf{B}^2)\mathbf{H}(\mathbf{B}^2)(\mathbf{B}^2)^-,$$

where $\mathbf{H}$ is *any* $m \times m$ matrix.

By virtue of the fact that $\mathbf{B}$ is symmetric, and hence by Theorem 6.4.4, $(\mathbf{B}^2)^- = (\mathbf{BB})^- = \mathbf{B}^-\mathbf{B}^-$, the solution for $\mathbf{K}$ reduces to (note also that, by Theorem 6.4.4, $\mathbf{BB}^- = \mathbf{B}^-\mathbf{B}$)

$$\mathbf{K} = \mathbf{B}^-\mathbf{B}^- + \mathbf{H} - \mathbf{B}^-\mathbf{BHBB}^-.$$

Notice that for any solution $\mathbf{K}$ to the equation $\mathbf{B}^2\mathbf{K}\mathbf{B}^2 = \mathbf{B}^2$, the quantity $\mathbf{BKB} = \mathbf{BB}^-\mathbf{B}^-\mathbf{B} = \mathbf{BB}^- = \mathbf{B}^-\mathbf{B}$ and hence is invariant. Thus the quantity $(\mathbf{BK})^2\mathbf{B}$ can be written as

$$(\mathbf{BK})^2\mathbf{B} = (\mathbf{BKB})\mathbf{KB} = \mathbf{B}^-(\mathbf{BKB}) = \mathbf{B}^-\mathbf{BB}^- = \mathbf{B}^-,$$

and the theorem is proved. ∎

The matrix $\mathbf{K}$ in this theorem can be found by using the procedure outlined in the next theorem.

Theorem 6.5.4

Let $\mathbf{B}$ *be an* $m \times m$ *symmetric matrix of rank* r, *and let* $\mathbf{P}$ *be a nonsingular* $m \times m$ *matrix such that*

$$\mathbf{P}'\mathbf{B}^2\mathbf{P} = \begin{bmatrix} \mathbf{I}_r & \mathbf{0} \\ \mathbf{0} & \mathbf{0} \end{bmatrix} = \mathbf{R}_r$$

(*see Theorems* 1.7.5 *and* 1.7.7); *then a matrix* $\mathbf{K}$ *that satisfies the equation*

$$\mathbf{B}^2\mathbf{K}\mathbf{B}^2 = \mathbf{B}^2$$

in Theorem 6.5.3 *is given by*

$$\mathbf{K} = \mathbf{P}\mathbf{R}_r\mathbf{P}'.$$

Proof: If we solve for $\mathbf{B}^2$ in the equation $\mathbf{P}'\mathbf{B}^2\mathbf{P} = \mathbf{R}_r$, we get

$$\mathbf{B}^2 = \mathbf{P}'^{-1}\mathbf{R}_r\mathbf{P}^{-1}.$$

Hence (note $\mathbf{R}_r\mathbf{R}_r = \mathbf{R}_r$),

$$\mathbf{B}^2\mathbf{K}\mathbf{B}^2 = (\mathbf{P}'^{-1}\mathbf{R}_r\mathbf{P}^{-1})(\mathbf{P}\mathbf{R}_r\mathbf{P}')(\mathbf{P}'^{-1}\mathbf{R}_r\mathbf{P}^{-1}) = \mathbf{P}^{-1}\mathbf{R}_r\mathbf{P}^{-1} = \mathbf{B}^2,$$

and the theorem is proved. ▮

Note that if $\mathbf{B}$ is nonsingular, then $\mathbf{K} = (\mathbf{B}^{-1})^2$ and $\mathbf{B}^- = \mathbf{B}^{-1} = \mathbf{K}\mathbf{B}$.

Example 6.5.4. We use the results of Theorem 6.5.3 to find the g-inverse of $\mathbf{B}$ where $\mathbf{B}$ is defined by

$$\mathbf{B} = \begin{bmatrix} 1 & 0 & -1 \\ 0 & 1 & 0 \\ -1 & 0 & 1 \end{bmatrix}.$$

Now $\mathbf{B}$ is a symmetric matrix; hence it satisfies the condition of Theorem 6.5.3. We get

$$\mathbf{B}^2 = \begin{bmatrix} 2 & 0 & -2 \\ 0 & 1 & 0 \\ -2 & 0 & 2 \end{bmatrix},$$

and the matrix $\mathbf{P}$ such that $\mathbf{P}'\mathbf{B}^2\mathbf{P} = \mathbf{R}_2$ is obtained by performing elementary row transformations on $\mathbf{B}^2$. We get

$$\mathbf{P}' = \begin{bmatrix} (1/2)\sqrt{2} & 0 & 0 \\ 0 & 1 & 0 \\ 1 & 0 & 1 \end{bmatrix},$$

and

$$\mathbf{P}'\mathbf{B}^2\mathbf{P} = \begin{bmatrix} 1 & 0 & 0 \\ 0 & 1 & 0 \\ 0 & 0 & 0 \end{bmatrix} = \mathbf{R}_2.$$

Also

$$\mathbf{K} = \mathbf{P}\mathbf{R}_2\mathbf{P}' = \begin{bmatrix} 1/2 & 0 & 0 \\ 0 & 1 & 0 \\ 0 & 0 & 0 \end{bmatrix},$$

and

$$\mathbf{B}^- = (\mathbf{BK})^2\mathbf{B} = \begin{bmatrix} 1/4 & 0 & -1/4 \\ 0 & 1 & 0 \\ -1/4 & 0 & 1/4 \end{bmatrix}.$$

The results of Theorem 6.5.3 can also be used to find the g-inverse of a non-symmetric matrix. This is the context of the next theorem.

Theorem 6.5.5

Let $\mathbf{A}$ *be an* $m \times n$ *matrix. The g-inverse of* $\mathbf{A}$ *is given by*

$$\mathbf{A}^- = (\mathbf{A}'\mathbf{A}\mathbf{K})^2\mathbf{A}'\mathbf{A}\mathbf{A}'$$

where $\mathbf{K}$ *is any solution to the equation*

$$(\mathbf{A}'\mathbf{A})^2\mathbf{K}(\mathbf{A}'\mathbf{A})^2 = (\mathbf{A}'\mathbf{A})^2.$$

Proof: If we set $\mathbf{B} = \mathbf{A}'\mathbf{A}$, then $\mathbf{B}$ satisfies the condition of Theorem 6.5.3. We get $\mathbf{B}^- = (\mathbf{BK})^2\mathbf{B}$ where $\mathbf{B}$ is any solution to $\mathbf{B}^2\mathbf{K}\mathbf{B}^2 = \mathbf{B}^2$; or, in other words, by substituting $\mathbf{A}'\mathbf{A}$ for $\mathbf{B}$ we obtain

$$(\mathbf{A}'\mathbf{A})^- = (\mathbf{A}'\mathbf{A}\mathbf{K})^2\mathbf{A}'\mathbf{A}$$

where $\mathbf{K}$ is any solution to $(\mathbf{A}'\mathbf{A})^2\mathbf{K}(\mathbf{A}'\mathbf{A})^2 = (\mathbf{A}'\mathbf{A})^2$. After finding $(\mathbf{A}'\mathbf{A})^-$, we multiply on the right by $\mathbf{A}'$, since $(\mathbf{A}'\mathbf{A})^-\mathbf{A}' = \mathbf{A}^-$. This proves the theorem. ∎

The next theorem can sometimes be used in applications of partitioned matrices.

Theorem 6.5.6

Let $\mathbf{A}$ *be an* $n \times m$ *matrix of rank* r *where* $r > 0$ *and let* $\mathbf{A}$ *be partitioned as*

$$\mathbf{A} = \begin{bmatrix} \mathbf{A}_{11} & \mathbf{A}_{12} \\ \mathbf{A}_{21} & \mathbf{A}_{22} \end{bmatrix},$$

where $\mathbf{A}_{11}$ *is an* $r \times r$ *matrix of rank* r *and the sizes of the other submatrices are determined. (We assume* $r < m$ *and* $r < n$.) *The matrix* $\mathbf{A}$ *can be written as*

$$\begin{bmatrix} \mathbf{A}_{11} & \mathbf{A}_{12} \\ \mathbf{A}_{21} & \mathbf{A}_{21}\mathbf{A}_{11}^{-1}\mathbf{A}_{12} \end{bmatrix}.$$

Proof: Multiply $\mathbf{A}$ on the left by the $n \times n$ matrix $\mathbf{B}$ where

$$\mathbf{B} = \begin{bmatrix} \mathbf{I}_1 & \mathbf{0} \\ \mathbf{A}_{21}\mathbf{A}_{11}^{-1} & \mathbf{I}_2 \end{bmatrix},$$

where $\mathbf{I}_1$ is the $r \times r$ identity matrix and $\mathbf{I}_2$ is the $(n-r) \times (n-r)$ identity matrix. Clearly $\mathbf{B}$ is nonsingular and hence has rank n; hence $\mathbf{BA}$ has the same rank as $\mathbf{A}$, which is r. But

$$\mathbf{BA} = \begin{bmatrix} \mathbf{I}_1 & \mathbf{0} \\ -\mathbf{A}_{21}\mathbf{A}_{11}^{-1} & \mathbf{I}_2 \end{bmatrix} \begin{bmatrix} \mathbf{A}_{11} & \mathbf{A}_{12} \\ \mathbf{A}_{21} & \mathbf{A}_{22} \end{bmatrix} = \begin{bmatrix} \mathbf{A}_{11} & \mathbf{A}_{12} \\ \mathbf{0} & \mathbf{A}_{22} - \mathbf{A}_{21}\mathbf{A}_{11}^{-1}\mathbf{A}_{12} \end{bmatrix}.$$

But since $\mathbf{BA}$ is triangular in blocks and the rank of $\mathbf{BA}$ is equal to the rank of $\mathbf{A}_{11}$, the rank of $(\mathbf{A}_{22} - \mathbf{A}_{21}\mathbf{A}_{11}^{-1}\mathbf{A}_{12})$ is zero. Thus

$$\mathbf{A}_{22} - \mathbf{A}_{21}\mathbf{A}_{11}^{-1}\mathbf{A}_{12} = \mathbf{0},$$

or, in other words,

$$\mathbf{A}_{22} = \mathbf{A}_{21}\mathbf{A}_{11}^{-1}\mathbf{A}_{12},$$

and the theorem is proved. ∎

Theorem 6.5.7

Let $\mathbf{A}$ be an $n \times m$ matrix of rank r where $0 < r$, $r < m$, and $r < n$. Let $\mathbf{A}$ be partitioned as

$$\mathbf{A} = \begin{bmatrix} \mathbf{A}_{11} & \mathbf{A}_{12} \\ \mathbf{A}_{21} & \mathbf{A}_{22} \end{bmatrix},$$

where $\mathbf{A}_{11}$ is an $r \times r$ matrix of rank r. The sizes of the remaining submatrices are determined. The g-inverse of $\mathbf{A}$ is given by

$$\mathbf{A}^{-} = \begin{bmatrix} \mathbf{A}'_{11}\mathbf{B}\mathbf{A}'_{11} & \mathbf{A}'_{11}\mathbf{B}\mathbf{A}'_{21} \\ \mathbf{A}'_{12}\mathbf{B}\mathbf{A}'_{11} & \mathbf{A}'_{12}\mathbf{B}\mathbf{A}'_{21} \end{bmatrix},$$

where

$$\mathbf{B} = (\mathbf{A}_{11}\mathbf{A}'_{11} + \mathbf{A}_{12}\mathbf{A}'_{12})^{-1}\mathbf{A}_{11}(\mathbf{A}'_{11}\mathbf{A}_{11} + \mathbf{A}'_{21}\mathbf{A}_{21})^{-1}.$$

Proof: By Theorem 1.7.7, $\mathbf{A}_{11}\mathbf{A}'_{11}$ is positive definite of rank r, since $\mathbf{A}_{11}$ is an $r \times r$ matrix of rank r; hence, $\mathbf{A}_{11}\mathbf{A}'_{11} + \mathbf{A}_{12}\mathbf{A}'_{12}$ is also positive definite and of rank r. Similarly $\mathbf{A}'_{11}\mathbf{A}_{11} + \mathbf{A}'_{21}\mathbf{A}_{21}$ is an $r \times r$ matrix of rank r. Thus the indicated inverses exist and $\mathbf{B}$ has rank r. By Theorem 6.5.6, we can write

$$\mathbf{A} = \begin{bmatrix} \mathbf{A}_{11} & \mathbf{A}_{12} \\ \mathbf{A}_{21} & \mathbf{A}_{21}\mathbf{A}_{11}^{-1}\mathbf{A}_{12} \end{bmatrix},$$

and by multiplication it can be shown that $\mathbf{A}^-$ satisfies the four conditions of Def. 6.2.1. ∎

Theorem 6.5.8

Let $\mathbf{A}$ *be an* $m \times n$ *matrix of rank* r*; then the g-inverse of* $\mathbf{A}$ *can be computed by the following steps.*

(1) *Compute* $\mathbf{B} = \mathbf{A}'\mathbf{A}$.
(2) *Let* $\mathbf{C}_1 = \mathbf{I}$.
(3) *Compute* $\mathbf{C}_{i+1} = \mathbf{I}(1/i)\operatorname{tr}(\mathbf{C}_i\mathbf{B}) - \mathbf{C}_i\mathbf{B}$, for $i = 1, 2, \ldots, r-1$.
(4) *Compute* $r\mathbf{C}_r\mathbf{A}'/\operatorname{tr}(\mathbf{C}_r\mathbf{B})$, and this is $\mathbf{A}^-$.

Also $\mathbf{C}_{r+1}\mathbf{B} = \mathbf{0}$ *and* $\operatorname{tr}(\mathbf{C}_r\mathbf{B}) \neq 0$.

The proof of this theorem is omitted here, but it can be found in [P–2]. Note that, due to the fact that $\mathbf{C}_{r+1}\ \mathbf{B} = \mathbf{0}$, one need not know the rank of $\mathbf{A}$ (which is r) in advance.

Example 6.5.5. We shall use the method of Theorem 6.5.8 to find the g-inverse of the matrix $\mathbf{A}$ in Example 6.5.1, where

$$\mathbf{A} = \begin{bmatrix} 1 & 0 & -2 \\ 0 & 1 & -1 \\ -1 & 1 & 1 \\ 2 & -1 & 2 \end{bmatrix}.$$

Now

(1) $$\mathbf{B} = \mathbf{A}'\mathbf{A} = \begin{bmatrix} 6 & -3 & 1 \\ -3 & 3 & -2 \\ 1 & -2 & 10 \end{bmatrix},$$

(2) $$\mathbf{C}_1 = \begin{bmatrix} 1 & 0 & 0 \\ 0 & 1 & 0 \\ 0 & 0 & 1 \end{bmatrix},$$

$$(3)\quad \mathbf{C}_2 = \mathbf{I}\,\mathrm{tr}\,(\mathbf{C}_1\mathbf{B}) - \mathbf{C}_1\mathbf{B} = \begin{bmatrix} 19 & 0 & 0 \\ 0 & 19 & 0 \\ 0 & 0 & 19 \end{bmatrix} - \begin{bmatrix} 6 & -3 & 1 \\ -3 & 3 & -2 \\ 1 & -2 & 10 \end{bmatrix}$$

$$= \begin{bmatrix} 13 & 3 & -1 \\ 3 & 16 & 2 \\ -1 & 2 & 9 \end{bmatrix},$$

$$\mathbf{C}_3 = \mathbf{I}\tfrac{1}{2}\,\mathrm{tr}\,(\mathbf{C}_2\mathbf{B}) - \mathbf{C}_2\mathbf{B} = \begin{bmatrix} 94 & 0 & 0 \\ 0 & 94 & 0 \\ 0 & 0 & 94 \end{bmatrix} - \begin{bmatrix} 68 & -28 & -3 \\ -28 & 35 & -9 \\ -3 & -9 & 85 \end{bmatrix}$$

$$= \begin{bmatrix} 26 & 28 & 3 \\ 28 & 59 & 9 \\ 3 & 9 & 9 \end{bmatrix}.$$

Clearly r cannot be greater than 3 and $\mathbf{C}_3\mathbf{B} \neq \mathbf{0}$, so $r = 3$. We get

$$(4)\quad \mathbf{A}^- = \frac{3\mathbf{C}_3\mathbf{A}'}{\mathrm{tr}\,(\mathbf{C}_3\mathbf{B})} = \frac{3}{225}\begin{bmatrix} 20 & 25 & 5 & 30 \\ 10 & 50 & 40 & 15 \\ -15 & 0 & 15 & 15 \end{bmatrix},$$

which of course is the same result as that obtained in Example 6.5.1.

6.6 Conditional Inverse

In the previous sections of this chapter we have shown that the generalized inverse of any matrix possesses many of the properties that the inverse of nonsingular matrices possesses. These properties can be very useful in many areas of statistics, especially in solving systems of linear equations. Since the theory of systems of linear equations plays such an extremely important role in statistics as well as in most other scientific fields, we shall discuss another type of inverse, which we call a *conditional* inverse, that is often valuable in these situations.

A conditional inverse of a matrix is generally easier to compute than the generalized inverse of **A**, and, if the end result of some theoretical work is to do computation that involves solutions of systems of linear equations, then it may be desirable to use a conditional inverse rather than the generalized inverse.

Definition 6.6.1

Conditional Inverse. *Let* $\mathbf{A}$ *be an* $m \times n$ *matrix. A matrix* $\mathbf{A}^c$ *is defined to be a conditional inverse of* $\mathbf{A}$ *if and only if it satisfies*

$$\mathbf{A}\mathbf{A}^c\mathbf{A} = \mathbf{A}. \tag{6.6.1}$$

We now state some theorems that are obvious consequences of the definition.

Theorem 6.6.1

The generalized inverse of a matrix $\mathbf{A}$ *is also a conditional inverse of* $\mathbf{A}$, *but a conditional inverse of* $\mathbf{A}$ *may not be the generalized inverse.*

Theorem 6.6.2

A conditional inverse exists for each matrix, but it may not be unique.

Theorem 6.6.3

If $\mathbf{A}$ *is an* $m \times n$ *matrix, a conditional inverse is an* $n \times m$ *matrix.*

Note: A conditional inverse must satisfy only (3) of Def. 6.2.1 of a generalized inverse. Hence any theorem concerning generalized inverses that involves only (3) of Def. 6.2.1 in the proof is also true for conditional inverses. We sometimes abbreviate conditional inverse as "c-inverse."

The Hermite (canonical) form of a matrix is useful in many areas of matrix theory, and it is particularly useful in a discussion of c-inverses, so we shall define it and state some important theorems.

Definition 6.6.2

Hermite Form. *An* $n \times n$ *matrix* $\mathbf{H}$ *is defined to be in (upper) Hermite form if and only if it satisfies the following four conditions.*

(1) $\mathbf{H}$ *is upper triangular.*
(2) *Only zeros and ones are on the diagonal.*
(3) *If a row has a zero on the diagonal, then every element in that row is zero.*
(4) *If a row has a one on the diagonal, then every off-diagonal element is zero in the column in which the one appears.*

We shall state two theorems here that will aid us in the discussion of c-inverses. Some additional theorems on Hermite forms are stated in Sec. 6.7.

Theorem 6.6.4

If $\mathbf{H}$ *is in Hermite form, then* $\mathbf{H} = \mathbf{H}^2$.

Proof: The proof is obtained by multiplication. ∎

Theorem 6.6.5

For any $n \times n$ *matrix* $\mathbf{A}$ *there exists a nonsingular matrix* $\mathbf{B}$ *such that* $\mathbf{BA} = \mathbf{H}$ *where* $\mathbf{H}$ *is in Hermite form.*

This theorem states that any $n \times n$ matrix $\mathbf{A}$ can be reduced to a Hermite-form matrix by elementary row operations.

Proof: The proof is obtained by row operations on $\mathbf{A}$. ∎

Example 6.6.1. Reduce the matrix $\mathbf{A}$ to Hermite form by elementary row operations where

$$\mathbf{A} = \begin{bmatrix} 1 & 2 & 1 \\ 2 & 3 & 1 \\ 1 & 1 & 0 \end{bmatrix}.$$

Perform the following operations on $\mathbf{A}$:

(1) Multiply row 1 by -2 and add the result to the second row.
(2) Multiply row 1 by -1 and add the result to the third row.

The result of these two operations is

$$\mathbf{A}_1 = \begin{bmatrix} 1 & 2 & 1 \\ 0 & -1 & -1 \\ 0 & -1 & -1 \end{bmatrix}.$$

Perform the following row operations on $\mathbf{A}_1$:

(3) Multiply row 2 by $+2$ and add the result to the first row.
(4) Multiply row 2 by -1 and add the result to the third row.

The result of these two operations on $\mathbf{A}_1$ is

$$\mathbf{A}_2 = \begin{bmatrix} 1 & 0 & -1 \\ 0 & -1 & -1 \\ 0 & 0 & 0 \end{bmatrix}.$$

(5) Multiply the second row of $\mathbf{A}_2$ by -1, and the result is

$$\mathbf{A}_3 = \begin{bmatrix} 1 & 0 & -1 \\ 0 & 1 & 1 \\ 0 & 0 & 0 \end{bmatrix} = \mathbf{H},$$

and by Def. 6.6.2, $\mathbf{A}_3$ is in Hermite form. The matrix $\mathbf{B}$, such that $\mathbf{BA} = \mathbf{H}$, is

$$\mathbf{B} = \begin{bmatrix} -3 & 2 & 0 \\ 2 & -1 & 0 \\ 1 & -1 & 1 \end{bmatrix},$$

and it is easily verified that $\mathbf{H}^2 = \mathbf{H}$.

Theorem 6.6.6

Let $\mathbf{A}$ *be an* $n \times n$ *matrix. Let* $\mathbf{B}$ *be a nonsingular matrix such that* $\mathbf{BA} = \mathbf{H}$ *where* $\mathbf{H}$ *is in Hermite form. Then* $\mathbf{B}$ *is a conditional inverse of* $\mathbf{A}$.

Proof: Since $\mathbf{BA} = \mathbf{H}$, $\mathbf{H}^2 = \mathbf{H}$ and $\mathbf{B}$ is nonsingular, we get

$$(\mathbf{BA})(\mathbf{BA}) = \mathbf{H}^2 = \mathbf{H} = \mathbf{BA},$$

or

$$\mathbf{BABA} = \mathbf{BA}.$$

Multiply by $\mathbf{B}^{-1}$ and the result is

$$\mathbf{ABA} = \mathbf{A},$$

so

$$\mathbf{B} = \mathbf{A}^c. \blacksquare$$

Corollary 6.6.6

Let $\mathbf{A}$ *be an* $m \times n$ *matrix with* $m > n$ *and let* $\mathbf{A}_0 = [\mathbf{A}, \mathbf{0}]$ *where* $\mathbf{0}$ *is the* $m \times m - n$ *zero matrix. Let* $\mathbf{B}_0$ *be a nonsingular matrix such that* $\mathbf{B}_0 \mathbf{A}_0 = \mathbf{H}$ *where* $\mathbf{H}$ *is in Hermite form. Let* $\mathbf{B}_0$ *be partitioned as*

$$\mathbf{B}_0 = \begin{bmatrix} \mathbf{B} \\ \mathbf{B}_1 \end{bmatrix},$$

where $\mathbf{B}$ *is an* $n \times m$ *matrix. Then* $\mathbf{B}$ *is a c-inverse of* $\mathbf{A}$.

A similar corollary can be obtained for the situation that $m < n$.

Example 6.6.2. Find $\mathbf{A}^c$, a c-inverse of the matrix $\mathbf{A}$ in Ex. 6.6.1. In this example we computed a matrix $\mathbf{B}$ such that $\mathbf{BA} = \mathbf{H}$ where $\mathbf{H}$ is in Hermite form. The reader can verify that $\mathbf{ABA} = \mathbf{A}$ and hence $\mathbf{A}^c = \mathbf{B}$. Note that $\mathbf{CA}$ is also a Hermite matrix where

$$\mathbf{C} = \begin{bmatrix} 0 & -1 & 3 \\ 0 & 1 & -2 \\ 1 & -1 & 1 \end{bmatrix},$$

and hence $\mathbf{C}$ is another c-inverse of the matrix $\mathbf{A}$.

Example 6.6.3. Find a c-inverse of the 3×2 matrix $\mathbf{A}$ where

$$\mathbf{A} = \begin{bmatrix} 1 & -1 \\ 2 & -1 \\ 0 & 1 \end{bmatrix}.$$

We shall join a column of zeros to $\mathbf{A}$ and obtain

$$\mathbf{A}_0 = \begin{bmatrix} 1 & -1 & 0 \\ 2 & -1 & 0 \\ 0 & 1 & 0 \end{bmatrix}.$$

We shall now reduce $\mathbf{A}_0$ to a Hermite form by a nonsingular matrix $\mathbf{B}_0$. Clearly $\mathbf{B}_0$ defined below will work

$$\mathbf{B}_0 = \begin{bmatrix} -1 & 1 & 0 \\ -2 & 1 & 0 \\ 2 & -1 & 1 \end{bmatrix}.$$

Hence

$$\mathbf{B} = \begin{bmatrix} -1 & 1 & 0 \\ -2 & 1 & 0 \end{bmatrix}$$

is a c-inverse of $\mathbf{A}$.

Theorem 6.6.7

For any c-inverse $\mathbf{A}^c$ of an $m \times n$ matrix $\mathbf{A}$, the matrices $\mathbf{A}^c\mathbf{A}$ and $\mathbf{AA}^c$ are each idempotent.

Proof: Since $\mathbf{AA}^c\mathbf{A} = \mathbf{A}$ the result $(\mathbf{A}^c\mathbf{A})^2 = \mathbf{A}^c\mathbf{A}$ is obtained by premultiplication by $\mathbf{A}^c$, and the result $(\mathbf{AA}^c)^2 = \mathbf{AA}^c$ is obtained by postmultiplication by $\mathbf{A}^c$. ▮

Theorem 6.6.8

For any c-inverse $\mathbf{A}^c$ *of an* $m \times n$ *matrix* $\mathbf{A}$,

$$\textit{rank}\ (\mathbf{A}) = \textit{rank}\ (\mathbf{A}^c\mathbf{A}) = \textit{rank}\ (\mathbf{AA}^c) \leqq \textit{rank}\ (\mathbf{A}^c).$$

Proof: Rank $(\mathbf{A})$ = rank $(\mathbf{AA}^c\mathbf{A}) \le$ rank $(\mathbf{A}^c\mathbf{A}) \le$ rank $(\mathbf{A})$, and hence rank $(\mathbf{A}^c\mathbf{A})$ = rank $(\mathbf{A})$. The proofs of the remaining results are similar. ▮

In Sec. 6.3 we briefly discussed systems of linear equations and showed how the g-inverse of the coefficient matrix could be used to determine whether or not the system is consistent and to find *all* solutions if they are consistent. In Chapter 7 we state similar theorems, using c-inverses instead of g-inverses.

Theorem 6.6.9, which follows, has very important applications in statistics. In the model $\mathbf{y} = \mathbf{X}\boldsymbol{\beta} + \mathbf{e}$ referred to in the Introduction, it is often the case that the $n \times p$ matrix $\mathbf{X}$ has rank k where $k < p$ and $k < n$. In this case the normal equations $\mathbf{X}'\mathbf{X}\hat{\boldsymbol{\beta}} = \mathbf{X}'\mathbf{y}$ have no unique solution for $\hat{\boldsymbol{\beta}}$. In this situation it is shown in statistical theory that there exists no unbiased estimator for every element in the vector $\boldsymbol{\beta}$. In these cases our interest is generally in certain linear combinations of $\boldsymbol{\beta}$, say γ, where $\gamma = \mathbf{c}'\boldsymbol{\beta}$ and where $\mathbf{c}$ is a known $p \times 1$ vector. An unbiased estimator of γ exists if and only if the vector $\mathbf{c}$ is in the column space of $\mathbf{X}'$ (the row space of $\mathbf{X}$)—that is, if and only if there is a vector $\mathbf{k}$ such that $\mathbf{X}'\mathbf{k} = \mathbf{c}$ or $\mathbf{k}'\mathbf{X} = \mathbf{c}'$. The general solution to the normal equations is $\hat{\boldsymbol{\beta}} = (\mathbf{X}'\mathbf{X})^-\mathbf{X}'\mathbf{y} + (\mathbf{I} - \mathbf{X}^-\mathbf{X})\mathbf{h}$ where $\mathbf{h}$ is any $p \times 1$ vector. If we use $\mathbf{c}'\hat{\boldsymbol{\beta}}$ as the estimator of γ, we note that $\mathbf{c}'\hat{\boldsymbol{\beta}}$ is unique for any solution $\hat{\boldsymbol{\beta}}$ to the normal equations if $\mathbf{c}$ is in the column space of $\mathbf{X}'$. It can be shown that the variance of $\mathbf{c}'\hat{\boldsymbol{\beta}}$ in this case is $\sigma^2\mathbf{c}'(\mathbf{X}'\mathbf{X})^-\mathbf{c}$. It is in general easier to compute $(\mathbf{X}'\mathbf{X})^c$ than to compute $(\mathbf{X}'\mathbf{X})^-$, and it can be shown that $\mathbf{c}'(\mathbf{X}'\mathbf{X})^c\mathbf{c} = \mathbf{c}'(\mathbf{X}'\mathbf{X})^-\mathbf{c}$ for any c-inverse of $\mathbf{X}'\mathbf{X}$ if $\mathbf{c}$ is in the column space of $\mathbf{X}'$. This has very important consequences in the theory of linear models in statistics. The next theorem is a statement of this result.

Theorem 6.6.9

Let $\mathbf{A}$ *be any* $p \times p$ *matrix. If* $\mathbf{c}$ *is a* $p \times 1$ *vector, then* $\mathbf{c}'\mathbf{A}^c\mathbf{c}$ *is invariant for any c-inverse of* $\mathbf{A}$ *if* $\mathbf{c}$ *is in the column space of* $\mathbf{A}$ *and of* $\mathbf{A}'$.

Proof: Since $\mathbf{A}^-$ is unique and since it is also a c-inverse of $\mathbf{A}$, we must find conditions on $\mathbf{c}$ such that $\mathbf{c}'\mathbf{A}^c\mathbf{c} = \mathbf{c}'\mathbf{A}^-\mathbf{c}$ for all matrices $\mathbf{A}^c$ that satisfy

$\mathbf{AA}^c\mathbf{A} = \mathbf{A}$. If we solve this last matrix equation, we obtain

$$\mathbf{A}^c = \mathbf{A}^- + \mathbf{K} - \mathbf{A}^-\mathbf{AKAA}^-$$

for all $p \times p$ matrices $\mathbf{K}$. If we multiply by $\mathbf{c}'$ and $\mathbf{c}$, we obtain

$$\mathbf{c}'\mathbf{A}^c\mathbf{c} = \mathbf{c}'\mathbf{A}^-\mathbf{c} + \mathbf{c}'\mathbf{Kc} - \mathbf{c}'\mathbf{A}^-\mathbf{AKAA}^-\mathbf{c}.$$

If $\mathbf{c}$ is in the column space of $\mathbf{A}$ and $\mathbf{A}'$, there exist vectors $\mathbf{b}$ and $\mathbf{d}$ such that $\mathbf{c} = \mathbf{Ad}$ and $\mathbf{c} = \mathbf{A}'\mathbf{b}$. If we substitute these values for $\mathbf{c}$, we get $\mathbf{c}'\mathbf{A}^c\mathbf{c} = \mathbf{c}'\mathbf{A}^-\mathbf{c}$ for all $p \times p$ matrices $\mathbf{K}$. This completes the proof of the theorem. ∎

Corollary 6.6.9.1

If $\mathbf{A}$ *is an* $n \times n$ *symmetric matrix, then* $\mathbf{c}'\mathbf{A}^c\mathbf{c}$ *is invariant for any* **c***-inverse of* $\mathbf{A}$ *if* $\mathbf{c}$ *is in the column space of* $\mathbf{A}$.

This theorem can be extended and generalized by using matrices $\mathbf{C}$ and $\mathbf{B}$ instead of a vector $\mathbf{c}$. This is the context of the next corollary.

Corollary 6.6.9.2

Let $\mathbf{A}$ *be any* $n \times m$ *matrix and let* $\mathbf{B}$ *and* $\mathbf{C}$ *be any* $p \times m$ *and* $n \times q$ *matrices, respectively. The matrix* $\mathbf{BA}^c\mathbf{C}$ *is invariant for any c-inverse of* $\mathbf{A}$ *if the column space of* $\mathbf{B}'$ *is in the column space of* $\mathbf{A}'$ *and the column space of* $\mathbf{C}$ *is in the column space of* $\mathbf{A}$.

We conclude this section with some additional special cases of Theorem 6.6.9.

Corollary 6.6.9.3

Let $\mathbf{c}$ *be any* $p \times 1$ *vector that is in the column space of the* $p \times n$ *matrix* $\mathbf{X}'$. *The vector* $\mathbf{c}'(\mathbf{X}'\mathbf{X})^c\mathbf{X}'$ *is invariant for any c-inverse of* $\mathbf{X}'\mathbf{X}$ *and* $\mathbf{X}(\mathbf{X}'\mathbf{X})^c\mathbf{X}' = \mathbf{XX}^-$.

Note: From the normal equations discussed above, if $\mathbf{c}$ is in the column space of $\mathbf{X}'$, then $\mathbf{c}'\hat{\boldsymbol{\beta}} = \mathbf{c}'(\mathbf{X}'\mathbf{X})^c\mathbf{X}'\mathbf{y}$ for any solution $\hat{\boldsymbol{\beta}}$ to the normal equations, and this is invariant for any c-inverse of $\mathbf{X}'\mathbf{X}$. Also $\mathbf{c}'\hat{\boldsymbol{\beta}}$ is an unbiased estimator of $\mathbf{c}'\boldsymbol{\beta}$.

Corollary 6.6.9.4

If $\mathbf{c}$ *is any* $p \times 1$ *vector that is in the column space of the* $p \times n$ *matrix* $\mathbf{X}'$, *then* $[\mathbf{c}'(\mathbf{X}'\mathbf{X})^c\mathbf{X}']' = \mathbf{X}(\mathbf{X}'\mathbf{X})^c\mathbf{c}$ *for any c-inverse of* $\mathbf{X}'\mathbf{X}$.

The next theorem is similar to Theorem 6.2.16.

Theorem 6.6.10

Let $\mathbf{A}$ *be any* $m \times n$ *matrix. Then* $\mathbf{A}^c = (\mathbf{A}'\mathbf{A})^c\mathbf{A}'$ *where* $(\mathbf{A}'\mathbf{A})^c$ *is any c-inverse of* $\mathbf{A}'\mathbf{A}$.

Proof: Let $\mathbf{B} = (\mathbf{A}'\mathbf{A})^c\mathbf{A}'$. Then $\mathbf{AB} = \mathbf{AA}^-$ and $\mathbf{ABA} = \mathbf{AA}^-\mathbf{A} = \mathbf{A}$, so $\mathbf{B} = \mathbf{A}^c$ and the theorem is proved. ▮

Theorem 6.6.11

Let $\mathbf{A}$ *be any* $m \times n$ *matrix of rank* r. *Let* $\mathbf{P}$ *and* $\mathbf{Q}$ *be any nonsingular matrices of appropriate sizes such that*

$$\mathbf{PAQ} = \begin{bmatrix} \mathbf{I}_r & \mathbf{0} \\ \mathbf{0} & \mathbf{0} \end{bmatrix}.$$

The matrix $\mathbf{B}$ *is a c-inverse of* $\mathbf{A}$, *where*

$$\mathbf{B} = \mathbf{Q}\begin{bmatrix} \mathbf{I}_r & \mathbf{B}_{12} \\ \mathbf{B}_{21} & \mathbf{B}_{22} \end{bmatrix}\mathbf{P} = \mathbf{QCP},$$

where $\mathbf{B}_{ij}$ *are any matrices of appropriate sizes.*

Proof: $$(\mathbf{PAQ})\mathbf{C}(\mathbf{PAQ}) = \begin{bmatrix} \mathbf{I}_r & \mathbf{0} \\ \mathbf{0} & \mathbf{0} \end{bmatrix}\begin{bmatrix} \mathbf{I}_r & \mathbf{B}_{12} \\ \mathbf{B}_{21} & \mathbf{B}_{22} \end{bmatrix}\begin{bmatrix} \mathbf{I}_r & \mathbf{0} \\ \mathbf{0} & \mathbf{0} \end{bmatrix} = \begin{bmatrix} \mathbf{I}_r & \mathbf{0} \\ \mathbf{0} & \mathbf{0} \end{bmatrix} = \mathbf{PAQ}$$

and hence $\mathbf{B}$ is a c-inverse of $\mathbf{A}$. ▮

Theorem 6.6.12

Let $\mathbf{A}$ *and* $\mathbf{B}$ *be any* $m \times n$ *matrices. The columns of* $\mathbf{B}$ *are in the column space of* $\mathbf{A}$ *if and only if* $\mathbf{AA}^c\mathbf{B} = \mathbf{B}$ *where* $\mathbf{A}^c$ *is any c-inverse of* $\mathbf{A}$.

Proof: The columns of $\mathbf{B}$ are in the column space of $\mathbf{A}$ if and only if there is a matrix $\mathbf{C}$ such that $\mathbf{B} = \mathbf{AC}$. Multiply both sides by $\mathbf{AA}^c$ and get $\mathbf{AA}^c\mathbf{B}$

$= \mathbf{AA}^c\mathbf{AC} = \mathbf{AC} = \mathbf{B}$. Next suppose $\mathbf{AA}^c\mathbf{B} = \mathbf{B}$. Then if we denote $\mathbf{A}^c\mathbf{B}$ by $\mathbf{C}$, we get $\mathbf{AC} = \mathbf{B}$ and the theorem is proved. ∎

Theorem 6.6.13

Let $\mathbf{A}^{c1}$, $\mathbf{A}^{c2}$, $\mathbf{A}^{c3}$ be any c-inverses of $\mathbf{A}$. Then $\mathbf{B}_1$ *and* $\mathbf{B}_2$ *are also c-inverses of* $\mathbf{A}$, *where*

$$\mathbf{B}_1 = \mathbf{A}^{c1} + \mathbf{H} - \mathbf{A}^{c2}\mathbf{AHAA}^{c3},$$

$$\mathbf{B}_2 = \mathbf{A}^{c1} + \mathbf{H}(\mathbf{I} - \mathbf{AA}^{c2}) + (\mathbf{I} - \mathbf{A}^{c3}\mathbf{A})\mathbf{G},$$

and where $\mathbf{H}$ *and* $\mathbf{G}$ *are any matrices of appropriate sizes. Also any c-inverse of* $\mathbf{A}$ *can be written as* $\mathbf{B}_1$ *and as* $\mathbf{B}_2$ *for some matrices* $\mathbf{H}$ *and* $\mathbf{G}$.

The proof of this theorem will be asked for in the problems.

6.7 Hermite Form of Matrices

This section gives some additional theorems about the Hermite form of matrices which are sometimes useful in theoretical work on matrices.

Definition 6.7.1

***Hermite Form of an $n \times n$ Matrix* $\mathbf{A}$.** *Let* $\mathbf{B}$ *be any nonsingular matrix such that* $\mathbf{BA} = \mathbf{H_A}$. *The matrix* $\mathbf{H_A}$ *is defined to be a Hermite form of* $\mathbf{A}$ *if and only if* $\mathbf{H_A}$ *is in Hermite form, that is, if it satisfies* Def. 6.6.2.

By Theorem 6.6.5, for every $n \times n$ matrix $\mathbf{A}$ there exists a Hermite form $\mathbf{H_A}$. In fact every $n \times n$ matrix has exactly one Hermite form, and this is the context of the next theorem.

Theorem 6.7.1

Let $\mathbf{A}$ *be any $n \times n$ matrix. There exists exactly one Hermite form of* $\mathbf{A}$.

The proof of this theorem is left as a problem.

Note: There may be many nonsingular matrices, say **B** and **C**, such that **BA** and **CA** are each in Hermite form, but the Hermite matrices are the same. For instance, in Example 6.6.1, we note that $\mathbf{BA} = \mathbf{H}_1$ gives us

$$\begin{bmatrix} -3 & 2 & 0 \\ 2 & -1 & 0 \\ 1 & -1 & 1 \end{bmatrix} \begin{bmatrix} 1 & 2 & 1 \\ 2 & 3 & 1 \\ 1 & 1 & 0 \end{bmatrix} = \begin{bmatrix} 1 & 0 & -1 \\ 0 & 1 & 1 \\ 0 & 0 & 0 \end{bmatrix} = \mathbf{H}_1,$$

and in Example 6.6.2 $\mathbf{CA} = \mathbf{H}_2$ gives us

$$\begin{bmatrix} 0 & -1 & 3 \\ 0 & 1 & -2 \\ 1 & -1 & 1 \end{bmatrix} \begin{bmatrix} 1 & 2 & 1 \\ 2 & 3 & 1 \\ 1 & 1 & 0 \end{bmatrix} = \begin{bmatrix} 1 & 0 & -1 \\ 0 & 1 & 1 \\ 0 & 0 & 0 \end{bmatrix} = \mathbf{H}_2.$$

$\mathbf{H}_1$ and $\mathbf{H}_2$ are Hermite forms for **A** and $\mathbf{H}_1 = \mathbf{H}_2$ but $\mathbf{B} \neq \mathbf{C}$.

Theorem 6.7.2

The Hermite form $\mathbf{H}_\mathbf{A}$ of an $n \times n$ matrix **A** *has the same rank as* **A**.

Proof: There exists a nonsingular matrix **B** such that $\mathbf{BA} = \mathbf{H}_\mathbf{A}$, and by Theorem 1.6.7, $\mathbf{H}_\mathbf{A}$ and **A** have the same rank. ▮

Theorem 6.7.3

The $n \times n$ identity matrix, **I**, *is the only $n \times n$ matrix in Hermite form that is nonsingular.*

Proof: Since $\mathbf{H}^2 = \mathbf{H}$ for any matrix in Hermite form and since **H** is nonsingular, we can multiply by $\mathbf{H}^{-1}$ and obtain $\mathbf{H} = \mathbf{I}$. ▮

Theorem 6.7.4

Let **A** *be any $n \times n$ nonsingular matrix. The Hermite form of* **A** *is the $n \times n$ identity matrix* **I**.

Proof: We have $\mathbf{BA} = \mathbf{H}_\mathbf{A}$, but since **B** and **A** are nonsingular, it follows that $\mathbf{H}_\mathbf{A}$ is also nonsingular and, by the previous theorem, $\mathbf{H}_\mathbf{A} = \mathbf{I}$. ▮

Theorem 6.7.5

Two $n \times n$ matrices **A** *and* **B** *have the same Hermite form if and only if the column space of* **A′** *is the same as the column space of* **B′**.

Proof: First assume that **A′** and **B′** have the same column space. By Theorem 5.4.5, there exists a nonsingular matrix **C′** such that $\mathbf{A}'\mathbf{C}' = \mathbf{B}'$ or $\mathbf{CA} = \mathbf{B}$. Now let **K** be a nonsingular matrix such that $\mathbf{KB} = \mathbf{H}$, where **H** is the Hermite form of **B**; but $\mathbf{KCA} = \mathbf{H}$ and **KC** is a nonsingular matrix; hence **H** is also the Hermite form of **A**. Next assume **A** and **B** have the same Hermite form **H**. This means there exist nonsingular matrices **F** and **G** such that $\mathbf{FA} = \mathbf{H} = \mathbf{GB}$ or $\mathbf{FA} = \mathbf{GB}$, and hence

$$\mathbf{A}' = \mathbf{B}'\mathbf{G}'\mathbf{F}^{-1\prime} = \mathbf{B}'\mathbf{K}'$$

where $\mathbf{K}' = \mathbf{G}'\mathbf{F}^{-1\prime}$ is nonsingular. By Theorem 5.4.5, **A′** and **B′** have the same column space, and the theorem is proved. ▮

Corollary 6.7.5.1

Let **A** *be any $n \times n$ matrix. The matrices* **A′A** *and* **A** *have the same Hermite form. Also,* $\mathbf{A}^-\mathbf{A}$ *and* **A** *have the same Hermite form.*

Corollary 6.7.5.2

Let **A** *be any $n \times n$ matrix and let* **B** *be any $n \times n$ nonsingular matrix. The matrices* **BA** *and* **A** *have the same Hermite form.*

Theorem 6.7.6

Let **A** *be an $n \times n$ matrix. Let* **B** *be a nonsingular matrix such that* $\mathbf{BA} = \mathbf{H_A}$ *where* $\mathbf{H_A}$ *is in Hermite form. Then* $\mathbf{AH_A} = \mathbf{A}$.

Proof: Since $\mathbf{BA} = \mathbf{H_A}$ and $\mathbf{H_A} = \mathbf{H_A^2}$, it follows that

$$\mathbf{AH_A} = \mathbf{ABA} = \mathbf{A}. \quad ▮$$

The Hermite form of a matrix **A** can be used to determine the rank of **A** and a set of linearly independent columns of **A**.

Theorem 6.7.7

Let $\mathbf{H_A}$ *be the Hermite form of* **A**. *Suppose that the $i_1, i_2, \ldots, i_k$ diagonal* *elements of* $\mathbf{H_A}$ *are each equal to one, and the remaining diagonal elements of* $\mathbf{H_A}$ *are equal to zero. Then the* $i_1, i_2, \ldots, i_k$ *columns of* $\mathbf{A}$ *are linearly independent.*

Proof: Since $\mathbf{AH_A} = \mathbf{A}$, the rank of $\mathbf{A}$ is k and we can write

$$\left[\sum_i \mathbf{a}_i h_{i1}, \sum_i \mathbf{a}_i h_{i2}, \ldots, \sum_i \mathbf{a}_i h_{in}\right] = [\mathbf{a}_1, \mathbf{a}_2, \ldots, \mathbf{a}_n].$$

However, if $h_{tt} = 0$, then $\mathbf{a}_t$ does not occur in any term on the left side of the equation. But if $h_{tt} = 1$ then $\mathbf{a}_t$ does occur. Hence all the columns of $\mathbf{A}$ can be written as a linear combination of those columns of $\mathbf{A}$ corresponding to the nonzero diagonal elements of $\mathbf{H_A}$, and the theorem is proved. ▮

From this discussion we can state the following two theorems.

Theorem 6.7.8

The rank of $\mathbf{A}$ *is equal to the number of diagonal elements of* $\mathbf{H_A}$ *that are equal to one.*

Theorem 6.7.9

Consider the t-th column of $\mathbf{A}$ *denoted by* $\mathbf{a}_t$. *This column is a linear combination of the set of linearly independent columns of* $\mathbf{A}$ *described in Theorem* 6.7.7. *The coefficients of the linear combination are the nonzero elements of the t-th column of* $\mathbf{H_A}$. *(Assume* $h_{tt} = 0$ *in* $\mathbf{H_A}$.*)*

Example 6.7.1. Consider the matrix $\mathbf{A}$ in Example 6.6.1 and the Hermite form $\mathbf{H_A}$ where

$$\mathbf{A} = \begin{bmatrix} 1 & 2 & 1 \\ 2 & 3 & 1 \\ 1 & 1 & 0 \end{bmatrix}; \qquad \mathbf{H_A} = \begin{bmatrix} 1 & 0 & -1 \\ 0 & 1 & 1 \\ 0 & 0 & 0 \end{bmatrix}.$$

$\mathbf{H_A}$ has exactly two nonzero diagonal elements, so the rank of $\mathbf{A}$ is 2. Also, the first and second diagonal elements of $\mathbf{H_A}$ are nonzero, so $\mathbf{a}_1$ and $\mathbf{a}_2$ are linearly independent columns of $\mathbf{A}$. Also $\mathbf{a}_3$ is equal to $(-1)\mathbf{a}_1 + (1)\mathbf{a}_2$ where the coefficients -1 and 1 are the elements of column 3 of $\mathbf{H_A}$.

Theorem 6.7.10

If $\mathbf{A}$ *is an* $n \times n$ *matrix, and* $\mathbf{H_A}$ *is the Hermite form of* $\mathbf{A}$, *then* $\mathbf{A}$ *is idempotent if and only if the Hermite form of* **A** *is also a c-inverse of* **A**. *Also* **A** *is idempotent if and only if* $\mathbf{H_A A} = \mathbf{H_A}$.

The proof of this theorem is left as a problem.

Problems

1. Find the g-inverse of the vector **a** where

$$\mathbf{a} = \begin{bmatrix} 1 \\ 3 \\ 1 \\ 5 \\ 2 \end{bmatrix}.$$

Use Theorem 6.4.8.

2. Find the g-inverse of the 2×2 matrix **A** where

$$\mathbf{A} = \begin{bmatrix} -1 & -1 \\ -1 & -1 \end{bmatrix}.$$

Use Theorems 6.4.1 and 6.4.9.

3. Find the g-inverse of the 6×2 matrix **A** where

$$\mathbf{A} = \begin{bmatrix} 1 & 1 \\ 3 & 3 \\ 5 & 2 \\ 2 & 1 \\ 0 & 6 \\ 1 & 5 \end{bmatrix}.$$

Use Theorem 6.2.16.

4. Find the g-inverse of the 5×2 matrix **A** where

$$\mathbf{A} = \begin{bmatrix} 2 & 4 \\ 1 & 2 \\ 3 & 6 \\ 5 & 10 \\ 2 & 4 \end{bmatrix}.$$

Use Theorem 6.5.1.

5. Show that the g-inverse of a general 2×2 symmetric matrix $\mathbf{A}$ of rank 1 defined by

$$\mathbf{A} = \begin{bmatrix} a_{11} & a_{12} \\ a_{21} & a_{22} \end{bmatrix}$$

is given by

$$\mathbf{A}^{-} = \begin{bmatrix} \frac{a_{11}}{T} & \frac{a_{12}}{T} \\ \frac{a_{21}}{T} & \frac{a_{22}}{T} \end{bmatrix},$$

where $T = a_{11}^2 + a_{12}^2 + a_{21}^2 + a_{22}^2 = \text{tr}\,(\mathbf{A}'\mathbf{A})$.

6. Find the g-inverse of the 3×3 matrix $\mathbf{A}$ where

$$\mathbf{A} = \begin{bmatrix} 3 & 2 & 1 \\ 1 & 1 & 1 \\ 3 & 1 & -1 \end{bmatrix},$$

using the methods presented in each of the Theorems 6.5.1, 6.5.2, 6.5.5, and 6.5.8.

7. Use Theorem 6.3.1 to show that the system of equations given below is consistent.

$$3x_1 - 2x_2 + x_3 = 3,$$

$$3x_1 + x_2 + 2x_3 = 5,$$

$$3x_1 + 10x_2 + 5x_3 = 11.$$

8. Find a solution to the system of equations in Prob. 7.
9. Find the general solution to the system of equations in Prob. 7.
10. Show that the system of equations given below is not consistent.

$$6x_1 + x_2 - 3x_3 + x_4 = 0,$$

$$4x_1 - x_2 + x_3 - 2x_4 = 5,$$

$$x_1 + 3x_2 + 4x_3 - x_4 = -6,$$

$$x_1 - x_2 - 8x_3 + 4x_4 = 3.$$

11. In Prob. 10, show that the first three equations are consistent.

12. Let the $m \times 2$ $(m \geq 2)$ matrix **A** of rank 1 be defined by

$$\mathbf{A} = [\mathbf{a}, c\mathbf{a}]$$

where c is a scalar and **a** is an $m \times 1$ nonzero vector. Find the g-inverse of **A** in terms of **a** and c.

13. Prove the following: Let **A** be an $m \times n$ matrix, **X** be an $n \times r$ matrix, **C** be an $m \times r$ matrix, **B** be an $r \times g$ matrix, and **D** be an $n \times g$ matrix. A necessary and sufficient condition that the two equations $\mathbf{AX} = \mathbf{C}$ and $\mathbf{XB} = \mathbf{D}$ have a common solution **X** is (1) each equation has a solution and (2) $\mathbf{AD} = \mathbf{CB}$.

14. Find the g-inverse of the matrix **A** where

$$\mathbf{A} = \begin{bmatrix} 6 & 1 & 2 & 4 & 9 \\ -3 & 1 & 5 & 2 & 7 \\ 1 & 0 & 3 & -4 & 1 \\ 1 & 3 & 17 & 4 & 24 \\ 1 & -1 & -13 & 6 & -9 \end{bmatrix}.$$

15. Find the g-inverse of the symmetric matrix **A** where

$$\mathbf{A} = \begin{bmatrix} 3 & 1 & 0 & 1 \\ 1 & 4 & -1 & 2 \\ 0 & -1 & 6 & 2 \\ 1 & 2 & 2 & 4 \end{bmatrix}.$$

16. Find the g-inverse of the matrix **A** where

$$\mathbf{A} = \begin{bmatrix} 1 & 1 & 1 & 0 & 0 \\ 1 & 1 & 1 & 0 & 0 \\ 1 & 1 & 1 & 0 & 0 \\ 0 & 0 & 0 & 2 & 2 \\ 0 & 0 & 0 & 2 & 2 \end{bmatrix}.$$

17. Find the g-inverse of the matrix **A** where

$$\mathbf{A} = \begin{bmatrix} 3 & 0 & 0 & 0 & 0 \\ 0 & 0 & 0 & 0 & 0 \\ 0 & 0 & 1 & 0 & 0 \\ 0 & 0 & 0 & 4 & 2 \\ 0 & 0 & 0 & 2 & 1 \end{bmatrix}.$$

18. Find the g-inverse of the matrix $\mathbf{A}$ where

$$\mathbf{A} = \begin{bmatrix} 1 & 1 & 1 \\ 2 & -2 & 0 \\ 3 & 3 & -3 \end{bmatrix}.$$

19. Let $\mathbf{A}$ be an $m \times n$ given matrix, and let $\mathbf{X}$ be any $n \times m$ matrix such that

$$\mathbf{A'AX} = \mathbf{A'}$$

is satisfied, and let $\mathbf{Y}$ be any $n \times m$ matrix such that

$$\mathbf{YAA'} = \mathbf{A'}$$

is satisfied. Show that the g-inverse of $\mathbf{A}$ is given by

$$\mathbf{A}^- = \mathbf{YAX}.$$

20. Find a solution to the system of equations

$$2x_1 - x_2 + x_3 = 8,$$

$$x_1 + 2x_2 - x_3 = -5.$$

21. If $\mathbf{A}$ is a given $m \times n$ matrix, find conditions on a matrix $\mathbf{X}$ so that the system $\mathbf{AX} = \mathbf{I}$ is consistent.
22. Prove Theorem 6.3.3.
23. Let $\mathbf{A}$ be an $m \times m$ symmetric matrix and $\mathbf{P}$ be an orthogonal matrix such that $\mathbf{P'AP} = \mathbf{D}$, where $\mathbf{D}$ is a diagonal matrix with the characteristic roots of $\mathbf{A}$ on the diagonal. Show that $\mathbf{P'A^-P}$ is also a diagonal matrix.
24. Let λ_i $(i = 1, 2, \ldots, r)$ be the nonzero characteristic roots of an $m \times m$ symmetric matrix $\mathbf{A}$. Show that λ_i^{-1} $(i = 1, 2, \ldots, r)$ are the nonzero characteristic roots of $\mathbf{A}^-$.
25. If $\mathbf{A}$ is an $m \times m$ symmetric matrix such that $\mathbf{a'A} = \mathbf{0}$, show that $\mathbf{a'A^-} = \mathbf{0}$ ($\mathbf{a}$ is an $m \times 1$ vector).
26. If $\mathbf{A}$ is an $m \times m$ symmetric matrix such that $\mathbf{1'A} = \mathbf{0}$, show that

$$\begin{bmatrix} \mathbf{A} \\ \mathbf{1'} \end{bmatrix}^- = \begin{bmatrix} \mathbf{A}^-, & \dfrac{1}{m}\,\mathbf{1} \end{bmatrix}.$$

27. If $\mathbf{A}$ is an $m \times n$ matrix, $\mathbf{B}$ is an $m \times n$ matrix and $\mathbf{AB}' = \mathbf{0}$, and $\mathbf{B}'\mathbf{A} = \mathbf{0}$, show that

(1) $\mathbf{A}^-\mathbf{B} = \mathbf{0}$, (2) $\mathbf{B}^-\mathbf{A} = \mathbf{0}$, (3) $\mathbf{AB}^- = \mathbf{0}$,

(4) $\mathbf{BA}^- = \mathbf{0}$, (5) $\mathbf{B}'^-\mathbf{A}^- = \mathbf{0}$, (6) $\mathbf{A}'^-\mathbf{B}^- = \mathbf{0}$.

28. Let $\mathbf{A}$ be an $m \times n$ matrix and let $\mathbf{P}$ be any $n \times k$ matrix of rank n. Show that $\mathbf{P}^c\mathbf{A}^c$ is a c-inverse of $\mathbf{AP}$, where $\mathbf{A}^c$ and $\mathbf{P}^c$ are any c-inverses of $\mathbf{A}$ and $\mathbf{P}$, respectively.
29. In Prob. 28, if $k = n$ and $\mathbf{P}$ is nonsingular, show that $\mathbf{P}^{-1}\mathbf{A}^c$ is a c-inverse of $\mathbf{AP}$.
30. In Prob. 28, if $\mathbf{PP}' = \mathbf{I}$, show that $\mathbf{P}'\mathbf{A}^c$ is a c-inverse of $\mathbf{AP}$.
31. Let $\mathbf{A}$ be an $m \times n$ matrix and let $\mathbf{B}$ be an $n \times k$ matrix. Define $\mathbf{F}$ and $\mathbf{G}$ by

$$\mathbf{G} = \mathbf{A}^-\mathbf{AB}; \quad \mathbf{F} = \mathbf{AGG}^-,$$

and show that

$$\mathbf{AB} = \mathbf{FG} \quad \text{and} \quad (\mathbf{FG})^- = \mathbf{G}^-\mathbf{F}^-.$$

32. Let $\mathbf{A}$ be an $m \times n$ matrix of rank m such that $\mathbf{A} = \mathbf{BC}$ where $\mathbf{B}$ and $\mathbf{C}$ each has rank m. Show that $(\mathbf{BC})^- = \mathbf{C}^-\mathbf{B}^-$.
33. If $\mathbf{A}$ is a positive semidefinite matrix, show that $\mathbf{A}^-$ is also a positive semidefinite matrix.
34. Let $\mathbf{A}$ be an $m \times m$ matrix and let $\mathbf{P}$ and $\mathbf{Q}$ be orthogonal matrices such that $\mathbf{PAQ} = \mathbf{D}$ where $\mathbf{D}$ is a diagonal matrix. Show that $\mathbf{QD}^-\mathbf{P} = \mathbf{A}^-$.
35. Let $\mathbf{A}_1^c$ and $\mathbf{A}_2^c$ be any two c-inverses of the $m \times n$ matrix $\mathbf{A}$, and let $\mathbf{g}$ be any $n \times 1$ vector such that $\mathbf{AA}_1^c\mathbf{g} = \mathbf{g}$. Show that $\mathbf{AA}_2^c\mathbf{g} = \mathbf{g}$.
36. Find a c-inverse of the matrix $\mathbf{A}$ in Prob. 6.
37. Show rank $(\mathbf{A}^c) \geq$ rank $(\mathbf{A})$ for any c-inverse of $\mathbf{A}$. (See Theorem 6.6.8.)
38. For the matrix

$$\mathbf{A} = \begin{bmatrix} 1 & 2 \\ 1 & 1 \\ -1 & 0 \end{bmatrix},$$

find a c-inverse.
39. For the matrix in Prob. 15, find a nonsingular matrix $\mathbf{B}$ such that $\mathbf{BA} = \mathbf{H}$, where $\mathbf{H}$ is in Hermite form.

40. Show that there does not always exist a c-inverse of a matrix $\mathbf{A}^c$ that equals $\mathbf{A}$.
41. Does there ever exist a c-inverse of $\mathbf{A}^c$ that is equal to $\mathbf{A}$ if $\mathbf{A}$ is singular?
42. If $\mathbf{A}$ is nonsingular, show that a c-inverse of $\mathbf{A}$ is unique and $\mathbf{A}^c = \mathbf{A}^{-1}$.
43. If $\mathbf{A}$ is defined by

$$\mathbf{A} = \begin{bmatrix} \mathbf{B} & \mathbf{0} \\ \mathbf{0} & \mathbf{C} \end{bmatrix},$$

show that $\mathbf{A}^c$ is a c-inverse of $\mathbf{A}$ where

$$\mathbf{A}^c = \begin{bmatrix} \mathbf{B}^c & \mathbf{0} \\ \mathbf{0} & \mathbf{C}^c \end{bmatrix}$$

where $\mathbf{B}^c$ and $\mathbf{C}^c$ are any c-inverses of $\mathbf{B}$ and $\mathbf{C}$, respectively.
44. Show that a c-inverse of a singular diagonal matrix is not unique.
45. If $\mathbf{A}^c$ is any c-inverse of a matrix $\mathbf{A}$, show that $(\mathbf{A}^c)'$ is a c-inverse of $\mathbf{A}'$.
46. Let $\mathbf{P}$ and $\mathbf{Q}$ be respectively $m \times m$ and $n \times n$ nonsingular matrices and let $\mathbf{A}$ be any $m \times n$ matrix. Show that there exists a c-inverse of $\mathbf{PAQ}$ denoted by $(\mathbf{PAQ})^c$ such that $(\mathbf{PAQ})^c = \mathbf{Q}^{-1}\mathbf{A}^c\mathbf{P}^{-1}$ where $\mathbf{A}^c$ is any c-inverse of $\mathbf{A}$.
47. If $\mathbf{B}$ is any c-inverse of $\mathbf{A}$, show that $\mathbf{BAB}$ is also a c-inverse of $\mathbf{A}$.
48. In Prob. 47, show that $\mathbf{BAB}$ has the same rank as $\mathbf{A}$.
49. In Prob. 47, let $\mathbf{C} = \mathbf{BAB}$. Show that $\mathbf{CAC} = \mathbf{C}$.
50. Find the Hermite form of the matrix $\mathbf{A}$ in Prob. 6.
51. In Prob. 6, show by using the Hermite form of $\mathbf{A}$ that the first two columns of $\mathbf{A}$ are linearly independent and find the linear combination of these two columns that is equal to the third column.
52. In Prob. 6, find the Hermite form of $\mathbf{BA}$ where

$$\mathbf{B} = \begin{bmatrix} 1 & 2 & 1 \\ 1 & 1 & -1 \\ 1 & -1 & 1 \end{bmatrix},$$

and show that it is the same as the Hermite form of $\mathbf{A}$.
53. Prove Theorem 6.7.1.
54. Prove Theorem 6.7.10.
55. Let $\mathbf{A}$ be an $m \times n$ matrix. Show that $\mathbf{B}$ is an orthogonal left identity for $\mathbf{A}$ where $\mathbf{B} = 2\mathbf{AA}^- - \mathbf{I}$.
56. If $\mathbf{A}$ is an $m \times m$ symmetric matrix of rank $k < m$, show that there exists an $m \times m - k$ matrix $\mathbf{B}$ of rank $m - k$ such that $\mathbf{B}'\mathbf{A} = \mathbf{0}$.

57. Let $\mathbf{A}$ and $\mathbf{B}$ be as given in Prob. 56. Show that $\mathbf{A} + \mathbf{BB}'$ is nonsingular.

58. In Prob. 56, show that $\mathbf{A}^-\mathbf{A} + \mathbf{BB}^- = \mathbf{I}$.

59. In Prob. 56, show that $\mathbf{A} + \mathbf{BB}^-$ is nonsingular and $(\mathbf{A} + \mathbf{BB}^-)^{-1} = \mathbf{A}^- + \mathbf{BB}^-$.

60. If $\mathbf{A}$ and $\mathbf{B}$ are defined as in Prob. 56, show that $\mathbf{C}$ is nonsingular, where

$$\mathbf{C} = \begin{bmatrix} \mathbf{A} & \mathbf{B} \\ \mathbf{B}' & \mathbf{0} \end{bmatrix},$$

and show that

$$\mathbf{C}^{-1} = \begin{bmatrix} \mathbf{A}^- & \mathbf{B}'^- \\ \mathbf{B}^- & \mathbf{0} \end{bmatrix}.$$

61. For $\mathbf{A}$ and $\mathbf{B}$ in Prob. 56, show that $\mathbf{F}$ is nonsingular, where

$$\mathbf{F} = \begin{bmatrix} \mathbf{A} + \mathbf{BB}' & \mathbf{B} \\ \mathbf{B}' & \mathbf{0} \end{bmatrix}.$$

62. Show that $\mathbf{A}'\mathbf{AB} = \mathbf{0}$ if and only if $\mathbf{AB} = \mathbf{0}$ for any matrices $\mathbf{A}$ and $\mathbf{B}$ such that the multiplications are defined.

63. Let $\mathbf{A}$ be any symmetric $n \times n$ matrix. Show that $\mathbf{B}$ is a symmetric c-inverse of $\mathbf{A}$, where $\mathbf{B} = \frac{1}{2}[\mathbf{A}^c + (\mathbf{A}^c)']$, where $\mathbf{A}^c$ is any c-inverse of $\mathbf{A}$.

64. Let $\mathbf{A} \neq \mathbf{0}$ be an $m \times n$ matrix. Show that there exist matrices $\mathbf{B}$ and $\mathbf{C}$ such that $\mathbf{BA}^c\mathbf{C} = \mathbf{I}$, where $\mathbf{A}^c$ is any c-inverse of $\mathbf{A}$.

65. Let $\mathbf{A}$ be an $n \times n$ symmetric matrix such that $\mathbf{A}^2 = m\mathbf{A}$. Show that $\mathbf{B} = \dfrac{1}{m}\mathbf{A}$ is a g-inverse of $\dfrac{1}{m}\mathbf{A}$.

66. Let $\mathbf{A}$ be an $m \times n$ matrix and let $\mathbf{B}$ be an $n \times k$ matrix. Show that $(\mathbf{AB})^c = \mathbf{B}^c\mathbf{A}^c$ (where $(\mathbf{AB})^c$, $\mathbf{A}^c$, $\mathbf{B}^c$ are any c-inverses of the respective matrices) if and only if $\mathbf{A}^c\mathbf{ABB}^c$ is idempotent.

67. If $\mathbf{A}^c$ is a c-inverse of $\mathbf{A}$, show that $\mathbf{B}$ is also a c-inverse of $\mathbf{A}$, where $\mathbf{B} = \mathbf{A}^c\mathbf{AA}^c + (\mathbf{I} - \mathbf{A}^c\mathbf{A})\mathbf{P} + \mathbf{Q}(\mathbf{I} - \mathbf{AA}^c)$, where $\mathbf{P}$ and $\mathbf{Q}$ are any matrices of appropriate sizes.

68. If $\mathbf{ABA} = k\mathbf{A}$ for $k \neq 0$, show that $(1/k)\mathbf{B}$ is a c-inverse of $\mathbf{A}$.

69. If $\mathbf{a}$ is an $m \times 1$ vector and $\mathbf{b}$ is an $n \times 1$ vector, find a c-inverse of $\mathbf{ab}'$ in terms of $\mathbf{a}$ and $\mathbf{b}$.

70. Let $\mathbf{A}$ be an $m \times n$ matrix and let $\mathbf{B}$ be an $n \times p$ matrix of rank n. Show that $(\mathbf{AB})(\mathbf{AB})^- = \mathbf{AA}^-$.

71. Let $\mathbf{A}$ be an $m \times n$ matrix and let $\mathbf{B}$ be an $n \times p$ matrix. Show that $(\mathbf{AB})^- = \mathbf{B}^-\mathbf{A}^-$ if either (1) or (2) below is true.

    (1) $\mathbf{A}'\mathbf{A} = \mathbf{I}$.
    (2) $\mathbf{BB}' = \mathbf{I}$.

72. Let $\mathbf{A}$ be an $m \times n$ matrix. Show that $\mathbf{A}^- = \mathbf{A}'(\mathbf{AA}')^c\mathbf{A}(\mathbf{A}'\mathbf{A})^c\mathbf{A}'$, where $(\mathbf{A}'\mathbf{A})^c$ and $(\mathbf{AA}')^c$ are any c-inverses of the respective matrices.
73. Let $\mathbf{A}$ be an $m \times n$ matrix of rank m and let $\mathbf{B}$ be an $m \times m$ matrix of rank m. Show that $(\mathbf{A}'\mathbf{BA})^- = \mathbf{A}^-\mathbf{B}^{-1}\mathbf{A}'^-$.
74. If $\mathbf{A}$, $\mathbf{B}$, and $\mathbf{X}$ are $m \times n$, $k \times n$, and $m \times k$ matrices, respectively, show that $\mathbf{XX}^- = \mathbf{AA}^-$ if $\mathbf{A} = \mathbf{XB}$ and $\mathbf{X} = \mathbf{AC}$.
75. Let $\mathbf{A}$ be an $m \times n$ matrix and $\mathbf{B}$ be an $n \times m$ matrix. Show that if $\mathbf{ABB}^- = \mathbf{A}$ and $\mathbf{B}^-\mathbf{B} = \mathbf{AB}$, then $\mathbf{A} = \mathbf{B}^-$.
76. Let $\mathbf{A}$ be any $n \times n$ matrix and let $\mathbf{H}$ be its Hermite form. Show that $\mathbf{A}^-\mathbf{A} = \mathbf{H}^-\mathbf{H}$.

Systems of Linear Equations

7

7.1 Introduction

There is perhaps no field of mathematical inquiry in which systems of linear equations do not play an important role, and the field of statistics is certainly no exception. This chapter assumes that the reader is well acquainted with some of the elementary theorems for solving systems of linear equations.

We shall write a system of m equations in n unknowns as

$$\mathbf{Ax} = \mathbf{g}, \tag{7.1.1}$$

where $\mathbf{A}$ is an $m \times n$ real matrix, $\mathbf{g}$ is an $m \times 1$ real vector, and $\mathbf{x}$ is an $n \times 1$ real vector. The important problems for our concern are:

(1) For a given $m \times n$ real matrix $\mathbf{A}$ and a given $m \times 1$ real vector $\mathbf{g}$, does there exist an $n \times 1$ real vector $\mathbf{x}$ that satisfies Eq. (7.1.1)?

(2) If the answer to (1) is "yes," the next question is, "how many solution vectors $\mathbf{x}$ are there?"

(3) If the answer to (1) is "no," the next question is, "does there exist a vector $\mathbf{x}$ so that Eq. (7.1.1) is approximately satisfied for a suitable definition of approximate?"

If the answer to (1) is "yes," the system is said to be *consistent*; if the answer is "no," the system is said to be *inconsistent*.

Reminder. As usual, we shall consider that every matrix and vector has real elements unless explicitly stated otherwise; however, in some cases the word "real" will be included for emphasis.

7.2 Existence of Solutions to $\mathbf{Ax} = \mathbf{g}$

In this section we state and prove a number of theorems concerning the existence of a solution vector $\mathbf{x}$ to the system of equations $\mathbf{Ax} = \mathbf{g}$.

Theorem 7.2.1

If $\mathbf{A}$ *is an* $n \times n$ *nonsingular matrix, then the system* $\mathbf{Ax} = \mathbf{g}$ *has a solution.*

Proof: Since $\mathbf{A}^{-1}$ exists, $\mathbf{x} = \mathbf{A}^{-1}\mathbf{g}$ is clearly a solution. It is evident that if $\mathbf{A}$ is real, then $\mathbf{A}^{-1}$ is real, and, since $\mathbf{g}$ is real, it follows that $\mathbf{x}$ is real. ▮

In the system of equations $\mathbf{Ax} = \mathbf{g}$, the matrix $\mathbf{A}$ is called the *coefficient* matrix, and the matrix $\mathbf{B}$ is called the *augmented* matrix where $\mathbf{B} = [\mathbf{A}, \mathbf{g}]$; that is, if the vector $\mathbf{g}$ is appended to the matrix $\mathbf{A}$ as the $(n + 1)$-st column, the resulting matrix has size $m \times (n + 1)$ and is called the augmented matrix.

Example 7.2.1. Consider the system

$$2x_1 + 3x_2 + x_3 = 1$$
$$2x_1 - x_2 = 1.$$

We obtain

$$\mathbf{A} = \begin{bmatrix} 2 & 3 & 1 \\ 2 & -1 & 0 \end{bmatrix}, \qquad \mathbf{g} = \begin{bmatrix} 1 \\ 1 \end{bmatrix}.$$

The coefficient matrix is

$$\mathbf{A} = \begin{bmatrix} 2 & 3 & 1 \\ 2 & -1 & 0 \end{bmatrix},$$

and the augmented matrix is

$$\mathbf{B} = [\mathbf{A}, \mathbf{g}] = \begin{bmatrix} 2 & 3 & 1 & 1 \\ 2 & -1 & 0 & 1 \end{bmatrix}.$$

The augmented and coefficient matrices can be used to determine whether or not a solution to the system exists.

Theorem 7.2.2

A solution to the system $\mathbf{Ax} = \mathbf{g}$ *exists if and only if the rank of the coefficient matrix* $\mathbf{A}$ *is equal to the rank of the augmented matrix* $[\mathbf{A}, \mathbf{g}]$.

Proof: First we assume that the rank of $\mathbf{A}$ is equal to the rank of $[\mathbf{A}, \mathbf{g}]$ and show that this implies that a solution to $\mathbf{Ax} = \mathbf{g}$ exists. Let the common rank be r. Then there are exactly r columns of $\mathbf{A}$ that are linearly independent; denote them by $\mathbf{a}_{i_1}, \mathbf{a}_{i_2}, \ldots, \mathbf{a}_{i_r}$. Every other column of $\mathbf{A}$ must be a linear combination of these r columns. Now the rank of $[\mathbf{A}, \mathbf{g}]$ is assumed to be r also, so there are exactly r linearly independent columns in $[\mathbf{A}, \mathbf{g}]$. Thus the columns $\mathbf{a}_{i_1}, \mathbf{a}_{i_2}, \ldots, \mathbf{a}_{i_r}$ are linearly independent, and every other column in $[\mathbf{A}, \mathbf{g}]$ can be obtained as a linear combination of these r columns. Specifically, $\mathbf{g}$ can be so obtained. Thus for some (real) scalars $x_{i_1}, x_{i_2}, \ldots, x_{i_r}$, we get

$$\mathbf{g} = \sum_{t=1}^{r} \mathbf{a}_{i_t} x_{i_t}.$$

If we let $x_j = 0$ for $j \neq i_1; j \neq i_2, \ldots; j \neq i_r$, then we can write

$$\mathbf{g} = \sum_{i=1}^{n} \mathbf{a}_i x_i, \quad \text{or} \quad \mathbf{g} = \mathbf{Ax};$$

and the sufficiency part of the theorem is proved. To prove the necessary part of the theorem, we assume that a solution to the system $\mathbf{Ax} = \mathbf{g}$ exists. We can write the system as

$$[\mathbf{a}_1, \mathbf{a}_2, \ldots, \mathbf{a}_n] \begin{bmatrix} x_1 \\ \vdots \\ x_n \end{bmatrix} = \mathbf{g},$$

and from this we obtain

$$\sum_{i=1}^{n} \mathbf{a}_i x_i = \mathbf{g}.$$

Thus the vector space spanned by the columns of **A** is the same as the vector space spanned by the columns of [**A**, **g**], and the dimension of this vector space is the rank of the matrix **A**; but the dimension is also the rank of the matrix [**A**, **g**]. Hence the two matrices must have the same rank, and the theorem is proved. ∎

Corollary 7.2.2

The system $\mathbf{Ax} = \mathbf{g}$ *has a solution if and only if the vector* $\mathbf{g}$ *is in the column space of* **A**.

Next we shall show how the *g*-inverse and a *c*-inverse of **A** can be used to examine the existence of a solution to $\mathbf{Ax} = \mathbf{g}$.

Theorem 7.2.3

A necessary and sufficient condition for a solution to exist to the system $\mathbf{Ax} = \mathbf{g}$ *is that there is a c-inverse,* $\mathbf{A}^c$, *of* **A** *such that* $\mathbf{AA}^c\mathbf{g} = \mathbf{g}$.

Proof: First we assume that $\mathbf{Ax} = \mathbf{g}$ has a solution denoted by $\mathbf{x}_0$; hence $\mathbf{g} = \mathbf{Ax}_0$. Let $\mathbf{A}^c$ be any *c*-inverse of **A**. Multiply both sides by $\mathbf{AA}^c$ and obtain

$$\mathbf{AA}^c\mathbf{g} = \mathbf{AA}^c\mathbf{Ax}_0 = \mathbf{Ax}_0 = \mathbf{g},$$

and thus $\mathbf{AA}^c\mathbf{g} = \mathbf{g}$. Next we assume that $\mathbf{AA}^c\mathbf{g} = \mathbf{g}$. Let $\mathbf{x}_0 = \mathbf{A}^c\mathbf{g}$; then $\mathbf{Ax}_0 = \mathbf{AA}^c\mathbf{g} = \mathbf{g}$ so $\mathbf{Ax}_0 = \mathbf{g}$, and hence a solution exists. This completes the proof. ∎

Note: From the results of Prob. 35, Chap 6, if $\mathbf{AA}^c\mathbf{g} = \mathbf{g}$ for any *c*-inverse of **A**, the equation holds for all *c*-inverses of **A**.

Corollary 7.2.3

The system $\mathbf{Ax} = \mathbf{g}$ *has a solution if and only if* $\mathbf{AA}^-\mathbf{g} = \mathbf{g}$.

Proof: The proof is obtained from the fact that $\mathbf{A}^-$ is a conditional inverse of **A**. ∎

Theorem 7.2.4

If $\mathbf{A}$ *is an* $m \times n$ *matrix of rank* m, *then the system* $\mathbf{Ax} = \mathbf{g}$ *has a solution.*

Proof: By Theorem 6.2.16, if $\mathbf{A}$ is $m \times n$ of rank m, then $\mathbf{AA}^- = \mathbf{I}$ and hence $\mathbf{AA}^-\mathbf{g} = \mathbf{g}$, so a solution exists. ∎

If $\mathbf{A}$ is an $m \times n$ matrix, the system $\mathbf{Ax} = \mathbf{0}$ is called a linear *homogeneous* system. Clearly, $\mathbf{x} = \mathbf{0}$ is always a solution to this system. We now state and prove a theorem about other solutions of these systems.

Theorem 7.2.5

Let $\mathbf{A}$ *be an* $m \times n$ *matrix. The linear homogeneous system* $\mathbf{Ax} = \mathbf{0}$ *has a solution other than* $\mathbf{x} = \mathbf{0}$ *if and only if rank* $(\mathbf{A}) < n$.

Proof: If $\mathbf{x} \neq \mathbf{0}$ is a solution, then rank $(\mathbf{A})$ must be less than n; for if it were not, then we could multiply $\mathbf{Ax} = \mathbf{0}$ on the left by $\mathbf{A}^-$ and obtain $\mathbf{x} = \mathbf{A}^-\mathbf{0} = \mathbf{0}$, which is a contradiction. Next assume rank $(\mathbf{A}) < n$; then the row vectors of $\mathbf{A}$ span a space of dimension less than n (say r). Thus the orthogonal complement of the row space of $\mathbf{A}$ has dimension $n - r$ (let $n - r = s$, $s > 0$, since $r < n$). Therefore, there is a nonzero vector $\mathbf{x}$ that is orthogonal to the rows of $\mathbf{A}$; that is, $\mathbf{Ax} = \mathbf{0}$, and the proof is complete. ∎

7.3 The Number of Solutions of the System $Ax = g$

In this section we assume that the system of equations $\mathbf{Ax} = \mathbf{g}$ has at least one solution, and we discuss the number of solutions and find the form of the general solution. The following theorem is extremely useful in the theory of systems of linear equations, since it gives every solution to the system.

Theorem 7.3.1

Let $\mathbf{A}$ *be an* $m \times n$ *matrix and let* $\mathbf{A}^c$ *be any c-inverse of* $\mathbf{A}$ *and suppose a solution exists to the system* $\mathbf{Ax} = \mathbf{g}$. *For each and every* $n \times 1$ *vector* $\mathbf{h}$, *the vector* $\mathbf{x}_0$ *is a solution where*

$$\mathbf{x}_0 = \mathbf{A}^c\mathbf{g} + (\mathbf{I} - \mathbf{A}^c\mathbf{A})\mathbf{h}. \tag{7.3.1}$$

Also, every solution to the system can be written in the form of Eq. (7.3.1) *for some* $n \times 1$ *vector* $\mathbf{h}$.

Proof: Since we assume that there is a solution to the system, Theorem 7.2.3 tells us that $\mathbf{AA}^c\mathbf{g} = \mathbf{g}$. Hence, to prove that $\mathbf{x}_0$ in Eq. (7.3.1) is a solution, we multiply it on the left by $\mathbf{A}$ and obtain

$$\mathbf{Ax}_0 = \mathbf{AA}^c\mathbf{g} + \mathbf{A}(\mathbf{I} - \mathbf{A}^c\mathbf{A})\mathbf{h};$$

but since $\mathbf{A}(\mathbf{I} - \mathbf{A}^c\mathbf{A}) = \mathbf{0}$ and $\mathbf{AA}^c\mathbf{g} = \mathbf{g}$, this reduces to $\mathbf{Ax}_0 = \mathbf{g}$, and hence $\mathbf{x}_0$ in Eq. (7.3.1) is a solution. Next we assume that $\mathbf{x}_0$ is any solution to $\mathbf{Ax} = \mathbf{g}$, and show that there exists a vector $\mathbf{h}$ such that $\mathbf{x}_0$ can be written in the form of Eq. (7.3.1). Since $\mathbf{x}_0$ is a solution, we have $\mathbf{Ax}_0 = \mathbf{g}$, and multiplying on the left by $\mathbf{A}^c$ gives us

$$\mathbf{A}^c\mathbf{Ax}_0 = \mathbf{A}^c\mathbf{g} \quad \text{or} \quad \mathbf{0} = \mathbf{A}^c\mathbf{g} - \mathbf{A}^c\mathbf{Ax}_0 .$$

If we add $\mathbf{x}_0$ to both sides, we get

$$\mathbf{x}_0 = \mathbf{A}^c\mathbf{g} + \mathbf{x}_0 - \mathbf{A}^c\mathbf{Ax}_0 = \mathbf{A}^c\mathbf{g} + (\mathbf{I} - \mathbf{A}^c\mathbf{A})\mathbf{x}_0 ,$$

which is of the form of Eq. (7.3.1) with $\mathbf{h} = \mathbf{x}_0$, and the theorem is proved. ▮

From this theorem we get immediately a number of useful corollaries.

Corollary 7.3.1.1

Let the linear system of equations $\mathbf{Ax} = \mathbf{g}$ *have a solution where* $\mathbf{A}$ *is an* $m \times n$ *matrix. For each and every* $n \times 1$ *vector* $\mathbf{h}$, *the vector* $\mathbf{x}_0$ *is a solution where*

$$\mathbf{x}_0 = \mathbf{A}^-\mathbf{g} + (\mathbf{I} - \mathbf{A}^-\mathbf{A})\mathbf{h}. \tag{7.3.2}$$

Also, every solution to the system can be written in the form of Eq. (7.3.2) *for some* $n \times 1$ *vector* $\mathbf{h}$.

Corollary 7.3.1.2

If the system $\mathbf{Ax} = \mathbf{g}$ *is consistent, the vector* $\mathbf{x}_0 = \mathbf{A}^-\mathbf{g}$, *and the vector* $\mathbf{x}_1 = \mathbf{A}^c\mathbf{g}$ *are solutions.*

The proof of this corollary is obtained by putting $\mathbf{h} = \mathbf{0}$ in Eqs. (7.3.2) and (7.3.1), respectively.

Corollary 7.3.1.3

If the system $\mathbf{Ax} = \mathbf{g}$ *is consistent, then the solution* $\mathbf{x}_0 = \mathbf{A}^-\mathbf{g}$ *is unique if and only if* $\mathbf{A}^-\mathbf{A} = \mathbf{I}$.

Proof: $\mathbf{A}^-\mathbf{A} = \mathbf{I}$ if and only if $\mathbf{A}^c\mathbf{A} = \mathbf{I}$ for any c-inverse, $\mathbf{A}^c$, of $\mathbf{A}$. The proof of this corollary follows from the fact that for every $n \times 1$ vector $\mathbf{h}$,

$$\mathbf{x}_0 = \mathbf{A}^-\mathbf{g} + (\mathbf{I} - \mathbf{A}^-\mathbf{A})\mathbf{h}$$

is a solution of $\mathbf{Ax} = \mathbf{g}$, and clearly this is equal to $\mathbf{A}^-\mathbf{g}$ for every vector $\mathbf{h}$ if and only if $\mathbf{I} - \mathbf{A}^-\mathbf{A} = \mathbf{0}$. ▮

Corollary 7.3.1.4

If the system $\mathbf{Ax} = \mathbf{g}$ is consistent (where $\mathbf{A}$ is an $m \times n$ matrix), then the system has a unique solution if and only if the rank of $\mathbf{A}$ is equal to n.

Proof: Clearly the solution is unique if and only if $\mathbf{A}^-\mathbf{A} = \mathbf{I}$, but in this case the rank of $\mathbf{I}$, and hence of $\mathbf{A}^-\mathbf{A}$, is n. But rank $(\mathbf{A})$ = rank $(\mathbf{A}^-\mathbf{A}) = n$; so the rank must equal n. Also if $\mathbf{A}$ is $m \times n$ of rank n, then $\mathbf{A}^-\mathbf{A} = \mathbf{I}$, by Theorem 6.2.16. ▮

Corollary 7.3.1.5

If a unique solution exists to the system $\mathbf{Ax} = \mathbf{g}$, then it is $\mathbf{A}^-\mathbf{g}$. In this case $\mathbf{A}^-\mathbf{g} = \mathbf{A}^c\mathbf{g}$ for any c-inverse $\mathbf{A}^c$.

If $\mathbf{x}_1$ and $\mathbf{x}_2$ are solutions to the system $\mathbf{Ax} = \mathbf{g}$, then $\mathbf{y} = c_1\mathbf{x}_1 + c_2\mathbf{x}_2$ is a solution if $c_1 + c_2 = 1$. This can be shown by the fact that

$$\mathbf{Ay} = \mathbf{A}(c_1\mathbf{x}_1 + c_2\mathbf{x}_2) = c_1\mathbf{g} + c_2\mathbf{g} = \mathbf{g}(c_1 + c_2) = \mathbf{g}.$$

This result can be generalized: if $\mathbf{x}_1, \mathbf{x}_2, \ldots, \mathbf{x}_t$ are solutions to the system $\mathbf{Ax} = \mathbf{g}$, $\mathbf{g} \neq \mathbf{0}$, then $\mathbf{y} = \sum_{i=1}^{t} c_i\mathbf{x}_i$ is a solution for any set of scalars c_i if and only if $\sum_{i=1}^{t} c_i = 1$.

When a set of equations is consistent, we may want to determine the number of solutions that exist for the system. By the above, if there is more than one solution, there are an infinite number. We may, however, be able to determine the number of linearly independent vectors that are solutions to the system. This is the context of the next theorem.

Theorem 7.3.2

Let the system $\mathbf{Ax} = \mathbf{g}$ be consistent where $\mathbf{A}$ is an $m \times n$ matrix of rank $r > 0$ and $\mathbf{g} \neq \mathbf{0}$. Then there are exactly $n - r + 1$ linearly independent vectors that satisfy the system.

Proof: Consider the $n \times n$ matrix $\mathbf{B}$ defined by

$$\mathbf{I} - \mathbf{A}^{-}\mathbf{A} = \mathbf{B} = [\mathbf{b}_1, \mathbf{b}_2, \ldots, \mathbf{b}_n].$$

Now $\mathbf{x}_0, \mathbf{x}_1, \ldots, \mathbf{x}_n$ are solutions to the system $\mathbf{Ax} = \mathbf{g}$ where

$$\mathbf{A}^{-}\mathbf{g} = \mathbf{x}_0, \mathbf{A}^{-}\mathbf{g} + \mathbf{b}_1 = \mathbf{x}_1, \ldots, \mathbf{A}^{-}\mathbf{g} + \mathbf{b}_n = \mathbf{x}_n.$$

This result is obtained by setting $\mathbf{h} = \mathbf{0}$, $\mathbf{h} = \mathbf{e}_1, \ldots;$ $\mathbf{h} = \mathbf{e}_n$ in Corollary 7.3.1.1, where $\mathbf{e}_i$ is an $n \times 1$ vector with i-th element equal to unity and the remaining elements equal to zero. We can write these vectors as

$$\mathbf{X} = [\mathbf{x}_0, \mathbf{x}_1, \ldots, \mathbf{x}_n] = [\mathbf{A}^{-}\mathbf{g}, \mathbf{I} - \mathbf{A}^{-}\mathbf{A}]\begin{bmatrix} 1 & \mathbf{1}' \\ \mathbf{0} & \mathbf{I} \end{bmatrix},$$

and the rank of $\mathbf{X}$ is the rank of $[\mathbf{A}^{-}\mathbf{g}, \mathbf{I} - \mathbf{A}^{-}\mathbf{A}]$, which is equal to

$$1 + \text{rank}\,(\mathbf{I} - \mathbf{A}^{-}\mathbf{A}) = 1 + n - r.$$

Thus there are at least $n - r + 1$ linearly independent vectors that satisfy the system. Assume that there is a maximum of t linearly independent vectors $\mathbf{x}_i$ that satisfy the system. Then there must exist t distinct vectors $\mathbf{h}_i$ such that

$$\mathbf{x}_i = \mathbf{A}^{-}\mathbf{g} + (\mathbf{I} - \mathbf{A}^{-}\mathbf{A})\mathbf{h}_i, \quad i = 1, 2, \ldots, t$$

or

$$\mathbf{X}^* = [\mathbf{x}_1, \ldots, \mathbf{x}_t] = [\mathbf{A}^{-}\mathbf{g}, \mathbf{I} - \mathbf{A}^{-}\mathbf{A}]\begin{bmatrix} 1, & 1, \ldots, & 1 \\ \mathbf{h}_1, & \mathbf{h}_2, \ldots, & \mathbf{h}_t \end{bmatrix};$$

but rank of $\mathbf{X}^* \leq$ rank $[\mathbf{A}^{-}\mathbf{g}, \mathbf{I} - \mathbf{A}^{-}\mathbf{A}]$, which is $n - r + 1$, so t cannot be greater than $n - r + 1$. Thus there are exactly $n - r + 1$ linearly independent vectors that satisfy the system $\mathbf{Ax} = \mathbf{g}$, and the theorem is proved. ▮

Note: The vector $\mathbf{0}$ satisfies the system $\mathbf{Ax} = \mathbf{g}$ if and only if $\mathbf{g} = \mathbf{0}$; hence if $\mathbf{g} \neq \mathbf{0}$, the vectors that satisfy the system cannot form a vector space.

Theorem 7.3.3.

Consider the system $\mathbf{Ax} = \mathbf{0}$ *where* $\mathbf{A}$ *is an* $m \times n$ *matrix of rank* $r > 0$. *The vectors that satisfy this system form a vector subspace of* E_n *of dimension* $n - r$.

Proof: If $\mathbf{x}_1$ and $\mathbf{x}_2$ are any two vectors that satisfy $\mathbf{Ax} = \mathbf{0}$, then clearly $c_1\mathbf{x}_1 + c_2\mathbf{x}_2$ also satisfies the system for any scalars c_1 and c_2; so the set of solutions is a vector subspace. By Corollary 7.3.1.1, every solution must be of the form $(\mathbf{I} - \mathbf{A}^-\mathbf{A})\mathbf{h}$, so clearly the number of linearly independent solution vectors must be the rank of $(\mathbf{I} - \mathbf{A}^-\mathbf{A})$, which is $n - r$. ∎

Corollary 7.3.3

The set of solutions of $\mathbf{Ax} = \mathbf{0}$ *is the orthogonal complement of the column space of* $\mathbf{A}'$.

7.4 Approximate Solutions to Inconsistent Systems of Linear Equations

Suppose that the system of equations in Eq. (7.1.1) is inconsistent—that is, there is no vector $\mathbf{x}$ that satisfies the system. Then we write Eq. (7.1.1) as

$$\mathbf{Ax} - \mathbf{g} = \mathbf{e}(\mathbf{x}), \tag{7.4.1}$$

where $\mathbf{e}(\mathbf{x})$ is a remainder vector or a vector of deviations. If there did exist a vector $\mathbf{x}_0$ that satisfied the system $\mathbf{Ax} = \mathbf{g}$, this would mean there exists a vector $\mathbf{x}_0$ such that $\mathbf{e}(\mathbf{x}_0) = \mathbf{0}$. If there is no vector $\mathbf{x}$ such that $\mathbf{e}(\mathbf{x}) = \mathbf{0}$ (that is ,such that $\mathbf{Ax} = \mathbf{g}$), then it may be desirable to seek a vector $\mathbf{x}_0$ such that $\mathbf{e}(\mathbf{x}_0)$ is "small." If $\mathbf{x}_0$ is such a vector, then we may wish to call $\mathbf{x}_0$ an "approximate" solution to the system $\mathbf{Ax} = \mathbf{g}$. If $\mathbf{x}_0$ is a vector such that it produces a "smaller" $\mathbf{e}(\mathbf{x})$ in Eq. (7.4.1) than any other vector $\mathbf{x}$, then we might call $\mathbf{x}_0$ the *best approximate* "solution" (BAS) to the system of equations $\mathbf{Ax} = \mathbf{g}$.

Note: The reader should notice that sometimes we write $\mathbf{Ax} = \mathbf{g}$ instead of $\mathbf{Ax} - \mathbf{g} = \mathbf{e}(\mathbf{x})$ even though there may exist no solution to $\mathbf{Ax} = \mathbf{g}$.

Example 7.4.1. For a simple example, suppose the system $\mathbf{Ax} = \mathbf{g}$ is

$$\begin{aligned} x_1 + x_2 &= 2 \\ 2x_1 + 2x_2 &= 2 \\ 3x_1 + 3x_2 &= 3. \end{aligned} \tag{7.4.2}$$

Clearly there is no solution to this system of equations. As an alternative to a solution, suppose we decide to determine x_1 and x_2 such that $f(x_1, x_2)$ is a minimum, where $f(x_1, x_2)$ is the sum of squares of deviations; that is,

$$f(x_1, x_2) = (x_1 + x_2 - 2)^2 + (2x_1 + 2x_2 - 2)^2 + (3x_1 + 3x_2 - 3)^2.$$

The system in Eq. (7.4.2) has a solution $x_1 = x_1^0$, $x_2 = x_2^0$, if and only if $f(x_1^0, x_2^0) = 0$, and for values of x_1 and x_2 that are not solutions, it is clear that $f(x_1, x_2) > 0$. We shall determine values x_1 and x_2 such that $f(x_1, x_2)$ is a minimum and call this an approximate solution. If we use calculus to determine the values, we obtain

$$\frac{\partial f(x_1, x_2)}{\partial x_1} = 0, \qquad \frac{\partial f(x_1, x_2)}{\partial x_2} = 0,$$

which yield the two identical equations

$$14x_1 + 14x_2 = 15,$$
$$14x_1 + 14x_2 = 15.$$

Thus $f(x_1, x_2)$ is a minimum for any x_1 and x_2 that satisfies $14x_1 + 14x_2 = 15$. Thus the criterion of minimizing the sum of squares of deviations $f(x_1, x_2)$ does not give a unique solution. We must have an additional criterion such as: choose those values that minimize $f(x_1, x_2)$ and, among all of the values of x_1 and x_2 that minimize $f(x_1, x_2)$, choose those values that minimize $x_1^2 + x_2^2$. In the example we must minimize $g(x_1, x_2) = x_1^2 + x_2^2$ subject to the restriction $14x_1 + 14x_2 = 15$. Using the calculus again, we obtain $x_1^0 = x_2^0 = 15/28$ as the answer. We shall now show that the vector $\mathbf{x}_0 = \mathbf{A}^-\mathbf{g}$ gives the same solution. If we evaluate $\mathbf{A}^-$, we obtain

$$\mathbf{A}^- = \frac{1}{28}\begin{bmatrix} 1 & 2 & 3 \\ 1 & 2 & 3 \end{bmatrix},$$

and if we compute $\mathbf{A}^-\mathbf{g}$, we get

$$\mathbf{A}^-\mathbf{g} = \frac{1}{28}\begin{bmatrix} 15 \\ 15 \end{bmatrix},$$

which has 15/28 for each entry, and these are the same values obtained earlier for the approximate solution to the system in Eq. (7.4.2).

Next we formulate a definition of an approximate solution based on the results of the example, and then prove a theorem demonstrating how the g-inverse of $\mathbf{A}$ can be used to obtain the approximate solution. We seek an $\mathbf{x}_0$ that minimizes the sum of squares of deviations $\sum e_i^2(x)$. We can write this sum of squares as

$$\mathbf{e}'(\mathbf{x})\mathbf{e}(\mathbf{x}) \quad \text{or} \quad (\mathbf{Ax} - \mathbf{g})'(\mathbf{Ax} - \mathbf{g}).$$

Definition 7.4.1

Best Approximate Solution. *The vector* $\mathbf{x}_0$ *is defined to be the best approximate solution* (BAS) *to the system of equations* ($\mathbf{A}$ *is an* $m \times n$ *matrix*)

$$\mathbf{Ax} - \mathbf{g} = \mathbf{e}(\mathbf{x})$$

if and only if

(1) *for all* $\mathbf{x}$ *in* E_n, *the relationship* $(\mathbf{Ax} - \mathbf{g})'(\mathbf{Ax} - \mathbf{g}) \geq (\mathbf{Ax}_0 - \mathbf{g})'(\mathbf{Ax}_0 - \mathbf{g})$ *obtains;*

(2) *and for those* $\mathbf{x} \neq \mathbf{x}_0$ *such that* $(\mathbf{Ax} - \mathbf{g})'(\mathbf{Ax} - \mathbf{g}) = (\mathbf{Ax}_0 - \mathbf{g})'(\mathbf{Ax}_0 - \mathbf{g})$, *the relationship* $\mathbf{x}'\mathbf{x} > \mathbf{x}_0'\mathbf{x}_0$ *obtains.*

The definition essentially states that the vector $\mathbf{x}_0$ minimizes the sum of squares of deviations; and, if there is a set S of vectors such that each member in the set gives the minimum sum of squares of deviations, then the vector $\mathbf{x}_0$ in S is chosen as BAS if for all other vectors $\mathbf{x}$ in S the sum of squares $\mathbf{x}'\mathbf{x}$ is larger than $\mathbf{x}_0'\mathbf{x}_0$.

The following theorem states that the BAS exists and the g-inverse of the coefficient matrix can be used to find it.

Theorem 7.4.1

The BAS *to the system of equations* $\mathbf{Ax} = \mathbf{g}$ *is* $\mathbf{x}_0$ *where*

$$\mathbf{x}_0 = \mathbf{A}^-\mathbf{g}.$$

Proof: We must show that for $\mathbf{x}_0 = \mathbf{A}^-\mathbf{g}$ we get

$$(\mathbf{Ax} - \mathbf{g})'(\mathbf{Ax} - \mathbf{g}) \geq (\mathbf{Ax}_0 - \mathbf{g})'(\mathbf{Ax}_0 - \mathbf{g})$$

for all vectors $\mathbf{x}$ in E_n; and for those vectors such that the equality holds, we have $\mathbf{x}'\mathbf{x} > \mathbf{x}_0\mathbf{x}_0$ if $\mathbf{x} \neq \mathbf{x}_0$. Add and subtract $\mathbf{AA}^-\mathbf{g}$ to the quantity $\mathbf{Ax} - \mathbf{g}$

and obtain

$$\begin{aligned}(\mathbf{Ax}-\mathbf{g})'(\mathbf{Ax}-\mathbf{g}) &= (\mathbf{Ax}-\mathbf{AA}^-\mathbf{g}+\mathbf{AA}^-\mathbf{g}-\mathbf{g})'(\mathbf{Ax}-\mathbf{AA}^-\mathbf{g}+\mathbf{AA}^-\mathbf{g}-\mathbf{g})\\ &= [\mathbf{A}(\mathbf{x}-\mathbf{A}^-\mathbf{g})+(\mathbf{AA}^- - \mathbf{I})\mathbf{g}]'\\ &\quad\times[\mathbf{A}(\mathbf{x}-\mathbf{A}^-\mathbf{g})+(\mathbf{AA}^- - \mathbf{I})\mathbf{g}]\\ &= [\mathbf{A}(\mathbf{x}-\mathbf{A}^-\mathbf{g})]'[\mathbf{A}(\mathbf{x}-\mathbf{A}^-\mathbf{g})]+[(\mathbf{AA}^- - \mathbf{I})\mathbf{g}]'\\ &\quad\times[(\mathbf{AA}^- - \mathbf{I})\mathbf{g}] \geq [(\mathbf{AA}^- - \mathbf{I})\mathbf{g}]'[(\mathbf{AA}^- - \mathbf{I})\mathbf{g}],\end{aligned}$$

since the cross-product terms are equal to zero. This inequality holds for all $\mathbf{x}$ in E_n. If we let $\mathbf{x}_0 = \mathbf{A}^-\mathbf{g}$, we obtain

$$(\mathbf{Ax}-\mathbf{g})'(\mathbf{Ax}-\mathbf{g}) \geq [(\mathbf{AA}^- - \mathbf{I})\mathbf{g}]'[(\mathbf{AA}^- - \mathbf{I})\mathbf{g}] = (\mathbf{Ax}_0-\mathbf{g})'(\mathbf{Ax}_0-\mathbf{g}) \tag{7.4.3}$$

for all $\mathbf{x}$ in E_n, and the equality holds if and only if $[\mathbf{A}(\mathbf{x}-\mathbf{A}^-\mathbf{g})]' \times [\mathbf{A}(\mathbf{x}-\mathbf{A}^-\mathbf{g})] = 0$; that is, if and only if $\mathbf{Ax} = \mathbf{AA}^-\mathbf{g}$.

Now we must show that, for the set of $\mathbf{x}$'s such that $\mathbf{Ax} = \mathbf{AA}^-\mathbf{g}$, the relationships

$$\mathbf{x}'\mathbf{x} \geq (\mathbf{A}^-\mathbf{g})'(\mathbf{A}^-\mathbf{g}) = \mathbf{x}_0'\mathbf{x}_0$$

obtain. The following holds for all vectors $\mathbf{x}$ in E_n:

$$\begin{aligned}&[\mathbf{A}^-\mathbf{g}+(\mathbf{I}-\mathbf{A}^-\mathbf{A})\mathbf{x}]'[\mathbf{A}^-\mathbf{g}+(\mathbf{I}-\mathbf{A}^-\mathbf{A})\mathbf{x}]\\ &\quad= (\mathbf{A}^-\mathbf{g})'(\mathbf{A}^-\mathbf{g})+[(\mathbf{I}-\mathbf{A}^-\mathbf{A})\mathbf{x}]'[(\mathbf{I}-\mathbf{A}^-\mathbf{A})\mathbf{x}].\end{aligned} \tag{7.4.4}$$

If we substitute $\mathbf{AA}^-\mathbf{g}$ for $\mathbf{Ax}$, or equivalently, $\mathbf{A}^-\mathbf{g}$ for $\mathbf{A}^-\mathbf{Ax}$ (the equality in Eq. (7.4.3) holds in this case), the identity in Eq. (7.4.4) becomes

$$\mathbf{x}'\mathbf{x} = (\mathbf{A}^-\mathbf{g})'(\mathbf{A}^-\mathbf{g})+(\mathbf{x}-\mathbf{A}^-\mathbf{g})'(\mathbf{x}-\mathbf{A}^-\mathbf{g})$$

or

$$\mathbf{x}'\mathbf{x} > \mathbf{x}_0'\mathbf{x}_0 \quad \text{if} \quad \mathbf{x} \neq \mathbf{x}_0,$$

and the theorem is proved. ∎

Note: This proof demonstrates that a BAS always exists and is unique.

Corollary 7.4.1

For a given $m \times n$ matrix $\mathbf{A}$ *and a given $m \times 1$ vector* $\mathbf{g}$, *the minimum of the quantity* $(\mathbf{Ax} - \mathbf{g})'(\mathbf{Ax} - \mathbf{g})$ *as* $\mathbf{x}$ *varies over* E_n *is* $\mathbf{g}'(\mathbf{I} - \mathbf{AA}^-)\mathbf{g}$.

7.5 Statistical Applications

In the Introduction we discussed the fact that in the theory of the linear model, the model can be written as

$$\mathbf{y} = \mathbf{X}\boldsymbol{\beta} + \mathbf{e},$$

where $\mathbf{y}$ is an $n \times 1$ vector of observations, $\mathbf{X}$ is a $n \times p$ known matrix, $\boldsymbol{\beta}$ is a $p \times 1$ vector of unknown parameters, and $\mathbf{e}$ is a vector of errors which are errors of deviation of observation $\mathbf{y}$ from the expected value; that is, $\mathbf{e} = \mathbf{y} - \mathbf{X}\boldsymbol{\beta}$. The normal equations are

$$\mathbf{X}'\mathbf{X}\hat{\boldsymbol{\beta}} = \mathbf{X}'\mathbf{y}. \tag{7.5.1}$$

To put this set of equations in the notation of this chapter, we identify $\mathbf{X}'\mathbf{X}$ as $\mathbf{A}$; $\hat{\boldsymbol{\beta}}$ as $\mathbf{x}$; and $\mathbf{X}'\mathbf{y}$ as $\mathbf{g}$. We assume that the $p \times p$ symmetric matrix, $\mathbf{X}'\mathbf{X}$ is known, and the $p \times 1$ vector $\mathbf{X}'\mathbf{y}$ is also known. We want to do the following:

(1) Show that the system in Eq. (7.5.1) is consistent.
(2) Find the general solution to Eq. (7.5.1).
(3) Find an $m \times p$ matrix $\mathbf{G}$ such that the vector $\mathbf{G}\hat{\boldsymbol{\beta}}$ is unique for any $\hat{\boldsymbol{\beta}}$ that satisfies Eq. (7.5.1).
(4) Show that if the vectors $\hat{\boldsymbol{\beta}}_1$ and $\hat{\boldsymbol{\beta}}_2$ satisfy Eq. (7.5.1), then $\hat{\boldsymbol{\beta}}_1'\mathbf{X}'\mathbf{y} = \hat{\boldsymbol{\beta}}_2'\mathbf{X}'\mathbf{y}$ (that is, any vector $\hat{\boldsymbol{\beta}}$ that satisfies Eq. (7.5.1) leaves $\hat{\boldsymbol{\beta}}'\mathbf{X}'\mathbf{y}$ invariant).
(5) Show that $(\mathbf{y} - \mathbf{X}\hat{\boldsymbol{\beta}})'(\mathbf{y} - \mathbf{X}\hat{\boldsymbol{\beta}})$ is invariant for any solution $\hat{\boldsymbol{\beta}}$ to the normal equations given in Eq. (7.5.1).
(6) Show that, for any $\hat{\boldsymbol{\beta}}$ that satisfies Eq. (7.5.1), the two quadratic forms

$$\hat{\boldsymbol{\beta}}'\mathbf{X}'\mathbf{y} = \mathbf{y}'\mathbf{C}\mathbf{y}$$

and

$$(\mathbf{y} - \mathbf{X}\hat{\boldsymbol{\beta}})'(\mathbf{y} - \mathbf{X}\hat{\boldsymbol{\beta}}) = \mathbf{y}'\mathbf{B}\mathbf{y}$$

are such that $\mathbf{CB} = \mathbf{0}$.

(7) In (6) show that $\mathbf{C}$ and $\mathbf{B}$ are each idempotent and also that $\mathbf{C} + \mathbf{B} = \mathbf{I}$.

The theorems in this chapter can be used to accomplish (1) through (7), as demonstrated below.

(1) We use Corollary 7.2.3 to show that the system in Eq. (7.5.1) is consistent. We must show that $(\mathbf{X}'\mathbf{X})(\mathbf{X}'\mathbf{X})^{-}\mathbf{X}'\mathbf{y} = \mathbf{X}'\mathbf{y}$. We get

$$(\mathbf{X}'\mathbf{X})(\mathbf{X}'\mathbf{X})^{-}\mathbf{X}'\mathbf{y} = (\mathbf{X}'\mathbf{X})[\mathbf{X}^{-}(\mathbf{X}')^{-}]\mathbf{X}'\mathbf{y}.$$

But by the definition of a g-inverse of $\mathbf{X}$, we note that $(\mathbf{X}')^{-}\mathbf{X}'$ is symmetric, and hence $[(\mathbf{X}')^{-}\mathbf{X}'] = [(\mathbf{X}')^{-}\mathbf{X}']' = \mathbf{X}\mathbf{X}^{-}$. If we substitute this, we get

$$\begin{aligned}(\mathbf{X}'\mathbf{X})(\mathbf{X}'\mathbf{X})^{-}\mathbf{X}'\mathbf{y} &= (\mathbf{X}'\mathbf{X})[\mathbf{X}^{-}(\mathbf{X}')^{-}]\mathbf{X}'\mathbf{y} = \mathbf{X}'\mathbf{X}\mathbf{X}^{-}\mathbf{X}\mathbf{X}^{-}\mathbf{y} \\ &= \mathbf{X}'\mathbf{X}\mathbf{X}^{-}\mathbf{y} = \mathbf{X}'(\mathbf{X}\mathbf{X}^{-})'\mathbf{y} = \mathbf{X}'(\mathbf{X}')^{-}\mathbf{X}'\mathbf{y} = \mathbf{X}'\mathbf{y},\end{aligned}$$

and the system of equations in Eq. (7.5.1) is consistent.

(2) The general solution

$$\hat{\boldsymbol{\beta}} = (\mathbf{X}'\mathbf{X})^{-}\mathbf{X}'\mathbf{y} + [\mathbf{I} - (\mathbf{X}'\mathbf{X})^{-}(\mathbf{X}'\mathbf{X})]\mathbf{h}$$

can be simplified to

$$\hat{\boldsymbol{\beta}} = \mathbf{X}^{-}\mathbf{y} + [\mathbf{I} - \mathbf{X}^{-}\mathbf{X}]\mathbf{h},$$

where $\mathbf{h}$ is *any* $p \times 1$ vector.

(3) If $\mathbf{G}$ is a matrix such that $\mathbf{G}\hat{\boldsymbol{\beta}}_1 = \mathbf{G}\hat{\boldsymbol{\beta}}_2$, where $\hat{\boldsymbol{\beta}}_1$ and $\hat{\boldsymbol{\beta}}_2$ are any solutions to the normal equations, then by (2) we get

$$\mathbf{G}[\mathbf{X}^{-}\mathbf{y} + (\mathbf{I} - \mathbf{X}^{-}\mathbf{X})\mathbf{h}_1] = \mathbf{G}[\mathbf{X}^{-}\mathbf{y} + (\mathbf{I} - \mathbf{X}^{-}\mathbf{X})\mathbf{h}_2],$$

where $\mathbf{h}_1$ and $\mathbf{h}_2$ are any $p \times 1$ vectors. From this we get

$$\mathbf{G}(\mathbf{I} - \mathbf{X}^{-}\mathbf{X})\mathbf{h}_1 = \mathbf{G}(\mathbf{I} - \mathbf{X}^{-}\mathbf{X})\mathbf{h}_2.$$

This equation is satisfied for all $p \times 1$ vectors $\mathbf{h}_1$ and $\mathbf{h}_2$, if and only if

$$\mathbf{G}(\mathbf{I} - \mathbf{X}^{-}\mathbf{X}) = \mathbf{0} \quad \text{or, equivalently,} \quad \mathbf{G} = \mathbf{G}\mathbf{X}^{-}\mathbf{X}.$$

So a necessary and sufficient condition that the vector $\mathbf{G}\hat{\boldsymbol{\beta}}$ is unique, where $\hat{\boldsymbol{\beta}}$ satisfies Eq. (7.5.1), is

$$\mathbf{G}\mathbf{X}^{-}\mathbf{X} = \mathbf{G}.$$

(4) Clearly if $\hat{\boldsymbol{\beta}}_1$ and $\hat{\boldsymbol{\beta}}_2$ satisfy Eq. (7.5.1), then for any vectors $\mathbf{h}_1$ and $\mathbf{h}_2$, we get

$$\hat{\boldsymbol{\beta}}_1' = \mathbf{y}'(\mathbf{X}^-)' + \mathbf{h}_1'[\mathbf{I} - \mathbf{X}'(\mathbf{X}')^-] \quad \text{and} \quad \hat{\boldsymbol{\beta}}_1'\mathbf{X}'\mathbf{y} = \mathbf{y}'(\mathbf{X}^-)'\mathbf{X}'\mathbf{y} = \mathbf{y}'\mathbf{X}\mathbf{X}^-\mathbf{y}.$$

Also

$$\hat{\boldsymbol{\beta}}_2' = \mathbf{y}'(\mathbf{X}^-)' + \mathbf{h}_2'[\mathbf{I} - \mathbf{X}'(\mathbf{X}')^-] \quad \text{and} \quad \hat{\boldsymbol{\beta}}_2\mathbf{X}'\mathbf{y} = \mathbf{y}'\mathbf{X}\mathbf{X}^-\mathbf{y}.$$

So $\hat{\boldsymbol{\beta}}'\mathbf{X}'\mathbf{y}$ is invariant for any solution $\hat{\boldsymbol{\beta}}$ to the normal equations in Eq. (7.5.1).

(5) We can write $(\mathbf{y} - \mathbf{X}\hat{\boldsymbol{\beta}})'(\mathbf{y} - \mathbf{X}\hat{\boldsymbol{\beta}})$ as

$$\mathbf{y}'\mathbf{y} - 2\hat{\boldsymbol{\beta}}'\mathbf{X}'\mathbf{y} + \hat{\boldsymbol{\beta}}'\mathbf{X}'\mathbf{X}\hat{\boldsymbol{\beta}},$$

but by Eq. (7.5.1), $\mathbf{X}'\mathbf{X}\hat{\boldsymbol{\beta}} = \mathbf{X}'\mathbf{y}$, so we get

$$(\mathbf{y} - \mathbf{X}\hat{\boldsymbol{\beta}})'(\mathbf{y} - \mathbf{X}\hat{\boldsymbol{\beta}}) = \mathbf{y}'\mathbf{y} - \hat{\boldsymbol{\beta}}'\mathbf{X}'\mathbf{y}.$$

By (4) it was shown that $\hat{\boldsymbol{\beta}}'\mathbf{X}'\mathbf{y}$ is invariant for any vector $\hat{\boldsymbol{\beta}}$ that satisfies Eq. (7.5.1), and, since $\mathbf{y}'\mathbf{y}$ does not depend on $\hat{\boldsymbol{\beta}}$, it follows that $\mathbf{y}'\mathbf{y} - \hat{\boldsymbol{\beta}}'\mathbf{X}'\mathbf{y}$, and hence $(\mathbf{y}-\mathbf{X}\hat{\boldsymbol{\beta}})'(\mathbf{y} - \mathbf{X}\hat{\boldsymbol{\beta}})$ is invariant for any vector $\hat{\boldsymbol{\beta}}$ that satisfies Eq. (7.5.1).

(6) We can write

$$\hat{\boldsymbol{\beta}}'\mathbf{X}'\mathbf{y} = \mathbf{y}'(\mathbf{X}')^-\mathbf{X}'\mathbf{y} = \mathbf{y}'\mathbf{X}\mathbf{X}^-\mathbf{y} = \mathbf{y}'\mathbf{C}\mathbf{y} \quad \text{and} \quad \mathbf{C} = \mathbf{X}\mathbf{X}^-.$$

Also

$$(\mathbf{y} - \mathbf{X}\hat{\boldsymbol{\beta}})'(\mathbf{y} - \mathbf{X}\hat{\boldsymbol{\beta}}) = \mathbf{y}'\mathbf{y} - \hat{\boldsymbol{\beta}}'\mathbf{X}'\mathbf{y} = \mathbf{y}'\mathbf{y} - \mathbf{y}'\mathbf{X}\mathbf{X}^-\mathbf{y} = \mathbf{y}'(\mathbf{I} - \mathbf{X}\mathbf{X}^-)\mathbf{y} = \mathbf{y}'\mathbf{B}\mathbf{y}$$

and

$$\mathbf{B} = \mathbf{I} - \mathbf{X}\mathbf{X}^-.$$

So, clearly, $\mathbf{CB} = \mathbf{0}$.

(7) By multiplication it is clear that $\mathbf{A}$ and $\mathbf{B}$ are each idempotent.
Note that

$$\text{rank}\,(\mathbf{X}\mathbf{X}^-) = \text{rank}\,\mathbf{X} = \text{tr}\,(\mathbf{X}\mathbf{X}^-)$$

and

$$\text{rank}\,(\mathbf{I} - \mathbf{X}\mathbf{X}^-) = \text{tr}\,(\mathbf{I} - \mathbf{X}\mathbf{X}^-) = \text{tr}\,(\mathbf{I}) - \text{tr}\,(\mathbf{X}\mathbf{X}^-) = n - \text{rank}\,(\mathbf{X}).$$

For further information about similar problems, see [G–7] and [R–3].

7.6 Least Squares

Probably no procedure in applied statistics is used more often than the theory of least squares. It is closely associated with the BAS of Sec. 7.4 in that the starting point is generally a set of equations that is inconsistent, such as Eq. (7.4.1),

$$\mathbf{Ax} - \mathbf{g} = \mathbf{e}(\mathbf{x}), \tag{7.6.1}$$

and the problem is to find a vector $\mathbf{x}_0$ such that $\mathbf{e}'(\mathbf{x})\mathbf{e}(\mathbf{x})$ is a minimum, and any vector that satisfies the requirement is called a least squares solution to the system in Eq. (7.4.1). In this section we define least squares and prove some theorems that may be useful.

Definition 7.6.1

Least Squares Solutions. *The vector* $\mathbf{x}_0$ *is defined to be a least squares solution* (LSS) *of the system* $\mathbf{Ax} - \mathbf{g} = \mathbf{e}(\mathbf{x})$ (*where* $\mathbf{A}$ *is* $m \times n$) *if and only if for all* $\mathbf{x}$ *in* E_n *the following relationship obtains:*

$$(\mathbf{Ax} - \mathbf{g})'(\mathbf{Ax} - \mathbf{g}) \geq (\mathbf{Ax}_0 - \mathbf{g})'(\mathbf{Ax}_0 - \mathbf{g}). \tag{7.6.2}$$

Note: The difference between a BAS and an LSS is the fact that if there is a set of $\mathbf{x}$'s such that the equality holds in Eq. (7.6.2), there is no further restriction (namely, $\mathbf{x}'\mathbf{x} > \mathbf{x}_0'\mathbf{x}_0$) for LSS as there is for BAS. Hence there may be many least squares solutions to a linear system. The BAS is always an LSS, but an LSS may not be the BAS.

Theorem 7.6.1

The vector $\mathbf{x}_0 = \mathbf{Bg}$ *is a least squares solution to the system* $\mathbf{Ax} - \mathbf{g} = \mathbf{e}(\mathbf{x})$, *where* $\mathbf{B}$ *is any matrix such that*

$$\begin{aligned} &(1)\ \mathbf{ABA} = \mathbf{A} \\ &(2)\ \mathbf{AB} \text{ is symmetric.} \end{aligned} \tag{7.6.3}$$

Proof: We must show that $\mathbf{x}_0 = \mathbf{Bg}$ is a minimum of $\mathbf{e}'(\mathbf{x})\mathbf{e}(\mathbf{x})$; that is, we must show that $\mathbf{x}_0 = \mathbf{Bg}$ is a minimum of

$$(\mathbf{Ax} - \mathbf{g})'(\mathbf{Ax} - \mathbf{g}).$$

Now let $\mathbf{B}$ be any matrix such that $\mathbf{ABA} = \mathbf{A}$ and $\mathbf{AB}$ is symmetric. These relationships imply

$$\begin{aligned} \mathbf{A}'\mathbf{B}'\mathbf{A}' &= \mathbf{A}', \\ \mathbf{B}'\mathbf{A}' &= \mathbf{AB}. \end{aligned} \tag{7.6.4}$$

Now

$$\begin{aligned} (\mathbf{Ax} - \mathbf{g})'(\mathbf{Ax} - \mathbf{g}) &= [\mathbf{Ax} - \mathbf{ABg} + \mathbf{ABg} - \mathbf{g}]'[\mathbf{Ax} - \mathbf{ABg} + \mathbf{ABg} - \mathbf{g}] \\ &= [\mathbf{A}(\mathbf{x} - \mathbf{Bg}) + (\mathbf{AB} - \mathbf{I})\mathbf{g}]'[\mathbf{A}(\mathbf{x} - \mathbf{Bg}) + (\mathbf{AB} - \mathbf{I})\mathbf{g}] \\ &= [\mathbf{A}(\mathbf{x} - \mathbf{Bg})]'[\mathbf{A}(\mathbf{x} - \mathbf{Bg})] + [(\mathbf{AB} - \mathbf{I})\mathbf{g}]'[(\mathbf{AB} - \mathbf{I})\mathbf{g}], \end{aligned} \tag{7.6.5}$$

since the cross products vanish; that is,

$$\begin{aligned} [\mathbf{A}(\mathbf{x} - \mathbf{Bg})]'[(\mathbf{AB} - \mathbf{I})\mathbf{g}] &= (\mathbf{x} - \mathbf{Bg})'\mathbf{A}'(\mathbf{AB} - \mathbf{I})\mathbf{g} \\ &= (\mathbf{x} - \mathbf{Bg})'\mathbf{A}'(\mathbf{B}'\mathbf{A}' - \mathbf{I})\mathbf{g} \\ &= (\mathbf{x} - \mathbf{Bg})'[(\mathbf{A}'\mathbf{B}'\mathbf{A}' - \mathbf{A}')\mathbf{g}] \\ &= \mathbf{0}. \end{aligned}$$

Thus from Eq. (7.6.5), for all $\mathbf{x}$ in E_n, the following relationship obtains:

$$(\mathbf{Ax} - \mathbf{g})'(\mathbf{Ax} - \mathbf{g}) \geq [(\mathbf{AB} - \mathbf{I})\mathbf{g}]'[(\mathbf{AB} - \mathbf{I})\mathbf{g}];$$

so $[(\mathbf{AB} - \mathbf{I})\mathbf{g}]'[(\mathbf{AB} - \mathbf{I})\mathbf{g}]$ *is a lower bound for* $(\mathbf{Ax} - \mathbf{g})'(\mathbf{Ax} - \mathbf{g})$.

When $\mathbf{x} = \mathbf{Bg}$, the quantity $(\mathbf{Ax} - \mathbf{g})'(\mathbf{Ax} - \mathbf{g})$ achieves its lower bound, and hence $\mathbf{x}_0 = \mathbf{Bg}$ is a least squares solution to the system $\mathbf{Ax} - \mathbf{g} = \mathbf{e}(\mathbf{x})$, and the theorem is proved.

Note: The vector $\mathbf{x}_0 = \mathbf{A}^-\mathbf{g}$ is a least squares solution, but $\mathbf{x}_0 = \mathbf{A}^c\mathbf{g}$ may *not* be a least squares solution for all conditional inverses $\mathbf{A}^c$. However, if $\mathbf{A}^c$ is such that $\mathbf{AA}^c$ is symmetric, then it is a least squares solution.

Corollary 7.6.1

If $\mathbf{A}$ *is an* $m \times n$ *matrix and* $\mathbf{B}$ *is such that* $\mathbf{ABA} = \mathbf{A}$ *and* $\mathbf{AB}$ *is symmetric, then* $\mathbf{AB} = \mathbf{AA}^-$.

Proof: $\mathbf{AB} = \mathbf{AA}^-\mathbf{AB} = (\mathbf{AA}^-)'(\mathbf{AB})' = \mathbf{A}'^-\mathbf{A}'\mathbf{B}'\mathbf{A}' = \mathbf{A}'^-\mathbf{A}' = \mathbf{AA}^-$. ▌

Example 7.6.1. Consider the system in Example 7.4.1. We have

$$\mathbf{A} = \begin{bmatrix} 1 & 1 \\ 2 & 2 \\ 3 & 3 \end{bmatrix},$$

and we can easily compute a c-inverse,

$$\mathbf{A}^c = \begin{bmatrix} 1 & 0 & 0 \\ -2 & 1 & 0 \end{bmatrix},$$

and if we let $\mathbf{x}_1 = \mathbf{A}^c\mathbf{g}$, we get

$$\mathbf{x}_1 = \begin{bmatrix} 2 \\ -2 \end{bmatrix}.$$

Now $(\mathbf{Ax}_1 - \mathbf{g})'(\mathbf{Ax}_1 - \mathbf{g}) = 18$. If we let $\mathbf{x}_0 = \mathbf{A}^-\mathbf{g}$, then

$$\mathbf{x}_0 = \begin{bmatrix} 15/28 \\ 15/28 \end{bmatrix}$$

and $(\mathbf{Ax}_0 - \mathbf{g})'(\mathbf{Ax}_0 - \mathbf{g}) = 91/98$, so this c-inverse $\mathbf{A}^c$ certainly does not give us a least squares solution.

Note: There may exist an LSS, $\mathbf{x}_0 = \mathbf{Bg}$, where $\mathbf{B}$ does not satisfy Eq. (7.6.3). However there always exists a matrix $\mathbf{B}$ such that any LSS can be written as $\mathbf{x}_0 = \mathbf{Bg}$ where $\mathbf{B}$ does satisfy Eq. (7.6.3). In Example 7.6.1, suppose we let

$$\mathbf{B} = \frac{1}{112}\begin{bmatrix} 15 & 15 & 0 \\ 30 & 0 & 0 \end{bmatrix}.$$

Clearly $\mathbf{x}_0 = \mathbf{Bg} = \begin{bmatrix} 15/28 \\ 15/28 \end{bmatrix}$, and hence $\mathbf{x}_0$ is an LSS, but it is easy to verify that this $\mathbf{B}$ does not satisfy Eq. (7.6.3).

Theorem 7.6.2

The $n \times 1$ vector $\mathbf{x}_0$ is an LSS *to the system $\mathbf{Ax} - \mathbf{g} = \mathbf{e}(\mathbf{x})$ if and only if*

$$(\mathbf{Ax}_0 - \mathbf{g})'(\mathbf{Ax}_0 - \mathbf{g}) = \mathbf{g}'(\mathbf{I} - \mathbf{AA}^-)\mathbf{g}.$$

Proof: By Corollary 7.4.1 a lower bound of $(\mathbf{Ax} - \mathbf{g})'(\mathbf{Ax} - \mathbf{g})$ is equal to $\mathbf{g}'(\mathbf{I} - \mathbf{AA}^-)\mathbf{g}$ and is always attainable (for example, $\mathbf{x}_0 = \mathbf{A}^-\mathbf{g}$). ∎

We now consider various methods for determining whether a given vector $\mathbf{x}_0$ is an LSS to a system $\mathbf{Ax} - \mathbf{g} = \mathbf{e}(\mathbf{x})$.

Theorem 7.6.3

An $n \times 1$ vector $\mathbf{x}_0$ is an LSS *to the system $\mathbf{Ax} - \mathbf{g} = \mathbf{e}(\mathbf{x})$ if and only if $\mathbf{x}_0$ satisfies the matrix equation*

$$\mathbf{Ax} = \mathbf{AA}^-\mathbf{g}. \tag{7.6.6}$$

Proof: Clearly the matrix equation $\mathbf{Ax} = \mathbf{AA}^-\mathbf{g}$ has a solution. If $\mathbf{x}_0$ satisfies $\mathbf{Ax}_0 = \mathbf{AA}^-\mathbf{g}$, we can solve for $\mathbf{x}_0$ and get the general solution

$$\mathbf{x}_0 = \mathbf{A}^-\mathbf{g} + (\mathbf{I} - \mathbf{A}^-\mathbf{A})\mathbf{h}.$$

Now if we substitute this value of $\mathbf{x}_0$ into $(\mathbf{Ax}_0 - \mathbf{g})'(\mathbf{Ax}_0 - \mathbf{g})$, we get

$$(\mathbf{Ax}_0 - \mathbf{g})'(\mathbf{Ax}_0 - \mathbf{g}) = \mathbf{g}'(\mathbf{I} - \mathbf{AA}^-)\mathbf{g},$$

and by Theorem 7.6.2, $\mathbf{x}_0$ is an LSS. Next we assume that $\mathbf{x}_0$ is an LSS and hence

$$(\mathbf{Ax}_0 - \mathbf{g})'(\mathbf{Ax}_0 - \mathbf{g}) = \mathbf{g}'(\mathbf{I} - \mathbf{AA}^-)\mathbf{g}.$$

We define a vector $\mathbf{q}$ by $\mathbf{q} = \mathbf{x}_0 - \mathbf{A}^-\mathbf{g}$ so $\mathbf{x}_0 = \mathbf{A}^-\mathbf{g} + \mathbf{q}$. If we substitute this value of $\mathbf{x}_0$ into the equation $(\mathbf{Ax}_0 - \mathbf{g})'(\mathbf{Ax}_0 - \mathbf{g}) = \mathbf{g}'(\mathbf{I} - \mathbf{AA}^-)\mathbf{g}$, we obtain

$$\mathbf{q}'\mathbf{A}'\mathbf{Aq} = 0,$$

which implies $\mathbf{Aq} = \mathbf{0}$ and hence $\mathbf{Ax}_0 = \mathbf{AA}^-\mathbf{g}$, and the theorem is proved. ∎

Corollary 7.6.3.1

An $n \times 1$ vector $\mathbf{x}_0$ is an LSS *to the system $\mathbf{Ax} - \mathbf{g} = \mathbf{e}(\mathbf{x})$ if and only if $\mathbf{x}_0$ satisfies the matrix equation*

$$\mathbf{A}'\mathbf{Ax} = \mathbf{A}'\mathbf{g}. \tag{7.6.7}$$

Note: The set of equations $\mathbf{A}'\mathbf{A}\mathbf{x} = \mathbf{A}'\mathbf{g}$ is the set of normal equations for the system $\mathbf{A}\mathbf{x} - \mathbf{g} = \mathbf{e}(\mathbf{x})$. See Sec. 7.5.

We shall extend Theorem 7.6.3 and Corollary 7.6.3.1 after first defining a least squares inverse of a matrix $\mathbf{A}$.

Definition 7.6.2

Least Squares Inverse. *Let* $\mathbf{A}$ *be any* $m \times n$ *matrix. A matrix denoted by* $\mathbf{A}^L$ *is defined to be a least squares inverse of* $\mathbf{A}$ *if and only if it satifies*

$$(1)\ \mathbf{A}\mathbf{A}^L\mathbf{A} = \mathbf{A},$$

$$(2)\ \mathbf{A}\mathbf{A}^L = (\mathbf{A}\mathbf{A}^L)'.$$

Note: The matrix $\mathbf{A}^L$ is the matrix $\mathbf{B}$ of Theorem 7.6.1, and hence $\mathbf{x}_0 = \mathbf{A}^L\mathbf{g}$ is a least squares solution to $\mathbf{A}\mathbf{x} - \mathbf{g} = \mathbf{e}(\mathbf{x})$. We shall sometimes refer to a least squares inverse as an L-inverse. We notice that $\mathbf{A}^L$ is a c-inverse of $\mathbf{A}$ and that $\mathbf{A}^-$ is both a c-inverse and an L-inverse of $\mathbf{A}$.

Corollary 7.6.3.2

An $n \times 1$ *vector* $\mathbf{x}_0$ *is an* LSS *to the system* $\mathbf{A}\mathbf{x} - \mathbf{g} = \mathbf{e}(\mathbf{x})$ *if and only if*

$$\mathbf{A}\mathbf{x}_0 = \mathbf{A}\mathbf{A}^L\mathbf{g} \tag{7.6.8}$$

for any L-inverse $\mathbf{A}^L$ of $\mathbf{A}$.

Corollary 7.6.3.3

An $n \times 1$ *vector* $\mathbf{x}_0$ *is an* LSS *to the system* $\mathbf{A}\mathbf{x} - \mathbf{g} = \mathbf{e}(\mathbf{x})$ *if and only if*

$$\mathbf{A}^-\mathbf{A}\mathbf{x}_0 = \mathbf{A}^-\mathbf{g}. \tag{7.6.9}$$

Next we consider the form of all LSS to the system $\mathbf{A}\mathbf{x} - \mathbf{g} = \mathbf{e}(\mathbf{x})$.

Theorem 7.6.4

Consider the system of equations $\mathbf{A}\mathbf{x} - \mathbf{g} = \mathbf{e}(\mathbf{x})$. *Let* $\mathbf{A}^L$ *be any L-inverse of* $\mathbf{A}$. *Then for any* $n \times 1$ *vector* $\mathbf{h}$ *the vector* $\mathbf{x}_0$ *is an* LSS *to the system, where*

$$\mathbf{x}_0 = \mathbf{A}^L\mathbf{g} + (\mathbf{I} - \mathbf{A}^L\mathbf{A})\mathbf{h}. \tag{7.6.10}$$

Also there exists a vector $\mathbf{h}$ *such that every* LSS, $\mathbf{x}_0$, *can be written in the form of Eq.* (7.6.10).

Proof: If we multiply Eq. (7.6.10) by $\mathbf{A}$ and use Corollary 7.6.3.2, we prove that $\mathbf{x}_0$ is an LSS for each and every $n \times 1$ vector $\mathbf{h}$. Next suppose that any vector $\mathbf{x}_0$ is an LSS. We can write $\mathbf{A}\mathbf{x}_0 = \mathbf{A}\mathbf{A}^L\mathbf{g}$. If we add $\mathbf{x}_0 - \mathbf{A}^L\mathbf{g}$ to both sides of this equation and simplify, we obtain

$$\mathbf{x}_0 = \mathbf{A}^L\mathbf{g} + (\mathbf{I} - \mathbf{A}^L\mathbf{A})(\mathbf{x}_0 - \mathbf{A}^L\mathbf{g}),$$

and if we let $\mathbf{h} = \mathbf{x}_0 - \mathbf{A}^L\mathbf{g}$, then $\mathbf{x}_0$ is of the form in Eq. (7.6.10), and the theorem is proved. ∎

A c-inverse of a matrix is relatively easy to compute, and the next theorem relates an L-inverse to a c-inverse.

Theorem 7.6.5

Let $\mathbf{A}$ *be any* $m \times n$ *matrix and let* $(\mathbf{A}'\mathbf{A})^c$ *be any c-inverse of* $\mathbf{A}'\mathbf{A}$. *Then* $\mathbf{B} = (\mathbf{A}'\mathbf{A})^c\mathbf{A}'$ *is an L-inverse of* $\mathbf{A}$.

Proof: Start with $(\mathbf{A}'\mathbf{A})(\mathbf{A}'\mathbf{A})^c\mathbf{A}'\mathbf{A} = \mathbf{A}'\mathbf{A}$ and multiply on the left by $\mathbf{A}'^-$ and get

$$\mathbf{A}(\mathbf{A}'\mathbf{A})^c\mathbf{A}'\mathbf{A} = \mathbf{A} \quad \text{or} \quad \mathbf{ABA} = \mathbf{A}.$$

Next multiply the result $\mathbf{A}(\mathbf{A}'\mathbf{A})^c\mathbf{A}'\mathbf{A} = \mathbf{A}$ on the right by $\mathbf{A}^-$ and obtain

$$\mathbf{A}(\mathbf{A}'\mathbf{A})^c\mathbf{A}' = \mathbf{A}\mathbf{A}^- \quad \text{or} \quad \mathbf{AB} = \mathbf{A}\mathbf{A}^-,$$

and hence $\mathbf{AB}$ is symmetric and the theorem is proved. ∎

Corollary 7.6.5

Let $\mathbf{A}^L$ *be any L-inverse of* $\mathbf{A}$. *Then* $\mathbf{A}\mathbf{A}^L = \mathbf{A}\mathbf{A}^-$, *and hence* $\mathbf{A}\mathbf{A}^L$ *is symmetric idempotent.*

This is a restatement of Corollary 7.6.1.1.

Theorem 7.6.6

The system $\mathbf{Ax} = \mathbf{g}$ *is consistent if and only if* $\mathbf{A}\mathbf{A}^L\mathbf{g} = \mathbf{g}$.

The proof of this theorem is similar to the proof of Theorem 7.2.3.

Theorem 7.6.7

If $\mathbf{A}$ *is an* $m \times n$ *matrix of rank* n, *the* LSS *solution of* $\mathbf{Ax} - \mathbf{g} = \mathbf{e(x)}$ *is unique.*

Proof: If $\mathbf{A}$ is $m \times n$ of rank n, then $\mathbf{A}^L\mathbf{A} = \mathbf{I}$ and by Theorem 7.6.4 the LSS is unique. ∎

Theorem 7.6.8

Consider the system of equations $\mathbf{Ax} - \mathbf{g} = \mathbf{e(x)}$ *where* $\mathbf{A}$ *is an* $m \times n$ *matrix. Then* $\mathbf{c}'\mathbf{x}_0$ *is unique for any* LSS *solution* $\mathbf{x}_0$ *if and only if* $\mathbf{c}$ *is in the column space of* $\mathbf{A}'$, *that is, if and only if there exists an* $n \times 1$ *vector* $\mathbf{k}$ *such that* $\mathbf{c} = \mathbf{A}'\mathbf{k}$.

Proof: Since every LSS solution can be written as $\mathbf{x}_0 = \mathbf{A}^-\mathbf{g} + (\mathbf{I} - \mathbf{A}^-\mathbf{A})\mathbf{h}$ as $\mathbf{h}$ takes on all values in E_n, $\mathbf{c}'\mathbf{x}_0$ is unique if and only if $\mathbf{c}'(\mathbf{I} - \mathbf{A}^-\mathbf{A}) = \mathbf{0}$. But if $\mathbf{c} = \mathbf{A}'\mathbf{k}$, then $\mathbf{c}'(\mathbf{I} - \mathbf{A}^-\mathbf{A}) = \mathbf{k}'\mathbf{A}(\mathbf{I} - \mathbf{A}^-\mathbf{A})$, which is the zero vector. If $\mathbf{c}'(\mathbf{I} - \mathbf{A}^-\mathbf{A}) = \mathbf{0}$, then $\mathbf{c} = \mathbf{A}'(\mathbf{c}'\mathbf{A}^-)' = \mathbf{A}'\mathbf{k}$. ∎

In statistics, as well as other fields, it is often desired to determine a least squares solution to a system of equations when the solution vector is restricted to a certain subset. This is the context of the next few theorems.

Theorem 7.6.9

Let $\mathbf{Ax} - \mathbf{a} = \mathbf{e(x)}$, *where* $\mathbf{A}$ *is an* $n \times p$ *matrix of rank* p, *and let* $\mathbf{Bx} = \mathbf{b}$ *be a consistent set of equations, where* $\mathbf{B}$ *is a* $q \times p$ *matrix of rank* q. *The vector that minimizes* $\mathbf{e}'(\mathbf{x})\mathbf{e}(\mathbf{x}) = (\mathbf{Ax} - \mathbf{a})'(\mathbf{Ax} - \mathbf{a})$ *subject to the condition* $\mathbf{Bx} = \mathbf{b}$ *is the vector* $\mathbf{x}_0$ *in the solution to the following system of equations.*

$$\begin{bmatrix} \mathbf{A}'\mathbf{A} & \mathbf{B}' \\ \mathbf{B} & \mathbf{0} \end{bmatrix} \begin{bmatrix} \mathbf{x}_0 \\ \mathbf{y}_0 \end{bmatrix} = \begin{bmatrix} \mathbf{A}'\mathbf{a} \\ \mathbf{b} \end{bmatrix}. \tag{7.6.11}$$

Proof: We shall use Lagrange multipliers to solve this problem. We want to minimize $(\mathbf{Ax} - \mathbf{a})'(\mathbf{Ax} - \mathbf{a})$ subject to the q constraints $\mathbf{Bx} = \mathbf{b}$. This leads to the function $f(\mathbf{x}, \mathbf{y})$, where

$$f(\mathbf{x}, \mathbf{y}) = (\mathbf{Ax} - \mathbf{a})'(\mathbf{Ax} - \mathbf{a}) + 2\mathbf{y}'(\mathbf{Bx} - \mathbf{b}),$$

where the $q \times 1$ vector $2\mathbf{y}$ is a vector of Lagrange multipliers. The set of $q + p$ partial derivatives of $f(\mathbf{x}, \mathbf{y})$ with respect to the variables in $\mathbf{x}$ and $\mathbf{y}$ gives

$$\left[\frac{\partial f(\mathbf{x}, \mathbf{y})}{\partial x_i}\right] = 2\mathbf{A}'\mathbf{A}\mathbf{x} - 2\mathbf{A}'\mathbf{a} + 2\mathbf{B}'\mathbf{y},$$

$$\left[\frac{\partial f(\mathbf{x}, \mathbf{y})}{\partial y_i}\right] = 2(\mathbf{B}\mathbf{x} - \mathbf{b}).$$

(In Chapter 10 we discuss in detail derivatives of functions with respect to vectors.) If we set these partial derivatives equal to zero and simplify, we get

$$\mathbf{A}'\mathbf{A}\mathbf{x}_0 + \mathbf{B}'\mathbf{y}_0 = \mathbf{A}'\mathbf{a},$$

$$\mathbf{B}\mathbf{x}_0 = \mathbf{b},$$

and these equations can be put in the system given in Eq. (7.6.11). The symbol

$$\left[\frac{\partial f(\mathbf{x}, \mathbf{y})}{\partial x_i}\right]$$

is used for a vector whose i-th element is $\partial f(\mathbf{x}, \mathbf{y})/\partial x_i$ (see Chapter 10 for details). To complete the proof of the theorem, we must show that the system in Eq. (7.6.11) does indeed have a solution. In the following corollary we exhibit the inverse of the coefficient matrix of the system in Eq. (7.6.11); hence this system has a unique solution. ∎

Corollary 7.6.9.1

The coefficient matrix in Eq. (7.6.11) *has an inverse, and it is given by*

$$\begin{bmatrix} \mathbf{F}^{-1} - \mathbf{F}^{-1}\mathbf{B}'\mathbf{K}^{-1}\mathbf{B}\mathbf{F}^{-1} & \mathbf{F}^{-1}\mathbf{B}'\mathbf{K}^{-1} \\ \mathbf{K}^{-1}\mathbf{B}\mathbf{F}^{-1} & -\mathbf{K}^{-1} \end{bmatrix},$$

where $\mathbf{K} = \mathbf{B}\mathbf{F}^{-1}\mathbf{B}'$ *and* $\mathbf{F} = \mathbf{A}'\mathbf{A}$.

Proof: The proof follows by direct multiplication after it is verified that $\mathbf{K}$ is nonsingular and hence has an inverse. To show that $\mathbf{K}$ is nonsingular we can write $\mathbf{C}'\mathbf{C} = \mathbf{F}$ (since $\mathbf{F}$ is assumed to be positive definite of rank p), where $\mathbf{C}$ is $p \times p$ of rank p. Hence $\mathbf{K} = \mathbf{BF}^{-1}\mathbf{B}' = \mathbf{B}(\mathbf{C}'\mathbf{C})^{-1}\mathbf{B}' = (\mathbf{BC}^{-1})(\mathbf{BC}^{-1})'$, and the rank of $\mathbf{K}$ is the rank of $\mathbf{BC}^{-1}$, which is the rank of $\mathbf{B}$, which is q. So $\mathbf{K}$ is $q \times q$ of rank q and $\mathbf{K}$ is nonsingular. ∎

Corollary 7.6.9.2

The solution for $\mathbf{x}_0$ *in Eq.* (7.6.11) *is*

$$\mathbf{x}_0 = \mathbf{A}^-\mathbf{a} - (\mathbf{A}'\mathbf{A})^{-1}\mathbf{B}'[\mathbf{B}(\mathbf{A}'\mathbf{A})^{-1}\mathbf{B}']^{-1}(\mathbf{BA}^-\mathbf{a} - \mathbf{b}).$$

Next we consider a more general matrix than the coefficient matrix in Eq. (7.6.11).

Theorem 7.6.10

Let $\mathbf{G}$ *be a* $(p + q) \times (p + q)$ *matrix given by*

$$\mathbf{G} = \begin{bmatrix} \mathbf{F} & \mathbf{B}' \\ \mathbf{B} & \mathbf{0} \end{bmatrix},$$

where $\mathbf{F} = \mathbf{A}'\mathbf{A}$ *is a* $p \times p$ *matrix of rank* $k \le p$, *and* $\mathbf{B}$ *is* $q \times p$ *of rank* $m \le q \le p$. *A c-inverse of* $\mathbf{G}$ *is*

$$\mathbf{G}^c = \begin{bmatrix} \mathbf{Q} & \mathbf{B}^c - \mathbf{QFB}^c \\ (\mathbf{B}^c)' - (\mathbf{B}^c)'\mathbf{FQ} & -(\mathbf{B}^c)'\mathbf{FB}^c + (\mathbf{B}^c)'\mathbf{FQFB}^c \end{bmatrix},$$

where $\mathbf{Q} = [\mathbf{I} - \mathbf{B}^c\mathbf{B}]\{[\mathbf{I} - \mathbf{B}'(\mathbf{B}^c)']\,\mathbf{F}\,[\mathbf{I} - \mathbf{B}^c\mathbf{B}]\}^c\,[\mathbf{I} - \mathbf{B}'(\mathbf{B}^c)']$, *where* $\mathbf{B}^c$ *is any c-inverse of* $\mathbf{B}$, *and where* $\{[\mathbf{I} - \mathbf{B}'(\mathbf{B}^c)']\,\mathbf{F}\,[\mathbf{I} - \mathbf{B}^c\mathbf{B}]\}^c$ *is any c-inverse of* $[\mathbf{I} - \mathbf{B}'(\mathbf{B}^c)']\,\mathbf{F}\,[\mathbf{I} - \mathbf{B}^c\mathbf{B}]$.

Proof: The proof is obtained by multiplication to demonstrate that $\mathbf{GG}^c\mathbf{G} = \mathbf{G}$. ∎

The next theorems are generalizations of Theorem 7.6.9; the matrices involved are not required to have full column (or row) rank.

Theorem 7.6.11

Let $\mathbf{Ax} - \mathbf{a} = \mathbf{e}(\mathbf{x})$, *where* $\mathbf{A}$ *is an* $n \times p$ *matrix, and let* $\mathbf{Bx} = \mathbf{b}$, *where* $\mathbf{B}$ *is a* $q \times p$ *matrix (the system* $\mathbf{Bx} = \mathbf{b}$ *is a consistent set, but the system*

$\mathbf{Ax} = \mathbf{a}$ *may not be a consistent set). Let Q be the set of vectors that satisfies* $\mathbf{Bx} = \mathbf{b}$. *Let* $\mathbf{x}_0$ *satisfy* $\min_{\mathbf{x} \text{ in } Q} [(\mathbf{Ax} - \mathbf{a})'(\mathbf{Ax} - \mathbf{a})]$. *Then* $\mathbf{x}_0$ *is given by*

$$\mathbf{x}_0 = \mathbf{B}^- \mathbf{b} + [\mathbf{A}(\mathbf{I} - \mathbf{B}^- \mathbf{B})]^- (\mathbf{a} - \mathbf{AB}^- \mathbf{b}) + (\mathbf{I} - \mathbf{B}^- \mathbf{B}) \times \{\mathbf{I} - [\mathbf{A}(\mathbf{I} - \mathbf{B}^- \mathbf{B})]^- [\mathbf{A}(\mathbf{I} - \mathbf{B}^- \mathbf{B})]\}\mathbf{h}$$

for any $p \times 1$ vector $\mathbf{h}$.

Proof: The general solution to the system $\mathbf{Bx} = \mathbf{b}$ is $\mathbf{x} = \mathbf{B}^- \mathbf{b} + (\mathbf{I} - \mathbf{B}^- \mathbf{B})\mathbf{t}$, where $\mathbf{t}$ is any vector in E_p. So we want to determine the vector $\mathbf{t}$ that minimizes

$$\begin{aligned}(\mathbf{Ax} - \mathbf{a})'(\mathbf{Ax} - \mathbf{a}) &= [\mathbf{AB}^- \mathbf{b} + \mathbf{A}(\mathbf{I} - \mathbf{B}^- \mathbf{B})\mathbf{t} - \mathbf{a}]' \times \\ &\quad [\mathbf{AB}^- \mathbf{b} + \mathbf{A}(\mathbf{I} - \mathbf{B}^- \mathbf{B})\mathbf{t} - \mathbf{a}] \\ &= [\mathbf{A}(\mathbf{I} - \mathbf{B}^- \mathbf{B})\mathbf{t} - (\mathbf{a} - \mathbf{AB}^- \mathbf{b})]' \times \\ &\quad [\mathbf{A}(\mathbf{I} - \mathbf{B}^- \mathbf{B})\mathbf{t} - (\mathbf{a} - \mathbf{AB}^- \mathbf{b})].\end{aligned}$$

By Theorem 7.6.3 this is a minimum if and only if $\mathbf{t}$ satisfies

$$\begin{aligned}\mathbf{t} &= [\mathbf{A}(\mathbf{I} - \mathbf{B}^- \mathbf{B})]^- (\mathbf{a} - \mathbf{AB}^- \mathbf{b}) \\ &\quad + \{\mathbf{I} - [\mathbf{A}(\mathbf{I} - \mathbf{B}^- \mathbf{B})]^- [\mathbf{A}(\mathbf{I} - \mathbf{B}^- \mathbf{B})]\}\mathbf{h}\end{aligned}$$

for $\mathbf{h}$ in E_p. If this value of $\mathbf{t}$ is substituted in $\mathbf{x} = \mathbf{B}^- \mathbf{b} + (\mathbf{I} - \mathbf{B}^- \mathbf{B})\mathbf{t}$ above, the theorem is proved if we use the fact that

$$\begin{aligned}(\mathbf{I} - \mathbf{B}^- \mathbf{B})[\mathbf{A}(\mathbf{I} - \mathbf{B}^- \mathbf{B})]^- &= (\mathbf{I} - \mathbf{B}^- \mathbf{B})[\mathbf{A}(\mathbf{I} - \mathbf{B}^- \mathbf{B})]^- \times \\ &\quad [\mathbf{A}(\mathbf{I} - \mathbf{B}^- \mathbf{B})][\mathbf{A}(\mathbf{I} - \mathbf{B}^- \mathbf{B})]^- \\ &= (\mathbf{I} - \mathbf{B}^- \mathbf{B})[\mathbf{A}(\mathbf{I} - \mathbf{B}^- \mathbf{B})]' \times \\ &\quad [\mathbf{A}(\mathbf{I} - \mathbf{B}^- \mathbf{B})]'^- [\mathbf{A}(\mathbf{I} - \mathbf{B}^- \mathbf{B})]^- \\ &= [\mathbf{A}(\mathbf{I} - \mathbf{B}^- \mathbf{B})]^-. \quad \blacksquare\end{aligned}$$

Corollary 7.6.11.1

If $\mathbf{A}$ *in Theorem* 7.6.11 *has rank p, then the vector* $\mathbf{x}_0$ *is given by* $\mathbf{x}_0 = \mathbf{B}^- \mathbf{b} + [\mathbf{A}(\mathbf{I} - \mathbf{B}^- \mathbf{B})]^- (\mathbf{a} - \mathbf{AB}^- \mathbf{b})$.

Proof: If $\mathbf{A}$ has rank p (full column rank), then

$$(\mathbf{I} - \mathbf{B}^- \mathbf{B}) \{\mathbf{I} - [\mathbf{A}(\mathbf{I} - \mathbf{B}^- \mathbf{B})]^- [\mathbf{A}(\mathbf{I} - \mathbf{B}^- \mathbf{B})]\} = \mathbf{0}$$

in Theorem 7.6.11. ∎

Corollary 7.6.11.2

In Theorem 7.6.11, *suppose* $\mathbf{Ax} - \mathbf{a} = \mathbf{e}_1(\mathbf{x})$ *and* $\mathbf{Bx} - \mathbf{b} = \mathbf{e}_2(\mathbf{x})$, *that is, both systems may be inconsistent. Then a vector* $\mathbf{x}_0$ *that minimizes* $(\mathbf{Ax} - \mathbf{a})'(\mathbf{Ax} - \mathbf{a})$ *subject to the condition that it minimizes* $(\mathbf{Bx} - \mathbf{b})'(\mathbf{Bx} - \mathbf{b})$ *is*

$$\mathbf{x}_0 = \mathbf{B}^-\mathbf{b} + [\mathbf{A}(\mathbf{I} - \mathbf{B}^-\mathbf{B})]^-(\mathbf{a} - \mathbf{AB}^-\mathbf{b}).$$

A problem that arises in experimental design models and is somewhat similar to the one discussed above is that of minimizing

$$\mathbf{e}'(\mathbf{x})\mathbf{e}(\mathbf{x}) = (\mathbf{Ax} - \mathbf{g})'(\mathbf{Ax} - \mathbf{g})$$

when $\mathbf{x}$ satisfies $\mathbf{Bx} = \mathbf{0}$ and no row of $\mathbf{B}$ is in the row space of $\mathbf{A}$. This is the context of the next theorem.

Theorem 7.6.12

Consider the (possibly inconsistent) system of equations $\mathbf{Ax} - \mathbf{a} = \mathbf{e}(\mathbf{x})$, *where* $\mathbf{A}$ *is an* $n \times p$ *matrix of rank* q. *Let* $\mathbf{B}$ *be a* $(p - q) \times p$ *matrix (where* $q < p < n$*) such that* $[\mathbf{A}', \mathbf{B}']$ *has rank* p. *A vector* $\mathbf{x}$ *that minimizes* $(\mathbf{Ax} - \mathbf{a})'(\mathbf{Ax} - \mathbf{a})$ *subject to the condition* $\mathbf{Bx} = \mathbf{0}$ *is the vector* $\mathbf{x}_0$ *in the solution to the following system of equations:*

$$\begin{bmatrix} \mathbf{A}'\mathbf{A} & \mathbf{B}' \\ \mathbf{B} & \mathbf{0} \end{bmatrix} \begin{bmatrix} \mathbf{x}_0 \\ \mathbf{y}_0 \end{bmatrix} = \begin{bmatrix} \mathbf{A}'\mathbf{a} \\ \mathbf{0} \end{bmatrix}. \tag{7.6.12}$$

Proof: The proof is similar to the proof for Theorem 7.6.9. ∎

In linear model, regression, and experimental design theory, Theorem 7.6.12 plays a significant role. The next theorem and corollaries contain some useful results about the system given in Eq. (7.6.12).

Theorem 7.6.13

Consider the matrices $\mathbf{A}$ *and* $\mathbf{B}$ *with the size and rank conditions given in Theorem* 7.6.12. *The results below follow.*

(1) $\mathbf{A'A} + \mathbf{B'B}$ *is positive definite.*
(2) $\mathbf{B}(\mathbf{I} - \mathbf{A}^{-}\mathbf{A})$ *is a* $(p - q) \times p$ *matrix of rank* $p - q$ *and hence* $[\mathbf{B}(\mathbf{I} - \mathbf{A}^{-}\mathbf{A})]\,[\mathbf{B}(\mathbf{I} - \mathbf{A}^{-}\mathbf{A})]^{-} = \mathbf{I}$.
(3) $\mathbf{A}(\mathbf{A'A} + \mathbf{B'B})^{-1}\,\mathbf{B'} = \mathbf{0}$.
(4) $\mathbf{B}(\mathbf{A'A} + \mathbf{B'B})^{-1}\,\mathbf{B'} = \mathbf{I}$.
(5) $(\mathbf{A'A} + \mathbf{B'B})^{-1}$ *is a c-inverse of* $\mathbf{A'A}$.

Proof: To prove this theorem, we denote $\mathbf{C}$ by $\mathbf{C'} = [\mathbf{A'}, \mathbf{B'}]$. Since $\mathbf{C}$ has rank p, it follows that $\mathbf{C}^{-}\mathbf{C} = \mathbf{I}$. To prove (1), we note that by the hypothesis of the theorem, $\mathbf{C}$ has column rank p, so it follows by Theorem 1.7.7 that $\mathbf{C'C}$ is positive definite. But $\mathbf{C'C} = \mathbf{A'A} + \mathbf{B'B}$ and (1) is proved. To prove (2), we get $p - q = \text{rank}\,[\mathbf{I} - \mathbf{A}^{-}\mathbf{A}] = \text{rank}\,[\mathbf{C}^{-}\mathbf{C}(\mathbf{I} - \mathbf{A}^{-}\mathbf{A})] \le \text{rank}\,[\mathbf{C}(\mathbf{I} - \mathbf{A}^{-}\mathbf{A})] \le \text{rank}\,[\mathbf{I} - \mathbf{A}^{-}\mathbf{A}] = p - q$; so $\text{rank}\,[\mathbf{C}(\mathbf{I} - \mathbf{A}^{-}\mathbf{A})] = p - q$. But

$$\mathbf{C}(\mathbf{I} - \mathbf{A}^{-}\mathbf{A}) = \begin{bmatrix} \mathbf{A} \\ \mathbf{B} \end{bmatrix} (\mathbf{I} - \mathbf{A}^{-}\mathbf{A}) = \begin{bmatrix} \mathbf{A}(\mathbf{I} - \mathbf{A}^{-}\mathbf{A}) \\ \mathbf{B}(\mathbf{I} - \mathbf{A}^{-}\mathbf{A}) \end{bmatrix} = \begin{bmatrix} \mathbf{0} \\ \mathbf{B}(\mathbf{I} - \mathbf{A}^{-}\mathbf{A}) \end{bmatrix}$$

and clearly the rank of $\mathbf{C}(\mathbf{I} - \mathbf{A}^{-}\mathbf{A})$, which is $p - q$, is the same as the rank of $\mathbf{B}(\mathbf{I} - \mathbf{A}^{-}\mathbf{A})$ and this proves (2). To prove (3), we get $\mathbf{0} = \mathbf{A}(\mathbf{I} - \mathbf{A}^{-}\mathbf{A}) = \mathbf{A}(\mathbf{A'A} + \mathbf{B'B})^{-1}\,(\mathbf{A'A} + \mathbf{B'B})(\mathbf{I} - \mathbf{A}^{-}\mathbf{A}) = \mathbf{A}(\mathbf{A'A} + \mathbf{B'B})^{-1} \times \mathbf{B'B}(\mathbf{I} - \mathbf{A}^{-}\mathbf{A}) = \mathbf{0}$. Multiply both sides on the right by $[\mathbf{B}(\mathbf{I} - \mathbf{A}^{-}\mathbf{A})]^{-}$ to get $\mathbf{A}(\mathbf{A'A} + \mathbf{B'B})^{-1}\,\mathbf{B'} = \mathbf{0}$ and this proves (3). To prove (4), we get $\mathbf{I} = (\mathbf{A'A} + \mathbf{B'B})(\mathbf{A'A} + \mathbf{B'B})^{-1}$. Multiply both sides on the right by $\mathbf{B'}$ to get $\mathbf{B'} = (\mathbf{A'A} + \mathbf{B'B})(\mathbf{A'A} + \mathbf{B'B})^{-1}\,\mathbf{B'}$. Use (3) to get $\mathbf{B'} = \mathbf{B'B}(\mathbf{A'A} + \mathbf{B'B})^{-1}\,\mathbf{B'}$ and multiply both sides on the left by $\mathbf{B}'^{-}$ to obtain the result in (4), since $\mathbf{B}'^{-}\,\mathbf{B'} = \mathbf{I}$. To prove (5), we get $\mathbf{A'A}(\mathbf{A'A} + \mathbf{B'B})^{-1}\,\mathbf{A'A} = \mathbf{A'A}(\mathbf{A'A} + \mathbf{B'B})^{-1}\,[(\mathbf{A'A} + \mathbf{B'B}) - \mathbf{B'B}] = \mathbf{A'A}$. This proves (5) and hence the theorem. ∎

Corollary 7.6.13.1

For the system in Eq. (7.6.12) *the inverse of the coefficient matrix is*

$$\begin{bmatrix} \mathbf{F}^{-1}(\mathbf{I} - \mathbf{B'BF}^{-1}) & \mathbf{F}^{-1}\,\mathbf{B'} \\ \mathbf{BF}^{-1} & \mathbf{0} \end{bmatrix},$$

where $\mathbf{F} = \mathbf{A'A} + \mathbf{B'B}$.

Proof: The proof is obtained by multiplication and by using some results of Theorem 7.6.13. ∎

Corollary 7.6.13.2

The solution of $\mathbf{y}$ *in the system in Eq.* (7.6.12) *is* $\mathbf{y}_0 = \mathbf{0}$, *and hence the unique vector* $\mathbf{x}_0$ *that minimizes* $(\mathbf{Ax} - \mathbf{a})'(\mathbf{Ax} - \mathbf{a})$ *subject to the constraint* $\mathbf{Bx} = \mathbf{0}$ *is the vector* $\mathbf{x}_0$ *given by* $\mathbf{x}_0 = (\mathbf{A}'\mathbf{A} + \mathbf{B}'\mathbf{B})^{-1}\mathbf{A}'\mathbf{a}$.

Note: Additional information on a matrix that closely resembles the one in Eq. (7.6.12) can be found in Chapter 6, Prob. 56–61.

Additional material on minimizing a possibly inconsistent set of equations subject to constraints can be found in [C–1], [R–3].

Another problem of interest in least squares is that of finding a vector $\mathbf{x}_0$ that minimizes $\mathbf{e}'(\mathbf{x})\mathbf{V}^{-1}\mathbf{e}(\mathbf{x})$ where $\mathbf{V}$ is a positive definite matrix and $\mathbf{Ax} - \mathbf{g} = \mathbf{e}(\mathbf{x})$. The problem is discussed in Theorem 7.7.1.

Two other problems of interest when considering systems of linear equations are: when do two systems have at least one common solution, and when are *all* solutions to the two systems the same? These problems are discussed in the next two theorems.

Theorem 7.6.14

Consider two systems of consistent equations

$$\text{(a)}\ \mathbf{A}_1\mathbf{x}_1 = \mathbf{g}_1,$$

$$\text{(b)}\ \mathbf{A}_2\mathbf{x}_2 = \mathbf{g}_2,$$

where $\mathbf{A}_i$ *for* $i = 1, 2$, *are* $m \times n$ *matrices. There is at least one common vector* $\mathbf{x}_0$ *that satisfies both systems of equations if: (1) the rows of* $\mathbf{B}_1$ *are in the row space of* $\mathbf{B}_2$, *or (2) the rows of* $\mathbf{B}_2$ *are in the row space of* $\mathbf{B}_1$, *where*

$$\mathbf{B}_1 = [\mathbf{A}_1, \mathbf{g}_1];\ \mathbf{B}_2 = [\mathbf{A}_2, \mathbf{g}_2].$$

Proof: Assume that the rows of $\mathbf{B}_1$ are in the row space of $\mathbf{B}_2$. By Theorem 5.4.6 there exists an $m \times m$ matrix $\mathbf{C}$ such that $\mathbf{B}_1 = \mathbf{CB}_2$, which implies $\mathbf{A}_1 = \mathbf{CA}_2$ and $\mathbf{g}_1 = \mathbf{Cg}_2$. Let $\mathbf{x}_0$ be a solution to the set of equations (b); thus $\mathbf{A}_2\mathbf{x}_0 = \mathbf{g}_2$. Multiply on the left by $\mathbf{C}$ and get $\mathbf{CA}_2\mathbf{x}_0 = \mathbf{Cg}_2$, which is $\mathbf{A}_1\mathbf{x}_0 = \mathbf{g}_1$, so $\mathbf{x}_0$ satisfies the systems of equations in (a) and in (b). A similar proof can be used if the rows $\mathbf{B}_2$ are in the row space of $\mathbf{B}_1$. ▮

Theorem 7.6.15

In Theorem 7.6.14 *the set of solutions for (a) and (b) are the same if there is a nonsingular matrix* $\mathbf{C}$ *such that* $\mathbf{CB}_1 = \mathbf{B}_2$.

Proof: If $\mathbf{CB}_1 = \mathbf{B}_2$, then multiply $\mathbf{A}_1\mathbf{x}_1 = \mathbf{g}_1$ by $\mathbf{C}$ to get $\mathbf{A}_2\mathbf{x}_1 = \mathbf{g}_2$, so any solution to (a) also satisfies (b). Use $\mathbf{A}_2\mathbf{x}_2 = \mathbf{g}_2$ and multiply by $\mathbf{C}^{-1}$ to get $\mathbf{A}_1\mathbf{x}_2 = \mathbf{g}_1$, so any solution to (b) also satisfies (a). Hence the set of solution vectors that satisfies (a) and (b) is the same. ∎

Additional material on solutions to sets of equations can be found in [C–1], [R–3].

7.7 Statistical Applications

Consider the linear model $\mathbf{y} = \mathbf{X\beta} + \mathbf{e}$ described in the Introduction and let us suppose that the objective is to estimate the vector $\boldsymbol{\beta}$. If the vector $\mathbf{e}$ is from a distribution with a mean of zero and covariance matrix $\sigma^2\mathbf{I}$, then least squares can be used to estimate $\boldsymbol{\beta}$. If we write this model as $\mathbf{X\beta} - \mathbf{y} = \mathbf{e}$, this is equivalent to Eq. (7.6.1), where we identify $\mathbf{y}$ with $\mathbf{g}$, $\mathbf{X}$ with $\mathbf{A}$, $\boldsymbol{\beta}$ with $\mathbf{x}$ and $\mathbf{e}$ with $\mathbf{e}(\mathbf{x})$. The value of $\boldsymbol{\beta}$ that is a least squares solution of $\mathbf{y} - \mathbf{X\beta} = \mathbf{e}$ is called a least squares estimator of $\boldsymbol{\beta}$. By Corollary 7.6.3.1, the least squares solution is any value $\hat{\boldsymbol{\beta}}$ that satisfies $\mathbf{X}'\mathbf{X}\hat{\boldsymbol{\beta}} = \mathbf{X}'\mathbf{y}$, and this is the set of normal equations given in Eq. (7.5.1). (See also Sec. 10.8.)

If the covariance matrix of $\mathbf{e}$ is not $\sigma^2\mathbf{I}$ but rather is a positive definite matrix $\mathbf{V}$, then the estimator $\hat{\boldsymbol{\beta}}$ of $\boldsymbol{\beta}$ is obtained by a method called *weighted least squares*. The weighted least squares, which is the value of $\boldsymbol{\beta}$ that minimizes $\mathbf{e}'\mathbf{V}^{-1}\mathbf{e}$, is the subject of the next theorem.

Theorem 7.7.1

In the model $\mathbf{y} - \mathbf{X\beta} = \mathbf{e}$, *a value of* $\boldsymbol{\beta}$ (*denoted by* $\hat{\boldsymbol{\beta}}$) *that minimizes* $\mathbf{e}'\mathbf{V}^{-1}\mathbf{e}$ *is*

$$\hat{\boldsymbol{\beta}} = (\mathbf{X}'\mathbf{V}^{-1}\mathbf{X})^c\mathbf{X}'\mathbf{V}^{-1}\mathbf{y}.$$

Proof: Since $\mathbf{V}$, and hence $\mathbf{V}^{-1}$, is positive definite, we can write $\mathbf{V}^{-1}$ as $\mathbf{B}'\mathbf{B}$. If we multiply $\mathbf{y} - \mathbf{X\beta} = \mathbf{e}$ by $\mathbf{B}$, we get $\mathbf{By} - \mathbf{BX\beta} = \mathbf{Be}$; observing that $(\mathbf{Be})'(\mathbf{Be}) = \mathbf{e}'\mathbf{V}^{-1}\mathbf{e}$, we must find the value of $\boldsymbol{\beta}$ that minimizes $(\mathbf{Be})'(\mathbf{Be})$. By Corollary 7.6.3.1, we get

$$\mathbf{X}'\mathbf{B}'\mathbf{BX}\hat{\boldsymbol{\beta}} = \mathbf{X}'\mathbf{B}'\mathbf{By} \quad \text{or} \quad \mathbf{X}'\mathbf{V}^{-1}\mathbf{X}\hat{\boldsymbol{\beta}} = \mathbf{X}'\mathbf{V}^{-1}\mathbf{y},$$

and a solution is

$$\hat{\boldsymbol{\beta}} = (\mathbf{X}'\mathbf{V}^{-1}\mathbf{X}')^c\mathbf{X}'\mathbf{V}^{-1}\mathbf{y}. \quad ∎$$

Problems

1. Show that the system of equations $\mathbf{Ax} = \mathbf{g}$ below is consistent by using Theorem 7.2.2.

$$\begin{aligned} x_1 - 2x_2 + 3x_3 - 2x_4 &= 2 \\ x_1 \qquad + x_3 - 3x_4 &= -4 \\ x_1 + 2x_2 - 3x_3 \qquad &= -4 \end{aligned}$$

2. In Prob. 1 find $\mathbf{A}^c$, a conditional inverse of $\mathbf{A}$.
3. In Prob. 1 show that the system is consistent by using Theorem 7.2.3.
4. In Prob. 1 find the number of linearly independent solution vectors.
5. In Prob. 1 find a linearly independent set of solution vectors.
6. Find any two distinct solutions $\mathbf{x}_1$ and $\mathbf{x}_2$ to the system in Prob. 1 and demonstrate that $\frac{1}{3}\mathbf{x}_1 + \frac{2}{3}\mathbf{x}_2$ is also a solution.
7. Consider the system $\mathbf{Ax} = \mathbf{g}$ and any c-inverse $\mathbf{A}^c$ of the $m \times n$ matrix $\mathbf{A}$. Let $\mathbf{h}_1, \mathbf{h}_2, \ldots, \mathbf{h}_t$ be any set of vectors from E_n, and let $c_1, c_2, \ldots, c_t$ be any set of scalars. Show that if the system is consistent the vector $\mathbf{y}$ is a solution to the system where

$$\mathbf{y} = \mathbf{A}^c\mathbf{g} + \sum_{i=1}^{t} c_i(\mathbf{I} - \mathbf{A}^c\mathbf{A})\mathbf{h}_i.$$

8. If $\mathbf{x}_1, \mathbf{x}_2, \ldots, \mathbf{x}_t$ are solutions to the system

$$\mathbf{Ax} = \mathbf{g},$$

show that

$$\mathbf{y} = \sum_{i=1}^{t} c_i \mathbf{x}_i$$

is also a solution, where the c_i are any scalars such that

$$\sum_{i=1}^{t} c_i = 1.$$

9. Prove that for any matrix $\mathbf{A}$ if there is a c-inverse $\mathbf{A}^c$, such that $\mathbf{AA}^c = \mathbf{I}$, then $\mathbf{AA}^- = \mathbf{I}$.

10. Prove that for any matrix $\mathbf{A}$ if $\mathbf{AA}^- = \mathbf{I}$, then $\mathbf{AA}^c = \mathbf{I}$ for each c-inverse of $\mathbf{A}$.
11. Let $\mathbf{x}_0$ be a solution to $\mathbf{Ax} = \mathbf{g}$ where $\mathbf{x}_0 = \mathbf{A}^-\mathbf{g}$. Show that $\mathbf{y}$ is a solution where $\mathbf{y} = \mathbf{x}_0 + \mathbf{z}$ for all vectors $\mathbf{z}$ that belong to the orthogonal complement of the column space of $\mathbf{A}'$.
12. Let $\mathbf{x}_1$ and $\mathbf{x}_2$ be any solutions to $\mathbf{Ax} = \mathbf{g}$. Show that the vector $\mathbf{x}_1 - \mathbf{x}_2$ is orthogonal to the rows of $\mathbf{A}$.
13. Show that the system below is inconsistent.

$$\begin{aligned} x_1 + x_2 + x_3 &= 3, \\ x_1 - x_2 + 2x_3 &= -3, \\ 3x_1 - x_2 + 5x_3 &= -2, \\ 2x_1 - x_2 - x_3 &= 4. \end{aligned}$$

14. In Prob. 13 find the BAS.
15. In Prob. 13 find an LSS.
16. Show that the system below is inconsistent.

$$\begin{aligned} x_1 + x_2 + x_4 &= 1, \\ x_1 + x_2 + x_5 &= 2, \\ x_1 + x_3 + x_4 &= 1, \\ x_1 + x_3 + x_5 &= 3. \end{aligned}$$

17. In Prob. 16 find a least squares solution by finding an L-inverse of the matrix.
18. Show that the system of equations below is consistent and that a unique solution does not exist.

$$\begin{aligned} 4x_1 + 2x_2 + 2x_3 &= 3, \\ 2x_1 + 2x_2 &= 0, \\ 2x_1 + 2x_3 &= 3. \end{aligned}$$

19. In Prob. 18 find a 2×3 matrix $\mathbf{G}$ such that $\mathbf{Gx}$ is unique for any solution vector $\mathbf{x}$.
20. For the system $\mathbf{X}'\mathbf{X}\hat{\boldsymbol{\beta}} = \mathbf{X}'\mathbf{y}$ in Sec. 7.5, prove that $\mathbf{G}\hat{\boldsymbol{\beta}}$ is unique for any solution $\hat{\boldsymbol{\beta}}$ if and only if the column space of $\mathbf{G}'$ is a subspace of the column space of $\mathbf{X}'$.

21. Find an L-inverse of the matrix $\mathbf{A}$ where

$$\mathbf{A} = \begin{bmatrix} 1 & 2 & 7 \\ 1 & 1 & 5 \\ -1 & 2 & 1 \\ 2 & 1 & 8 \end{bmatrix}.$$

22. Prove that for any matrix $\mathbf{A}$ if there is an L-inverse $\mathbf{A}^L$ such that $\mathbf{AA}^L = \mathbf{I}$ then $\mathbf{AA}^- = \mathbf{I}$.
23. Prove that if $\mathbf{AA}^- = \mathbf{I}$ then $\mathbf{AA}^L = \mathbf{I}$ for all L-inverses of $\mathbf{A}$.
24. Prove that if $\mathbf{A}$ is nonsingular, then $\mathbf{A}^c = \mathbf{A}^L = \mathbf{A}^- = \mathbf{A}^{-1}$.
25. Let $\mathbf{A}$ be any $m \times n$ matrix and define $\mathbf{B}$ by $\mathbf{B} = (\mathbf{A}'\mathbf{A})^c\mathbf{A}'$ where $(\mathbf{A}'\mathbf{A})^c$ is any c-inverse of $\mathbf{A}'\mathbf{A}$. Show that $\mathbf{ABA} = \mathbf{A}$; $\mathbf{BAB} = \mathbf{B}$; $\mathbf{AB}$ is symmetric.
26. In Prob. 25 show that $\mathbf{B}$ may not always be a g-inverse of $\mathbf{A}$, since $\mathbf{BA}$ may not always be symmetric.
27. If $\mathbf{A}$ is a symmetric matrix and $\mathbf{A}^c$ is any c-inverse of $\mathbf{A}$, show that $\mathbf{A}^c$ is not necessarily symmetric.
28. If $\mathbf{A}$ is symmetric show that for any c-inverse $\mathbf{A}^c$ of $\mathbf{A}$ the matrix $(\mathbf{A}^c)'$ is also a c-inverse of $\mathbf{A}$.
29. In Prob. 28 show that $\mathbf{B}$ is a symmetric c-inverse of $\mathbf{A}$ where

$$\mathbf{B} = \frac{1}{2}[\mathbf{A}^c + (\mathbf{A}^c)']$$

and where $\mathbf{A}^c$ is any c-inverse of $\mathbf{A}$.
30. Prove Theorems 6.6.1, 6.6.2, 6.6.3, 6.6.7, and 6.6.8 if the c-inverses are replaced by L-inverses.
31. Prove that $\mathbf{B}$ is symmetric idempotent where $\mathbf{B} = \mathbf{A}(\mathbf{A}'\mathbf{A})^c\mathbf{A}'$ and where $(\mathbf{A}'\mathbf{A})^c$ is any c-inverse of $\mathbf{A}'\mathbf{A}$.
32. In Prob. 31 show that $\mathbf{B} = \mathbf{AA}^-$.
33. If $\mathbf{A}$ is a symmetric matrix show that $\mathbf{A}^- = \mathbf{A}(\mathbf{A}^L)^2$ for any L-inverse of $\mathbf{A}$.
34. Let $\mathbf{A}$ be any matrix and let $(\mathbf{A}'\mathbf{A})^L$ be any L-inverse of $\mathbf{A}'\mathbf{A}$. Show that $\mathbf{A}'^- = \mathbf{A}(\mathbf{A}'\mathbf{A})^L$.
35. Prove Theorem 7.6.6.
36. Consider the linear model $\mathbf{y} = \mathbf{X}\boldsymbol{\beta} + \mathbf{e}$ and the normal equations

$$\mathbf{X}'\mathbf{X}\hat{\boldsymbol{\beta}} = \mathbf{X}'\mathbf{y}.$$

(1) For any solution $\hat{\boldsymbol{\beta}}$ show that $\mathbf{X}\hat{\boldsymbol{\beta}}$ is unique.
(2) In (1) show that $\hat{\boldsymbol{\beta}}'\mathbf{X}'$ is the projection of $\mathbf{y}'$ into the column space of $\mathbf{X}$.

37. If $\mathbf{A}$ is an $m \times n$ matrix and if the rank of $\mathbf{A}$ is n, show that $\mathbf{A}^L$ is unique.
38. Let $\mathbf{A}$ be a symmetric $n \times n$ matrix of rank $n - 1$ such that $\mathbf{1}'\mathbf{A} = \mathbf{0}$; that is, every column of $\mathbf{A}$ adds to zero. Show that $\mathbf{B} = \mathbf{A} + (1/n)\mathbf{1}\mathbf{1}'$ is nonsingular and the inverse is $\mathbf{A}^- + (1/n)\mathbf{J}$.
39. Show that $\mathbf{A}^k + \mathbf{J}$ is nonsingular where $\mathbf{A}$ is defined in Prob. 38 and k is any positive integer.
40. Show that $\mathbf{C}$ is nonsingular where $\mathbf{C}$ is defined by

$$\mathbf{C} = \begin{bmatrix} \mathbf{A} & \mathbf{1} \\ \mathbf{1}' & 0 \end{bmatrix}$$

and $\mathbf{A}$ is defined in Prob. 38.
41. In Prob. 40, let $\mathbf{B} = \mathbf{C}^{-1}$ and partition $\mathbf{B}$ as

$$\mathbf{B} = \begin{bmatrix} \mathbf{B}_{11} & \mathbf{b}_{12} \\ \mathbf{b}_{21} & b \end{bmatrix},$$

where $\mathbf{B}_{11}$ is an $n \times n$ submatrix. Show that
(1) $b = 0$,
(2) $\mathbf{b}_{12} = (1/n)\mathbf{1}$,
(3) $\mathbf{1}'\mathbf{B}_{11} = \mathbf{0}$,
(4) $\mathbf{A}\mathbf{B}_{11}$ and $\mathbf{B}_{11}\mathbf{A}$ are idempotent.

Patterned Matrices and Other Special Matrices

8

8.1 Introduction

In general, the task of finding the inverse of a matrix, the determinant of a matrix, or the characteristic roots of a matrix is very laborious and time consuming. However, by recognizing a particular structure, or pattern, in certain matrices, one may be able to significantly reduce the work required to invert a matrix or to find its determinant or characteristic roots. We shall call such matrices *patterned matrices.*

For a trivial case consider the $k \times k$ diagonal matrix $\mathbf{D}$ where $d_{ii} \neq 0$. We do not use the same procedure to find the inverse of this matrix that we would use to find the inverse of a general $k \times k$ matrix, since we know that $\mathbf{D}^{-1} = [c_{ij}]$, where $c_{ii} = d_{ii}^{-1}$, $c_{ij} = 0$ if $i \neq j$. In other words, we recognize $\mathbf{D}$ as having a special structure (diagonal), and we can write the inverse of this special matrix immediately.

Below are some other simple examples of patterned matrices.

1. If $\mathbf{A}$ is a symmetric matrix, then it is known that
   (a) if the inverse exists, the inverse is symmetric;
   (b) the characteristic roots are real.

   Thus, noticing a certain pattern of the matrix $\mathbf{A}$ (say symmetry) may be of value in finding its inverse or characteristic roots. Also if $\mathbf{A}$ is a positive definite matrix, it is known that the characteristic roots are real and positive and the determinant is positive.

2. If a matrix **B** is orthogonal, then the inverse is, of course, extremely easy to evaluate, and a great deal is known about the characteristic roots; also it is known that det (**B**) is either plus or minus one.
3. If a square matrix **C** is partitioned into blocks so that the diagonal blocks are square and the off-diagonal blocks are each equal to the null matrix, then these facts are very valuable if one wants to find the inverse, determinant, or characteristic roots of **C**.
4. The identity matrix, the null matrix, and a triangular matrix are also recognized as matrices with a special pattern.

The above matrices are certainly patterned matrices, and previous chapters contain a number of theorems that can be used to a great advantage when the pattern is recognized; but we are also interested in other types of patterned matrices. For example, the matrix

$$\begin{bmatrix} a & a & b & b \\ a & a & b & b \\ b & b & a & a \\ b & b & a & a \end{bmatrix}$$

has a recognizable pattern that one might take advantage of in finding the inverse, the determinant, or the characteristic roots. In many applications, an explicit formula (as a function of the elements of the matrix) for the determinant, the characteristic roots, or the elements of the inverse matrix is quite useful, especially if the matrix has a pattern such that the formula is quite simple.

Before discussing certain general theorems regarding patterned matrices in the light of special cases that occur in statistics, we shall state and prove some theorems on *partitioned* matrices.

8.2 Partitioned Matrices

It is sometimes convenient to find determinants and inverses of matrices in terms of submatrices. Our first theorem on this subject has many applications.

Theorem 8.2.1

Let **B** *be an* $n \times n$ *matrix that is partitioned as follows:*

$$\mathbf{B} = \begin{bmatrix} \mathbf{B}_{11} & \mathbf{B}_{12} \\ \mathbf{B}_{21} & \mathbf{B}_{22} \end{bmatrix}, \tag{8.2.1}$$

where $\mathbf{B}_{ij}$ *has size* $n_i \times n_j$, $i, j = 1, 2$, *and where* $n_1 + n_2 = n$.

(1) *Suppose* $|\mathbf{B}| \neq 0$, $|\mathbf{B}_{11}| \neq 0$, $|\mathbf{B}_{22}| \neq 0$ *and we let* $\mathbf{A} = \mathbf{B}^{-1}$ *and partition* $\mathbf{A}$ *as*

$$\mathbf{A} = \begin{bmatrix} \mathbf{A}_{11} & \mathbf{A}_{12} \\ \mathbf{A}_{21} & \mathbf{A}_{22} \end{bmatrix},$$

where $\mathbf{A}_{ij}$ *has size* $n_i \times n_j$ *for* $i, j = 1, 2$. *The results* 1(a) *through* 1(h) *follow:*

1(a) $\mathbf{A}_{11}^{-1}, \mathbf{A}_{22}^{-1}$ *exist.*

1(b) $[\mathbf{B}_{11} - \mathbf{B}_{12}\mathbf{B}_{22}^{-1}\mathbf{B}_{21}]^{-1}$ *and* $[\mathbf{B}_{22} - \mathbf{B}_{21}\mathbf{B}_{11}^{-1}\mathbf{B}_{12}]^{-1}$ *exist.*

1(c) $\mathbf{B}^{-1}$ *can be written as*

$$\mathbf{B}^{-1} = \begin{bmatrix} [\mathbf{B}_{11} - \mathbf{B}_{12}\mathbf{B}_{22}^{-1}\mathbf{B}_{21}]^{-1} & -\mathbf{B}_{11}^{-1}\mathbf{B}_{12}[\mathbf{B}_{22} - \mathbf{B}_{21}\mathbf{B}_{11}^{-1}\mathbf{B}_{12}]^{-1} \\ -\mathbf{B}_{22}^{-1}\mathbf{B}_{21}[\mathbf{B}_{11} - \mathbf{B}_{12}\mathbf{B}_{22}^{-1}\mathbf{B}_{21}]^{-1} & [\mathbf{B}_{22} - \mathbf{B}_{21}\mathbf{B}_{11}^{-1}\mathbf{B}_{12}]^{-1} \end{bmatrix}. \tag{8.2.2}$$

1(d) $\mathbf{A}_{11} = [\mathbf{B}_{11} - \mathbf{B}_{12}\mathbf{B}_{22}^{-1}\mathbf{B}_{21}]^{-1} = \mathbf{B}_{11}^{-1} + \mathbf{B}_{11}^{-1}\mathbf{B}_{12}\mathbf{A}_{22}\mathbf{B}_{21}\mathbf{B}_{11}^{-1}$.

1(e) $\mathbf{A}_{12} = -\mathbf{B}_{11}^{-1}\mathbf{B}_{12}[\mathbf{B}_{22} - \mathbf{B}_{21}\mathbf{B}_{11}^{-1}\mathbf{B}_{12}]^{-1} = -\mathbf{B}_{11}^{-1}\mathbf{B}_{12}\mathbf{A}_{22}$.

1(f) $\mathbf{A}_{22} = [\mathbf{B}_{22} - \mathbf{B}_{21}\mathbf{B}_{11}^{-1}\mathbf{B}_{12}]^{-1} = \mathbf{B}_{22}^{-1} + \mathbf{B}_{22}^{-1}\mathbf{B}_{21}\mathbf{A}_{11}\mathbf{B}_{12}\mathbf{B}_{22}^{-1}$.

1(g) $\mathbf{A}_{21} = -\mathbf{B}_{22}^{-1}\mathbf{B}_{21}[\mathbf{B}_{11} - \mathbf{B}_{12}\mathbf{B}_{22}^{-1}\mathbf{B}_{21}]^{-1} = -\mathbf{B}_{22}^{-1}\mathbf{B}_{21}\mathbf{A}_{11}$.

1(h) $|\mathbf{B}| = \dfrac{|\mathbf{B}_{11}|}{|\mathbf{A}_{22}|} = \dfrac{|\mathbf{B}_{22}|}{|\mathbf{A}_{11}|}$ *and* $|\mathbf{B}_{11}\mathbf{A}_{11}| = |\mathbf{B}_{22}\mathbf{A}_{22}|$.

(2) *If* $\mathbf{B}_{22}$ *is a nonsingular matrix, then the determinant of* $\mathbf{B}$ *can be written as*

$$|\mathbf{B}| = |\mathbf{B}_{22}| \cdot |\mathbf{B}_{11} - \mathbf{B}_{12}\mathbf{B}_{22}^{-1}\mathbf{B}_{21}|.$$

(3) *If* $\mathbf{B}_{11}$ *is a nonsingular matrix, the determimant of* $\mathbf{B}$ *can be written as*

$$|\mathbf{B}| = |\mathbf{B}_{11}| \cdot |\mathbf{B}_{22} - \mathbf{B}_{21}\mathbf{B}_{11}^{-1}\mathbf{B}_{12}|.$$

Proof: Statement 1(c) is proved by showing that $\mathbf{BB}^{-1} = \mathbf{I}$ after it is demonstrated that the matrices $[\mathbf{B}_{11} - \mathbf{B}_{12}\mathbf{B}_{22}^{-1}\mathbf{B}_{21}]$ and $[\mathbf{B}_{22} - \mathbf{B}_{21}\mathbf{B}_{11}^{-1}\mathbf{B}_{12}]$ are nonsingular. To prove (2) and parts of 1(a) and 1(b), multiply $|\mathbf{B}_{22}|$ by $|\mathbf{BC}|$,

where

$$\mathbf{C} = \begin{bmatrix} \mathbf{I} & \mathbf{0} \\ -\mathbf{B}_{22}^{-1}\mathbf{B}_{21} & \mathbf{B}_{22}^{-1} \end{bmatrix}.$$

It is clear that $|\mathbf{C}| = |\mathbf{I}| \cdot |\mathbf{B}_{22}^{-1}| = |\mathbf{B}_{22}^{-1}|$ and $|\mathbf{B}| = |\mathbf{B}_{22}||\mathbf{B}||\mathbf{B}_{22}^{-1}|$, or

$$\begin{aligned} |\mathbf{B}| &= |\mathbf{B}_{22}| \begin{vmatrix} \mathbf{B}_{11} & \mathbf{B}_{12} \\ \mathbf{B}_{21} & \mathbf{B}_{22} \end{vmatrix} \begin{vmatrix} \mathbf{I} & \mathbf{0} \\ -\mathbf{B}_{22}^{-1}\mathbf{B}_{21} & \mathbf{B}_{22}^{-1} \end{vmatrix} \\ &= |\mathbf{B}_{22}| \left| \begin{bmatrix} \mathbf{B}_{11} & \mathbf{B}_{12} \\ \mathbf{B}_{21} & \mathbf{B}_{22} \end{bmatrix} \begin{bmatrix} \mathbf{I} & \mathbf{0} \\ -\mathbf{B}_{22}^{-1}\mathbf{B}_{21} & \mathbf{B}_{22}^{-1} \end{bmatrix} \right| \\ &= |\mathbf{B}_{22}| \begin{vmatrix} \mathbf{B}_{11} - \mathbf{B}_{12}\mathbf{B}_{22}^{-1}\mathbf{B}_{21} & \mathbf{B}_{12}\mathbf{B}_{22}^{-1} \\ \mathbf{0} & \mathbf{I} \end{vmatrix} \\ &= |\mathbf{B}_{22}|\,|\mathbf{B}_{11} - \mathbf{B}_{12}\mathbf{B}_{22}^{-1}\mathbf{B}_{21}|. \end{aligned}$$

To prove (3) and the remaining parts of 1(a) and 1(b), we assume $\mathbf{B}_{11}$ is nonsingular and show that the determinant of $\mathbf{B}$ can be written as

$$|\mathbf{B}| = |\mathbf{B}_{11}| \cdot |\mathbf{B}_{22} - \mathbf{B}_{21}\mathbf{B}_{11}^{-1}\mathbf{B}_{12}|.$$

Statements 1(d), 1(e), 1(f), and 1(g) can be proved from 1(c). To prove statement 1(h), we use (2) and 1(d) to get

$$|\mathbf{B}| = |\mathbf{B}_{22}| \cdot |\mathbf{B}_{11} - \mathbf{B}_{12}\mathbf{B}_{22}^{-1}\mathbf{B}_{21}| = |\mathbf{B}_{22}| \cdot |\mathbf{A}_{11}^{-1}| = \frac{|\mathbf{B}_{22}|}{|\mathbf{A}_{11}|},$$

and we use (3) and 1(f) to obtain

$$|\mathbf{B}| = |\mathbf{B}_{11}| \cdot |\mathbf{B}_{22} - \mathbf{B}_{21}\mathbf{B}_{11}^{-1}\mathbf{B}_{12}| = |\mathbf{B}_{11}| \cdot |\mathbf{A}_{22}^{-1}| = \frac{|\mathbf{B}_{11}|}{|\mathbf{A}_{22}|}.$$

The details of the proofs are left for the reader. ∎

Example 8.2.1. Consider the matrix $\mathbf{B}$ where

$$\mathbf{B} = \begin{bmatrix} a\mathbf{I} & b\mathbf{I} \\ c\mathbf{I} & d\mathbf{I} \end{bmatrix} = \left[\begin{array}{cccc|cccc} a & 0 & \cdots & 0 & b & 0 & \cdots & 0 \\ 0 & a & \cdots & 0 & 0 & b & \cdots & 0 \\ \vdots & \vdots & & \vdots & \vdots & \vdots & & \vdots \\ 0 & 0 & \cdots & a & 0 & 0 & \cdots & b \\ \hline c & 0 & \cdots & 0 & d & 0 & \cdots & 0 \\ 0 & c & \cdots & 0 & 0 & d & \cdots & 0 \\ \vdots & \vdots & & \vdots & \vdots & \vdots & & \vdots \\ 0 & 0 & \cdots & c & 0 & 0 & \cdots & d \end{array}\right],$$

where I is an $m \times m$ identity matrix and a, b, c, d are scalars. Then if $d \neq 0$,

$$|\mathbf{B}| = |d\mathbf{I}| \cdot |a\mathbf{I} - b\mathbf{I}(d\mathbf{I})^{-1}c\mathbf{I}|$$

$$= |d\mathbf{I}|\left|\left(a - \frac{bc}{d}\right)\mathbf{I}\right| = d^m\left(a - \frac{bc}{d}\right)^m = (ad - bc)^m.$$

Note: If $\mathbf{B}$ is a positive definite matrix in Theorem 8.2.1, then all the assumptions are satisfied and, therefore, all of the results of the theorem follow.

8.3 The Inverse of Certain Patterned Matrices

In this section we shall give some theorems that are useful in finding the inverse of certain patterned matrices that occur quite often in statistical applications as well as other applications. Note that the theorems are not for general matrices, but each theorem is for a matrix with a specific form or pattern. However, most of the theorems allow for some generality in that they may involve arbitrary elements. In most cases, examples are given to illustrate the theorems.

Theorem 8.3.1

Let the $k \times k$ lower triangular matrix $\mathbf{C}$ be defined by

$$\mathbf{C} = \begin{bmatrix} a_1 b_1 & 0 & 0 & \cdots & 0 \\ a_2 b_1 & a_2 b_2 & 0 & \cdots & 0 \\ \vdots & \vdots & \vdots & & \vdots \\ a_k b_1 & a_k b_2 & a_k b_3 & \cdots & a_k b_k \end{bmatrix}, \tag{8.3.1}$$

where a_i and b_i are nonzero for all $i = 1, 2, \ldots, k$. The inverse of $\mathbf{C}$ is given by

$$\mathbf{C}^{-1} = \begin{bmatrix} (a_1 b_1)^{-1} & 0 & 0 & \cdots & 0 & 0 \\ -(b_2 a_1)^{-1} & (a_2 b_2)^{-1} & 0 & \cdots & 0 & 0 \\ 0 & -(b_3 a_2)^{-1} & (a_3 b_3)^{-1} & \cdots & 0 & 0 \\ \vdots & \vdots & \vdots & & \vdots & \vdots \\ 0 & 0 & 0 & \cdots & (b_{k-1} a_{k-1})^{-1} & 0 \\ 0 & 0 & 0 & \cdots & -(b_k a_{k-1})^{-1} & (a_k b_k)^{-1} \end{bmatrix}. \tag{8.3.2}$$

Proof: Multiply $\mathbf{C}$ by $\mathbf{C}^{-1}$ and observe that the result is the identity matrix $\mathbf{I}$. ▮

Theorem 8.3.2

Let the matrix $\mathbf{C}$ *be defined by*

$$\mathbf{C} = \begin{bmatrix} a_1^2 b_1 & a_1 a_2 b_1 & a_1 a_3 b_1 & \cdots & a_1 a_k b_1 \\ a_1 a_2 b_1 & a_2^2(b_1 + b_2) & a_2 a_3(b_1 + b_2) & \cdots & a_2 a_k(b_1 + b_2) \\ a_1 a_3 b_1 & a_2 a_3(b_1 + b_2) & a_3^2(b_1 + b_2 + b_3) & \cdots & a_3 a_k(b_1 + b_2 + b_3) \\ \vdots & \vdots & \vdots & & \vdots \\ a_1 a_k b_1 & a_2 a_k(b_1 + b_2) & a_3 a_k(b_1 + b_2 + b_3) & \cdots & a_k^2(b_1 + b_2 + \cdots + b_k) \end{bmatrix}, \tag{8.3.3}$$

where none of the numbers a_i, b_j, *is zero; then the inverse of* $\mathbf{C}$ *is given by*

$$\mathbf{C}^{-1} = \begin{bmatrix} \frac{1}{a_1^2}\left(\frac{1}{b_1} + \frac{1}{b_2}\right) & -\frac{1}{a_1 a_2 b_2} & 0 & \cdots & 0 \\ -\frac{1}{a_1 a_2 b_2} & \frac{1}{a_2^2}\left(\frac{1}{b_2} + \frac{1}{b_3}\right) & -\frac{1}{a_2 a_3 b_3} & \cdots & 0 \\ 0 & -\frac{1}{a_2 a_3 b_3} & \frac{1}{a_3^2}\left(\frac{1}{b_3} + \frac{1}{b_4}\right) & \cdots & 0 \\ \vdots & \vdots & \vdots & & \vdots \\ 0 & 0 & 0 & \cdots & -\frac{1}{a_{k-1} a_k b_k} \\ 0 & 0 & 0 & \cdots & \frac{1}{a_k^2 b_k} \end{bmatrix}. \tag{8.3.4}$$

Proof: Again the proof is given by showing that $\mathbf{CC}^{-1} = \mathbf{I}$. Notice that $\mathbf{C}$ can be written as $\mathbf{D}_1\mathbf{T}\mathbf{D}_2\mathbf{T}'\mathbf{D}_1$, where $\mathbf{D}_1$ is a diagonal matrix whose i-th diagonal element is a_i, $\mathbf{D}_2$ is a diagonal matrix whose i-th diagonal element is b_i, and $\mathbf{T}$ is a lower triangular matrix with each element on and below the diagonal equal to one. Of course, $\mathbf{C}^{-1} = \mathbf{D}_1^{-1}\mathbf{T}'^{-1}\mathbf{D}_2^{-1}\mathbf{T}^{-1}\mathbf{D}_1^{-1}$. ∎

The two examples that follow illustrate matrices for which Theorems 8.3.1 and 8.3.2 can be used.

Example 8.3.1. The following is the variance-covariance matrix of order-statistics for a random sample of size k from a population with an exponential density

$$\mathbf{V} = \begin{bmatrix} \frac{1}{k^2} & \frac{1}{k^2} & \frac{1}{k^2} & \cdots & \frac{1}{k^2} \\ \frac{1}{k^2} & \frac{1}{k^2} + \frac{1}{(k-1)^2} & \frac{1}{k^2} + \frac{1}{(k-1)^2} & \cdots & \frac{1}{k^2} + \frac{1}{(k-1)^2} \\ \frac{1}{k^2} & \frac{1}{k^2} + \frac{1}{(k-1)^2} & \sum_{j=1}^{3} \frac{1}{(k-j+1)^2} & \cdots & \sum_{j=1}^{3} \frac{1}{(k-j+1)^2} \\ \vdots & \vdots & \vdots & & \vdots \\ \frac{1}{k^2} & \frac{1}{k^2} + \frac{1}{(k-1)^2} & \sum_{j=1}^{3} \frac{1}{(k-j+1)^2} & \cdots & \sum_{j=1}^{k} \frac{1}{(k-j+1)^2} \end{bmatrix}. \tag{8.3.5}$$

This matrix is a special case of the matrix $\mathbf{C}$ in Theorem 8.3.2 if we let $a_1 = a_2 = \cdots = a_k = 1$ and $b_1 = 1/k^2$, $b_2 = 1/(k-1)^2, \cdots, b_k = 1$. Therefore, the inverse is given by $\mathbf{V}^{-1}$, where

$$\mathbf{V}^{-1} = \begin{bmatrix} k^2 + (k-1)^2 & -(k-1)^2 & 0 & 0 & \cdots & 0 & 0 \\ -(k-1)^2 & (k-1)^2 + (k-2)^2 & -(k-2)^2 & 0 & \cdots & 0 & 0 \\ 0 & -(k-2)^2 & (k-2)^2 + (k-3)^2 & -(k-3)^2 & \cdots & 0 & 0 \\ \vdots & \vdots & \vdots & \vdots & & \vdots & \vdots \\ 0 & 0 & 0 & 0 & \cdots & -1 & 1 \end{bmatrix} \tag{8.3.6}$$

Example 8.3.2. By applying Theorem 8.3.1 to the following matrix $\mathbf{C}$, we get $\mathbf{C}^{-1}$ as given below. (Note from Eq. (8.3.1) that $a_i = i$, $b_j = j$ if $i \geq j$.) Thus

$$\mathbf{C} = \begin{bmatrix} 1 & 0 & 0 & 0 \\ 2 & 4 & 0 & 0 \\ 3 & 6 & 9 & 0 \\ 4 & 8 & 12 & 16 \end{bmatrix}; \quad \mathbf{C}^{-1} = \begin{bmatrix} 1 & 0 & 0 & 0 \\ -1/2 & 1/4 & 0 & 0 \\ 0 & -1/6 & 1/9 & 0 \\ 0 & 0 & -1/12 & 1/16 \end{bmatrix}.$$

Theorem 8.3.3

Let the $k \times k$ matrix $\mathbf{C}$ be given by

$$\mathbf{C} = \mathbf{D} + \alpha \mathbf{a}\mathbf{b}', \tag{8.3.7}$$

where $\mathbf{D}$ is a nonsingular diagonal matrix, $\mathbf{a}$ and $\mathbf{b}$ are each $k \times 1$ vectors, and α is a scalar such that

$$\alpha \neq -\left[\sum_{i=1}^{k} a_i b_i / d_{ii}\right]^{-1}.$$

The inverse of $\mathbf{C}$ is

$$\mathbf{C}^{-1} = \mathbf{D}^{-1} + \gamma \mathbf{a}^* \mathbf{b}^{*\prime}, \tag{8.3.8}$$

where $\gamma = -\alpha(1 + \alpha\sum_{i=1}^{k} a_i b_i d_{ii}^{-1})^{-1}$; $a_i^* = a_i/d_{ii}$; $b_i^* = b_i/d_{ii}$; and d_{ii} is the i-th diagonal element of $\mathbf{D}$.

Proof: The proof is given by showing that $\mathbf{C}\mathbf{C}^{-1} = \mathbf{I}$. ∎

Example 8.3.3. The matrix $\mathbf{V}$ given below is the variance-covariance matrix of a k-dimensional random variable that has the multinomial density (assume $p_i > 0$, $\sum_{i=1}^{k} p_i < 1$)

$$\mathbf{V} = \begin{bmatrix} p_1(1-p_1) & -p_1 p_2 & -p_1 p_3 & \cdots & -p_1 p_k \\ -p_1 p_2 & p_2(1-p_2) & -p_2 p_3 & \cdots & -p_2 p_k \\ -p_1 p_3 & -p_2 p_3 & p_3(1-p_3) & \cdots & -p_3 p_k \\ \vdots & \vdots & \vdots & & \vdots \\ -p_1 p_k & -p_2 p_k & -p_3 p_k & \cdots & p_k(1-p_k) \end{bmatrix}. \tag{8.3.9}$$

If we let $\mathbf{D}$, α, $\mathbf{a}$, and $\mathbf{b}$ in Theorem 8.3.3 be given by

$$\mathbf{D} = \begin{bmatrix} p_1 & 0 & 0 & \cdots & 0 \\ 0 & p_2 & 0 & \cdots & 0 \\ 0 & 0 & p_3 & \cdots & 0 \\ \vdots & \vdots & \vdots & & \vdots \\ 0 & 0 & 0 & \cdots & p_k \end{bmatrix}; \qquad \alpha = -1; \qquad \mathbf{a} = \mathbf{b} = \begin{bmatrix} p_1 \\ p_2 \\ \vdots \\ p_k \end{bmatrix}, \tag{8.3.10}$$

then we see that $\mathbf{V} = \mathbf{D} + \alpha \mathbf{a}\mathbf{b}'$ and $\mathbf{V}^{-1} = \mathbf{D}^{-1} + \mathbf{b}\mathbf{a}^*\mathbf{b}^{*\prime}$. But $a_i^* = b_i^* = 1$, and hence $\mathbf{a}^*\mathbf{b}^{*\prime} = (\mathbf{1})(\mathbf{1})' = \mathbf{J}$; also

$$\gamma = \left[1 - \sum_{i=1}^{k} p_i\right]^{-1};$$

so if we let $\gamma = 1/p$, we can write

$$\mathbf{V}^{-1} = \begin{bmatrix} \frac{1}{p_1} + \frac{1}{p} & \frac{1}{p} & \frac{1}{p} & \cdots & \frac{1}{p} \\ \frac{1}{p} & \frac{1}{p_2} + \frac{1}{p} & \frac{1}{p} & \cdots & \frac{1}{p} \\ \frac{1}{p} & \frac{1}{p} & \frac{1}{p_3} + \frac{1}{p} & \cdots & \frac{1}{p} \\ \vdots & \vdots & \vdots & & \vdots \\ \frac{1}{p} & \frac{1}{p} & \frac{1}{p} & \cdots & \frac{1}{p_k} + \frac{1}{p} \end{bmatrix}. \tag{8.3.11}$$

Another patterned matrix that occurs quite frequently in probability and statistics, as well as other areas of interest, is the $k \times k$ matrix $\mathbf{C}$ defined by

$$\mathbf{C} = \begin{bmatrix} a & b & b & \cdots & b \\ b & a & b & \cdots & b \\ b & b & a & \cdots & b \\ \vdots & \vdots & \vdots & & \vdots \\ b & b & b & \cdots & a \end{bmatrix}. \tag{8.3.12}$$

That is, the diagonal elements of $\mathbf{C}$ are each equal to a and the off-diagonal elements are each equal to b. Another way to write this matrix is

$$\mathbf{C} = (a - b)\mathbf{I} + b\mathbf{J}. \tag{8.3.13}$$

This matrix does not have an inverse for all values of a and b.

Theorem 8.3.4

Let the $k \times k$ matrix $\mathbf{C}$ be defined by

$$\mathbf{C} = (a - b)\mathbf{I} + b\mathbf{J}.$$

The matrix **C** *has an inverse if and only if* $a \neq b$ *and* $a \neq -(k-1)b$. *If* $\mathbf{C}^{-1}$ *exists, it is given by*

$$\mathbf{C}^{-1} = \frac{1}{a-b}\left[\mathbf{I} - \frac{b}{a+(k-1)b}\mathbf{J}\right]. \tag{8.3.14}$$

Proof: To show that $\mathbf{C}^{-1}$ exists if and only if $a \neq b$ and $a \neq -(k-1)b$, we shall evaluate the determinant of **C**. Perform the following operations:

1. Subtract the second row from the first row; subtract the third row from the second row; subtract the fourth row from the third row; etc. The result is

$$\mathbf{C}^* = \begin{bmatrix} a-b & b-a & 0 & 0 & \cdots & 0 \\ 0 & a-b & b-a & 0 & \cdots & 0 \\ 0 & 0 & a-b & b-a & \cdots & 0 \\ \vdots & \vdots & \vdots & \vdots & & \vdots \\ b & b & b & b & \cdots & a \end{bmatrix},$$

but the value of the determinant is unchanged; that is, $|\mathbf{C}^*| = |\mathbf{C}|$.

2. Now add the first column to the second column, add the second column to the third column, add the third column to the fourth column, and so on. The result is

$$\mathbf{C}^{**} = \begin{bmatrix} a-b & 0 & 0 & \cdots & 0 \\ 0 & a-b & 0 & \cdots & 0 \\ 0 & 0 & a-b & \cdots & 0 \\ \vdots & \vdots & \vdots & & \vdots \\ b & 2b & 3b & \cdots & a+(k-1)b \end{bmatrix},$$

and the value of the determinant is unchanged; that is, $|\mathbf{C}| = |\mathbf{C}^*| = |\mathbf{C}^{**}|$. But $\mathbf{C}^{**}$ is a lower triangular matrix; hence,

$$|\mathbf{C}| = |\mathbf{C}^{**}| = (a-b)^{k-1}[a+(k-1)b].$$

So **C** has an inverse if and only if the determinant of $\mathbf{C}^{**}$ is not zero—that is, if and only if $a \neq b$ and $a \neq -(k-1)b$. It is easily shown that $\mathbf{C}^{-1}\mathbf{C} = \mathbf{I}$ if $\mathbf{C}^{-1}$ exists. ▮

Example 8.3.4. Let the 4×4 matrix $\mathbf{C}$ be defined by

$$\mathbf{C} = \begin{bmatrix} 2 & 4 & 4 & 4 \\ 4 & 2 & 4 & 4 \\ 4 & 4 & 2 & 4 \\ 4 & 4 & 4 & 2 \end{bmatrix};$$

then $\mathbf{C}^{-1}$ exists, since $b = 4$, $a = 2$, and $k = 4$, and is readily determined to be

$$\mathbf{C}^{-1} = \begin{bmatrix} -\frac{5}{14} & \frac{1}{7} & \frac{1}{7} & \frac{1}{7} \\ \frac{1}{7} & -\frac{5}{14} & \frac{1}{7} & \frac{1}{7} \\ \frac{1}{7} & \frac{1}{7} & -\frac{5}{14} & \frac{1}{7} \\ \frac{1}{7} & \frac{1}{7} & \frac{1}{7} & -\frac{5}{14} \end{bmatrix}.$$

Notice that the matrix $\mathbf{C}$ in Eq. (8.3.13) is a special case of the matrix $\mathbf{C}$ in Theorem 8.3.3. In fact, if we let

$$\mathbf{D} = (a - b)\mathbf{I},$$

$$\mathbf{a} = \mathbf{b} = \mathbf{1},$$

and

$$\alpha = b,$$

we obtain

$$d_{ii} = a - b,$$

$$\gamma = -b\left[1 + \frac{bk}{a-b}\right]^{-1},$$

$$\mathbf{a}^* = \frac{1}{a-b}\mathbf{1},$$

$$\mathbf{b}^* = \frac{1}{a-b}\mathbf{1},$$

and, by Theorem 8.3.3, we get

$$\mathbf{C}^{-1} = \frac{1}{a-b}\mathbf{I} - \frac{b}{1 + \dfrac{bk}{a-b}}\mathbf{a}^*\mathbf{b}^{*\prime}$$

or

$$\mathbf{C}^{-1} = \frac{1}{a-b}\left[\mathbf{I} - \frac{b}{a+b(k-1)}\mathbf{J}\right],$$

which of course is the same result as given in Theorem 8.3.4.

A matrix of a more general character and its inverse are discussed in the next theorem.

Theorem 8.3.5

Let the $(m+n) \times (m+n)$ matrix $\mathbf{C}$ be defined by

$$\mathbf{C} = \begin{bmatrix} a_1\mathbf{I}_1 & a_2\mathbf{J}_1 \\ a_2\mathbf{J}_1' & a_3\mathbf{I}_2 \end{bmatrix}, \tag{8.3.15}$$

where $a_3 \neq 0$, and $n > 0$, and where $\mathbf{I}_1$ is the $m \times m$ identity matrix, $\mathbf{I}_2$ is the $n \times n$ identity matrix, $\mathbf{J}_1$ is an $m \times n$ matrix of ones. The inverse exists if and only if $a_1 \neq 0$ and $a_1 \neq mna_2^2/a_3$. If $\mathbf{C}^{-1}$ exists, it is given by

$$\mathbf{C}^{-1} = \begin{bmatrix} \dfrac{1}{a_1}\mathbf{I}_1 + b_1\mathbf{J}_3 & b_2\mathbf{J}_1 \\ b_2\mathbf{J}_1' & \dfrac{1}{a_3}\mathbf{I}_2 + b_3\mathbf{J}_4 \end{bmatrix}, \tag{8.3.16}$$

where $\mathbf{J}_3$ is an $m \times m$ matrix of ones and $\mathbf{J}_4$ is an $n \times n$ matrix of ones and where

$$b_1 = -\frac{na_2^2}{a_1(mna_2^2 - a_3a_1)}, \quad b_2 = \frac{a_2}{mna_2^2 - a_3a_1},$$

$$\text{and} \quad b_3 = -\frac{ma_2^2}{a_3(mna_2^2 - a_3a_1)}.$$

Proof: We shall use Statement (2) of Theorem 8.2.1. We identify $\mathbf{B}_{11} = a_1\mathbf{I}_1$,

$\mathbf{B}_{12} = a_2\mathbf{J}_1$, $\mathbf{B}_{21} = \mathbf{B}'_{12} = a_2\mathbf{J}'_1$, $\mathbf{B}_{22} = a_3\mathbf{I}_2$. Since in the theorem we assume that $a_3 \neq 0$, this implies that $\mathbf{B}_{22}$ is nonsingular. Hence we can write

$$|\mathbf{B}| = |\mathbf{B}_{22}| \cdot |\mathbf{B}_{11} - \mathbf{B}_{12}\mathbf{B}_{22}^{-1}\mathbf{B}_{21}|,$$

as

$$|\mathbf{C}| = |a_3\mathbf{I}_2| \cdot |a_1\mathbf{I}_1 - (a_2\mathbf{J}_1)(a_3\mathbf{I}_2)^{-1}(a_2\mathbf{J}'_1)|$$
$$= (a_3^n)\left|a_1\mathbf{I}_1 - \frac{a_2^2 n}{a_3}\mathbf{J}\right|,$$

where $n\mathbf{J} = \mathbf{J}_1\mathbf{J}'_1$, and $\mathbf{J}$ is an $m \times m$ matrix of ones. We must now evaluate the determinant of $a_1\mathbf{I}_1 - (na_2^2/a_3)\mathbf{J}$, but this can be done by using the results of Theorem 8.3.4 where we identify $a = a_1 - na_2^2/a_3$ and $b = -na_2^2/a_3$. Thus $|\mathbf{C}|$ is not zero if and only if $a \neq b$ and $a \neq -(m-1)b$, which implies $a_1 \neq 0$ and $a_1 \neq mna_2^2/a_3$. To show that $\mathbf{C}^{-1}$ in Eq. (8.3.16) is actually the inverse of $\mathbf{C}$ given in Eq. (8.3.15), if the inverse exists, one can show that $\mathbf{C}^{-1}\mathbf{C} = \mathbf{I}$. ▌

Example 8.3.5. In the two-way classification model with equal numbers in the subcells in Experimental Design theory, the matrix $\mathbf{B}$ plays an important role where

$$\mathbf{B} = \begin{bmatrix} n\mathbf{I}_1 & \mathbf{J}_1 \\ \mathbf{J}'_1 & m\mathbf{I}_2 \end{bmatrix}, \tag{8.3.17}$$

$\mathbf{I}_1$ is the $m \times m$ identity matrix, $\mathbf{I}_2$ is the $(n-1) \times (n-1)$ identity matrix, and $\mathbf{J}_1$ is an $m \times (n-1)$ matrix of ones. By Theorem 8.3.5, the inverse is

$$\mathbf{B}^{-1} = \begin{bmatrix} \dfrac{1}{n}\mathbf{I}_1 + \dfrac{n-1}{mn}\mathbf{J}_3 & -\dfrac{1}{m}\mathbf{J}_1 \\ -\dfrac{1}{m}\mathbf{J}'_1 & \dfrac{1}{m}\mathbf{I}_2 + \dfrac{1}{m}\mathbf{J}_4 \end{bmatrix}, \tag{8.3.18}$$

where $\mathbf{J}_3$ is an $m \times m$ matrix and $\mathbf{J}_4$ is an $(n-1) \times (n-1)$ matrix.

In the study of Experimental Design theory of Latin square models, response surface models, and many other models, a matrix of the following general structure is encountered:

$$\mathbf{C} = \begin{bmatrix} \alpha_1 & \alpha_2\mathbf{1}' & \alpha_3\mathbf{1}' & \cdots & \alpha_t\mathbf{1}' \\ \alpha_2\mathbf{1} & \beta_2\mathbf{I}+\gamma_2\mathbf{J} & \beta_3\mathbf{I}+\gamma_3\mathbf{J} & \cdots & \beta_t\mathbf{I}+\gamma_t\mathbf{J} \\ \alpha_3\mathbf{1} & \beta_3\mathbf{I}+\gamma_3\mathbf{J} & \delta_3\mathbf{I}+\theta_3\mathbf{J} & \cdots & \delta_t\mathbf{I}+\theta_t\mathbf{J} \\ \vdots & \vdots & \vdots & & \vdots \\ \alpha_t\mathbf{1} & \beta_t\mathbf{I}+\gamma_t\mathbf{J} & \delta_t\mathbf{I}+\theta_t\mathbf{J} & \cdots & \xi_t\mathbf{I}+\mu_t\mathbf{J} \end{bmatrix}, \tag{8.3.19}$$

where the $\alpha, \beta, \ldots, \mu$ are scalars, $\mathbf{1}$ is the $k \times 1$ vector of ones, $\mathbf{I}$ is the $k \times k$ identity matrix, and $\mathbf{J}$ is a $k \times k$ matrix of ones. If the inverse of $\mathbf{C}$ exists, then it can be shown that it has the same pattern as $\mathbf{C}$ and we need only evaluate the scalars.

We can illustrate a method of determining the inverse of a patterned matrix $\mathbf{C}$ as in Eq. (8.3.19) above by an example.

Example 8.3.6. Find the inverse of the matrix $\mathbf{C}$ given by

$$\mathbf{C} = \left[\begin{array}{c|ccc|ccc} 8 & 2 & 2 & 2 & 3 & 3 & 3 \\ \hline 2 & 4 & 1 & 1 & 3 & 2 & 2 \\ 2 & 1 & 4 & 1 & 2 & 3 & 2 \\ 2 & 1 & 1 & 4 & 2 & 2 & 3 \\ \hline 3 & 3 & 2 & 2 & 1 & 2 & 2 \\ 3 & 2 & 3 & 2 & 2 & 1 & 2 \\ 3 & 2 & 2 & 3 & 2 & 2 & 1 \end{array}\right]. \tag{8.3.20}$$

Notice that if we identify the matrix in Eq. (8.3.20) with the one in Eq. (8.3.19), we have $\alpha_1 = 8$, $\alpha_2 = 2$, $\alpha_3 = 3$, $\beta_2 = 3$, $\gamma_2 = 1$, $\beta_3 = 1$, $\gamma_3 = 2$, $\delta_3 = -1$, $\theta_3 = 2$, $k = 3$, and $\mathbf{C}$ is a 7×7 matrix. If $\mathbf{C}^{-1}$ exists, it can be shown that it has the same pattern and we can write $\mathbf{CC}^{-1} = \mathbf{I}$ as

$$\begin{bmatrix} 8 & 2\mathbf{1}' & 3\mathbf{1}' \\ 2\mathbf{1} & 3\mathbf{I}+\mathbf{J} & \mathbf{I}+2\mathbf{J} \\ 3\mathbf{1} & \mathbf{I}+2\mathbf{J} & -\mathbf{I}+2\mathbf{J} \end{bmatrix} \begin{bmatrix} a_1 & a_2\mathbf{1}' & a_3\mathbf{1}' \\ a_2\mathbf{1} & b_2\mathbf{I}+c_2\mathbf{J} & b_3\mathbf{I}+c_3\mathbf{J} \\ a_3\mathbf{1} & b_3\mathbf{I}+c_3\mathbf{J} & d_3\mathbf{I}+e_3\mathbf{J} \end{bmatrix} = \begin{bmatrix} 1 & \mathbf{0} & \mathbf{0} \\ \mathbf{0} & \mathbf{I} & \mathbf{0} \\ \mathbf{0} & \mathbf{0} & \mathbf{I} \end{bmatrix}, \tag{8.3.21}$$

where the constants $a_1, a_2, a_3, b_2, b_3, c_2, c_3, d_3$, and e_3 are to be determined.

Upon multiplying the three block-rows of $\mathbf{C}$ (each element is a matrix) by the first block-column of $\mathbf{C}^{-1}$ in turn, using the fact that $\mathbf{1}'\mathbf{1} = 3 = k$ and

$\mathbf{J1} = 3\mathbf{1}$, we get that

$$\begin{aligned} 8a_1 + 6a_2 + 9a_3 &= 1, \\ 2a_1\mathbf{1} + 3a_2\mathbf{1} + 3a_2\mathbf{1} + a_3\mathbf{1} + 6a_3\mathbf{1} &= \mathbf{0}, \\ 3a_1\mathbf{1} + a_2\mathbf{1} + 6a_2\mathbf{1} - a_3\mathbf{1} + 6a_3\mathbf{1} &= \mathbf{0}. \end{aligned} \tag{8.3.22}$$

From these equations we get

$$\begin{aligned} 8a_1 + 6a_2 + 9a_3 &= 1, \\ 2a_1 + 6a_2 + 7a_3 &= 0, \\ 3a_1 + 7a_2 + 5a_3 &= 0. \end{aligned} \tag{8.3.23}$$

The solution is $a_1 = \dfrac{57}{366}$, $a_2 = -\dfrac{33}{366}$, $a_3 = \dfrac{12}{366}$.

Next we shall multiply the three block-rows of $\mathbf{C}$ by the second block-column of $\mathbf{C}^{-1}$ in turn and (notice that $\mathbf{11}' = \mathbf{J}$, $\mathbf{JJ} = 3\mathbf{J}$), we obtain

$$\begin{aligned} &\text{(a)} \quad 8a_2\mathbf{1}' + 2b_2\mathbf{1}' + 6c_2\mathbf{1}' + 3b_3\mathbf{1}' + 9c_3\mathbf{1}' = \mathbf{0}, \\ &\text{(b)} \quad 2a_2\mathbf{J} + 3b_2\mathbf{I} + (3c_2 + b_2 + 3c_2)\mathbf{J} + b_3\mathbf{I} + (2b_3 + c_3 + 6c_3)\mathbf{J} = \mathbf{I}, \\ &\text{(c)} \quad 3a_2\mathbf{J} + b_2\mathbf{I} + (2b_2 + c_2 + 6c_2)\mathbf{J} - b_3\mathbf{I} + (6c_3 - c_3 + 2b_3)\mathbf{J} = \mathbf{0}. \end{aligned} \tag{8.3.24}$$

Notice that two equations in the a_i, b_i, and c_i can be obtained from the second matrix equation above, that is, from Eq. (8.3.24b), one equation by using the off-diagonal elements, and one by using the diagonal elements. This is also true of the third matrix equation. Hence we obtain the five equations given in Eq. (8.3.25) below using the values of a_1, a_2, a_3 from Eq. (8.3.23). Equation (8.3.25a) is obtained from Eq. (8.3.24a); Eq. (8.3.25b) is obtained from the off-diagonal elements of Eq. (8.3.24b); Eq. (8.3.25c) is obtained from the diagonal elements of Eq. (8.3.24b); Eqs. (8.3.25d) and (8.3.25e) are obtained from the off-diagonal and diagonal elements, respectively, of Eq. (8.3.24c).

$$\begin{aligned} &\text{(a)} \quad 2b_2 + 6c_2 + 3b_3 + 9c_3 = 264/366 \\ &\text{(b)} \quad b_2 + 6c_2 + 2b_3 + 7c_3 = 66/366 \\ &\text{(c)} \quad 3b_2 + b_3 = 1 \\ &\text{(d)} \quad 2b_2 + 7c_2 + 2b_3 + 5c_3 = 99/366 \\ &\text{(e)} \quad b_2 - b_3 = 0 \end{aligned} \tag{8.3.25}$$

We have a system of five equations and four unknowns. We do not need the first equation to solve for b_2, c_2, b_3, and c_3. The solution is

$$b_2 = \frac{1}{4}, \qquad b_3 = \frac{1}{4}, \qquad c_2 = -\frac{87}{732}, \qquad c_3 = \frac{15}{732}.$$

By using the third column of $\mathbf{C}^{-1}$ in a similar manner, we get

$$\begin{aligned} &8a_3\mathbf{1}' + 2b_3\mathbf{1}' + 6c_3\mathbf{1}' + 3d_3\mathbf{1}' + 9e_3\mathbf{1}' = \mathbf{0}, \\ &2a_3\mathbf{J} + 3b_3\mathbf{I} + (3c_3 + b_3 + 3c_3)\mathbf{J} + d_3\mathbf{I} + (e_3 + 2d_3 + 6e_3)\mathbf{J} = \mathbf{0}, \\ &3a_3\mathbf{J} + b_3\mathbf{I} + (2b_3 + 6c_3 + c_3)\mathbf{J} - d_3\mathbf{I} + (2d_3 + 6e_3 - e_3)\mathbf{J} = \mathbf{I}. \end{aligned} \tag{8.3.26}$$

From these we can obtain the following equations:

$$\begin{aligned} 8a_3 + 2b_3 + 6c_3 + 3d_3 + 9e_3 &= 0, \\ 3b_3 \qquad\quad + d_3 \qquad\quad &= 0, \\ 2a_3 + b_3 + 6c_3 + 2d_3 + 7e_3 &= 0, \\ b_3 \qquad\quad - d_3 \qquad\quad &= 1, \\ 3a_3 + 2b_3 + 7c_3 + 2d_3 + 5e_3 &= 0. \end{aligned} \tag{8.3.27}$$

Since we need to solve only for d_3 and e_3, we can use the first two equations only. The solution is

$$d_3 = -\frac{3}{4}, \qquad e_3 = \frac{111}{732}.$$

Thus $\mathbf{C}^{-1}$ is given by

$$\mathbf{C}^{-1} = \frac{1}{732}\begin{bmatrix} 114 & -66 & -66 & -66 & 24 & 24 & 24 \\ -66 & 96 & -87 & -87 & 198 & 15 & 15 \\ -66 & -87 & 96 & -87 & 15 & 198 & 15 \\ -66 & -87 & -87 & 96 & 15 & 15 & 198 \\ 24 & 198 & 15 & 15 & 438 & 111 & 111 \\ 24 & 15 & 198 & 15 & 111 & 438 & 111 \\ 24 & 15 & 15 & 198 & 111 & 111 & 438 \end{bmatrix}.$$

Thus we have found the inverse of a 7×7 matrix $\mathbf{C}$ given in Eq. (8.3.20) by solving (1) a set of three equations and three unknowns in Eq. (8.3.23); (2) a set of four equations and four unknowns in Eq. (8.3.25); (3) a set of two equations and two unknowns in Eq. (8.3.27). For a discussion of matrices of the type given in Eq. (8.3.19), see [G–8].

Before discussing the next theorem, we define diagonal matrices of type 2.

Definition 8.3.1

Matrices of Type 2. *A $k \times k$ matrix $\mathbf{A}$ is defined to be a diagonal matrix of type 2 if and only if (1) the diagonal elements are nonzero; (2) each element immediately above and immediately below the diagonal is nonzero; (3) the remaining elements are zero. Another way to state this is $a_{ij} \neq 0$ if $|i-j| \leq 1$, $a_{ij} = 0$ if $|i-j| > 1$.*

As an example, the matrix given below is a diagonal matrix of type 2.

$$\mathbf{A} = \begin{bmatrix} 1 & 2 & 0 & 0 \\ 8 & 3 & 4 & 0 \\ 0 & 3 & 2 & 7 \\ 0 & 0 & 1 & 4 \end{bmatrix}.$$

We state the following theorem without giving the proof.

Theorem 8.3.6

Let $\mathbf{B}$ be a $k \times k$ symmetric nonsingular matrix with $b_{1j} \neq 0$ for $j = 2, \ldots, k$. A necessary and sufficient condition for the inverse of the matrix $\mathbf{B}$ to be a diagonal matrix of type 2 is the following:

$$\begin{aligned}
\frac{b_{2j}}{b_{1j}} &= \theta_2, \quad 2 \leq j \leq k, \\
\frac{b_{3j}}{b_{1j}} &= \theta_3, \quad 3 \leq j \leq k, \\
&\vdots \\
\frac{b_{tj}}{b_{1j}} &= \theta_t, \quad t \leq j \leq k, \\
&\vdots \\
\frac{b_{kk}}{b_{1k}} &= \theta_k, \quad j = k.
\end{aligned} \tag{8.3.28}$$

Equation (8.3.28) means that all the elements on and to the right of the diagonal in the t-th row of **B** have a constant relation to the corresponding column elements of the first row; θ_t is the proportionality factor. That is, if the elements in the first row are given by $b_{11}, b_{12}, \ldots, b_{1k}$, then the elements on the diagonal and to the right of the diagonal in the t-th row are given by

$$b_{tj} = \theta_t b_{1j}; \quad j = t, t+1, \ldots, k, \quad t = 2, 3, \ldots, k.$$

Notice that the matrix in Theorem 8.3.2 satisfies the conditions of Theorem 8.3.6.

The following matrix **B** is another example of a matrix that satisfies Theorem 8.3.6, where

$$\mathbf{B} = \begin{bmatrix} 3 & 2 & 1 \\ 2 & 4 & 2 \\ 1 & 2 & 3 \end{bmatrix},$$

and therefore $\mathbf{B}^{-1}$ must be a diagonal matrix of type 2. Actually $\mathbf{B}^{-1}$ is given by

$$\mathbf{B}^{-1} = \frac{1}{4}\begin{bmatrix} 2 & -1 & 0 \\ -1 & 2 & -1 \\ 0 & -1 & 2 \end{bmatrix},$$

which is a diagonal matrix of type 2.

Our next theorem gives the elements of the inverse of the matrix **B** in Theorem 8.3.6.

Theorem 8.3.7

Let **B** *be a matrix that satisfies the conditions of Theorem* 8.3.6. *Then the inverse* $\mathbf{C} = \mathbf{B}^{-1}$ *is a diagonal matrix of type* 2 *and the elements are given by*

$$\begin{aligned}
c_{11} &= -\theta_2(b_{12} - \theta_2 b_{11})^{-1}, \\
c_{tt} &= -\frac{b_{t-1,t+1} - \theta_{t+1} b_{1,t-1}}{(b_{t-1,t} - \theta_t b_{1,t-1})(b_{t,t+1} - \theta_{t+1} b_{1t})}, \quad \text{for} \quad t = 2, 3, \ldots, k-1, \\
c_{kk} &= -\frac{b_{1,k-1}}{b_{1k}(b_{k-1,k} - \theta_k b_{1,k-1})}, \qquad (8.3.29) \\
c_{t,t-1} &= c_{t-1,t} = (b_{t-1,t} - \theta_t b_{1,t-1})^{-1}, \quad \text{for} \quad t = 2, 3, \ldots, k. \\
c_{ij} &= 0, \quad |i-j| > 1.
\end{aligned}$$

This theorem can be proved by showing that $\mathbf{CB} = \mathbf{I}$.

Example 8.3.7. The following matrix $\mathbf{B}$ arises in the theory of statistics in the variance-covariance matrix of ordered observations of random samples of size k from rectangular distributions.

$$\mathbf{B} = \begin{bmatrix} k & k-1 & k-2 & k-3 & \cdots & 1 \\ k-1 & 2(k-1) & 2(k-2) & 2(k-3) & \cdots & 2 \\ k-2 & 2(k-2) & 3(k-2) & 3(k-3) & \cdots & 3 \\ k-3 & 2(k-3) & 3(k-3) & 4(k-3) & \cdots & 4 \\ \vdots & \vdots & \vdots & \vdots & & \vdots \\ 1 & 2 & 3 & 4 & \cdots & k \end{bmatrix}. \tag{8.3.30}$$

Matrix $\mathbf{B}$ in Eq. (8.3.30) satisfies the conditions of Theorem 8.3.6 with

$$\theta_2 = 2, \quad \theta_3 = 3, \ldots, \quad \theta_t = t, \ldots, \quad \theta_k = k.$$

By Theorem 8.3 the inverse of $\mathbf{B}$ has elements

$$c_{11} = -2[(k-1) - 2k]^{-1} = \frac{2}{k+1},$$

$$c_{tt} = -\frac{(t-1)(k-t) - (t+1)(k-t+2)}{[(t-1)(k-t+1) - t(k-t+2)][t(k-t) - (t+1)(k-t+1)]}$$

$$= \frac{2}{k+1},$$

$$c_{kk} = -\frac{2}{(k-1) - 2k} = \frac{2}{k+1},$$

$$c_{t-1,t} = [(t-1)(k-t+1) - t(k-t+2)]^{-1} = -\frac{1}{k+1}, \quad t = 2, \ldots, k-1. \tag{8.3.31}$$

Notice that $b_{ij} = i(k - j + 1)$ if $i \le j$. Thus we get for $\mathbf{B}^{-1}$

$$\mathbf{B}^{-1} =$$

$$\begin{bmatrix} 2(k+1)^{-1} & -(k+1)^{-1} & 0 & 0 & \cdots & 0 \\ -(k+1)^{-1} & 2(k+1)^{-1} & -(k+1)^{-1} & 0 & \cdots & 0 \\ 0 & -(k+1)^{-1} & 2(k+1)^{-1} & -(k+1)^{-1} & \cdots & 0 \\ 0 & 0 & -(k+1)^{-1} & 2(k+1)^{-1} & \cdots & 0 \\ \vdots & \vdots & \vdots & \vdots & & \vdots \\ 0 & 0 & 0 & 0 & \cdots & 2(k+1)^{-1} \end{bmatrix}. \tag{8.3.32}$$

Example 8.3.8. In the theory of stationary time series, the $n \times n$ covariance matrix $\mathbf{V}$ given below occurs.

$$\mathbf{V} = \sigma^2 \begin{bmatrix} 1 & \rho & \rho^2 & \rho^3 & \cdots & \rho^{n-1} \\ \rho & 1 & \rho & \rho^2 & \cdots & \rho^{n-2} \\ \rho^2 & \rho & 1 & \rho & \cdots & \rho^{n-3} \\ \rho^3 & \rho^2 & \rho & 1 & \cdots & \rho^{n-4} \\ \vdots & \vdots & \vdots & \vdots & & \vdots \\ \rho^{n-1} & \rho^{n-2} & \rho^{n-3} & \rho^{n-4} & \cdots & 1 \end{bmatrix},$$

where σ^2 is any positive constant and $|\rho| < 1$. Another way to define $\mathbf{V}$ is $v_{ij} = \sigma^2 \rho^{|i-j|}$. It is easy to see that the conditions in Eq. (8.3.28) are satisfied. Hence $\mathbf{V}^{-1}$ is a diagonal matrix of type 2. The inverse elements are given in Theorem 8.3.7, and $\mathbf{V}^{-1}$ is given by

$$\mathbf{V}^{-1} = [(1-\rho^2)\sigma^2]^{-1} \begin{bmatrix} 1 & -\rho & 0 & \cdots & 0 & 0 \\ -\rho & 1+\rho^2 & -\rho & \cdots & 0 & 0 \\ 0 & -\rho & 1+\rho^2 & \cdots & 0 & 0 \\ \vdots & \vdots & \vdots & & \vdots & \vdots \\ 0 & 0 & 0 & \cdots & -\rho & 1 \end{bmatrix}.$$

8.4 Determinants of Certain Patterned Matrices

If one can recognize a certain structure of a matrix, it may be possible to obtain a simple formula for the determinant. We shall illustrate by using the patterned matrices in Sec. 8.3.

Theorem 8.4.1

The determinant of the matrix in Theorem 8.3.1 *is*

$$a_1 a_2 \cdots a_k b_1 b_2 \cdots b_k. \tag{8.4.1}$$

The proof of this theorem is obvious, since the determinant of a triangular matrix is equal to the product of the diagonal elements.

Theorem 8.4.2

The determinant of the matrix in Theorem 8.3.2 *is*

$$a_1^2 a_2^2 \cdots a_k^2 b_1 b_2 \cdots b_k. \tag{8.4.2}$$

Proof: For the proof we shall find the determinant of $\mathbf{C}^{-1}$ and use the fact that $\det(\mathbf{C}) = 1/\det(\mathbf{C}^{-1})$. Perform the following elementary operations on the rows of $\mathbf{C}^{-1}$:

(1) Multiply the k-th row by a_k/a_{k-1} and add the result to the $(k-1)$-st row.
(2) Multiply the resulting $(k-1)$-st row by a_{k-1}/a_{k-2} and add the result to the $(k-2)$-nd row.
(3) Continue in this fashion until the last operation results in multiplying the second row by a_2/a_1 and adding the result to the first row.

The resulting matrix is

$$\begin{bmatrix} \dfrac{1}{a_1^2 b_1} & 0 & 0 & \cdots & 0 \\ -\dfrac{1}{a_1 a_2 b_2} & \dfrac{1}{a_2^2 b_2} & 0 & \cdots & 0 \\ 0 & -\dfrac{1}{a_2 a_3 b_3} & \dfrac{1}{a_3^2 b_3} & \cdots & 0 \\ \vdots & \vdots & \vdots & & \vdots \\ 0 & 0 & 0 & \cdots & \dfrac{1}{a_k^2 b_k} \end{bmatrix}. \tag{8.4.3}$$

This matrix is triangular and the determinant is equal to $(a_1^2 a_2^2 \cdots a_k^2 b_1 b_2 \cdots b_k)^{-1}$, but the determinant of this matrix is equal to $\det(\mathbf{C}^{-1})$, since it was obtained by a series of transformations that leave the value of the determinant unchanged. Hence the theorem is proved. A very simple proof is also obtained by using the fact that $\mathbf{C} = \mathbf{D}_1\mathbf{T}'\mathbf{D}_2\mathbf{T}\mathbf{D}_1$ and $|\mathbf{C}| = |\mathbf{D}_1||\mathbf{T}'||\mathbf{D}_2||\mathbf{T}||\mathbf{D}_1|$. ▮

Theorem 8.4.3

The determinant of the matrix given in Theorem 8.3.3 *is equal to*

$$\det(\mathbf{C}) = \alpha\Big[a_1 b_1 \prod_{i\neq 1} d_{ii} + a_2 b_2 \prod_{i\neq 2} d_{ii} + \cdots + a_k b_k \prod_{i\neq k} d_{ii}\Big] + \prod d_{ii} \tag{8.4.4}$$

and can also be written as

$$\det(\mathbf{C}) = \left[1 + \alpha \sum_j \frac{a_j b_j}{d_{jj}}\right] \prod_i d_{ii}. \tag{8.4.5}$$

Proof: We shall assume that none of the b_i is equal to zero. If some of the b_i are equal to zero, then only a slight modification is necessary. We shall also assume that $\alpha \neq 0$, since if it is zero, the determinant is obviously equal to $\prod_{i=1}^{k} d_{ii}$.

The matrix in Eq. (8.3.7) in expanded form is

$$\mathbf{C} = \begin{bmatrix} d_{11} + \alpha a_1 b_1 & \alpha a_1 b_2 & \alpha a_1 b_3 & \cdots & \alpha a_1 b_k \\ \alpha a_2 b_1 & d_{22} + \alpha a_2 b_2 & \alpha a_2 b_3 & \cdots & \alpha a_2 b_k \\ \alpha a_3 b_1 & \alpha a_3 b_2 & d_{33} + \alpha a_3 b_3 & \cdots & \alpha a_3 b_k \\ \vdots & \vdots & \vdots & & \vdots \\ \alpha a_k b_1 & \alpha a_k b_2 & \alpha a_k b_3 & \cdots & d_{kk} + \alpha a_k b_k \end{bmatrix}. \tag{8.4.6}$$

Factor αb_1 from the first column, αb_2 from the second column, and so forth, and get $\det(\mathbf{C}) = \alpha^k b_1 b_2 \cdots b_k \det(\mathbf{G})$, where

$$\mathbf{G} = \begin{bmatrix} \dfrac{d_{11}}{b_1\alpha} + a_1 & a_1 & a_1 & \cdots & a_1 \\ a_2 & \dfrac{d_{22}}{b_2\alpha} + a_2 & a_2 & \cdots & a_2 \\ a_3 & a_3 & \dfrac{d_{33}}{b_3\alpha} + a_3 & \cdots & a_3 \\ \vdots & \vdots & \vdots & & \vdots \\ a_k & a_k & a_k & \cdots & \dfrac{d_{kk}}{\alpha_k b} + a_k \end{bmatrix}. \tag{8.4.7}$$

Next perform the following operations on **G**:

(1) Subtract the second column from the first column.
(2) Subtract the third column from the second column.
(3) Continue until the k-th column is subtracted from the $(k - 1)$-st column.

Call the resulting matrix **H** (clearly the det (**H**) = det (**G**) by virtue of the type of operation performed on **G** to obtain **H**).

$$\mathbf{H} = \begin{bmatrix} \dfrac{d_{11}}{b_1\alpha} & 0 & 0 & \cdots & a_1 \\ -\dfrac{d_{22}}{b_2\alpha} & \dfrac{d_{22}}{b_2\alpha} & 0 & \cdots & a_2 \\ 0 & -\dfrac{d_{33}}{b_3\alpha} & \dfrac{d_{33}}{b_3\alpha} & \cdots & a_3 \\ \vdots & \vdots & \vdots & & \vdots \\ 0 & 0 & 0 & \cdots & \dfrac{d_{kk}}{b_k\alpha} + a_k \end{bmatrix}. \tag{8.4.8}$$

Next evaluate the det (**H**) by expanding on the last column. The result is

$$\det(\mathbf{H}) = \sum_{j=1}^{k-1} a_j A_j + \left(\frac{d_{kk}}{b_k\alpha} + a_k\right) A_k, \tag{8.4.9}$$

where A_j is the cofactor of the j-th element in the last column of **H**. The cofactor can be easily evaluated, since each reduces to a triangular matrix. It is seen to be

$$A_j = \prod_{\substack{i=1 \\ i \neq j}}^{k-1} \frac{d_{ii}}{b_i\alpha}; \quad j = 1, 2, \ldots k. \tag{8.4.10}$$

If det (**G**) is substituted into the formula for det (**C**), the theorem is proved. ∎

Theorem 8.4.4

The determinant of the matrix given in Theorem 8.3.4 *is equal to*

$$(a - b)^{k-1}[a + (k - 1)b]. \tag{8.4.11}$$

Proof: This theorem is proved in the proof of Theorem 8.3.4 and is stated here for completeness. ▮

Theorem 8.4.5

The determinant of the matrix given in Theorem 8.3.5 is equal to

$$a_3^{n-1}a_1^{m-1}(a_1a_3 - mna_2^2). \tag{8.4.12}$$

Proof: The proof is obtained by using the results of the proof of Theorem 8.3.5 along with Theorem 8.4.4. The results of Theorem 8.4.5 hold regardless of the values of a_1, a_2, and a_3. ▮

Sometimes one can evaluate the determinant of a matrix by first evaluating the inverse and then finding the determinant of the inverse. Of course this generally requires more work than evaluating the determinant directly. However, sometimes by discerning a certain pattern, one may be able to write the inverse directly, and from this it may be simple to evaluate the determinant.

8.5 Characteristic Equations and Roots of Some Patterned Matrices

In this section we shall state some theorems that can be used to evaluate the characteristic equations and, hence, the characteristic roots of certain matrices that occur in statistics.

Theorem 8.5.1

If $\mathbf{A}$ *is a* $k \times k$ *triangular matrix, then the characteristic equation is*

$$\prod_{i=1}^{k}(a_{ii} - \lambda) = 0, \tag{8.5.1}$$

and the characteristic roots are $a_{11}, a_{22}, \ldots, a_{kk}$.

Proof: The proof of this theorem follows from the fact that the matrix $\mathbf{A} - \lambda\mathbf{I}$

is also triangular with diagonal elements $a_{ii} - \lambda$, and the determinant of a triangular matrix is the product of the diagonal elements. ▮

The characteristic equation and roots of the matrix **C** given in Theorem 8.3.2 are difficult to evaluate in general, but sometimes one can take advantage of the fact that **C** is the product of five matrices, each of which has a special pattern; that is,

$$\mathbf{C} = \mathbf{D}_1\mathbf{T}\mathbf{D}_2\mathbf{T}'\mathbf{D}_1 \quad \text{and} \quad |\mathbf{C}| = |\mathbf{D}_1||\mathbf{T}||\mathbf{D}_2||\mathbf{T}'||\mathbf{D}_1|.$$

Theorem 8.5.2

If the matrix **C** *is as defined in Theorem* 8.3.3, that is, *if* $\mathbf{C} = \mathbf{D} + \alpha\mathbf{ab}'$, *then the characteristic equation is*

$$\left(1 + \alpha\sum_{i=1}^{k} \frac{a_i b_i}{d_{ii} - \lambda}\right) \prod_{i=1}^{k}(d_{ii} - \lambda) = 0. \tag{8.5.2}$$

Proof: The characteristic equation is obtained from evaluating the determinant of $\mathbf{C} - \lambda\mathbf{I} = \mathbf{D} - \lambda\mathbf{I} + \alpha\mathbf{ab}'$, but this has the same pattern as **C** and the determinant is given in Theorem 8.4.3. ▮

Theorem 8.5.3

In Theorem 8.5.2, *suppose that* $\mathbf{D} = d\mathbf{I}$; *then the characteristic equation is*

$$\left(d + \alpha\sum_{i=1}^{k} a_i b_i - \lambda\right)(d - \lambda)^{k-1} = 0, \tag{8.5.3}$$

and hence $k - 1$ *characteristic roots are equal to* d *and one root is equal to*

$$d + \alpha\sum_{i=1}^{k} a_i b_i. \tag{8.5.4}$$

Proof: Substitute d for d_{ii} in Theorem 8.5.2 and simplify. ▮

As a final remark in this section, we note that sometimes a matrix **C** has a pattern such that $\mathbf{C} - \lambda\mathbf{I}$ has the same pattern. This fact can often be used to find the characteristic equation and sometimes the roots from the evaluation of the determinant of **C**. For example, if **C** is the matrix in Theorem 8.3.4, then $\mathbf{C} - \lambda\mathbf{I}$ has the same pattern with a replaced by $a - \lambda$. A similar situation holds for the matrices in Eqs. (8.3.15) and (8.3.19).

8.6 Triangular Matrices

Triangular matrices play a significant role in the theories of multivariate analysis and regression. A triangular matrix is a special patterned matrix that is easily recognized, and it has many special properties that will be discussed in this section.

Theorem 8.6.1

Let $\mathbf{A}$ *be a* $k \times k$ *(real) matrix such that every leading principal minor is nonzero. Then* $\mathbf{A}$ *can be written as the product of a lower (real) and an upper (real) triangular matrix; that is,*

$$\mathbf{A} = \mathbf{RT}, \tag{8.6.1}$$

where $\mathbf{R}$ *is a lower (real) triangular matrix and* $\mathbf{T}$ *is an upper (real) triangular matrix. Further, if each of the diagonal elements of* $\mathbf{T}$ *(or* $\mathbf{R}$*) is set equal to unity, then the two triangular matrices are unique.*

Proof: For the proof of this theorem, we can use mathematical induction. If $k = 1$, then $a_{11} = r_{11}t_{11}$, and the theorem is certainly true. Suppose the theorem is true for every $(k-1) \times (k-1)$ matrix $\mathbf{A}_{11}$ that satisfies the hypothesis of the theorem; that is, there exist $(k-1) \times (k-1)$ lower and upper triangular matrices $\mathbf{R}_{11}$ and $\mathbf{T}_{11}$, respectively, such that

$$\mathbf{A}_{11} = \mathbf{R}_{11}\mathbf{T}_{11}.$$

We must then show that any $k \times k$ matrix $\mathbf{A}$ that satisfies the hypothesis of the theorem can also be written as the product of two triangular matrices. Let $\mathbf{A}$ be defined by

$$\mathbf{A} = \begin{bmatrix} \mathbf{A}_{11} & \mathbf{a}_{12} \\ \mathbf{a}_{21} & a_{kk} \end{bmatrix}$$

and, where $\mathbf{A}$ is partitioned, so that $\mathbf{A}_{11}$ is $(k-1) \times (k-1)$, etc. Since by hypothesis $\det(\mathbf{A}_{11}) \neq 0$, it follows that $\det(\mathbf{R}_{11}) \neq 0$ and $\det(\mathbf{T}_{11}) \neq 0$, so both $\mathbf{R}_{11}$ and $\mathbf{T}_{11}$ have inverses, and we can write

$$\mathbf{A} = \begin{bmatrix} \mathbf{A}_{11} & \mathbf{a}_{12} \\ \mathbf{a}_{21} & a_{kk} \end{bmatrix} = \begin{bmatrix} \mathbf{R}_{11} & \mathbf{0} \\ \mathbf{a}_{21}\mathbf{T}_{11}^{-1} & a_{kk} - \mathbf{a}_{21}\mathbf{T}_{11}^{-1}\mathbf{R}_{11}^{-1}\mathbf{a}_{12} \end{bmatrix} \begin{bmatrix} \mathbf{T}_{11} & \mathbf{R}_{11}^{-1}\mathbf{a}_{12} \\ \mathbf{0} & 1 \end{bmatrix} = \mathbf{RT}.$$

Thus we have shown that if the $(k-1) \times (k-1)$ matrix $\mathbf{A}_{11}$ can be written as a product of a lower and an upper triangular matrix, then any $k \times k$ matrix $\mathbf{A}$ with $\mathbf{A}_{11}$ in the upper left corner can also be so written. This result with the result for $k = 1$ completes the proof for the first part of the theorem. The proof for uniqueness is similar. ▮

We have stressed the fact that real matrices **R** and **T** exist by placing the word real in parentheses in the theorem. We shall continue to do this even though we remind the reader that all matrices and scalars in this book are real unless explicitly stated otherwise.

Theorem 8.6.2

If the $k \times k$ matrix **A** *is positive definite, then* **A** *can be written as*

$$\mathbf{A} = \mathbf{T}'\mathbf{T}, \tag{8.6.2}$$

where **T** *is an upper triangular matrix* (**T** *is unique except for signs*).

Proof: The proof follows the line of proof for the previous theorem except that **A** is assumed to be symmetric and positive definite. ▮

In the theorem it was stated that **T** is unique except for signs. By this we mean that for any matrix **T** such that $\mathbf{A} = \mathbf{T}'\mathbf{T}$, if any set of rows is multiplied by (-1) and a matrix $\mathbf{T}_1$ is obtained, then $\mathbf{A} = \mathbf{T}_1'\mathbf{T}_1$. There are thus 2^k distinct matrices that satisfy Eq. (8.6.2) for a given matrix **A**. If **A** is symmetric but *not* positive definite, such that the leading principal minors are nonzero, then there exists an upper triangular matrix **T** such that $\mathbf{A} = \mathbf{T}'\mathbf{T}$, but the elements of **T** may not be real numbers.

Example 8.6.1. Let **A** be defined by

$$\mathbf{A} = \begin{bmatrix} 1 & 2 & 0 \\ 2 & 5 & 1 \\ 0 & 1 & 17 \end{bmatrix}.$$

Clearly **A** is positive definite, and if we define **T** by

$$\mathbf{T} = \begin{bmatrix} 1 & 2 & 0 \\ 0 & 1 & 1 \\ 0 & 0 & 4 \end{bmatrix},$$

then $\mathbf{T}'\mathbf{T} = \mathbf{A}$. If we multiply the first two rows of $\mathbf{T}$ by (-1), we get

$$\mathbf{T}_1 = \begin{bmatrix} -1 & -2 & 0 \\ 0 & -1 & -1 \\ 0 & 0 & 4 \end{bmatrix},$$

and clearly $\mathbf{T}_1'\mathbf{T}_1 = \mathbf{A}$.

Corollary 8.6.2

If $\mathbf{A}$ is a positive definite $n \times n$ matrix, there exists a unique upper triangular (real) matrix $\mathbf{T}$ with $t_{ii} = 1$ such that $\mathbf{A} = \mathbf{T}'\mathbf{D}\mathbf{T}$, where $\mathbf{D}$ is a (real) diagonal matrix.

Theorem 8.6.3

The product of a finite number of lower (upper) triangular $k \times k$ matrices is a lower (upper) triangular matrix.

Proof: We shall prove the theorem for the case of two lower triangular matrices and the extension can be made by induction. Let $\mathbf{R}$ and $\mathbf{S}$ be two $k \times k$ lower triangular matrices, and we write $\mathbf{T} = \mathbf{RS}$. We want to show that $\mathbf{T}$ is a lower triangular matrix; that is, $t_{ij} = 0$ for $j > i$. We get

$$t_{ij} = \sum_{p=1}^{k} r_{ip} s_{pj} = \sum_{p=1}^{i} r_{ip} s_{pj},$$

since $r_{ip} = 0$ if $p > i$. But $s_{pj} = 0$ if p is less than j. In the second summation the subscript p goes from 1 to i, so if i is less than j, then the subscript p is always less than j and $s_{pj} = 0$. Thus $t_{ij} = 0$ if $i < j$. ▮

Theorem 8.6.4

If the inverse of a lower (upper) triangular $k \times k$ matrix exists, it is a lower (upper) triangular matrix.

Proof: If $\mathbf{T}$ is a lower (upper) triangular matrix, the cofactor of t_{ij} is clearly equal to zero if $i > j$ $(j > i)$; and therefore, $\mathbf{T}^{-1}$ is a lower (upper) triangular matrix. ▮

Theorem 8.6.5

The determinant of a lower (upper) triangular $k \times k$ matrix $\mathbf{T}$ *is equal to the product of the diagonal elements; that is,* $\det(\mathbf{T}) = \prod_{i=1}^{k} t_{ii}$.

Proof: Evaluate the determinant of a lower (upper) triangular matrix by the method of cofactors on the first row (column) and repeat the process. ∎

Theorem 8.6.6

The characteristic roots of a triangular $k \times k$ matrix $\mathbf{T}$ *are equal to* $t_{11}, t_{22}, \ldots, t_{kk}$.

Proof: The characteristic equation is clearly $\det(\mathbf{T} - \lambda\mathbf{I}) = 0$; but $\mathbf{T} - \lambda\mathbf{I}$ is a triangular matrix and, hence, by Theorem 8.6.5 we get

$$\det(\mathbf{T} - \lambda\mathbf{I}) = \prod_{i=1}^{k} (t_{ii} - \lambda).$$

Set this result equal to zero and the conclusion follows. ∎

Theorem 8.6.7

If $\mathbf{A}$ *is a $k \times k$ (real) symmetric matrix such that every leading principal minor is nonzero, there exists a (real) upper triangular matrix* $\mathbf{T}$ *and a diagonal matrix* $\mathbf{D}$ *with diagonal elements equal to plus and minus unity such that* $\mathbf{A} = \mathbf{T}'\mathbf{D}\mathbf{T}$.

Proof: The proof of this theorem follows from Theorem 8.6.1. ∎

Theorem 8.6.8

If $\mathbf{A}$ *is a $k \times k$ (real) matrix, there exists an (real) orthogonal matrix* $\mathbf{P}$ *such that* $\mathbf{PA} = \mathbf{T}$, *where* $\mathbf{T}$ *is an upper (real) triangular matrix and* $t_{ii} \geq 0$ *for each* $i = 1, 2, \ldots, k$.

Proof: We shall use induction to prove the theorem. If $k = 1$, the proof is trivial since we define $\mathbf{P} = 1$ if $a_{11} \geq 0$ and $\mathbf{P} = -1$ if $a_{11} < 0$ and $\mathbf{P}$ is an orthogonal 1×1 matrix. Also $\mathbf{PA} = \mathbf{T}$ gives $t_{11} = |a_{11}|$ and t_{11} is nonnegative and is a 1×1 upper triangular matrix. Let us assume that the theorem is true for any $(k-1) \times (k-1)$ matrix $\mathbf{B}$; that is, for a $(k-1) \times (k-1)$ matrix $\mathbf{B}$, let $\mathbf{R}$ be an orthogonal matrix such that $\mathbf{RB} = \mathbf{T}_1$, where $\mathbf{T}_1$ is an upper triangular $(k-1) \times (k-1)$ matrix with non-negative diagonal elements. Now form an orthogonal matrix $\mathbf{Q}$ where the first row $\mathbf{q}_1'$ is equal to $w\mathbf{a}_1'$ if $\mathbf{a}_1 \neq \mathbf{0}$ and the first row is equal to $(1, 0, \ldots, 0)$ if $\mathbf{a}_1 = \mathbf{0}$, where $\mathbf{a}_1$ is the first column of the matrix $\mathbf{A}$ and w is a scalar such that

$$w = \begin{cases} (\mathbf{a}_1'\mathbf{a}_1)^{-1/2}, & \text{if } \mathbf{a}_1 \neq \mathbf{0} \\ 0, & \text{if } \mathbf{a}_1 = \mathbf{0}. \end{cases}$$

We get (define w^{-1} to be zero if w is zero)

$$\mathbf{Q} = \begin{bmatrix} \mathbf{q}_1' \\ \mathbf{Q}_2 \end{bmatrix},$$

and notice that since $\mathbf{Q}$ is orthogonal we get $\mathbf{Q}_2\,\mathbf{q}_1 = \mathbf{0}$.

$$\mathbf{PA} = \begin{bmatrix} 1 & \mathbf{0} \\ \mathbf{0} & \mathbf{R} \end{bmatrix}\mathbf{QA} = \begin{bmatrix} 1 & \mathbf{0} \\ \mathbf{0} & \mathbf{R} \end{bmatrix}\begin{bmatrix} \mathbf{q}_1' \\ \mathbf{Q}_2 \end{bmatrix}[\mathbf{a}_1, \quad \mathbf{A}_2]$$

$$= \begin{bmatrix} 1 & \mathbf{0} \\ \mathbf{0} & \mathbf{R} \end{bmatrix}\begin{bmatrix} w^{-1} & \mathbf{q}_1'\mathbf{A}_2 \\ \mathbf{0} & \mathbf{Q}_2\,\mathbf{A}_2 \end{bmatrix} = \begin{bmatrix} w^{-1} & \mathbf{q}_1'\mathbf{A}_2 \\ \mathbf{0} & \mathbf{RQ}_2\,\mathbf{A}_2 \end{bmatrix} = \mathbf{T},$$

since, by the hypothesis of the induction, $\mathbf{R}$ is an orthogonal matrix such that $\mathbf{R}(\mathbf{Q}_2\mathbf{A}_2) = \mathbf{RB}$ is an upper triangular matrix with non-negative diagonal elements, and hence $\mathbf{T}$ is an upper triangular matrix with non-negative diagonal elements. But notice that

$$\begin{bmatrix} 1 & \mathbf{0} \\ \mathbf{0} & \mathbf{R} \end{bmatrix}$$

is an orthogonal matrix and $\mathbf{Q}$ is an orthogonal matrix; hence

$$\mathbf{P} = \begin{bmatrix} 1 & \mathbf{0} \\ \mathbf{0} & \mathbf{R} \end{bmatrix}\mathbf{Q}$$

is an orthogonal matrix and, by induction, the proof is complete. Note that each matrix involved can be taken to be real, since $\mathbf{A}$ is assumed to be real. ∎

Theorem 8.6.9

If $\mathbf{A}$ is a $k \times k$ (real) matrix, there exists a nonsingular matrix $\mathbf{P}$ (not necessarily real) such that $\mathbf{P}^{-1}\mathbf{AP}$ is an upper triangular matrix, $\mathbf{T}$ (not necessarily real).

The proofs of this theorem and the following two corollaries are left for the reader. The proofs can be found in [S–8].

Corollary 8.6.9.1

In Theorem 8.6.9, *the diagonal elements of the triangular matrix* $\mathbf{T}$ *are the characteristic roots of* $\mathbf{A}$.

Corollary 8.6.9.2

In Theorem 8.6.9, *if* $\mathbf{A}$ *and the roots of* $\mathbf{A}$ *are real, then there exists a real non-singular matrix* $\mathbf{P}$ *such that* $\mathbf{P}^{-1}\mathbf{AP}$ *is a real upper triangular matrix* $\mathbf{T}$ *and the diagonal elements of* $\mathbf{T}$ *are the characteristic roots of* $\mathbf{A}$.

Theorem 8.6.10

If $\mathbf{T}$ *is a (real)* $k \times k$ *upper (lower) triangular matrix and if* $\mathbf{T'T} = \mathbf{TT'}$, *then* $\mathbf{T}$ *is a diagonal matrix.*

Proof: Let $\mathbf{T}$ be an upper triangular matrix ($t_{pq} = 0$ if $p > q$). Set $\mathbf{A} = \mathbf{T'T}$, $\mathbf{B} = \mathbf{TT'}$. Then

$$a_{ij} = \sum_{m=1}^{k} t_{mi} t_{mj}, \qquad b_{ij} = \sum_{n=1}^{k} t_{in} t_{jn}$$

and

$$a_{ii} = \sum_{m=1}^{i} t_{mi}^2 = b_{ii} = \sum_{n=i}^{k} t_{in}^2.$$

So when $i = 1$, we get $t_{11}^2 = \sum_{n=1}^{k} t_{1n}^2$ or $t_{12} = t_{13} = \cdots = t_{1k} = 0$. When $i = 2$, we get

$$t_{12}^2 + t_{22}^2 = t_{22}^2 + t_{23}^2 + \cdots + t_{2k}^2,$$

which gives

$$t_{23} = t_{24} = \cdots = t_{2k} = 0.$$

By continuing in this fashion we find that $t_{pq} = 0$ if $p < q$. Hence, since $\mathbf{T}$ was assumed to be an upper triangular matrix, the result is that $\mathbf{T}$ is diagonal. ▌

The proofs of the next three theorems are left for the reader.

Theorem 8.6.11

Let $\mathbf{T}$ *be an upper (lower) triangular* $k \times k$ *matrix with* i-th *diagonal element equal to* t_{ii}. *The* i-th *diagonal element of* $\mathbf{T}^n$ *is* t_{ii}^n.

Theorem 8.6.12

Let $\mathbf{T}$ *be a nonsingular* $k \times k$ *triangular matrix and denote* $\mathbf{T}^{-1}$ *by* $\mathbf{B}$; *then* $t_{ii} b_{ii} = 1$ *for* $i = 1, 2, \ldots, k$.

Theorem 8.6.13

Let $\mathbf{A}$ *be a* $k \times k$ *matrix with real characteristic roots. There exists an orthogonal matrix* $\mathbf{P}$ *such that* $\mathbf{P}'\mathbf{A}\mathbf{P} = \mathbf{T}$ *where* $\mathbf{T}$ *is a real upper triangular matrix with the characteristic roots of* $\mathbf{A}$ *on the diagonal of* $\mathbf{T}$.

The final theorem in this section is an extension of Theorem 8.6.2.

Theorem 8.6.14

Let $\mathbf{A}$ *be a* $k \times k$ *positive semidefinite matrix. There exists an upper triangular (real) matrix* $\mathbf{T}$ *such that* $\mathbf{A} = \mathbf{T}'\mathbf{T}$.

Proof: Since $\mathbf{A}$ is positive semidefinite, there exists a matrix $\mathbf{B}$ such that $\mathbf{A} = \mathbf{B}'\mathbf{B}$; and by Theorem 8.6.8 there exists an orthogonal matrix $\mathbf{P}$ such that $\mathbf{P}\mathbf{B} = \mathbf{T}$, where $\mathbf{T}$ is upper triangular. Hence we have $\mathbf{A} = \mathbf{B}'\mathbf{P}'\mathbf{P}\mathbf{B} = \mathbf{T}'\mathbf{T}$, and the theorem is proved. ▮

8.7 Correlation Matrix

Let y_1 and y_2 be two random variables. We denote the covariance of y_1 and y_2 by v_{12}, the variance of y_1 by v_{11}, and the variance of y_2 by v_{22}. The correlation between the two random variables is denoted by ρ_{12} and defined by $\rho_{12} = v_{12}/\sqrt{v_{11}v_{22}}$. To generalize these ideas, let $\mathbf{y}$ be an $n \times 1$ random vector and let the $n \times n$ positive definite matrix $\mathbf{V}$ denote the covariance matrix of the vector $\mathbf{y}$. This merely states that v_{ij} is the covariance of y_i and y_j for $i \neq j$, and v_{ii} is the variance of y_i. We can also define an $n \times n$ matrix $\mathbf{R} = [\rho_{ij}]$ where ρ_{ij}, which is the correlation between y_i and y_j, is defined above. We note from above that ρ_{ii} represents the correlation of y_i with itself and is equal to 1.

Definition 8.7.1

Correlation Matrix. *Let* $\mathbf{y}$ *be an* $n \times 1$ *random vector with positive definite covariance matrix denoted by* $\mathbf{V}$. *The correlation matrix of* $\mathbf{y}$ *is denoted by* $\mathbf{R} = [\rho_{ij}]$ *where* ρ_{ij} *is defined by*

$$\rho_{ij} = \frac{v_{ij}}{\sqrt{v_{ii} v_{jj}}}$$

for all i and j.

Note: It may be desirable to discuss the correlation matrix of a random vector that has a positive semidefinite covariance matrix. However, we shall be concerned here with only a positive definite covariance matrix.

We shall let $\mathbf{D}_v$ denote a diagonal matrix with the i-th diagonal element equal to v_{ii}, the i-th diagonal element of $\mathbf{V}$. We note that $\mathbf{D}_v^{-1}$ exists, since $\mathbf{V}$ is assumed to be positive definite, and hence $v_{ii} > 0$ for all i.

Theorem 8.7.1

The correlation matrix $\mathbf{R}$ *of a random vector* $\mathbf{y}$ *with positive definite covariance matrix* $\mathbf{V}$ *is determined from* $\mathbf{V}$ *by*

$$\mathbf{R} = \mathbf{D}_v^{-1/2}\mathbf{V}\mathbf{D}_v^{-1/2},$$

where $\mathbf{D}_v$ *is a diagonal matrix with* i-th *diagonal element* v_{ii}.

Proof: The ij-th element of $\mathbf{D}_v^{-1/2}\mathbf{V}\mathbf{D}_v^{-1/2}$ is clearly equal to $v_{ij}/\sqrt{v_{ii}v_{jj}}$, which is the definition of ρ_{ij}. ▌

Theorem 8.7.2

A correlation matrix is positive definite.

Proof: Since $\mathbf{V}$ is positive definite and since $\mathbf{D}_v^{-1/2}$ is nonsingular and symmetric, the result follows. ▌

Theorem 8.7.3

In a correlation matrix the following relationships hold:

(a) $\rho_{ii} = 1; \quad i = 1, 2, \ldots, n,$
(b) $-1 < \rho_{ij} < 1; \quad$ *all* $i \neq j$.

Proof: The relationship (a) is obtained by evaluating the i-th diagonal element of $\mathbf{D}_v^{-1/2}\mathbf{V}\mathbf{D}_v^{-1/2}$. The relationship (b) is obtained by setting the i-th element of a vector $\mathbf{x}$ equal to $+1$, the j-th element equal to $+1$, and the remaining elements equal to zero. From $\mathbf{x}'\mathbf{R}\mathbf{x} > 0$, we obtain $\rho_{ii} + \rho_{ij} + \rho_{ji} + \rho_{jj} > 0$ or $\rho_{ij} > -1$. By changing x_j to -1 we get $\rho_{ij} < 1$, and the proof is complete. ∎

Theorem 8.7.4

Let $\mathbf{R}$ *be an* $n \times n$ *correlation matrix. The largest characteristic root of* $\mathbf{R}$ *is less than* n.

Proof: Clearly $n = \sum_{i=1}^{n} \lambda_i$ but each λ_i is positive, and hence the result follows. ∎

Theorem 8.7.5

Let the $n \times n$ *matrix* $\mathbf{R}$ *be defined by* $\mathbf{R} = (1 - \rho)\mathbf{I} + \rho\mathbf{J}$. $\mathbf{R}$ *is a correlation matrix if and only if* $-1/(n-1) < \rho < 1$.

The proof of this theorem is left for the reader.

Theorem 8.7.6

If $\mathbf{R}$ *is any correlation matrix, then*

$$0 < |\mathbf{R}| \leq 1.$$

The proof of this theorem is left for the reader.

8.8 Direct Product and Sum of Matrices

In the theory of the design of experiments it is sometimes advantageous to use a method of multiplication of two matrices that is different from the one we have used so far. This method, called the *direct product* or sometimes referred to as the Kronecker product of matrices, is especially useful in some situations when one is working with blocks of submatrices, as when matrices are partitioned. References [C–10] and [S–5] may be consulted to find out more about the importance of the direct product of matrices in the design of experiments.

Definition 8.8.1

Direct Product. *Let* **A** *be an* $m_2 \times n_2$ *matrix and let* **B** *be an* $m_1 \times n_1$ *matrix; then the direct product of* **A** *and* **B**, *which we write as* **A × B**, *is a matrix* **C** *of size* $m_1 m_2 \times n_1 n_2$ *defined by*

$$\mathbf{C} = \begin{bmatrix} \mathbf{A}b_{11} & \mathbf{A}b_{12} & \cdots & \mathbf{A}b_{1n_1} \\ \mathbf{A}b_{21} & \mathbf{A}b_{22} & \cdots & \mathbf{A}b_{2n_1} \\ \vdots & \vdots & & \vdots \\ \mathbf{A}b_{m_11} & \mathbf{A}b_{m_12} & \cdots & \mathbf{A}b_{m_1n_1} \end{bmatrix} = \begin{bmatrix} b_{11}\mathbf{A} & b_{12}\mathbf{A} & \cdots & b_{1n_1}\mathbf{A} \\ b_{21}\mathbf{A} & b_{22}\mathbf{A} & \cdots & b_{2n_1}\mathbf{A} \\ \vdots & \vdots & & \vdots \\ b_{m_11}\mathbf{A} & b_{m_12}\mathbf{A} & \cdots & b_{m_1n_1}\mathbf{A} \end{bmatrix}.$$

Actually the definition is a *left* direct product. One could also define a *right* direct product.

Notice that **C** contains $m_1 n_1$ submatrices each of the size $m_2 \times n_2$, and the ij-th submatrix, denoted by $\mathbf{C}_{ij}$, is $\mathbf{A}b_{ij}$. We sometimes write

$$\mathbf{C} = [\mathbf{C}_{ij}] = [\mathbf{A}b_{ij}], \qquad i = 1, 2, \ldots, m_1; \quad j = 1, 2, \ldots, n_1.$$

Note also that the direct product of two matrices is defined for any size matrices.

Example 8.8.1. The matrices **A** and **B** are defined by

$$\mathbf{A} = \begin{bmatrix} 3 & 1 \\ -1 & 0 \end{bmatrix}; \qquad \mathbf{B} = [1, 4].$$

Then the direct product **A × B** is

$$\mathbf{A} \times \mathbf{B} = [1\mathbf{A}, 4\mathbf{A}] = \begin{bmatrix} 3 & 1 & 12 & 4 \\ -1 & 0 & -4 & 0 \end{bmatrix}.$$

Also the direct product **B × A** is

$$\mathbf{B} \times \mathbf{A} = \begin{bmatrix} 3\mathbf{B} & 1\mathbf{B} \\ -1\mathbf{B} & 0\mathbf{B} \end{bmatrix} = \begin{bmatrix} 3 & 12 & 1 & 4 \\ -1 & -4 & 0 & 0 \end{bmatrix};$$

note that $\mathbf{B} \times \mathbf{A} \neq \mathbf{A} \times \mathbf{B}$.

Theorem 8.8.1

The direct product of two matrices exists for any two matrices **A** *and* **B**, *and in general* $\mathbf{A} \times \mathbf{B} \neq \mathbf{B} \times \mathbf{A}$.

Example 8.8.2. Let $\mathbf{I}$ be the $m_1 \times m_1$ identity matrix and let $\mathbf{A}$ be any $m_2 \times n_2$ matrix; then

$$\mathbf{A} \times \mathbf{I} = \begin{bmatrix} \mathbf{A} & \mathbf{0} & \cdots & \mathbf{0} \\ \mathbf{0} & \mathbf{A} & \cdots & \mathbf{0} \\ \vdots & & & \vdots \\ \mathbf{0} & \mathbf{0} & \cdots & \mathbf{A} \end{bmatrix}$$

is an $m_1m_2 \times m_1n_2$ matrix that is a block diagonal matrix. If $\mathbf{A} = \mathbf{I}$, then of course $\mathbf{I} \times \mathbf{I}$ is the $m_1m_2 \times m_1m_2$ identity matrix.

Before proceeding we shall develop some notation for submatrices, and sub-submatrices, etc., that will be useful in working with the direct product of matrices.

Suppose that $\mathbf{G}$ is a matrix composed of submatrices $\mathbf{G}_{i_1, j_1}$ such that there are m_1 rows of submatrices and n_1 columns; that is,

$$\mathbf{G} = \begin{bmatrix} \mathbf{G}_{11} & \mathbf{G}_{12} & \cdots & \mathbf{G}_{1n_1} \\ \mathbf{G}_{21} & \mathbf{G}_{22} & \cdots & \mathbf{G}_{2n_1} \\ \vdots & \vdots & & \vdots \\ \mathbf{G}_{m_11} & \mathbf{G}_{m_12} & \cdots & \mathbf{G}_{m_1n_1} \end{bmatrix}.$$

Suppose that each matrix $\mathbf{G}_{i_1, j_1}$ is composed of submatrices such that

$$\mathbf{G}_{11} = \begin{bmatrix} \mathbf{G}^*_{11} & \mathbf{G}^*_{12} & \cdots & \mathbf{G}^*_{1n_2} \\ \mathbf{G}^*_{21} & \mathbf{G}^*_{22} & \cdots & \mathbf{G}^*_{2n_2} \\ \vdots & \vdots & & \vdots \\ \mathbf{G}^*_{m_21} & \mathbf{G}^*_{m_2\,2} & \cdots & \mathbf{G}^*_{m_2n_2} \end{bmatrix},$$

where $\mathbf{G}_{22}$, and so on, are defined similarly and where, say, each $\mathbf{G}^*_{i_2, j_2}$ has size $m_3 \times n_3$. Of course $\mathbf{G}^*_{i_2, j_2}$ could be thought of as consisting of submatrices, etc., but we shall have occasion to use only submatrices and sub-submatrices. We shall use the notation

$$g_3(i_1, j_1 : i_2, j_2 : i_3, j_3)$$

to represent a certain element in a matrix $\mathbf{G}$. The subscript 3 on g means that $\mathbf{G}$ has 3 levels of submatrices, the final level is composed of scalars, and g represents a certain one of these scalars. Of course, the fact that the argument of g has three sets of subscripts $i_1, j_1 : i_2, j_2 : i_3, j_3$ indicates this also. The indicated subscripts will always be defined to range over certain values as follows:

$$\begin{aligned} i_1 &= 1, 2, \ldots, m_1; & j_1 &= 1, 2, \ldots, n_1 \\ i_2 &= 1, 2, \ldots, m_2; & j_2 &= 1, 2, \ldots, n_2 \\ i_3 &= 1, 2, \ldots, m_3; & j_3 &= 1, 2, \ldots, n_3. \end{aligned}$$

So by writing

$$\mathbf{G} = [g_3(i_1, j_1 : i_2, j_2 : i_3, j_3)],$$

we mean **G** has size $m_1 m_2 m_3 \times n_1 n_2 n_3$ and **G** is composed of m_1 rows and n_1 columns of block matrices and each submatrix $\mathbf{G}_{i_1, j_1}$ has size $m_2 m_3 \times n_2 n_3$. Also, each submatrix is composed of m_2 rows and n_2 columns of block matrices (called sub-submatrices of **G**), and each has size $m_3 \times n_3$. Further,

$$g_3(i_1, j_1 : i_2, j_2 : i_3, j_3)$$

is an element in a sub-submatrix. More specifically, it is the (i_3, j_3)-th element in the (i_2, j_2)-th submatrix of $\mathbf{G}_{i_1, j_1}$. To state this another way it is in the i_1, j_1 submatrix of **G**, and the i_2, j_2 submatrix of the resulting matrix, and the (i_3, j_3)-th element in the sub-submatrix. For a simple illustration, let $m_1 = m_2 = m_3 = n_1 = n_2 = n_3 = 2$. We get

$$\mathbf{G} = \left[\begin{array}{cc:cc|cc:cc} 2 & 1 & -1 & 3 & -2 & 0 & 1 & 4 \\ 1 & 3 & -2 & 4 & 5 & -9 & 8 & 7 \\ \hdashline -2 & 1 & 4 & 0 & 6 & 18 & 4 & -9 \\ -3 & 1 & 0 & 4 & 7 & 2 & 1 & 8 \\ \hline 4 & 10 & -9 & 3 & 2 & 1 & 16 & 8 \\ 1 & 0 & -5 & 4 & 3 & 2 & 0 & 7 \\ \hdashline -2 & -1 & 4 & 5 & 16 & 2 & -10 & -7 \\ 9 & -4 & 1 & 2 & -3 & 0 & -6 & 61 \end{array}\right];$$

$$\mathbf{G}_{11} = \left[\begin{array}{cc:cc} 2 & 1 & -1 & 3 \\ 1 & 3 & -2 & 4 \\ \hdashline -2 & 1 & 4 & 0 \\ -3 & 1 & 0 & 4 \end{array}\right]; \quad \mathbf{G}_{12} = \left[\begin{array}{cc:cc} -2 & 0 & 1 & 4 \\ 5 & -9 & 8 & 7 \\ \hdashline 6 & 18 & 4 & -9 \\ 7 & 2 & 1 & 8 \end{array}\right];$$

and so on. Also

$$g_3(1, 2 : 2, 1 : 1, 1) = 6,$$

since the first pair of numbers 1, 2 indicates block $\mathbf{G}_{12}$ and the second pair of numbers 2, 1 indicates subblock 2, 1, which is

$$\begin{bmatrix} 6 & 18 \\ 7 & 2 \end{bmatrix},$$

and the third pair of numbers 1, 1 indicates the element in this block—that is, the element in the first row and first column which is equal to 6. Similarly,

$$g_3(2, 2 : 1, 1 : 1, 2) = 1,$$

$$g_3(1, 1 : 2, 1 : 1, 1) = -2,$$

and so on. Also,

$$\mathbf{G}_2(1, 1 : 2, 2) = \begin{bmatrix} 4 & 0 \\ 0 & 4 \end{bmatrix},$$

$$\mathbf{G}_2(1, 2 : 1, 1) = \begin{bmatrix} -2 & 0 \\ 5 & -9 \end{bmatrix},$$

and

$$\mathbf{G}_1(1, 2) = \mathbf{G}_{12}, \text{ etc.}$$

In this example g_4, g_5 and so forth are not defined. Consider an example in which a matrix $\mathbf{F}$ is divided into submatrices, but the submatrices are not further divided;

$$\mathbf{F} = \left[\begin{array}{cc|cc|cc} 1 & 3 & 2 & 1 & 3 & 1 \\ -1 & 4 & 5 & 0 & -1 & 2 \\ \hline 6 & 1 & 8 & 7 & -1 & 4 \\ 2 & 9 & 1 & 5 & -4 & 0 \end{array}\right].$$

Then

$$\mathbf{F}_1(2, 2) = \begin{bmatrix} 8 & 7 \\ 1 & 5 \end{bmatrix}, \quad \mathbf{F}_1(1, 3) = \begin{bmatrix} 3 & 1 \\ -1 & 2 \end{bmatrix}, \quad \text{etc;}$$

$f_2(1, 2 : 2, 1) = 5$; $f_2(2, 3 : 1, 2) = 4$; $f_2(1, 3 : 1, 1) = 3$, etc., and $f_3, f_4, \ldots$ are not defined. Note that if we define **F** by

$$\mathbf{F} = \mathbf{A} \times \mathbf{B},$$

then

$$\mathbf{F}_1(i_1, j_1) = b_{i_1 j_1}\mathbf{A}$$

and

$$f_2(i_1, j_1 : i_2, j_2) = b_{i_1 j_1}\, a_{i_2 j_2}, \text{ etc.}$$

In $\mathbf{A} \times \mathbf{B}$ in Ex, 8.8.1, we get $m_1 = 1, n_1 = 2; m_2 = 2, n_2 = 2$;

$$\mathbf{F}_1(1, 2) = \begin{bmatrix} 12 & 4 \\ -4 & 0 \end{bmatrix},$$

$$f_2(1, 1 : 2, 1) = b_{11}a_{21} = -1; \quad f_2(1, 2 : 2, 2) = b_{12}a_{22} = 0.$$

Theorem 8.8.2

*Let a be any scalar and **A** and **B** be any matrices; then*

$$(a\mathbf{A}) \times \mathbf{B} = \mathbf{A} \times (a\mathbf{B}) = a(\mathbf{A} \times \mathbf{B}).$$

Proof: The result follows directly from Def. 8.8.1. ∎

Theorem 8.8.3

*Let **A**, **B**, and **C** be any matrices; then*

$$(\mathbf{A} \times \mathbf{B}) \times \mathbf{C} = \mathbf{A} \times (\mathbf{B} \times \mathbf{C}).$$

Proof: We shall use the following notation:

$$\mathbf{B} \times \mathbf{C} = \mathbf{E}, \qquad \mathbf{A} \times \mathbf{B} = \mathbf{F},$$

$$(\mathbf{A} \times \mathbf{B}) \times \mathbf{C} = \mathbf{G}, \qquad \mathbf{A} \times (\mathbf{B} \times \mathbf{C}) = \mathbf{H}.$$

We must show that $\mathbf{G} = \mathbf{H}$. Now

$$\mathbf{G} = \mathbf{F} \times \mathbf{C}; \qquad \mathbf{G}_1(i_1, j_1) = \mathbf{F}c_{i_1, j_1},$$

$$\mathbf{G}_2(i_1, j_1 : i_2, j_2) = \mathbf{A}b_{i_2, j_2} c_{i_1, j_1},$$

and

$$g_3(i_1, j_1 : i_2, j_2 : i_3, j_3) = a_{i_3, j_3} b_{i_2, j_2} c_{i_1, j_1}.$$

Also

$$\mathbf{H} = \mathbf{A} \times \mathbf{E}; \qquad \mathbf{E}_1(i_1, j_1) = \mathbf{B}c_{i_1, j_1};$$

$$e_2(i_1, j_1 : i_2, j_2) = b_{i_2, j_2} c_{i_1, j_1};$$

and

$$h_3(i_1, j_1 : i_2, j_2 : i_3, j_3) = a_{i_3, j_3} e_2(i_1, j_1 : i_2, j_2) = a_{i_3, j_3} b_{i_2, j_2} c_{i_1, j_1};$$

hence $(\mathbf{A} \times \mathbf{B}) \times \mathbf{C} = \mathbf{A} \times (\mathbf{B} \times \mathbf{C})$. In general we shall write each expression as $\mathbf{A} \times \mathbf{B} \times \mathbf{C}$. ▮

Theorem 8.8.4

Let $\mathbf{A}$ *and* $\mathbf{B}$ *be any matrices; then*

$$(\mathbf{A} \times \mathbf{B})' = \mathbf{A}' \times \mathbf{B}'.$$

Proof: Let $\mathbf{C} = \mathbf{A} \times \mathbf{B}$; then

$$\mathbf{C} = [\mathbf{C}_{ij}] = [\mathbf{A}b_{ij}], \quad \text{but} \quad \mathbf{C}' = [\mathbf{C}'_{ji}] = [\mathbf{A}'b_{ji}] = \mathbf{A}' \times \mathbf{B}'. \quad ▮$$

Example 8.8.3. Let $\mathbf{A}$ and $\mathbf{B}$ be defined as in Example 8.8.1.

$$\mathbf{A}' \times \mathbf{B}' = \begin{bmatrix} 3 & -1 \\ 1 & 0 \end{bmatrix} \times \begin{bmatrix} 1 \\ 4 \end{bmatrix} = \begin{bmatrix} 3 & -1 \\ 1 & 0 \\ 12 & -4 \\ 4 & 0 \end{bmatrix} = (\mathbf{A} \times \mathbf{B})'$$

Theorem 8.8.5

Let **A** *and* **B** *each be square matrices; then*

$$\operatorname{tr}(\mathbf{A} \times \mathbf{B}) = [\operatorname{tr}(\mathbf{A})][\operatorname{tr}(\mathbf{B})].$$

Proof: Let $\mathbf{C} = \mathbf{A} \times \mathbf{B}$; then $\mathbf{C}_{ii} = \mathbf{A}b_{ii}$ and

$$\operatorname{tr}(\mathbf{C}) = \sum_{i=1}^{m_1} \operatorname{tr}\mathbf{C}_{ii} = \sum_{i=1}^{m_1} \operatorname{tr}(\mathbf{A}b_{ii}) = \operatorname{tr}(\mathbf{A}) \sum_{i=1}^{m_1} b_{ii} = [\operatorname{tr}(\mathbf{A})][\operatorname{tr}(\mathbf{B})]. \quad \blacksquare$$

Theorem 8.8.6

Let **A** *be an* $m_1 \times n_1$ *matrix,* **B** *an* $m_2 \times n_2$ *matrix,* **F** *an* $n_1 \times k_1$ *matrix, and* **G** *an* $n_2 \times k_2$ *matrix. Then*

$$(\mathbf{A} \times \mathbf{B})(\mathbf{F} \times \mathbf{G}) = (\mathbf{AF}) \times (\mathbf{BG}).$$

Proof: The matrix $\mathbf{A} \times \mathbf{B}$ has size $m_1 m_2 \times n_1 n_2$ and can be written as a block matrix with the *it*-th block equal to $\mathbf{A}b_{it}$. The matrix $\mathbf{F} \times \mathbf{G}$ has size $n_1 n_2 \times k_1 k_2$ and can be written as a block matrix with the *tj*-th block equal to $\mathbf{F}g_{tj}$. If we perform the multiplication by blocks, we get for the *ij*-th block of $(\mathbf{A} \times \mathbf{B})$ $(\mathbf{F} \times \mathbf{G})$ the following:

$$\sum_{t=1}^{n_2} \mathbf{A}b_{it} \mathbf{F}g_{tj} = \mathbf{AF} \sum_{t=1}^{n_2} b_{it} g_{tj}.$$

But $\sum_{t=1}^{n_2} b_{it} g_{tj}$ is the *ij*-th element of the product **BG**; hence $(\mathbf{A} \times \mathbf{B})(\mathbf{F} \times \mathbf{G}) = (\mathbf{AF}) \times (\mathbf{BG})$.
It may be useful to write these matrices in detail. We get

$$\mathbf{A} \times \mathbf{B} = \begin{bmatrix} \mathbf{A}b_{11} & \mathbf{A}b_{12} & \cdots & \mathbf{A}b_{1n_2} \\ \vdots & \vdots & & \vdots \\ \mathbf{A}b_{m_2 1} & \mathbf{A}b_{m_2 2} & \cdots & \mathbf{A}b_{m_2 n_2} \end{bmatrix},$$

$$\mathbf{F} \times \mathbf{G} = \begin{bmatrix} \mathbf{F}g_{11} & \mathbf{F}g_{12} & \cdots & \mathbf{F}g_{1k_2} \\ \vdots & \vdots & & \vdots \\ \mathbf{F}g_{n_2 1} & \mathbf{F}g_{n_2 2} & \cdots & \mathbf{F}g_{n_2 k_2} \end{bmatrix},$$

and clearly $(\mathbf{A} \times \mathbf{B})(\mathbf{F} \times \mathbf{G}) = (\mathbf{AF}) \times (\mathbf{BG})$. ▮

Example 8.8.4. Find $(\mathbf{A} \times \mathbf{B})(\mathbf{F} \times \mathbf{G})$ and $(\mathbf{AF}) \times (\mathbf{BG})$ for the matrices defined below.

$$\mathbf{A} = \begin{bmatrix} 2 & 1 & 1 \\ -1 & 0 & 1 \end{bmatrix}; \quad \mathbf{B} = \begin{bmatrix} 1 & 4 \\ -3 & 2 \end{bmatrix}; \quad \mathbf{F} = \begin{bmatrix} 1 \\ -1 \\ 5 \end{bmatrix}; \quad \mathbf{G} = \begin{bmatrix} 3 & 2 \\ 1 & 4 \end{bmatrix};$$

$$\mathbf{A} \times \mathbf{B} = \begin{bmatrix} 2 & 1 & 1 & 8 & 4 & 4 \\ -1 & 0 & 1 & -4 & 0 & 4 \\ -6 & -3 & -3 & 4 & 2 & 2 \\ 3 & 0 & -3 & -2 & 0 & 2 \end{bmatrix}; \quad \mathbf{F} \times \mathbf{G} = \begin{bmatrix} 3 & 2 \\ -3 & -2 \\ 15 & 10 \\ 1 & 4 \\ -1 & -4 \\ 5 & 20 \end{bmatrix};$$

$$(\mathbf{A} \times \mathbf{B})(\mathbf{F} \times \mathbf{G}) = \begin{bmatrix} 42 & 108 \\ 28 & 72 \\ -42 & 12 \\ -28 & 8 \end{bmatrix};$$

$$\mathbf{AF} = \begin{bmatrix} 6 \\ 4 \end{bmatrix}; \quad \mathbf{BG} = \begin{bmatrix} 7 & 18 \\ -7 & 2 \end{bmatrix};$$

and

$$(\mathbf{AF}) \times (\mathbf{BG}) = (\mathbf{A} \times \mathbf{B})(\mathbf{F} \times \mathbf{G}).$$

Theorem 8.8.7

Let $\mathbf{A}$ *be an* $m_1 \times m_1$ *nonsingular matrix and* $\mathbf{B}$ *an* $m_2 \times m_2$ *nonsingular matrix. Then* $\mathbf{A} \times \mathbf{B}$ *is nonsingular and the inverse is given by*

$$(\mathbf{A} \times \mathbf{B})^{-1} = \mathbf{A}^{-1} \times \mathbf{B}^{-1}.$$

Proof: $\mathbf{A} \times \mathbf{B}$ has dimension $m_1 m_2 \times m_1 m_2$ and hence is a square matrix. By Theorem 8.8.6 we get

$$(\mathbf{A} \times \mathbf{B})(\mathbf{A}^{-1} \times \mathbf{B}^{-1}) = (\mathbf{AA}^{-1}) \times (\mathbf{BB}^{-1}) = \mathbf{I} \times \mathbf{I} = \mathbf{I},$$

since $\mathbf{I} \times \mathbf{I}$ is the $m_1 m_2 \times m_1 m_2$ identity matrix. ∎

Theorem 8.8.8

If $\mathbf{P}$ *and* $\mathbf{Q}$ *are orthogonal matrices, then* $\mathbf{P} \times \mathbf{Q}$ *is an orthogonal matrix.*

Proof: We shall show that $(\mathbf{P} \times \mathbf{Q})'(\mathbf{P} \times \mathbf{Q})$ is equal to the identity matrix, and since $\mathbf{P} \times \mathbf{Q}$ is a square matrix, this implies that $\mathbf{P} \times \mathbf{Q}$ is an orthogonal matrix. By Theorem 8.8.4 we get $(\mathbf{P} \times \mathbf{Q})' = \mathbf{P}' \times \mathbf{Q}'$, and hence

$$(\mathbf{P} \times \mathbf{Q})'(\mathbf{P} \times \mathbf{Q}) = (\mathbf{P}' \times \mathbf{Q}')(\mathbf{P} \times \mathbf{Q}) = (\mathbf{P}'\mathbf{P}) \times (\mathbf{Q}'\mathbf{Q}) = \mathbf{I} \times \mathbf{I} = \mathbf{I}. \quad \blacksquare$$

Theorem 8.8.9

The quantity $\mathbf{A} \times \mathbf{I}$ *can be written as*

$$\mathbf{A} \times \mathbf{I} = \begin{bmatrix} \mathbf{A} & \mathbf{0} & \cdots & \mathbf{0} \\ \mathbf{0} & \mathbf{A} & \cdots & \mathbf{0} \\ \vdots & \vdots & & \vdots \\ \mathbf{0} & \mathbf{0} & \cdots & \mathbf{A} \end{bmatrix} = \operatorname{diag}(\mathbf{A}),$$

and $\mathbf{A} \times \mathbf{B} = (\mathbf{A} \times \mathbf{I})(\mathbf{I} \times \mathbf{B})$.

Proof: The result is obtained by performing the indicated multiplications. ▮

Theorem 8.8.10

Let $\mathbf{A}$ *be any* $m \times m$ *matrix and* $\mathbf{B}$ *be any* $n \times n$ *matrix; then*

$$\det(\mathbf{A} \times \mathbf{B}) = \det(\mathbf{B} \times \mathbf{A}) = |\mathbf{A}|^n |\mathbf{B}|^m.$$

Proof: By Theorem 8.6.9 there exists a nonsingular matrix $\mathbf{P}$ such that $\mathbf{PBP}^{-1} = \mathbf{T}$ where $\mathbf{T}$ is an upper triangular matrix. By the results of Theorem 8.8.9, we can write ($\mathbf{P}$ and $\mathbf{T}$ may not be real matrices)

$$\mathbf{A} \times \mathbf{T} = (\mathbf{A} \times \mathbf{I})(\mathbf{I} \times \mathbf{T}) = \begin{bmatrix} \mathbf{A} & \mathbf{0} & \mathbf{0} & \cdots & \mathbf{0} \\ \mathbf{0} & \mathbf{A} & \mathbf{0} & \cdots & \mathbf{0} \\ \mathbf{0} & \mathbf{0} & \mathbf{A} & \cdots & \mathbf{0} \\ \vdots & \vdots & \vdots & & \vdots \\ \mathbf{0} & \mathbf{0} & \mathbf{0} & \cdots & \mathbf{A} \end{bmatrix} \begin{bmatrix} \mathbf{I}t_{11} & \mathbf{I}t_{12} & \cdots & \mathbf{I}t_{1n} \\ \mathbf{0} & \mathbf{I}t_{22} & \cdots & \mathbf{I}t_{2n} \\ \vdots & \vdots & & \vdots \\ \mathbf{0} & \mathbf{0} & \cdots & \mathbf{I}t_{nn} \end{bmatrix}.$$

Clearly $\mathbf{I} \times \mathbf{T}$ is an upper triangular matrix. We can also write

$$(\mathbf{I} \times \mathbf{P})(\mathbf{I} \times \mathbf{B})(\mathbf{I} \times \mathbf{P}^{-1}) = \mathbf{I} \times (\mathbf{PBP}^{-1}) = \mathbf{I} \times \mathbf{T}.$$

But since $(\mathbf{I} \times \mathbf{P})^{-1} = \mathbf{I} \times \mathbf{P}^{-1}$, we get

$$(\mathbf{I} \times \mathbf{P})(\mathbf{I} \times \mathbf{B})(\mathbf{I} \times \mathbf{P})^{-1} = \mathbf{I} \times \mathbf{T}$$

and

$$\det(\mathbf{I} \times \mathbf{T}) = \det[(\mathbf{I} \times \mathbf{P})(\mathbf{I} \times \mathbf{B})(\mathbf{I} \times \mathbf{P})^{-1}] = \det(\mathbf{I} \times \mathbf{B}),$$

since $\det(\mathbf{I} \times \mathbf{P}) \det[(\mathbf{I} \times \mathbf{P})^{-1}] = 1$.
But since $\mathbf{I} \times \mathbf{T}$ is an upper triangular matrix, we get

$$\det(\mathbf{I} \times \mathbf{T}) = \prod_{i=1}^{n} t_{ii}^{m} = \left[\prod_{i=1}^{n} t_{ii}\right]^{m} = |\mathbf{T}|^{m} = |\mathbf{PBP}^{-1}|^{m} = |\mathbf{B}|^{m}.$$

Now we get

$$\det(\mathbf{A} \times \mathbf{B}) = \det[(\mathbf{A} \times \mathbf{I})(\mathbf{I} \times \mathbf{B})] = \det(\mathbf{A} \times \mathbf{I}) \det(\mathbf{I} \times \mathbf{B}) = |\mathbf{A}|^{n}|\mathbf{B}|^{m}.$$

By a similar procedure, it can be shown that $\det(\mathbf{B} \times \mathbf{A})$ also equals $|\mathbf{A}|^{n}|\mathbf{B}|^{m}$, and the theorem is proved. ▮

As a result of this theorem and the statements contained in the proof, we state some corollaries.

Corollary 8.8.10.1

Let $\mathbf{A}$ *be any* $m \times m$ *nonsingular matrix; then*

$$\det(\mathbf{A} \times \mathbf{A}^{-1}) = \det(\mathbf{A}^{-1} \times \mathbf{A}) = 1.$$

Corollary 8.8.10.2

Let $\mathbf{A}$ *be any* $m \times m$ *matrix and let* $\mathbf{P}$ *be any* $m \times m$ *matrix such that* $\mathbf{PA} = \mathbf{T}$, *where* $\mathbf{T}$ *is upper (lower) triangular; then*

$$(\mathbf{I} \times \mathbf{P})(\mathbf{I} \times \mathbf{A}) = \mathbf{I} \times \mathbf{T} = \mathbf{C},$$

and $\mathbf{C}$ *is an upper (lower) triangular matrix.*

Corollary 8.8.10.3

Let $\mathbf{A}$ *be any* $m \times m$ *matrix and let* $\mathbf{P}$ *and* $\mathbf{Q}$ *be any* $m \times m$ *matrices and let*

$\mathbf{PAQ} = \mathbf{B}$; *then*

$$(\mathbf{I} \times \mathbf{P})(\mathbf{I} \times \mathbf{A})(\mathbf{I} \times \mathbf{Q}) = \mathbf{I} \times \mathbf{B}.$$

Corollary 8.8.10.4

Let $\mathbf{A}$ *be any* $m \times m$ *matrix and let* $\mathbf{P}$ *and* $\mathbf{Q}$ *be any* $m \times m$ *matrices (nonsingular if the inverse is required) such that*

$$\mathbf{PAQ}^{-1} = \mathbf{B} \quad \text{and} \quad \mathbf{PAQ}' = \mathbf{C};$$

then

$$(\mathbf{I} \times \mathbf{P})(\mathbf{I} \times \mathbf{A})(\mathbf{I} \times \mathbf{Q})^{-1} = \mathbf{I} \times \mathbf{B}.$$

and

$$(\mathbf{I} \times \mathbf{P})(\mathbf{I} \times \mathbf{A})(\mathbf{I} \times \mathbf{Q})' = \mathbf{I} \times \mathbf{C}.$$

Corollary 8.8.10.5

Let $\mathbf{A}$ *be any* $m \times m$ *matrix with characteristic roots* $\lambda_1, \lambda_2, \ldots, \lambda_m$; *then the characteristic roots of* $\mathbf{I} \times \mathbf{A}$ *are the same as those of* $\mathbf{A} \times \mathbf{I}$ *and are* $\lambda_1, \lambda_2, \ldots, \lambda_m$, *each with multiplicity* n, *where* $\mathbf{I}$ *is an* $n \times n$ *identity matrix.*

Theorem 8.8.11

Let $\mathbf{A}$ *and* $\mathbf{B}$ *be* $m \times m$ *matrices, and* $\mathbf{C}$ *be an* $n \times n$ *matrix; then*

$$(\mathbf{A} + \mathbf{B}) \times \mathbf{C} = (\mathbf{A} \times \mathbf{C}) + (\mathbf{B} \times \mathbf{C}).$$

Proof: $(\mathbf{A} + \mathbf{B}) \times \mathbf{C} = [(\mathbf{A} + \mathbf{B})c_{ij}]$, $\mathbf{A} \times \mathbf{C} = [\mathbf{A}c_{ij}]$, and $\mathbf{B} \times \mathbf{C} = [\mathbf{B}c_{ij}]$.

Thus

$$(\mathbf{A} \times \mathbf{C}) + (\mathbf{B} \times \mathbf{C}) = [(\mathbf{A} + \mathbf{B})c_{ij}] = (\mathbf{A} + \mathbf{B}) \times \mathbf{C},$$

and the theorem is proved. ▮

Theorem 8.8.12

Let $\mathbf{D}_1$ *and* $\mathbf{D}_2$ *be diagonal matrices; then* $\mathbf{D}_1 \times \mathbf{D}_2$ *is a diagonal matrix. Let* $\mathbf{T}_1$

and $\mathbf{T}_2$ *be upper (lower) triangular matrices; then* $\mathbf{T}_1 \times \mathbf{T}_2$ *is an upper (lower) triangular matrix.*

The proof is left for the reader.

Theorem 8.8.13

Let $\mathbf{A}$ *be an* $m \times m$ *matrix with characteristic roots* $a_1, a_2, \ldots, a_m$; *let* $\mathbf{B}$ *be an* $n \times n$ *matrix with characteristic roots* $b_1, b_2, \ldots, b_n$; *then the characteristic roots of* $\mathbf{A} \times \mathbf{B}$ *are* $a_i b_j$; $i = 1, 2, \ldots, m$; $j = 1, 2, \ldots, n$. *(These are also the characteristic roots of* $\mathbf{B} \times \mathbf{A}$.*)*

Proof: Let $\mathbf{P}$ be a nonsingular $m \times m$ matrix such that $\mathbf{PAP}^{-1} = \mathbf{T}_1$, where $\mathbf{T}_1$ is an upper triangular matrix with the characteristic roots a_i of $\mathbf{A}$ on the diagonal. Let $\mathbf{Q}$ be a nonsingular $n \times n$ matrix such that $\mathbf{QBQ}^{-1} = \mathbf{T}_2$, where $\mathbf{T}_2$ is an upper triangular matrix with the characteristic roots b_j of $\mathbf{B}$ on the diagonal. Then the characteristic roots of

$$(\mathbf{P} \times \mathbf{Q})(\mathbf{A} \times \mathbf{B})(\mathbf{P} \times \mathbf{Q})^{-1}$$

are the same as the characteristic roots of $\mathbf{A} \times \mathbf{B}$. But since $(\mathbf{P} \times \mathbf{Q})^{-1} = \mathbf{P}^{-1} \times \mathbf{Q}^{-1}$, we get

$$(\mathbf{P} \times \mathbf{Q})(\mathbf{A} \times \mathbf{B})(\mathbf{P} \times \mathbf{Q})^{-1} = (\mathbf{PAP}^{-1}) \times (\mathbf{QBQ}^{-1}) = \mathbf{T}_1 \times \mathbf{T}_2,$$

and the characteristic roots of $\mathbf{T}_1 \times \mathbf{T}_2$ are $a_i b_j$; $i = 1, 2, \ldots, m$; $j = 1, 2, \ldots, n$, and the theorem is proved. Note that $\mathbf{P}$, $\mathbf{Q}$, $\mathbf{T}_1$ and $\mathbf{T}_2$ may not be real matrices. ▌

Corollary 8.8.13

Let $\mathbf{A}$ *and* $\mathbf{B}$ *be positive (semi) definite matrices; then* $\mathbf{A} \times \mathbf{B}$ *is a positive (semi) definite matrix.*

Theorem 8.8.14

Let $\mathbf{A}$ *be an* $m_1 \times n_1$ *matrix of rank* r_1 *and let* $\mathbf{B}$ *be an* $m_2 \times n_2$ *matrix of rank* r_2; *then* $\mathbf{A} \times \mathbf{B}$ *has rank* $r_1 r_2$.

The proof of this theorem is left for the reader.

Theorem 8.8.15

Let $\mathbf{A}_i$ *be* $m_2 \times n_2$ *matrices and let* $\mathbf{B}_i$ *be* $m_1 \times n_1$ *matrices for* $i = 1, 2, \ldots, k$. *Then*

$$\left(\sum_{i=1}^{K} \mathbf{A}_i\right) \times \left(\sum_{j=1}^{K} \mathbf{B}_j\right) = \sum_{i=1}^{K} \sum_{j=1}^{K} (\mathbf{A}_i \times \mathbf{B}_j).$$

The proof will be left for the reader.
Next we discuss the direct sum of matrices.

Definition 8.8.2

Direct Sum of Matrices. *Let* $\mathbf{A}_1$ *be an* $n_1 \times n_1$ *matrix and* $\mathbf{A}_2$ *be an* $n_2 \times n_2$ *matrix. The direct sum of* $\mathbf{A}_1$ *and* $\mathbf{A}_2$*, denoted by* $\mathbf{A}_1 + \mathbf{A}_2$*, is defined to be the* $(n_1 + n_2) \times (n_1 + n_2)$ *block diagonal matrix* $\mathbf{A}$ *given by*

$$\mathbf{A} = \begin{bmatrix} \mathbf{A}_1 & \mathbf{0} \\ \mathbf{0} & \mathbf{A}_2 \end{bmatrix}.$$

The theorem below is an immediate result of the definition of direct sum of matrices. The proof will be left for the reader.

Theorem 8.8.16

Let $\mathbf{A}_i$ *be* $n_i \times n_i$ *matrices for* $i = 1, 2, \ldots, K$; *let* $\mathbf{A} = \mathbf{A}_1 + \mathbf{A}_2 + \ldots + \mathbf{A}_K$.

(1) *Then* $\mathbf{A}$ *is the* $n \times n$ *matrix* $\left(n = \sum_{k=1}^{K} n_i\right)$ *given by*

$$\mathbf{A} = \begin{bmatrix} \mathbf{A}_1 & \mathbf{0} & \ldots & \mathbf{0} \\ \mathbf{0} & \mathbf{A}_2 & \ldots & \mathbf{0} \\ \vdots & & & \vdots \\ \mathbf{0} & \ldots & & \mathbf{A}_K \end{bmatrix}.$$

(2) *In general,* $\mathbf{A}_1 + \mathbf{A}_2 \neq \mathbf{A}_2 + \mathbf{A}_1$.

(3) *If* a_i *for* $i = 1, \ldots, K$ *is a set of* K *scalars, then* $[a_1] + [a_2] + \ldots + [a_K] = \mathbf{D}$*, where* $\mathbf{D}$ *is a* $K \times K$ *diagonal matrix whose i-th diagonal element is* a_i.

(4) $(\mathbf{A}_1 + \mathbf{A}_2) + \mathbf{A}_3 = \mathbf{A}_1 + (\mathbf{A}_2 + \mathbf{A}_3)$.

(5) $\mathbf{A}_1 + \ldots + \mathbf{A}_K = \mathbf{A}_1' + \ldots + \mathbf{A}_K' = \mathbf{A}' = \mathbf{A}$ *if and only if each* $\mathbf{A}_i$ *is symmetric.*

(6) $\det(\mathbf{A}_1 + \ldots + \mathbf{A}_K) = \prod_{i=1}^{K} \det(\mathbf{A}_i)$.

(7) $\mathrm{tr}(\mathbf{A}_1 + \ldots + \mathbf{A}_K) = \sum_{i=1}^{K} \mathrm{tr}(\mathbf{A}_i)$.

(8) *If* $\mathbf{A}_i^{-1}$ *exists for each* $i = 1, \ldots, K$, *then* $\mathbf{A}^{-1}$ *exists and is given by* $\mathbf{A}^{-1} = \mathbf{A}_1^{-1} + \ldots + \mathbf{A}_K^{-1}$.

(9) If $\mathbf{A}_i$ *for* $i = 1, \ldots, K$ *are upper (or lower) triangular matrices, then* $\mathbf{A}_1 + \ldots + \mathbf{A}_K$ *is an upper (or lower) triangular matrix.*

(10) *If* $\mathbf{A}_i$ *for* $i = 1, \ldots, K$ *are orthogonal, then* $\mathbf{A}_1 + \ldots + \mathbf{A}_K$ *is orthogonal and* $\mathbf{A}'\mathbf{A} = \mathbf{A}\mathbf{A}' = \mathbf{I}$.

(11) $\mathbf{A}^M = \mathbf{A}_1^M + \ldots + \mathbf{A}_K^M$ *for any positive integer* M.

(12) *If* rank $(\mathbf{A}_i) = r_i$ *for* $i = 1, \ldots, K$, *then* rank $(\mathbf{A}_1 + \ldots + \mathbf{A}_K) = \sum_{i=1}^{K} r_i$.

(13) *If* $\lambda_j^{(i)}$ *for* $j = 1, \ldots, n_i$ *are the characteristic roots of* $\mathbf{A}_i$ *for* $i = 1, \ldots, K$, *then* $\lambda_j^{(i)}$ *for* $j = 1, \ldots, n_i$; $i = 1, \ldots, K$ *are the characteristic roots of* $\mathbf{A}_1 + \ldots + \mathbf{A}_K$.

(14) $(\mathbf{A}_1 + \mathbf{A}_2) + (\mathbf{A}_3 + \mathbf{A}_4) = (\mathbf{A}_1 + \mathbf{A}_3) + (\mathbf{A}_2 + \mathbf{A}_4) = (\mathbf{A}_1 + \mathbf{A}_4) + (\mathbf{A}_2 + \mathbf{A}_3)$, *where we assume all operations are defined.*

(15) $(\mathbf{A}_1 + \mathbf{A}_2)(\mathbf{A}_3 + \mathbf{A}_4) = (\mathbf{A}_1\mathbf{A}_3) + (\mathbf{A}_2\mathbf{A}_4)$, *where we assume all operations are defined.*

Theorem 8.8.17

Let $\mathbf{A}$ *be an* $n \times n$ *matrix and let* $\mathbf{I}$ *be the* $m \times m$ *identity; then* $\mathbf{A} \times \mathbf{I} = \mathbf{A} + \mathbf{A} + \ldots + \mathbf{A}$.

The proof will be left for the reader.

Example 8.8.5. The two-way classification model can be written as

$$y_{ij} = \mu + \tau_i + \gamma_j + e_{ij}, \qquad i = 1, 2, \ldots, t; \quad j = 1, 2, \ldots, g.$$

If we write this as

$$\mathbf{y} = \mathbf{X}\boldsymbol{\beta} + \mathbf{e},$$

where

$$\mathbf{y}' = [y_{11}, y_{12}, \ldots, y_{1g}, y_{21}, y_{22}, \ldots, y_{2g}, \ldots, y_{t1}, y_{t2}, \ldots, y_{tg}],$$

where

$$\mathbf{X} = [\mathbf{1}_g \times \mathbf{1}_t, \mathbf{1}_g \times \mathbf{I}_t, \mathbf{I}_g \times \mathbf{1}_t],$$

and where

$$\boldsymbol{\beta}' = [\mu, \tau_1, \tau_2, \ldots, \tau_t, \gamma_1, \gamma_2, \ldots, \gamma_g] = [\mu, \boldsymbol{\tau}', \boldsymbol{\gamma}'],$$

then the model can be written as

$$\mathbf{y} = (\mathbf{1}_g \times \mathbf{1}_t)\mu + (\mathbf{1}_g \times \mathbf{I}_t)\boldsymbol{\tau} + (\mathbf{I}_g \times \mathbf{1}_t)\boldsymbol{\gamma} + \mathbf{e}.$$

An analysis of variance includes the sum of squares due to τ, due to γ, and due to error, which are, respectively,

$$\sum_{j=1}^{g}\sum_{i=1}^{t}(y_{i.} - y_{..})^2 = \frac{1}{g}\mathbf{y}'(\mathbf{1}_g \times \mathbf{I}_t)\left(\mathbf{I}_t - \frac{1}{t}\mathbf{J}_t\right)(\mathbf{1}_g' \times \mathbf{I}_t)\mathbf{y},$$

$$\sum_{i=1}^{t}\sum_{j=1}^{g}(y_{.j} - y_{..})^2 = \frac{1}{t}\mathbf{y}'(\mathbf{I}_g \times \mathbf{1}_t)\left(\mathbf{I}_g - \frac{1}{g}\mathbf{J}_g\right)(\mathbf{I}_g \times \mathbf{1}_t')\mathbf{y}, \quad (8.8.1)$$

$$\sum_{i=1}^{t}\sum_{j=1}^{g}(y_{ij} - y_{i.} - y_{.j} + y_{..})^2 = \frac{1}{gt}\mathbf{y}'\left(\mathbf{I}_g - \frac{1}{g}\mathbf{J}_g\right) \times \left(\mathbf{I}_t - \frac{1}{t}\mathbf{J}_t\right)\mathbf{y}.$$

8.9 Additional Theorems

This section gives a number of miscellaneous theorems on patterned matrices. Some of the proofs are not given, but are requested in the problems.

Theorem 8.9.1

If $\mathbf{A}$ *and* $\mathbf{B}$ *are* $k \times k$ *matrices and*

$$\mathbf{C} = \begin{bmatrix} \mathbf{A} & \mathbf{B} & \cdots & \mathbf{B} \\ \mathbf{B} & \mathbf{A} & \cdots & \mathbf{B} \\ \vdots & \vdots & & \vdots \\ \mathbf{B} & \mathbf{B} & \cdots & \mathbf{A} \end{bmatrix} \quad (8.9.1)$$

and $\mathbf{C}$ *has dimension* $mk \times mk$, *then* $\det(\mathbf{C}) = |\mathbf{A} - \mathbf{B}|^{m-1}|\mathbf{A} + (m-1)\mathbf{B}|$. (Note that this is a generalization of Theorem 8.4.4.)

Theorem 8.9.2

If $\mathbf{I}$ *and* $\mathbf{J}$ *are* $k \times k$ *matrices and*

$$\mathbf{C} = \begin{bmatrix} \mathbf{I} & \mathbf{J} & \cdots & \mathbf{J} \\ \mathbf{J} & \mathbf{I} & \cdots & \mathbf{J} \\ \vdots & \vdots & & \vdots \\ \mathbf{J} & \mathbf{J} & \cdots & \mathbf{I} \end{bmatrix} \tag{8.9.2}$$

is an $mk \times mk$ *matrix with* $m > 1$, *then*

(a) $\det(\mathbf{C}) = (1-k)^{m-1}[1 + k(m-1)]$, and
(b) $\mathbf{C}$ *is nonsingular if and only if* $k > 1$.
(c) *If* $\mathbf{C}^{-1}$ *exists (that is, if* $k > 1$), *then*

$$\mathbf{C}^{-1} = \begin{bmatrix} \mathbf{I} + a\mathbf{J} & b\mathbf{J} & \cdots & b\mathbf{J} \\ b\mathbf{J} & \mathbf{I} + a\mathbf{J} & \cdots & b\mathbf{J} \\ \vdots & \vdots & & \vdots \\ b\mathbf{J} & b\mathbf{J} & \cdots & \mathbf{I} + a\mathbf{J} \end{bmatrix}, \tag{8.9.3}$$

where

$$a = \frac{k(k-1)(m-1)}{(m-1)k + 1}$$

and

$$b = \frac{-(k-1)}{(m-1)k + 1}(k-1).$$

Theorem 8.9.3

If $\mathbf{A}$ *is a* $k \times k$ *nonsingular matrix and* $\mathbf{c}$ *and* $\mathbf{d}$ *are* $k \times 1$ *vectors, then* $\det(\mathbf{A} + \mathbf{cd}') = |\mathbf{A}| \cdot (1 + \mathbf{d}'\mathbf{A}^{-1}\mathbf{c})$. *If the inverse of the matrix* $\mathbf{A} + \mathbf{cd}'$ *exists, the inverse is given by*

$$(\mathbf{A} + \mathbf{cd}')^{-1} = \mathbf{A}^{-1} - \frac{(\mathbf{A}^{-1}\mathbf{c})(\mathbf{d}'\mathbf{A}^{-1})}{1 + \mathbf{d}'\mathbf{A}^{-1}\mathbf{c}}. \tag{8.9.4}$$

Theorem 8.9.4

If $\mathbf{A}$ *is an idempotent* $k \times k$ *matrix and* a_1, a_2 *are nonzero constants, then* $\mathbf{B}$ *is nonsingular, where*

$$\mathbf{B} = a_1\mathbf{A} + a_2(\mathbf{I} - \mathbf{A}) \tag{8.9.5}$$

and

$$\mathbf{B}^{-1} = \frac{1}{a_1}\mathbf{A} + \frac{1}{a_2}(\mathbf{I} - \mathbf{A}). \tag{8.9.6}$$

Theorem 8.9.5

Let $\mathbf{A}$ *be a* $k \times k$ *nonsingular matrix, let* $\mathbf{b}$ *and* $\mathbf{c}$ *be* $k \times 1$ *vectors, and let* a *be a nonzero scalar; then*

$$\begin{vmatrix} \mathbf{A} & \mathbf{b} \\ \mathbf{c}' & a \end{vmatrix} = a \cdot \left| \mathbf{A} - \frac{1}{a}\mathbf{b}\mathbf{c}' \right| = |\mathbf{A}| \cdot (a - \mathbf{c}'\mathbf{A}^{-1}\mathbf{b}). \tag{8.9.7}$$

Theorem 8.9.6

Let $\mathbf{a}$ *be an* $n \times 1$ *vector and let* c *be a scalar; then* $|c\mathbf{I} - \mathbf{a}\mathbf{a}'| = c^{n-1}(c - \mathbf{a}'\mathbf{a})$.

Theorem 8.9.7

Let $\mathbf{A}$ *be an* $m \times n$ *matrix* $m \leq n$ *of rank* r*; then there exist orthogonal matrices* $\mathbf{P}$ *of order* $m \times m$ *and* $\mathbf{Q}$ *of order* $n \times n$ *such that*

$$\mathbf{P}'\mathbf{A}\mathbf{Q} = [\mathbf{D}, \mathbf{0}],$$

where $\mathbf{D}$ *is a (real)* $m \times m$ *diagonal matrix with diagonal elements* d_i *where* d_i^2, $i = 1, 2, \ldots, m$, *are the characteristic roots of* $\mathbf{A}\mathbf{A}'$ *(if* $m = n$, *then* $\mathbf{P}'\mathbf{A}\mathbf{Q} = \mathbf{D}$*).*

Proof: Assume $m < n$ and $m > r$ (the proof for the other cases is similar). Since $\mathbf{A}\mathbf{A}'$ is a symmetric (and positive semidefinite) matrix, let $\mathbf{P}$ be an orthogonal matrix such that $\mathbf{P}'\mathbf{A}\mathbf{A}'\mathbf{P} = \mathbf{D}^2$ where the diagonal elements of $\mathbf{D}^2$, denoted by d_i^2, are the characteristic roots of $\mathbf{A}\mathbf{A}'$. We can write the characteristic roots as d_i^2, since they are non-negative. Since the rank of $\mathbf{A}$ is assumed to be r, the rank of $\mathbf{A}\mathbf{A}'$ is also equal to r, so there are $m - r$ diagonal elements equal to zero in $\mathbf{D}^2$.

We choose $\mathbf{P}$ such that the last $(m - r)$ diagonal elements of $\mathbf{D}^2$ are zero and the first r diagonal elements are positive. Then we can partition $\mathbf{P}$ such that $\mathbf{P} = [\mathbf{P}_1, \mathbf{P}_2]$, where $\mathbf{P}_1$ has size $m \times r$ and $\mathbf{P}_2$ has size $m \times (m - r)$. We write

$$\mathbf{P}'\mathbf{A}\mathbf{A}'\mathbf{P} = \begin{bmatrix} \mathbf{P}_1' \\ \mathbf{P}_2' \end{bmatrix} \mathbf{A}\mathbf{A}'[\mathbf{P}_1, \mathbf{P}_2] = \begin{bmatrix} \mathbf{P}_1'\mathbf{A}\mathbf{A}'\mathbf{P}_1 & \mathbf{P}_1'\mathbf{A}\mathbf{A}'\mathbf{P}_2 \\ \mathbf{P}_2'\mathbf{A}\mathbf{A}'\mathbf{P}_1 & \mathbf{P}_2'\mathbf{A}\mathbf{A}'\mathbf{P}_2 \end{bmatrix} = \begin{bmatrix} \mathbf{D}_1^2 & \mathbf{0} \\ \mathbf{0} & \mathbf{0} \end{bmatrix},$$

and $\mathbf{D}_1^2$ is an $r \times r$ diagonal matrix with positive diagonal elements.

Now $\mathbf{P}_2'\mathbf{A}\mathbf{A}'\mathbf{P}_2 = \mathbf{0}$ or $(\mathbf{P}_2'\mathbf{A})(\mathbf{P}_2'\mathbf{A})' = \mathbf{0}$ and hence $\mathbf{P}_2'\mathbf{A} = \mathbf{0}$. Also

$$\mathbf{P}_1'\mathbf{A}\mathbf{A}'\mathbf{P}_1 = \mathbf{D}_1^2 \quad \text{or} \quad (\mathbf{D}_1^{-1}\mathbf{P}_1'\mathbf{A})(\mathbf{D}_1^{-1}\mathbf{P}_1'\mathbf{A})' = \mathbf{I}.$$

Let $\mathbf{Q}_1 = \mathbf{A}'\mathbf{P}_1\mathbf{D}_1^{-1}$ where $\mathbf{Q}_1$ is an $n \times r$ matrix whose columns are orthogonal and $\mathbf{Q}_1'\mathbf{Q}_1 = \mathbf{I}$. So $\mathbf{Q}_1$ contains r columns of an orthogonal $n \times n$ matrix. There exists a matrix $\mathbf{Q}_2$ of size $n \times (n - r)$ such that

$$\mathbf{Q} = [\mathbf{Q}_1, \mathbf{Q}_2],$$

where $\mathbf{Q}$ is an orthogonal matrix. Now

$$\mathbf{P}'\mathbf{A}\mathbf{Q} = \begin{bmatrix} \mathbf{P}_1' \\ \mathbf{P}_2' \end{bmatrix} \mathbf{A}[\mathbf{Q}_1, \mathbf{Q}_2] = \begin{bmatrix} \mathbf{P}_1'\mathbf{A}\mathbf{A}'\mathbf{P}_1\mathbf{D}_1^{-1} & \mathbf{P}_1'\mathbf{A}\mathbf{Q}_2 \\ \mathbf{P}_2'\mathbf{A}\mathbf{Q}_1 & \mathbf{P}_2'\mathbf{A}\mathbf{Q}_2 \end{bmatrix}.$$

But $\mathbf{P}_2'\mathbf{A} = \mathbf{0}$ and, since $\mathbf{Q}$ is an orthogonal matrix, we have that $\mathbf{Q}'\mathbf{Q} = \mathbf{I}$, which gives

$$\mathbf{I} = \mathbf{Q}'\mathbf{Q} = \begin{bmatrix} \mathbf{Q}_1' \\ \mathbf{Q}_2' \end{bmatrix} [\mathbf{Q}_1, \mathbf{Q}_2],$$

or $\mathbf{Q}_1'\mathbf{Q}_2 = \mathbf{0}$; but $\mathbf{Q}_1' = \mathbf{D}_1^{-1}\mathbf{P}_1'\mathbf{A}$, so $\mathbf{Q}_1'\mathbf{Q}_2 = \mathbf{0}$ implies that $\mathbf{D}_1^{-1}\mathbf{P}_1'\mathbf{A}\mathbf{Q}_2 = \mathbf{0}$

In other words, $\mathbf{P}_1'\mathbf{A}\mathbf{Q}_2 = \mathbf{0}$. Hence

$$\mathbf{P}'\mathbf{A}\mathbf{Q} = \begin{bmatrix} \mathbf{D}_1 & \mathbf{0} \\ \mathbf{0} & \mathbf{0} \end{bmatrix} = [\mathbf{D}, \mathbf{0}].$$

This proves the theorem. ∎

Note: The diagonal elements of $\mathbf{D}$ can be chosen to be non-negative, if we choose the diagonal elements of $\mathbf{D}_1$ to be the positive square roots of the diagonal elements of $\mathbf{D}_1^2$.

8.10 Circulants

A class of matrices that finds application in statistics and other fields is called circulants, and they have very special patterns. Three types of circulants (regular, symmetric regular, and symmetric) will be investigated.

To study these matrices, we will use the symbol $(j - i)|k$, where i, j, k are positive integers, to mean the *positive* remainder when $j - i$ is divided by k. Thus $(j - i)|k$ can have only the values $0, 1, 2, \ldots, k - 1$. Note that if $j - i$ is negative, the remainder is negative, and to this remainder we must add k to obtain $(j - i)|k$, which is defined as the *positive* remainder. For example, if $j - i = 17$ and $k = 5$, then $(j - i)|k = (17)|5 = 2$. If $j - i = -17$ and $k = 5$, then $(j - i)|k = (-17)|5$. But -17 divided by 5 has a negative remainder, -2, so to get $(-17)|5$ we must add $k = 5$ to -2. Thus $(-17)|5 = 5 - 2 = 3$. The symbol $(j - i)|k$ is sometimes referred to as $(j - i)$ modulo k. In our discussions i and j will always be integers between 1 and k inclusive, so it follows that

$$(j - i)|k = k + j - i \qquad \text{when } i > j,$$

$$(j - i)|k = j - i \qquad \text{when } i \leq j.$$

Definition 8.10.1

Regular Circulants. *A $k \times k$ matrix $\mathbf{A}$ with i, j-th element a_{ij} is defined to be a regular circulant if and only if $(j - i)|k = (q - p)|k$ implies that $a_{ij} = a_{pq}$.*

Note: From this definition we see that a regular circulant contains at most k distinct elements, $a_0, a_1, \ldots, a_{k-1}$, and

$$\mathbf{A} = \begin{bmatrix} a_0 & a_1 & a_2 \ldots a_{k-1} \\ a_{k-1} & a_0 & a_1 \ldots a_{k-2} \\ a_{k-2} & a_{k-1} & a_0 \ldots a_{k-3} \\ \vdots \cdots & & \\ a_1 & a_2 & a_3 \ldots a_0 \end{bmatrix},$$

so $\mathbf{A}$ is a $k \times k$ regular circulant if and only if $a_{ij} = a_{(j-i)|k}$. Another way to state this is: $a_{ij} = a_{j-i}$ if $j \geq i$ (elements on and above the diagonal), and $a_{ij} = a_{k+j-i}$ if $j < i$ (elements below the diagonal).

Note: The t-th row of a regular circulant for $t = 2, 3, \ldots, k$ is obtained from the $(t - 1)$-st row by moving each element of the $(t - 1)$-st row one column to the right and placing the element in the last column of the $(t - 1)$-st row in the first column of the t-th row. A similar arrangement applies for obtaining the $(t - 1)$-st column from the t-th column.

An example of a 4×4 regular circulant is

$$\mathbf{A} = \begin{bmatrix} 6 & 0 & -1 & 4 \\ 4 & 6 & 0 & -1 \\ -1 & 4 & 6 & 0 \\ 0 & -1 & 4 & 6 \end{bmatrix}.$$

In the following theorem and corollary, some results are stated that are a direct result of the definition.

Theorem 8.10.1

The following $k \times k$ matrices are regular circulants.

(1) *The zero matrix,* $\mathbf{0}$.
(2) *The identity matrix,* $\mathbf{I}$.
(3) *The matrix with every element equal to a constant b.*
(4) *The matrix $a\mathbf{A} + b\mathbf{B}$, where $\mathbf{A}$ and $\mathbf{B}$ are $k \times k$ regular circulants.*

Clearly the zero matrix in (1) is a special case of the matrices in (3) and (4). Also, the identity matrix in (2) is a special case of the matrix in (4).

Corollary 8.10.1

If $\mathbf{A}$ is a $k \times k$ regular circulant with first-row elements $a_0, a_1, \ldots, a_{k-1}$, then

(1) $\mathbf{A}'$ *is a regular circulant,*
(2) *The diagonal elements of $\mathbf{A}$ are equal,*
(3) *The elements are equal on each diagonal parallel to the main diagonal.*

Theorem 8.10.2

Let $\mathbf{A}$ *be a* $k \times k$ *matrix. Then* $\mathbf{A}$ *is a regular circulant if and only if* $\mathbf{P}'\mathbf{A}\mathbf{P} = \mathbf{A}$, *where* $\mathbf{P} = [\mathbf{e}_2, \mathbf{e}_3, \ldots, \mathbf{e}_k, \mathbf{e}_1]$, *where* $\mathbf{e}_i$ *is the i-th unit* $k \times 1$ *vector (that is, the i-th element is unity and the other elements are zeros).*

Proof: The proof is obtained by direct multiplication. ∎

Note: The matrix $\mathbf{P}$ in Theorem 8.10.2 is called a permutation matrix, and these matrices will be discussed in more detail in Sec. 8.13. Clearly $\mathbf{P}' = \mathbf{P}^{-1}$, so $\mathbf{P}$ is an orthogonal matrix.

Note: The results of Theorem 8.10.2 can be used as the definition of a regular circulant; that is, a $k \times k$ matrix is defined to be a regular circulant if and only if $\mathbf{A} = \mathbf{P}'\mathbf{A}\mathbf{P}$, where $\mathbf{P}$ is the permutation matrix in Theorem 8.10.2.

Theorem 8.10.3

Let $\mathbf{A}$ *and* $\mathbf{B}$ *be* $k \times k$ *regular circulants. Then the product* $\mathbf{AB}$ *is also a regular circulant.*

Proof: Let $\mathbf{AB} = \mathbf{D} = [d_{ij}]$. Then

$$d_{ij} = \sum_{s=1}^{k} a_{is}\, b_{sj} = \sum_{s=1}^{k} a_{(s-i)|k}\, b_{(j-s)|k}$$

$$= \sum_{u=0}^{k-1} a_u\, b_{(j-i-u)|k},$$

since $j - s = j - i - (s - i) = j - i - u$ for $u = s - i$, and

$$d_{pq} = \sum_{t=1}^{k} a_{pt}\, b_{tq} = \sum_{t=1}^{k} a_{(t-p)|k}\, b_{(q-t)|k}$$

$$= \sum_{u=0}^{k-1} a_u\, b_{(q-p-u)|k},$$

since $q - t = q - p - (t - p) = q - p - u$ for $u = t - p$. But $(q - p - u)|k = (j - i - u)|k$ if $(q - p)|k = (j - i)|k$. Therefore,

$d_{ij} = d_{pq}$ for $(j - i)|k = (q - p)|k$, which implies that **D**, and hence **AB**, is a regular circulant. ▮

We also exhibit a simple proof using the alternative definition. We must show $\mathbf{P'ABP} = \mathbf{AB}$ if $\mathbf{A} = \mathbf{P'AP}$ and $\mathbf{B} = \mathbf{P'BP}$, where **P** is given in Theorem 8.10.2. Clearly $\mathbf{AB} = \mathbf{P'APP'BP} = \mathbf{P'ABP}$, so **AB** is a regular circulant if **A** and **B** are regular circulants.

Theorem 8.10.4

If **C** *is a nonsingular* $k \times k$ *regular circulant, then* $\mathbf{C}^{-1}$ *is also a* $k \times k$ *regular circulant.*

Proof: Denote $\mathbf{C}^{-1}$ by **B**. Then $\mathbf{CB} = \mathbf{I}$, which may be rewritten as $\mathbf{C}[\mathbf{b}_1, \mathbf{b}_2, \ldots, \mathbf{b}_k] = [\mathbf{e}_1, \mathbf{e}_2, \ldots, \mathbf{e}_k]$, where $\mathbf{b}_i$ is the i-th column of **B** and $\mathbf{e}_j$ is the j-th column of **I**. Consider $\mathbf{Cb}_1 = \mathbf{e}_1$, where

$$\mathbf{b}_1 = \begin{bmatrix} b_0 \\ b_{k-1} \\ b_{k-2} \\ \vdots \\ b_1 \end{bmatrix}; \ \mathbf{e}_1 = \begin{bmatrix} 1 \\ 0 \\ \vdots \\ 0 \end{bmatrix}.$$

Writing out the system of equations, we get:

$$\begin{array}{r|l}
(1) & c_0b_0 + c_1b_{k-1} + c_2b_{k-2} + \cdots + c_{k-2}b_2 + c_{k-1}b_1 = 1. \\
(2) & c_{k-1}b_0 + c_0b_{k-1} + c_1b_{k-2} + \cdots + c_{k-3}b_2 + c_{k-2}b_1 = 0. \\
 & \cdots \\
(k-1) & c_2b_0 + c_3b_{k-1} + c_4b_{k-2} + \cdots + c_0b_2 + c_1b_1 = 0. \\
(k) & c_1b_0 + c_2b_{k-1} + c_3b_{k-2} + \cdots + c_{k-1}b_2 + c_0b_1 = 0.
\end{array}$$

Now, rearrange the above system in the following manner:

$$\begin{array}{r|l}
(k) & c_0b_1 + c_1b_0 + c_2b_{k-1} + c_3b_{k-2} + \cdots + c_{k-1}b_2 = 0. \\
(1) & c_{k-1}b_1 + c_0b_0 + c_1b_{k-1} + c_2b_{k-2} + \cdots + c_{k-2}b_2 = 1. \\
(2) & c_{k-2}b_1 + c_{k-1}b_0 + c_0b_{k-1} + c_1b_{k-2} + \cdots + c_{k-3}b_2 = 0. \\
 & \cdots \\
(k-1) & c_1b_1 + c_2b_0 + c_3b_{k-1} + c_4b_{k-2} + \cdots + c_0b_2 = 0.
\end{array}$$

The second system may now be rewritten as $\mathbf{Cb}_2 = \mathbf{e}_2$, where

$$\mathbf{b}_2 = \begin{bmatrix} b_1 \\ b_0 \\ b_{k-1} \\ \vdots \\ b_2 \end{bmatrix}; \ \mathbf{e}_2 = \begin{bmatrix} 0 \\ 1 \\ 0 \\ \vdots \\ 0 \end{bmatrix}.$$

By continuing this same process, we arrive at the result $\mathbf{Cb}_i = \mathbf{e}_i$ for each $i = 1, 2, \ldots, k$, where

$$\mathbf{b}_i = \begin{bmatrix} b_{(i-1)|k} \\ b_{(i-2)|k} \\ \vdots \\ b_{(i-k)|k} \end{bmatrix}.$$

Therefore, $\mathbf{B}$ (that is, $\mathbf{C}^{-1}$) is a regular circulant. ∎
An alternative proof is given below.

Proof: Since $\mathbf{C}$ is a regular circulant, this means that $\mathbf{C} = \mathbf{P'CP}$ and clearly $\mathbf{C}^{-1} = (\mathbf{P'CP})^{-1} = \mathbf{P'C}^{-1}\mathbf{P}$, so $\mathbf{C}^{-1}$ is a regular circulant. ∎

One use of this theorem is that if it is known that $\mathbf{C}$ is a regular circulant and is nonsingular, then to compute $\mathbf{C}^{-1}$ we only need to calculate the first row of $\mathbf{C}^{-1}$. The remaining rows of $\mathbf{C}^{-1}$ can be obtained from the first row, since $\mathbf{C}^{-1}$ is a regular circulant. In Theorem 8.10.23 we show that $\mathbf{C}^{-}$, the g-inverse of a regular circulant, is also a regular circulant.

Theorem 8.10.5

If $\mathbf{A}$ *and* $\mathbf{B}$ *are regular* $k \times k$ *circulants, then* $\mathbf{AB} = \mathbf{BA}$.

Proof: Let $\mathbf{AB} = \mathbf{X}$ and $\mathbf{BA} = \mathbf{Y}$. We will show that $x_{ij} = y_{ij}$ for all i, j.

$$x_{ij} = \sum_{s=1}^{k} a_{is} b_{sj} = \sum_{s=1}^{k} a_{(s-i)|k} b_{(j-s)|k}$$

$$= \sum_{u=0}^{k-1} a_{u|k} b_{(j-i-u)|k},$$

since $j - s = j - i - (s - i)$.

$$y_{ij} = \sum_{t=1}^{k} b_{it}\, a_{tj} = \sum_{t=1}^{k} b_{(t-i)|k}\, a_{(j-t)|k}$$

$$= \sum_{u=0}^{k-1} b_{(j-i-u)|k}\, a_{u|k} = \sum_{u=0}^{k-1} a_{u|k}\, b_{(j-i-u)|k},$$

since $t - i = j - i - (j - t)$. Therefore, $\mathbf{AB} = \mathbf{X} = \mathbf{Y} = \mathbf{BA}$ and the theorem is proved. ▮

Theorem 8.10.6

Let $\mathbf{A}$ *and* $\mathbf{C}$ *be* $k \times k$ *regular circulants, and suppose there exists a matrix* $\mathbf{X}$ *such that* $\mathbf{AX} = \mathbf{C}$. *Then there exists a regular circulant* $\mathbf{B}$ *such that* $\mathbf{AB} = \mathbf{C}$.

The proof of this theorem is straightforward and will be omitted. A simple proof can be found in [M–7]. Note that $\mathbf{A}$ may be singular.

Next we discuss characteristic roots, characteristic vectors, and determinants of regular circulants.

Theorem 8.10.7

Let $\mathbf{C}$ *be a* $k \times k$ *regular circulant with the first row equal to* $[c_0, c_1, c_2, \ldots, c_{k-1}]$. *Let* $\omega_1, \ldots, \omega_k$ *be the* k *roots of unity. The characteristic roots of* $\mathbf{C}$ *are given by* $\lambda_1, \lambda_2, \ldots, \lambda_k$, *where for each* $i = 1, 2, \ldots, k$

$$\lambda_i = c_0\omega_i^0 + c_1\omega_i + c_2\omega_i^2 + \cdots + c_{k-1}\omega_i^{k-1}$$

and $\mathbf{x}_1, \mathbf{x}_2, \ldots, \mathbf{x}_k$ *are the corresponding characteristic vectors, where*

$$\mathbf{x}_i = \begin{bmatrix} \omega_i^0 \\ \omega_i \\ \omega_i^2 \\ \vdots \\ \omega_i^{k-1} \end{bmatrix}.$$

Proof: By multiplying, one obtains $\mathbf{C}\mathbf{x}_i = \lambda_i\mathbf{x}_i$ for $i = 1, 2, \ldots, k$. ▮

Example 8.10.1. Find the characteristic roots and vectors of

$$\mathbf{C} = \begin{bmatrix} 3 & -1 & 2 & 0 \\ 0 & 3 & -1 & 2 \\ 2 & 0 & 3 & -1 \\ -1 & 2 & 0 & 3 \end{bmatrix}.$$

The four roots of unity are 1, -1, i, and $-i$, where $i = \sqrt{-1}$. Thus by theorem 8.10.7 we get

$$\lambda_1 = 3 - 1 + 2 + 0 = 4 \qquad ; \mathbf{x}_1' = [1, 1, 1, 1]$$

$$\lambda_2 = 3 - 1(-1) + 2(-1)^2 + 0(-1)^3 = 6 \qquad ; \mathbf{x}_2' = [1, -1, 1, -1]$$

$$\lambda_3 = 3 - 1(i) + 2(i)^2 + 0(i)^3 = 1 - i \qquad ; \mathbf{x}_3' = [1, i, -1, -i]$$

$$\lambda_4 = 3 - 1(-i) + 2(-i)^2 + 0(-i)^3 = 1 + i; \mathbf{x}_4' = [1, -i, -1, i]$$

To verify the above results, we may consider the characteristic equation $|\mathbf{C} - \lambda\mathbf{I}| = 0$, which reduces to

$$(3 - \lambda)^4 - 8(3 - \lambda)^2 + 8(3 - \lambda) + 15 = 0$$

or

$$\lambda^4 - 12\lambda^3 + 46\lambda^2 - 68\lambda + 48 = 0.$$

It can then be verified that the roots of this equation are 4, 6, $1 + i$, and $1 - i$.

Corollary 8.10.7.1

Let the sum of the elements in a row of a regular $k \times k$ circulant $\mathbf{C}$ *be denoted by s. Then s is a characteristic root of* $\mathbf{C}$.

Proof: Since 1 is always one of the k roots of unity, we see that

$$\lambda = c_0 1^0 + c_1 1 + c_2 1^2 + \cdots + c_{k-1} 1^{k-1} = \sum_{i=0}^{k-1} c_i = s$$

is always a characteristic root of $\mathbf{C}$. ∎

From matrix theory we know that the product of the characteristic roots of a matrix is always equal to the determinant of the matrix. Thus we have the following corollary.

Corollary 8.10.7.2

The determinant of a $k \times k$ regular circulant **C** *with first-row elements c_0, $c_1, \ldots, c_{k-1}$ is given by*

$$\prod_{i=1}^{k} \lambda_i = \prod_{i=1}^{k} (c_0 + c_1\omega_i + c_2\omega_i^2 + \cdots + c_{k-1}\,\omega_i^{k-1}).$$

The proof of the next theorem can be found in [D–1].

Theorem 8.10.8

For any $k \times k$ regular circulant, there exists a matrix **P** *such that* $\overline{\mathbf{P}}'\,\mathbf{AP} = \mathbf{D}$, *where* **D** *is a diagonal matrix where* $\overline{\mathbf{P}}$ *is the conjugate of the matrix* **P**.

Next we discuss a special type of $k \times k$ regular circulant—a $k \times k$ *symmetric* regular circulant.

Definition 8.10.2

Symmetric Regular Circulant. *A matrix* **A** *that is a $k \times k$ regular circulant and is also symmetric is defined to be a symmetric regular circulant.*

Example 8.10.2. If **A** and **B** are 3×3 and 4×4 symmetric regular circulants, respectively, they have the form given below.

$$\mathbf{A} = \begin{bmatrix} a_0 & a_1 & a_1 \\ a_1 & a_0 & a_1 \\ a_1 & a_1 & a_0 \end{bmatrix}; \mathbf{B} = \begin{bmatrix} b_0 & b_1 & b_2 & b_1 \\ b_1 & b_0 & b_1 & b_2 \\ b_2 & b_1 & b_0 & b_1 \\ b_1 & b_2 & b_1 & b_0 \end{bmatrix}.$$

The following two theorems are direct consequences of the definition of a symmetric regular circulant.

Theorem 8.10.9

Let **I**, **O**, *and* **J** *be the $k \times k$ identity matrix, null matrix, and matrix of ones, respectively. Then the following are symmetric regular circulants:*

$$a\mathbf{I},\ \mathbf{0},\ a\mathbf{J},\ a\mathbf{I} + b\mathbf{J},$$

where a and b are any real numbers.

Theorem 8.10.10

Let **A** *and* **B** *be* $k \times k$ *symmetric regular circulants. Then*

(1) **A** *has at most* $[k/2] + 1$ *distinct elements, where* $[k/2]$ *is the integral part of* $k/2$,
(2) $a\mathbf{A} + b\mathbf{B}$ *is a* $k \times k$ *symmetric regular circulant, where* a *and* b *are any real numbers,*
(3) $\mathbf{A}'$ *and* $\mathbf{B}'$ *are symmetric regular circulants.*

Theorem 8.10.11

Let **A** *and* **B** *be* $k \times k$ *symmetric regular circulants. Then*

(1) $\mathbf{AB} = \mathbf{BA}$,
(2) **AB** *is a symmetric regular circulant,*
(3) If **A** *is nonsingular, then* $\mathbf{A}^{-1}$ *is a symmetric regular circulant.*

Proof: The proof of (1) is a direct result of Theorem 8.10.5, since **A** and **B** are regular circulants. To prove (2), we use (1) and get $\mathbf{AB} = \mathbf{BA} = \mathbf{B}'\mathbf{A}' = (\mathbf{AB})'$, so **AB** is symmetric; but by Theorem 8.10.3, **AB** is a regular circulant, so **AB** is a symmetric regular circulant. To prove (3), we know by Theorem 8.10.4 that $\mathbf{A}^{-1}$ is a regular circulant, but the inverse of any nonsingular symmetric matrix is symmetric. This proves (3) and the theorem. ■

If **A** is a $k \times k$ symmetric regular circulant, the characteristic roots are real and Theorem 8.10.7 can be used to obtain these roots.

Another type of circulant that is useful in statistics is a symmetric circulant, which we now define and discuss.

Definition 8.10.3

Symmetric Circulant. *The* $k \times k$ *matrix* **A** *is a symmetric circulant if and only if* $(i + j - 2)|k = (p + q - 2)|k$ *implies* $a_{ij} = a_{pq}$.

We note that if **A** is a $k \times k$ symmetric matrix, then **A** has at most k distinct elements, which we denote by $a_0, a_1, \ldots, a_{k-1}$, and

$$\mathbf{A} = \begin{bmatrix} a_0 & a_1 & a_2 \ldots a_{k-1} \\ a_1 & a_2 & a_3 \ldots a_0 \\ a_2 & a_3 & a_4 \ldots a_1 \\ \vdots & & \\ a_{k-1} & a_0 & a_1 \ldots a_{k-2} \end{bmatrix}.$$

We note that $a_{ij} = a_{(i+j-2)|k}$, and so this can be used as the definition of a $k \times k$ symmetric circulant. Also we note that $\mathbf{A} = \mathbf{A}'$. In a symmetric circulant the t-th row for $t = 2, 3, \ldots, k$ is obtained from the $(t - 1)$-st row by moving each element of the $(t - 1)$-st row to the left by one column and placing the first element in the $(t - 1)$-st row in the last column of the t-th row. We point out that a symmetric circulant is not in general a symmetric *regular* circulant. The matrices below illustrate this: $\mathbf{A}_1$ and $\mathbf{B}_1$ are symmetric circulants; $\mathbf{A}_2$ and $\mathbf{B}_2$ are symmetric *regular* circulants.

$$\mathbf{A}_1 = \begin{bmatrix} a_0 & a_1 & a_2 \\ a_1 & a_2 & a_0 \\ a_2 & a_0 & a_1 \end{bmatrix}; \mathbf{B}_1 = \begin{bmatrix} b_0 & b_1 & b_2 & b_3 \\ b_1 & b_2 & b_3 & b_0 \\ b_2 & b_3 & b_0 & b_1 \\ b_3 & b_0 & b_1 & b_2 \end{bmatrix};$$

$$\mathbf{A}_2 = \begin{bmatrix} a_0 & a_1 & a_1 \\ a_1 & a_0 & a_1 \\ a_1 & a_1 & a_0 \end{bmatrix}; \mathbf{B}_2 = \begin{bmatrix} b_0 & b_1 & b_2 & b_1 \\ b_1 & b_0 & b_1 & b_2 \\ b_2 & b_1 & b_0 & b_1 \\ b_1 & b_2 & b_1 & b_0 \end{bmatrix}.$$

The following theorems are direct results of the definition of symmetric circulants.

Theorem 8.10.12

Let $\mathbf{A}$ and $\mathbf{B}$ be $k \times k$ symmetric circulants. Then $a\mathbf{A}$ and $b\mathbf{B}$ is a $k \times k$ symmetric circulant, where a and b are any real numbers.

Theorem 8.10.13

Let $\mathbf{I}$, $\mathbf{O}$, and $\mathbf{J}$ be the $k \times k$ identity matrix, the null matrix, and the matrix of ones, respectively. Then

(1) $\mathbf{I}$ *is* not *a symmetric circulant,*
(2) $\mathbf{0}$ *and $a\mathbf{J}$ are symmetric circulants for any real number a.*

It is of interest to determine if the product and inverse of symmetric circulants are symmetric circulants. These results are given in the next two theorems.

Theorem 8.10.14

If $\mathbf{A}$ is a $k \times k$ symmetric circulant and is nonsingular, then $\mathbf{A}^{-1}$ is a symmetric circulant.

Proof: This proof is almost identical to the first proof for Theorem 8.10.4. ∎

The next theorem demonstrates an interesting result between symmetric circulants and symmetric regular circulants.

Theorem 8.10.15

Let **A** *and* **B** *be* $k \times k$ *symmetric circulants. Then* **AB** *is a regular circulant. Also, in general,* $\mathbf{AB} \neq \mathbf{BA}$.

Proof: Let $\mathbf{AB} = \mathbf{C}$. Then

$$c_{ij} = \sum_{s=1}^{k} a_{is} b_{sj} = \sum_{s=1}^{k} a_{(s+i-2)|k}\, b_{(s+j-2)|k}$$

$$= \sum_{u=0}^{k-1} a_{u|k}\, b_{(j-i+u)|k},$$

since $s + j - 2 = j - i + (s + i - 2)$. Also

$$c_{pq} = \sum_{t=1}^{k} a_{pt} b_{tq} = \sum_{t=1}^{k} a_{(p+t-2)|k}\, b_{(t+q-2)|k}$$

$$= \sum_{u=0}^{k-1} a_{u|k}\, b_{(q-p+u)|k},$$

since $t + q - 2 = q - p + (p + t - 2)$. Thus $c_{ij} = c_{pq}$ if $b_{(j-i+u)|k} = b_{(q-p+u)|k}$, which is true if $(j - i + u)|k = (q - p + u)|k$, which is true if $(j - i)|k = (q - p)|k$. In summary, $c_{ij} = c_{pq}$ if $(j - i)|k = (q - p)|k$. Therefore, **C** is a regular circulant. One can multiply two symmetric circulants to demonstrate that in general, $\mathbf{AB} \neq \mathbf{BA}$. ▌

Theorem 8.10.16

If **B** *is a regular circulant and* **C** *is a symmetric circulant, then* **BC** *and* **CB** *are symmetric circulants. In general,* $\mathbf{BC} \neq \mathbf{CB}$.

Proof: To demonstrate $\mathbf{BC} \neq \mathbf{CB}$ one can multiply two appropriate matrices. We prove that **BC** is a symmetric circulant and the proof that **CB** is also a symmetric circulant is similar.

Let **BC** = **D**. Then

$$d_{ij} = \sum_{s=1}^{k} b_{is}c_{sj} = \sum_{s=1}^{k} b_{(s-i)|k}\, c_{(s+j-2)|k}$$

$$= \sum_{u=0}^{k-1} b_{u|k}\, c_{(i+j-2+u)|k},$$

where $s + j - 2 = i + j - 2 + (s - i)$. Similarly,

$$d_{pq} = \sum_{t=1}^{k} b_{pt}c_{tq} = \sum_{t=1}^{k} b_{(t-p)|k}\, c_{(t+q-2)|k}$$

$$= \sum_{u=0}^{k-1} b_{u|k}\, c_{(p+q-2+u)|k}.$$

Then $d_{ij} = d_{pq}$ if $c_{(i+j-2+u)|k} = c_{(p+q-2+u)|k}$, which is true if $(i + j - 2 + u)|k = (p + q - 2 + u)|k$, which is true if $(i + j - 2)|k = (p + q - 2)|k$. In summary, $d_{ij} = d_{pq}$ if $(i + j - 2)|k = (p + q - 2)|k$. Therefore, **D** is a symmetric circulant. ∎

The following theorem is a direct result of Theorems 8.10.15 and 8.10.16 and will be used in Theorem 8.10.23 concerning g-inverses of circulants.

Theorem 8.10.17

The product of an even number of symmetric circulants is a regular circulant, and the product of an odd number of symmetric circulants is a symmetric circulant.

Theorem 8.10.18

Let **A** *be a* $k \times k$ *regular circulant,* **C** *be a* $k \times k$ *symmetric circulant, and suppose there is a solution* **X** *to the matrix equation* **AX** = **C**. *There exists a* $k \times k$ *symmetric circulant* **B** *such that* **AB** = **C**.

The proof of Theorem 8.10.18 is straightforward and will be left for the reader. A simple proof can be found in [M–7].

In the next two theorems we discuss the characteristic roots and the determinant of symmetric circulants.

Theorem 8.10.19

Let **C** *be a* $k \times k$ *symmetric circulant where* $c_0, c_1, \ldots, c_{k-1}$ *are the* k *elements of* **C**. *Also let* $\omega_1, \omega_2, \ldots, \omega_k$ *be the* k *roots of unity. If the characteristic roots of* **C** *are denoted by* $\lambda_0, \lambda_1, \lambda_2, \ldots, \lambda_{k-1}$, *then*

$$\lambda_i^2 = \omega_i^0 \sum_{j=0}^{k-1} c_j^2 + \omega_i \sum_{j=0}^{k-1} c_j c_{(j+1)|k} + \omega_i^2 \sum_{j=0}^{k-1} c_j c_{(j+2)|k}$$

$$+ \cdots + \omega_i^{k-1} \sum_{j=0}^{k-1} c_j c_{(j+k-1)|k}.$$

Proof: Since $\mathbf{C} = \mathbf{C}'$, $\mathbf{C}^2 = (\mathbf{C}^2)'$. Thus both **C** and $\mathbf{C}^2$ have real eigenvalues. The eigenvalues of $\mathbf{C}^2$ are the squares of the eigenvalues of **C**, since $\mathbf{Cx} = \lambda\mathbf{x}$ implies $\mathbf{C}^2\mathbf{x} = \lambda\mathbf{Cx} = \lambda^2\mathbf{x}$. (Furthermore, the eigenvalues of $\mathbf{C}^2$ are non-negative, since they are the squares of the real eigenvalues of **C**.) By Theorem 8.10.15, $\mathbf{C}^2$ is a $k \times k$ regular circulant. The k elements in any row of $\mathbf{C}^2$ are

$$\sum_{j=0}^{k-1} c_j^2, \sum_{j=0}^{k-1} c_j c_{(j+1)|k}, \ldots, \sum_{j=0}^{k-1} c_j c_{(j+k-1)|k}.$$

So by Theorem 8.10.7,

$$\lambda_i^2 = \omega_i^0 \sum_{j=0}^{k-1} c_j^2 + \cdots + \omega_i^{k-1} \sum_{j=0}^{k-1} c_j c_{(j+k-1)|k}$$

and this proves the theorem. ∎

As the example below will show, this theorem is quite an aid in finding the eigenvalues of a symmetric circulant. However, in most cases more information will be needed.

Example 8.10.3. If **C**, a symmetric circulant, is defined by

$$\mathbf{C} = \begin{bmatrix} 2 & 1 & 0 & -1 \\ 1 & 0 & -1 & 2 \\ 0 & -1 & 2 & 1 \\ -1 & 2 & 1 & 0 \end{bmatrix},$$

then

$$\mathbf{C}^2 = \begin{bmatrix} 6 & 0 & -2 & 0 \\ 0 & 6 & 0 & -2 \\ -2 & 0 & 6 & 0 \\ 0 & -2 & 0 & 6 \end{bmatrix}.$$

The four roots of unity are 1, -1, i, and $-i$, where $i = \sqrt{-1}$. Thus

$$\lambda_1^2 = 6 - 2 = 4; \lambda_2^2 = 6 - 2 = 4; \lambda_3^2 = 6 + 2 = 8; \lambda_4^2 = 6 + 2 = 8.$$

Now using the fact that the trace of $\mathbf{C}$ is equal to the sum of the eigenvalues, we know that $\lambda_1 = \lambda_2 = 2$ and $\lambda_3 = -\lambda_4 = 2\sqrt{2}$.

A further aid in finding eigenvalues is given in the following theorem.

Theorem 8.10.20

Let $\mathbf{C}$ *be a* $k \times k$ *symmetric circulant with first-row elements* $c_0, c_1, \ldots,$ c_{k-1}. *Then* $s = c_0 + c_1 + \cdots + c_{k-1}$ *is an eigenvalue of* $\mathbf{C}$.

Proof: Consider the determinant $|\mathbf{C} - \lambda\mathbf{I}|$. Now add each of rows 2 through k to the first row. Then each element of the first row will be of the form $\Sigma_{i=0}^{k-1} c_i - \lambda$, which may then be factored out in the following manner:

$$|\mathbf{C} - \lambda\mathbf{I}| = \left(\sum_{i=0}^{k-1} c_i - \lambda\right) |\mathbf{C}_1|,$$

where $|\mathbf{C}_1|$ is the remaining determinant. Now we see that $|\mathbf{C} - \lambda\mathbf{I}| = 0$ for

$$\lambda = \sum_{i=0}^{k-1} c_i.$$

Thus $\Sigma_{i=0}^{k-1} c_i$ is an eigenvalue of $\mathbf{C}$. ∎

Theorem 8.10.21

Let **A** *and* **B** *be* $k \times k$ *regular and symmetric circulants, respectively, where* $c_0, c_1, \ldots, c_{k-1}$ *is the first row of both* **A** *and* **B**. *Then*

$$\det(\mathbf{A}) = (-1)^{\left[\frac{k-1}{2}\right]} \det(\mathbf{B}),$$

where $\left[\frac{k-1}{2}\right]$ *is the integral part of* $\frac{k-1}{2}$.

Proof: Let

$$\mathbf{A} = \begin{bmatrix} c_0 & c_1 & c_2 & \cdots & c_{k-1} \\ c_{k-1} & c_0 & c_1 & \cdots & c_{k-2} \\ c_{k-2} & c_{k-1} & c_0 & \cdots & c_{k-3} \\ \cdots & & & & \\ c_2 & c_3 & c_4 & \cdots & c_1 \\ c_1 & c_2 & c_3 & \cdots & c_0 \end{bmatrix}; \mathbf{B} = \begin{bmatrix} c_0 & c_1 & c_2 & \cdots & c_{k-1} \\ c_1 & c_2 & c_3 & \cdots & c_0 \\ c_2 & c_3 & c_4 & \cdots & c_1 \\ \cdots & & & & \\ c_{k-2} & c_{k-1} & c_0 & \cdots & c_{k-3} \\ c_{k-1} & c_0 & c_1 & \cdots & c_{k-2} \end{bmatrix}.$$

By elementary matrix theory we know that if two rows of a determinant are interchanged, the determinant value is changed by a factor of -1. Now note that row 2 of **A** is row k of **B**, row 3 of **A** is row $k - 1$ of **B**. Thus by interchanging rows 2 and k, 3 and $k - 1$, 4 and $k - 2$, and so forth, of **A**, we get **B**. Altogether we have a total of $\left[\frac{k-1}{2}\right]$ row interchanges of **A** to get **B**. Thus

$$\det(\mathbf{B}) = (-1)^{\left[\frac{k-1}{2}\right]} \det(\mathbf{A}). \quad \blacksquare$$

Referring back to Theorem 8.10.15, we are reminded that the product of two symmetric circulants is a regular circulant. This theorem brings to mind the question of whether the product of two symmetric circulants can ever be a symmetric circulant. This breaks down to two questions: (1) when can a symmetric circulant be a symmetric regular circulant and (2) when does the product of two symmetric circulants give a symmetric regular circulant? This is the context of the next theorem and corollary. The proof will be asked for in the problems.

Theorem 8.10.22

Let **C** *be a* $k \times k$ *matrix that is both a symmetric circulant and a symmetric regular circulant. If* k *is an odd integer, then* **C** *must be a matrix with each* *element equal to a constant, say c, and we denote this matrix by* $\mathbf{C}_1$; *if k is an even integer, then* **C** *must be of the form*

$$\begin{bmatrix} c_0 & c_1 & c_0 & c_1 & \cdots & c_0 & c_1 \\ c_1 & c_0 & c_1 & c_0 & \cdots & c_1 & c_0 \\ c_0 & c_1 & c_0 & c_1 & \cdots & c_0 & c_1 \\ \vdots & \vdots & \vdots & \vdots & & \vdots & \vdots \\ c_1 & c_0 & c_1 & c_0 & \cdots & c_1 & c_0 \end{bmatrix}.$$

We denote this matrix by $\mathbf{C}_0$.

Corollary 8.10.22

In Theorem 8.10.22 *the results below follow.*

(1) *The sum, difference, and product of two matrices of the form denoted by* $\mathbf{C}_1$ *is also a matrix of the form* $\mathbf{C}_1$. *The rank of a matrix of the form of* $\mathbf{C}_1$ *is zero if* $c = 0$; *it is* 1 *if* $c \neq 0$.

(2) *The sum, difference, and product of two matrices of the form denoted by* $\mathbf{C}_0$ *is also a matrix of the form* $\mathbf{C}_0$. *The rank of* $\mathbf{C}_0$ *is* (a) *zero if* $c_0 = c_1 = 0$; (b) 1 *if* $|c_0| = |c_1| \neq 0$; (c) 2 *if* $|c_0| \neq |c_1|$.

In Theorems 8.10.4, 8.10.11, and 8.10.14 we proved that the inverse of nonsingular circulants of a certain type were circulants of the same type. In the next theorem we discuss g-inverses of circulants.

Theorem 8.10.23

Let **A** *be a* $k \times k$ *regular circulant and* **B** *be a* $k \times k$ *symmetric circulant. Then* $\mathbf{A}^-$, *the g-inverse of* **A**, *is a regular circulant and* $\mathbf{B}^-$, *the g-inverse of* **B**, *is a symmetric circulant.*

Proof: We use Theorem 6.5.2, which states that the g-inverse of a matrix **C** is given by $\mathbf{C}^- = \mathbf{H}'\mathbf{C}'$, where **H** is any solution to $(\mathbf{C}'\mathbf{C})^2\mathbf{H} = \mathbf{C}'\mathbf{C}$. Let $\mathbf{A} = \mathbf{C}$; then $\mathbf{A}'\mathbf{A}$ is a regular circulant and $(\mathbf{A}'\mathbf{A})^2$ is also a regular circulant; hence by Theorem 8.10.6, there is a regular circulant **H** that satisfies $(\mathbf{A}'\mathbf{A})^2\mathbf{H} = \mathbf{A}'\mathbf{A}$ and $\mathbf{H}'\mathbf{A}$ (which is $\mathbf{A}^-$) is also a regular circulant. The proof that $\mathbf{B}^-$ is a symmetric circulant is similar. ∎

8.11 Dominant Diagonal Matrices

In the next few sections we present a brief discussion of some special types of matrices. The objective is to define various matrices that have special patterns and are useful in some areas of statistics and probability as well as other fields of study. None of the types will be discussed in great detail; only introductory material will be presented, along with some recent references, so that the interested reader can obtain further information.

In this section we discuss dominant diagonal matrices. We use d.d. for dominant diagonal and det(**A**) for determinant of **A**. The symbol $|a_{ij}|$ will denote absolute value of a_{ij} when a_{ij} is a real number. When a_{ij} is a complex number that is nonreal, the symbol $|a_{ij}|$ will denote the modulus of a_{ij}.

Definition 8.11.1

Dominant Diagonal Matrices. *Let* **A** *be an* $n \times n$ *(real),* $n \geq 2$, *matrix with elements* a_{ij}. *Let*

$$C_q = \sum_{\substack{i=1 \\ i \neq q}}^{n} |a_{iq}|;$$

that is, C_q *is the sum of the absolute values of the off-diagonal elements of the q-th column of* **A**. *Let*

$$R_p = \sum_{\substack{j=1 \\ j \neq p}}^{n} |a_{pj}|;$$

that is, R_p *is the sum of the absolute values of the off-diagonal elements of the p-th row of* **A**. *If* $|a_{qq}| > C_q$, *the q-th column of* **A** *is defined to have a d.d. If* $|a_{pp}| > R_p$, *the p-th row of* **A** *is defined to have a d.d. If* $|a_{qq}| > C_q$ *for all* $q = 1, 2, \ldots, n$, *then* **A** *is called column d.d. If* $|a_{pp}| > R_p$ *for all* $p = 1, 2, \ldots, n$, *then* **A** *is called row d.d. If* **A** *is either column or row d.d., it is defined to be a d.d. matrix.*

Note: When we state that **A** is not d.d., we mean that it is *neither* row *nor* column d.d.

The following theorem is a direct result of the definition of a diagonal dominant matrix.

Theorem 8.11.1

Let $\mathbf{A}$ *be an* $n \times n$ *matrix.*

(1) *If* $\mathbf{D}$ *is any* $n \times n$ *nonsingular diagonal matrix and* $\mathbf{A}$ *is row (column) dominant diagonal, then* $\mathbf{DA}(\mathbf{AD})$ *is row (column) dominant diagonal.*
(2) *If* $\mathbf{P}$ *is any* $n \times n$ *permutation matrix and* $\mathbf{A}$ *is row (column) dominant diagonal, then* $\mathbf{P'AP}$ *is row (column) dominant diagonal.*
(3) *If* $\mathbf{A}$ *is row (column) dominant diagonal and* $\tilde{\mathbf{A}}$ *is any principal submatrix of* $\mathbf{A}$, *then* $\tilde{\mathbf{A}}$ *is row (column) dominant diagonal. Note that* $\tilde{\mathbf{A}}$ *is the matrix resulting from eliminating any set of* $r(r \le n - 1)$ *rows and the corresponding columns of* $\mathbf{A}$.
(4) *If* $\mathbf{A}$ *is (is not) d.d. then* $\mathbf{A}'$ *is (is not) d.d.*
(5) *If* $\mathbf{A}$ *is d.d. and the signs of any of the elements of* $\mathbf{A}$ *are changed, then the resulting matrix is d.d.*
(6) *If* $\mathbf{A}$ *is any nonsingular diagonal matrix, it is row and column d.d.*
(7) *If any diagonal element of* $\mathbf{A}$ *is zero, then* $\mathbf{A}$ *is not d.d.*

Example 8.11.1. Let $\mathbf{A}$ and $\mathbf{B}$ be defined below.

$$\mathbf{A} = \begin{bmatrix} -6 & 1 & 0 & 2 \\ 1 & -5 & 1 & 2 \\ 1 & 5 & 9 & -2 \\ 3 & -2 & 4 & 12 \end{bmatrix}; \ \mathbf{B} = \begin{bmatrix} -6 & 1 & 0 & 2 \\ 1 & -4 & 1 & 2 \\ 1 & 5 & 9 & -2 \\ 3 & -2 & 4 & 12 \end{bmatrix}.$$

The matrix $\mathbf{A}$ is row d.d. but not column d.d. (note column 2), so $\mathbf{A}$ is d.d.

The matrix $\mathbf{B}$ is neither row d.d. nor column d.d., so $\mathbf{B}$ is *not* d.d.

We now state a theorem and some corollaries about the relationship of d.d. matrices with singular and nonsingular matrices.

Theorem 8.11.2

Let $\mathbf{A}$ *be an* $n \times n$ *d.d. matrix. Then* $\mathbf{A}$ *is nonsingular.*

Proof: Assume $\mathbf{A}$ is row d.d. The proof will be by contradiction. Assume $\mathbf{A}$ is singular; then there exists a nonzero vector $\mathbf{x}$ such that $\mathbf{Ax} = \mathbf{0}$ by Theorem 7.2.5. Let $\mathbf{x}_0$ be a vector ($\mathbf{x}_0 \ne \mathbf{0}$) such that $\mathbf{Ax}_0 = \mathbf{0}$, and suppose the elements x_i of the vector $\mathbf{x}_0$ are such that max $|x_i| \leq |x_q|$ for $i = 1, \ldots, n$; that is, no element of $\mathbf{x}_0$ has absolute value greater than $|x_q|$. Examine the q-th equation of $\mathbf{A}\mathbf{x}_0 = \mathbf{0}$. We obtain

$$0 = \sum_{j=1}^{n} a_{qj}x_j,$$

so

$$a_{qq}x_q = -\sum_{\substack{j=1 \\ j\neq q}}^{n} a_{qj}x_j$$

and

$$|a_{qq}x_q| = \left| -\sum_{\substack{j=1 \\ j\neq q}}^{n} a_{qj}x_j \right| = \left| \sum_{\substack{j=1 \\ j\neq q}}^{n} a_{qj}x_j \right| \leq \sum_{\substack{j=1 \\ j\neq q}}^{n} |a_{qj}x_j| = \sum_{\substack{j=1 \\ j\neq q}}^{n} |a_{qj}|\ |x_j|.$$

This results in

$$|a_{qq}|\ |x_q| \leq |x_q| \sum_{\substack{j=1 \\ j\neq q}}^{n} |a_{qj}|.$$

Divide both sides by $|x_q|$, which cannot be zero, since $\mathbf{x}_0 \neq \mathbf{0}$, and the result is

$$|a_{qq}| \leq \sum_{\substack{j=1 \\ j\neq q}}^{n} |a_{qj}|,$$

which contradicts the assumption that $\mathbf{A}$ is row d.d. Hence the theorem is established. If $\mathbf{A}$ is column d.d., use the above proof on $\mathbf{A}'$. ∎

Corollary 8.11.2.1

Let **P** *and* **Q** *be any nonsingular* $n \times n$ *matrices. If* **PAQ** *is a d.d. matrix, then* **A** *is nonsingular.*

Corollary 8.11.2.2

If **A** *is an* $n \times n$ *singular matrix, then* **A** *is not d.d.*

Corollary 8.11.2.3

If $\mathbf{A}$ *is an idempotent matrix and* $\mathbf{A} \neq \mathbf{I}$, *then* $\mathbf{A}$ *is not d.d.*

Corollary 8.11.2.4

If $\mathbf{A}$ *is d.d. and the signs of any set of the elements* a_{ij} *are changed, the resulting matrix is nonsingular.*

Corollary 8.11.2.5

If λ *is any characteristic root of* $\mathbf{A}$, *then* $\mathbf{A} - \lambda\mathbf{I}$ *is not d.d.*

Corollary 8.11.2.6

If $\mathbf{A}$ *is a* $k \times k$ *regular circulant with first-row elements* $a_0, a_1, \ldots, a_{k-1}$ *and if*

$$|a_q| > \sum_{\substack{i=0 \\ i \neq q}}^{k-1} |a_i|$$

for some q, *then* $\mathbf{A}$ *is nonsingular.*

Corollary 8.11.2.7

If $\mathbf{A}$ *is nonsingular, there exist nonsingular matrices* $\mathbf{P}$ *and* $\mathbf{Q}$ *such that* $\mathbf{PAQ}$ *is d.d.*

Example 8.11.2. To illustrate Corollary 8.11.2.1, consider the matrix $\mathbf{A}$ below.

$$\mathbf{A} = \begin{bmatrix} 6 & 2 & 4 & 1 \\ 2 & 8 & 1 & -1 \\ 3 & 2 & -11 & 1 \\ -2 & -1 & 1 & 7 \end{bmatrix}.$$

This matrix is not d.d., since neither the first row nor the first column is d.d. Multiply $\mathbf{A}$ on the left by a diagonal matrix $\mathbf{D}$ with diagonal elements 2, 1, 1, 1 and get

$$\mathbf{DA} = \begin{bmatrix} 2 & 0 & 0 & 0 \\ 0 & 1 & 0 & 0 \\ 0 & 0 & 1 & 0 \\ 0 & 0 & 0 & 1 \end{bmatrix} \begin{bmatrix} 6 & 2 & 4 & 1 \\ 2 & 8 & 1 & -1 \\ 3 & 2 & -11 & 1 \\ -2 & -1 & 1 & 7 \end{bmatrix} = \begin{bmatrix} 12 & 4 & 8 & 2 \\ 2 & 8 & 1 & -1 \\ 3 & 2 & -11 & 1 \\ -2 & -1 & 1 & 7 \end{bmatrix}.$$

The matrix **DA** is (column) d.d., so **DA** is nonsingular; but since **D** is nonsingular, it follows that **A** is also nonsingular.

The next theorem is a slight generalization of Theorem 8.11.2.

Theorem 8.11.3

*Consider an $n \times n$ matrix **A** such that each row, except one (say the s-th row), is d.d., and suppose the s-th row is such that $0 < |a_{ss}| = R_s$. Then **A** is nonsingular.*

Proof: By the hypothesis of the theorem, $|a_{ii}| > R_i$ for $i = 1, 2, \ldots, n$; $i \neq s$ and $0 < |a_{ss}| = R_s$. Multiply **A** on the right by a diagonal matrix **D** whose diagonal elements are all equal to 1 except the s-th diagonal element, which is equal to $1 + \varepsilon$, where $\varepsilon > 0$. The sum of the absolute values of the off-diagonal elements of the i-th row of **AD**, denoted by R_i^*, is

$$R_i^* = R_i + |a_{is}|\,|\varepsilon|, \qquad \text{for } i = 1, \ldots, n \text{ and } i \neq s; \text{ also } R_s^* = R_s.$$

We let a_{ii}^* denote the i-th diagonal element of **AD**, so $|a_{ii}^*| = |a_{ii}|$ for $i = 1, \ldots, n$ and $i \neq s$; also $|a_{ss}^*| = |a_{ss}(1 + \varepsilon)|$. Since $|a_{ii}| > R_i$ for all $i \neq s$, clearly $|a_{ii}^*| > R_i^*$ for all $i \neq s$ and for a small enough $\varepsilon > 0$. Also since $|a_{ss}| = R_{ss}$, it is clear that for any $\varepsilon > 0$, we obtain $|a_{ss}^*| > R_{ss}^*$. Hence **AD** is row d.d. for an appropriate value of $\varepsilon > 0$; so by Theorem 8.11.2 it follows that **AD** is nonsingular. But since **D** is nonsingular, **A** is also nonsingular, and this proves the theorem. ∎

A similar theorem is true if row d.d. of **A** is replaced by column d.d. Let **A** be an $n \times n$ matrix such that

$$|a_{ij}| > \sum_{\substack{k=1 \\ k \neq i}}^{n} |a_{kj}|. \tag{8.11.1}$$

Then the j-th column is said to have a dominant element and it is in the i-th row.

This leads to the following theorem, which is a slight generalization of Theorem 8.11.2.

Theorem 8.11.4

*Let **A** be an $n \times n$ matrix such that each column has a dominant element and such that each row contains one of the dominant elements. Then **A** is nonsingular.*

Proof: If each row contains one of the dominant elements, then by permuting the rows, the dominant elements can be transformed to the main diagonal. But this is equivalent to multiplying **A** on the left by a permutation matrix **P** (which is nonsingular). Thus the matrix **B**, where **B** = **PA**, satisfies the conditions of Theorem 8.11.2 and hence is nonsingular. So **A** is also nonsingular and the theorem is proved. ▮

Note: The above theorem is also true if each row has a dominant element and these are in distinct columns. The theorem is also true if each column (row) except one, say the t-th, has a dominant element and the equals sign holds for an element in the t-th column (row), where each of these elements is in distinct rows (columns).

Example 8.11.3. Let **A** be defined as below.

$$\mathbf{A} = \begin{bmatrix} -2 & -2 & 0 & -\underline{17} \\ 1 & \underline{4} & 6 & -3 \\ \underline{6} & 0 & -4 & 8 \\ 2 & -1 & \underline{10} & -5 \end{bmatrix}.$$

Each column in **A** except one has a dominant element, and in that column (column 3) an equals sign holds (we have underscored them), and there is one such element in each row. So **A** is nonsingular. We note that if we multiply **A** on the left by the appropriate permutation matrix **P**, we get **PA** = **B**. We illustrate **PA** = **B** below.

$$\begin{bmatrix} 0 & 0 & 1 & 0 \\ 0 & 1 & 0 & 0 \\ 0 & 0 & 0 & 1 \\ 1 & 0 & 0 & 0 \end{bmatrix} \begin{bmatrix} -2 & -2 & 0 & -\underline{17} \\ 1 & \underline{4} & 6 & -3 \\ \underline{6} & 0 & -4 & 8 \\ 2 & -1 & \underline{10} & -5 \end{bmatrix} = \begin{bmatrix} \underline{6} & 0 & -4 & 8 \\ 1 & \underline{4} & 6 & -3 \\ 2 & -1 & \underline{10} & -5 \\ -2 & -2 & 0 & -\underline{17} \end{bmatrix}.$$

So **B** satisfies Theorem 8.11.3 and is nonsingular. But a permutation matrix **P** is nonsingular and hence **A** is nonsingular.

The next theorem is a generalization of Theorem 8.11.3.

Theorem 8.11.5

Let **A** *be an* $n \times n$ *matrix such that for one value of* $j = 1, \ldots, n$ *(say* $j = t$*) either Eq.* (8.11.2) *or Eq.* (8.11.3) *holds.*

$$0 < |a_{tt}| < R_t \text{ and } |a_{ii}|\,|a_{tt}| > R_iR_t \text{ for } i = 1, \ldots, n;\ i \neq t, \tag{8.11.2}$$

$$0 < |a_{tt}| < C_t \text{ and } |a_{ii}|\,|a_{tt}| > C_iC_t \text{ for } i = 1, \ldots, n;\ i \neq t. \tag{8.11.3}$$

Then **A** *is nonsingular.*

Proof: We prove the theorem for $t = 1$ and the relations in Eq. (8.11.2). The proof for any other value of t is similar. Thus we assume $0 < |a_{11}| < R_1$ and $|a_{11}|\,|a_{ii}| > R_iR_1$ for $i = 2, \ldots, n$. This implies $|a_{ii}| > R_i$ for $i = 2, \ldots, n$; also $1 < R_1/|a_{11}| = b$ (say), so $|a_{ii}| > (R_1/|a_{11}|)R_i = bR_i$ for $i = 2, \ldots, n$. Multiply **A** on the right by **D**, where **D** is a diagonal matrix with $d_{11} = b$ and $d_{ii} = 1$ for $i = 2, \ldots, n$. We get **AD** = **B** where $b_{ij} = a_{ij}$ for $i = 1, \ldots, n$; $j = 2, \ldots, n$ and $b_{i1} = ba_{i1}$ for $i = 1, \ldots, n$ (that is, **B** is the same as **A** except that the first column of **A** is multiplied by b). We will show that **B** is nonsingular and hence **A** is nonsingular, since **D** is nonsingular. To show that **B** is nonsingular, we note that

$$|b_{11}| = \sum_{j=2}^{n} |b_{1j}|,$$

since $|b_{11}| = b|a_{11}| = R_1$ and

$$\sum_{j=2}^{n} |b_{1j}| = \sum_{j=2}^{n} |a_{1j}| = R_1.$$

Also for $i = 2, \ldots, n$ we get

$$|b_{ii}| > \sum_{\substack{j=1 \\ j\neq i}}^{n} |b_{ij}|,$$

since $|b_{ii}| = |a_{ii}|$ and

$$\sum_{\substack{j=1 \\ j\neq i}}^{n} |b_{ij}| = b|a_{i1}| + \sum_{\substack{j=2 \\ j\neq i}}^{n} |a_{ij}| \leq bR_i.$$

Now we use Theorem 8.11.3 to complete the proof of the theorem. ∎

Example 8.11.4. Consider the 3×3 matrix $\mathbf{A}$ below and suppose one wants to determine if $\mathbf{A}$ is nonsingular.

$$\mathbf{A} = \begin{bmatrix} 5 & 5 & 2 \\ 1 & 5 & -2 \\ -3 & 2 & 8 \end{bmatrix}.$$

This matrix is neither row nor column d.d. However, all the rows except row 1 are d.d., so there is a possibility of using either Theorem 8.11.3 or 8.11.5. Clearly Theorem 8.11.3 cannot be used, since $|a_{11}| \neq R_1$, but $|a_{11}| < R_1$, so it might be possible to use Theorem 8.11.5. We get

$\diagdown i$	1	2	3
$\lvert a_{ii}\rvert$	5	5	8
R_i	7	3	5

and $|a_{11}|\ |a_{22}| > R_1R_2$ (that is, $25 > 21$), and $|a_{11}|\ |a_{33}| > R_1R_3$ (that is, $40 > 35$). So $\mathbf{A}$ is nonsingular.

Theorem 8.11.6

If $\mathbf{A}$ *is row (column) d.d., then at least one column (row) must be d.d.*

Proof: The proof will be by contradiction. Assume that the theorem is false, that is, no column is d.d. If $\mathbf{A}$ is row d.d., then $|a_{ii}| > R_i$ for $i = 1, 2, \ldots, n$, and so

$$\sum_{i=1}^{n} |a_{ii}| > \sum_{i=1}^{n} R_i = \sum_{i=1}^{n} \sum_{\substack{j=1 \\ i \neq j}}^{n} |a_{ij}| = b \text{ (say)}.$$

Also if no column of $\mathbf{A}$ is d.d., then $|a_{ii}| \leq C_i$ for $i = 1, 2, \ldots, n$ and so

$$\sum_{i=1}^{n} |a_{ii}| \leq \sum_{i=1}^{n} C_i = \sum_{j=1}^{n} \sum_{\substack{i=1 \\ i \neq j}}^{n} |a_{ij}| = b,$$

which contradicts the fact that

$$\sum_{i=1}^{n} |a_{ii}| > b$$

if **A** is row d.d. This proves the theorem for row d.d., and a similar proof can be used for column d.d. ∎

In the next theorem we discuss how diagonal dominance might be used to determine if the determinant of a matrix is positive.

Theorem 8.11.7

Let **A** *be an* $n \times n$ *matrix that is d.d. and such that the diagonal elements of* **A** *are positive; that is,* $a_{ii} > 0$ *for* $i = 1, 2, \ldots, n$. *Then* $\det(\mathbf{A}) > 0$.

Proof: We assume that **A** is row d.d. The proof is similar if **A** is column d.d. The proof will be by induction. Clearly the theorem is true if $n = 1$ (and $n = 2$). Assume it is true if $n = k$. We must show it is also true when $n = k + 1$. Consider any $(k + 1) \times (k + 1)$ matrix **A** that satisfies the conditions of the theorem. Partition **A** as below where **B** is $k \times k$ and hence the size of **d**, **c**, and a are determined.

$$\mathbf{A} = \begin{bmatrix} \mathbf{B} & \mathbf{d} \\ \mathbf{c}' & a \end{bmatrix}.$$

By the hypothesis of the theorem,

$$a_{ii} = b_{ii} > 0; \; a_{ii} = b_{ii} > \sum_{\substack{j=1 \\ j \neq i}}^{k} |b_{ij}| + |d_i|$$

$$\text{for } i = 1, 2, \ldots, k; \; a > 0; \; a > \sum_{i=1}^{k} |c_i|.$$

By the induction hypothesis, $\det(\mathbf{B}) > 0$. We must show $\det(\mathbf{A}) > 0$. By Theorem 8.2.1, $\det(\mathbf{A}) = a\det(\mathbf{B} - \mathbf{dc}'/a)$.

We must show $\det(\mathbf{A}) > 0$, but by the hypothesis of the theorem, $a > 0$, so we must show $\det(\mathbf{B} - \mathbf{dc}'/a) > 0$. Denote $\mathbf{B} - \mathbf{dc}'/a$ by **G** and we must show $\det(\mathbf{G}) > 0$. But by the induction hypothesis, $\det(\mathbf{G}) > 0$ if $g_{ii} > 0$ for $i = 1, 2, \ldots, k$ (the induction hypothesis states that the theorem is true for all $k \times k$ matrices with positive diagonal elements, and **G** is $k \times k$). To show $g_{ii} > 0$ we note that $a_{ii} > 0$ implies

$$0 < a_{ii} = b_{ii} \text{ for } i = 1, \ldots, k; \; 0 < a_{k+1,\, k+1} = a.$$

Also since **A** is d.d., this implies

$$a_{ii} = b_{ii} > \sum_{\substack{j=1\\ j\neq i}}^{n} |b_{ij}| + |d_i| \text{ for } i = 1, \ldots, k \text{ and } a > \sum_{i=1}^{n} |c_i|.$$

We get $g_{ii} = b_{ii} - d_i\, c_i/a$ and for those values of i for which $d_i c_i \leq 0$, it is clear that $g_{ii} = b_{ii} - d_i c_i/a \geq b_{ii} > 0$. For those cases where $d_i c_i > 0$, it follows that $d_i c_i/a = |d_i|\ |c_i/a| < |d_i|$, since $|c_i|/a < |c_i|/\Sigma|c_i|$. Thus for those cases,

$$g_{ii} = b_{ii} - d_i c_i/a = b_{ii} - |d_i|\ |c_i|/a > b_{ii} - |d_i| > \sum_{\substack{j=1\\ j\neq i}}^{k} |b_{ij}| + |d_i| - |d_i|$$

$$= \sum_{\substack{j=1\\ j\neq i}}^{k} |b_{ij}| \geq 0.$$

Hence $g_{ii} > 0$ for all $i = 1, 2, \ldots, k$ and $\det(\mathbf{A}) > 0$. By the use of the induction hypothesis, this proves the theorem. ∎

Corollary 8.11.7.1

If **A** *is an* $n \times n$ *d.d. matrix and* $a_{ii} > 0$ *for* $i = 1, 2, \ldots, n$, *and if* $\tilde{\mathbf{A}}$ *is any principal submatrix of* **A**, *then* $\det(\tilde{\mathbf{A}}) > 0$.

Corollary 8.11.7.2

If **A** *is an* $n \times n$ *row d.d. matrix with* $a_{ii} > 0$ *for* $i = 1, 2, \ldots, n$, *and if* **D** *is any diagonal matrix with* $d_{ii} > 0$, *then* $\det(\mathbf{DA}) > 0$. *If* **A** *is as above except it is column d.d., then* $\det(\mathbf{AD}) > 0$.

Corollary 8.11.7.3

If **A** *is defined as in Theorem* 8.11.7 *and if the signs of any set of off-diagonal elements are changed, then* $\det(\mathbf{A}) > 0$.

Example 8.11.5. Consider the 3×3 matrix **A** given below.

$$\mathbf{A} = \begin{bmatrix} 6 & -2 & 3 \\ 4 & 7 & -2 \\ -1 & 2 & 4 \end{bmatrix}.$$

Clearly **A** is d.d. and $a_{ii} > 0$, so $\det(\mathbf{A}) > 0$, as can be verified by direct computation. Also it is easily shown that the determinant of any principal matrix is positive. If the signs of any off-diagonal elements are changed, it can easily be shown that the determinant of the resulting matrix is positive.

Corollary 8.11.7.4

If **A** *is any* $n \times n$ *matrix and* $\mathbf{P'AP}$ *is d.d. with positive diagonal elements, where* **P** *is any* $n \times n$ *orthogonal matrix, then* $\det(\mathbf{A}) > 0$. *Note that a permutation matrix is an orthogonal matrix, so the results apply if by permuting any rows and the corresponding columns of* **A** *the resulting matrix is d.d. with positive diagonal elements.*

Dominant diagonal matrices play a special role when we are determining bounds on characteristic roots of matrices and when we are examining convergence of matrices. In general, the characteristic roots of real matrices are complex (nonreal) numbers, since they are the roots of a polynomial with real coefficients. If the elements of a matrix are complex numbers (nonreal), then the characteristic roots are in general nonreal, complex numbers. To discuss the characteristic roots of a general matrix with complex elements, we note that any complex number z can be written as $z = x + iy$, where x and y are real numbers and $i = \sqrt{-1}$; $i^2 = -1$; $i^3 = -\sqrt{-1}$; $i^4 = 1$; and so forth. The modulus of a complex number z will be denoted by $|z|$ and defined by $|z| = (x^2 + y^2)^{1/2}$; note that $|z|$ is a real number. Also $|z| = |x|$, the absolute value of x, if z is a real number.

In the next theorem we generalize Theorem 8.11.2 to a matrix **A** that has complex numbers as elements.

Theorem 8.11.8

Let **A** *be an* $n \times n$ *complex matrix (a_{ij} are complex numbers). Then if*

$$|a_{ii}| > \sum_{\substack{j=1 \\ j \neq i}}^{n} |a_{ij}| \text{ for each } i = 1, 2, \ldots, n$$

or if

$$|a_{jj}| > \sum_{\substack{i=1 \\ i \neq j}}^{n} |a_{ij}| \text{ for each } j = 1, 2, \ldots, n,$$

then **A** *is nonsingular.*

Proof: The proof is the same as for Theorem 8.11.2 where $|a_{ij}|$ is the modulus of the complex number a_{ij}. ∎

Corollary 8.11.8

If $\mathbf{A}$ *is an* $n \times n$ *complex matrix, then all the results of Theorems* 8.11.1 *through* 8.11.6 *are true where the symbol* $|a_{ij}|$ *means the modulus of the complex number* a_{ij}.

Note: If a complex matrix $\mathbf{A} = [a_{ij}]$ satisfies Def. 8.11.1, where $|a_{ij}|$ denotes the modulus of the complex number a_{ij}, then $\mathbf{A}$ is defined to be d.d.

Next we prove a theorem that relates a d.d. matrix and the value of its characteristic roots.

Theorem 8.11.9

Let $\mathbf{A}$ *be an* $n \times n$ *(real) matrix that is d.d. and with positive diagonal elements. Then the real part of each characteristic root of* $\mathbf{A}$ *is positive.*

Proof: Consider $\mathbf{A} - \lambda\mathbf{I} = \mathbf{B}$. The values of λ (which may be complex and nonreal) that satisfy $\det(\mathbf{B}) = 0$ are the characteristic roots of $\mathbf{A}$. Denote λ by $\lambda = \alpha + i\beta$ and suppose $\alpha \leq 0$. Then $b_{tt} = a_{tt} - \alpha - i\beta = (a_{tt} - \alpha) - i\beta$. But $|b_{tt}| = |(a_{tt} - \alpha) - i\beta| = [(a_{tt} - \alpha)^2 + \beta^2]^{1/2} \geq a_{tt} > 0$ for $t = 1, \ldots, n$ if $\alpha \leq 0$. Hence if $\alpha \leq 0$, $\det(\mathbf{A} - \lambda\mathbf{I}) = \det(\mathbf{B})$ cannot be zero, since $\mathbf{B}$ is d.d.; so λ cannot be a characteristic root of $\mathbf{A}$. Thus α, the real part of λ, cannot be negative or zero and must be positive. This proves the theorem. ∎

Corollary 8.11.9.1

In Theorem 8.11.9 *all real characteristic roots are positive.*

Corollary 8.11.9.2

In Theorem 8.11.9 *if the diagonal elements* a_{tt} *of* $\mathbf{A}$ *are negative and* $\mathbf{A}$ *is d.d., then the real parts of all characteristic roots of* $\mathbf{A}$ *are negative.*

For a real (or complex) $n \times n$ matrix $\mathbf{A}$, denote the n characteristic roots by $\lambda_1, \lambda_2, \ldots, \lambda_n$, where $\lambda_t = \alpha_t + i\beta_t$, and denote the moduli of these roots by $|\lambda_1|, \ldots, |\lambda_n|$, where $|\lambda_t| = (\alpha_t^2 + \beta_t^2)^{1/2}$. We define the root of maximum modulus as the largest of the real numbers $|\lambda_1|, \ldots, |\lambda_n|$ and denote this largest number by $\rho(\mathbf{A})$; that is,

$$\rho(\mathbf{A}) = \max |\lambda_t|, \tag{8.11.4}$$

where $\lambda_1, \ldots, \lambda_n$ are the characteristic roots of $\mathbf{A}$.

Note: $\rho(\mathbf{A})$ is sometimes called the spectral radius of $\mathbf{A}$.

We now state and prove a theorem about $\rho(\mathbf{A})$ for a matrix with d.d.

Theorem 8.11.10

Let $\mathbf{A}$ be an $n \times n$ matrix with d.d., and let $\mathbf{D_A}$ denote a diagonal matrix with diagonal elements a_{ii} (that is, $\mathbf{A}$ and $\mathbf{D_A}$ have the same diagonal elements). The maximum of the moduli of the matrix $\mathbf{I} - \mathbf{D_A^{-1}A}$ is less than 1; that is, $\rho(\mathbf{B}) < 1$, where $\mathbf{B} = \mathbf{I} - \mathbf{D_A^{-1}A}$.

Proof: Let λ be any characteristic root of $\mathbf{B}$. Then there is a vector $\mathbf{x} \neq \mathbf{0}$ such that $\mathbf{Bx} = \lambda\mathbf{x}$, which we can write as $(\mathbf{I} - \mathbf{D_A^{-1}A})\mathbf{x} = \lambda\mathbf{x}$. For any vector $\mathbf{x} \neq \mathbf{0}$ satisfying $\mathbf{Bx} = \lambda\mathbf{x}$, let x_q be an element of $\mathbf{x}$ such that $|x_q| \geq |x_i|$ for $i = 1, \ldots, n$. Since $\mathbf{x} \neq \mathbf{0}$, we also get $|x_q| > 0$. The q-th equation of $\mathbf{Bx} = \lambda\mathbf{x}$ can be written as

$$\lambda x_q = \sum_{j=1}^{n} b_{qj}x_j = x_q - \sum_{j=1}^{n} (a_{qj}/a_{qq})x_j = \sum_{\substack{j=1 \\ j\neq q}}^{n} (a_{qj}/a_{qq})x_j.$$

This can be written as

$$|\lambda|\,|x_q| = |\lambda x_q| = \left|\sum_{\substack{j=1 \\ j\neq q}}^{n} (a_{qj}/a_{qq})x_j\right| \leq \sum_{\substack{j=1 \\ j\neq q}}^{n} |(a_{qj}/a_{qq})|\,|x_j|$$

$$\leq \sum_{\substack{j=1 \\ j\neq q}}^{n} (|a_{qj}|/|a_{qq}|) \max |x_j|$$

$$= |x_q| \sum_{\substack{j=1 \\ j\neq q}}^{n} |a_{qj}|/|a_{qq}| < |x_q|.$$

So $|\lambda| < 1$ and the theorem is proved. Note that d.d. of $\mathbf{A}$ is used in stating

$$\sum_{\substack{j=1 \\ j \neq q}}^{n} |a_{qj}|/|a_{qq}| < 1. \quad \blacksquare$$

Before we consider the next theorem, we define a type of matrix that will play a special role in many contexts—a reducible matrix.

Definition 8.11.2

Reducible and Irreducible Matrices. *An $n \times n$ $(n > 1)$ matrix $\mathbf{A}$ is defined to be reducible if and only if by permuting a set of rows and the corresponding set of columns it can be transformed to a matrix of the form*

$$\mathbf{A} = \begin{bmatrix} \mathbf{A}_{11} & \mathbf{A}_{12} \\ \mathbf{0} & \mathbf{A}_{22} \end{bmatrix}, \tag{8.11.5}$$

where $\mathbf{A}_{11}$ and $\mathbf{A}_{22}$ are square matrices. If a matrix is not reducible, it is defined to be irreducible.

One use of a reducible matrix $\mathbf{A}$ is that a system of equations $\mathbf{Ax} = \mathbf{g}$ can be written as

$$\begin{aligned} \mathbf{A}_{11}\mathbf{x}_1 + \mathbf{A}_{12}\mathbf{x}_2 &= \mathbf{g}_1, \\ \mathbf{A}_{22}\mathbf{x}_2 &= \mathbf{g}_2, \end{aligned}$$

where $\mathbf{x} = [\mathbf{x}_1, \mathbf{x}_2]'$ and $\mathbf{g} = [\mathbf{g}_1, \mathbf{g}_2]'$. This system is ordinarily easier to solve than the general irreducible system $\mathbf{Ax} = \mathbf{g}$, since it can be solved as two smaller systems.

If $\mathbf{A}$ can be written in the form

$$\mathbf{A} = \begin{bmatrix} \mathbf{B} & \mathbf{0} \\ \mathbf{C} & \mathbf{E} \end{bmatrix},$$

where $\mathbf{B}$ and $\mathbf{E}$ are square matrices, it is reducible, since we can transform it to a form given in Eq. (8.11.5) by permuting a set of rows and the corresponding set of columns.

The following theorem follows directly from the definition of reducible matrices.

Theorem 8.11.11

Let **A** *be an* $n \times n$ *matrix.*

(1) *If* $a_{ij} \neq 0$ *for* $i = 1, \ldots, n; j = 1, \ldots, n$ *(that is,* **A** *has no zero elements), then* **A** *is irreducible.*
(2) *If* $a_{ii} = 0$ *for* $i = 1, \ldots, n$ *and* $a_{ij} \neq 0$ *for all* $i \neq j$, *then* **A** *is irreducible.*
(3) *If* **A** *is reducible, it must have at least* $n - 1$ *elements equal to zero.*
(4) *If* **A** *has at least one row (column) of zeros, then* **A** *is reducible.*

The next theorem gives a condition for which dominant diagonal can be used to determine the nonsingularity of an irreducible matrix.

Theorem 8.11.12

Let **A** *be an* $n \times n$ *irreducible matrix such that either* $|a_{ii}| \geq R_i$ *for* $i = 1, \ldots, n$ *with* $|a_{ii}| > R_i$ *for at least one value of* i, *or* $|a_{ii}| \geq C_i$ *for* $i = 1, \ldots, n$ *with* $|a_{ii}| > C_i$ *for at least one value of* i. *Then* **A** *is nonsingular.*

Proof: The proof of this theorem is similar to the proof of Theorem 8.11.2, and the details will be omitted. ∎

Example 8.11.6. Consider the 4×4 matrix **A** given below.

$$\mathbf{A} = \begin{bmatrix} 4 & 0 & 3 & 1 \\ 0 & 3 & 2 & 1 \\ 1 & -1 & 3 & 1 \\ -2 & -3 & 1 & 7 \end{bmatrix}.$$

Clearly this matrix is irreducible (it contains only $n - 2 = 2$ zero elements). Also $|a_{ii}| = R_i$ for $i = 1, 2, 3$ and $|a_{44}| > R_4$, so by Theorem 8.11.12, **A** is nonsingular.

Next we discuss how d.d. matrices can be used to determine whether solutions exist to a certain type of linear equations. First we state some notation that will be used.

Notation: Let $\mathbf{A} = [a_{ij}]$ be any real $m \times n$ matrix. We use the notation $\mathbf{A} \geq \mathbf{0}$ to mean $a_{ij} \geq 0$ for $i = 1, \ldots, m$ and $j = 1, \ldots, n$. We use the notation $\mathbf{A} > \mathbf{0}$ to mean $a_{ij} > 0$ for $i = 1, \ldots, m$ and $j = 1, \ldots, n$. The notation $\mathbf{A} \geq \mathbf{B}$ means $\mathbf{A} - \mathbf{B} \geq \mathbf{0}$, and $\mathbf{A} > \mathbf{B}$ means $\mathbf{A} - \mathbf{B} > 0$. A similar notation will be used for vectors, that is, when $n = 1$ or $m = 1$.

Theorem 8.11.13

Let **A** *be an* $n \times n$ *matrix with positive diagonal elements and nonpositive off-diagonal elements. For any* $n \times 1$ *vector* **g**, *where* $\mathbf{g} \geq \mathbf{0}$, *there exists a unique vector* **x** *such that* $\mathbf{x} \geq \mathbf{0}$ *that is a solution to* $\mathbf{Ax} = \mathbf{g}$ *if* **A** *is d.d.*

Proof: Assume **A** is d.d. Then $\mathbf{A}^{-1}$ exists and $\mathbf{x} = \mathbf{A}^{-1}\mathbf{g}$ is a unique solution to $\mathbf{Ax} = \mathbf{g}$. We must show $x_i \geq 0$ for $i = 1, \ldots, n$ if $g_i \geq 0$. We assume the contrary and show that it leads to a contradiction. Without loss of generality, assume $x_j < 0$ for $j = 1, \ldots, m$ and $x_j \geq 0$ for $j = m + 1, \ldots, n$. For the i-th equation in $\mathbf{Ax} = \mathbf{g}$ we get

$$\sum_{j=1}^{m} a_{ij}x_j + \sum_{j=m+1}^{n} a_{ij}x_j = g_i \geq 0.$$

Summing over $i = 1, \ldots, m$, we get

$$\sum_{i=1}^{m}\sum_{j=1}^{m} a_{ij}x_j + \sum_{i=1}^{m}\sum_{j=m+1}^{n} a_{ij}x_j = \sum_{i=1}^{m} g_i.$$

Clearly $\Sigma_{i=1}^{m} \Sigma_{j=m+1}^{n} a_{ij}x_j \leq 0$, since $a_{ij} \leq 0$ for $i \neq j$ and $x_j \geq 0$ for $j = m + 1, \ldots, n$. Next we show that $\Sigma_{i=1}^{m} \Sigma_{j=1}^{m} a_{ij}x_j$ is negative. Since **A** is d.d. and $a_{jj} > 0$, we get $\Sigma_{i=1}^{m} a_{ij} > 0$ for $j = 1, \ldots, m$. Thus since $x_j < 0$ for $j = 1, \ldots, m$, we get $\Sigma_{j=1}^{m} (\Sigma_{i=1}^{m} a_{ij})x_j < 0$, so this together with $\Sigma_{i=1}^{m} \Sigma_{j=m+1}^{n} a_{ij}x_j \leq 0$ contradicts the fact that $\mathbf{g} \geq \mathbf{0}$, that is, that $\Sigma_{i=1}^{m} g_i \geq 0$, so this proves that if **A** is d.d., then $\mathbf{x} \geq \mathbf{0}$. ∎

Corollary 8.11.13

Let **A** *be an* $n \times n$ *matrix such that* $a_{ii} > 0$ *for all* i *and* $a_{ij} \leq 0$ *for all* $i \neq j$. *If there exists a diagonal matrix* **D** *with positive diagonal elements such that* **AD** *is d.d., then for any* $\mathbf{g} \geq \mathbf{0}$ *there exists a unique* $\mathbf{x} \geq \mathbf{0}$ *such that* $\mathbf{Ax} = \mathbf{g}$.

8.12 Vandermonde and Fourier Matrices

A patterned matrix that has many uses in statistics, as well as in many other applied fields, is a Vandermonde matrix, which we shall define and discuss in this section.

Definition 8.12.1

Vandermonde Matrix. *Let $x_1, x_2, \ldots, x_k$ be a set of k real numbers and let* **X** *be a $k \times k$ matrix defined by*

$$\mathbf{X} = \begin{bmatrix} 1 & 1 & 1 & \cdots & 1 \\ x_1 & x_2 & x_3 & \cdots & x_k \\ x_1^2 & x_2^2 & x_3^2 & \cdots & x_k^2 \\ x_1^3 & x_2^3 & x_3^3 & \cdots & x_k^3 \\ \vdots & \vdots & \vdots & \vdots & \vdots \\ x_1^{k-1} & x_2^{k-1} & x_3^{k-1} & \cdots & x_k^{k-1} \end{bmatrix}. \tag{8.12.1}$$

Then **X** *and* **X**$'$ *are defined to be $k \times k$ Vandermonde matrices.*

Example 8.12.1. The following 4×4 matrix is a Vandermonde matrix, where $x_1 = -1, x_2 = 3, x_3 = 2, x_4 = -2$.

$$\mathbf{X} = \begin{bmatrix} 1 & 1 & 1 & 1 \\ -1 & 3 & 2 & -2 \\ 1 & 9 & 4 & 4 \\ -1 & 27 & 8 & -8 \end{bmatrix}.$$

The following theorem follows directly from the definition.

Theorem 8.12.1

Let **X** *be a $k \times k$ Vandermonde matrix. Then every leading principal submatrix of size $n \times n$, where $n \le k$, is an $n \times n$ Vandermonde matrix.*

Theorem 8.12.2

Let **X** *be a $k \times k$ Vandermonde matrix given in Eq.* (8.12.1). *Then the determinant of* **X** *is*

$$\det(\mathbf{X}) = \prod_{t=2}^{k} \prod_{i=1}^{t-1} (x_t - x_i).$$

Proof: Multiply $\mathbf{X}$ on the left by the matrix $\mathbf{A}$ to get $\mathbf{AX} = \mathbf{B}$.

$$\begin{bmatrix} 1 & 0 & 0 & \dots & 0 & 0 \\ -x_1 & 1 & 0 & \dots & 0 & 0 \\ 0 & -x_1 & 1 & \dots & 0 & 0 \\ \vdots & \vdots & \vdots & & \vdots & \vdots \\ 0 & 0 & 0 & \dots & -x_1 & 1 \end{bmatrix} \begin{bmatrix} 1 & 1 & 1 & \dots & 1 \\ x_1 & x_2 & x_3 & \dots & x_k \\ x_1^2 & x_2^2 & x_3^2 & \dots & x_k^2 \\ \vdots & \vdots & \vdots & & \vdots \\ x_1^{k-1} & x_2^{k-1} & x_3^{k-1} & \dots & x_k^{k-1} \end{bmatrix}$$

$$= \begin{bmatrix} 1 & 1 & 1 & \dots & 1 \\ 0 & x_2-x_1 & x_3-x_1 & \dots & x_k-x_1 \\ 0 & x_2(x_2-x_1) & x_3(x_3-x_1) & \dots & x_k(x_k-x_1) \\ \vdots & & & & \\ 0 & x_2^{k-2}(x_2-x_1) & x_3^{k-2}(x_3-x_1) & \dots & x_k^{k-2}(x_k-x_1) \end{bmatrix}.$$

Now $\det(\mathbf{A}) = 1$, so $\det(\mathbf{AX}) = \det(\mathbf{A})\det(\mathbf{X}) = \det(\mathbf{X}) = \det(\mathbf{B})$. To find $\det(\mathbf{B})$, expand on the first column of $\mathbf{B}$ and get

$$\det(\mathbf{B}) = 1 \cdot \det \begin{bmatrix} x_2-x_1 & x_3-x_1 & \dots & x_k-x_1 \\ x_2(x_2-x_1) & x_3(x_3-x_1) & \dots & x_k(x_k-x_1) \\ \vdots & \vdots & & \vdots \\ x_2^{k-2}(x_2-x_1) & x_3^{k-2}(x_3-x_1) & \dots & x_k^{k-2}(x_k-x_1) \end{bmatrix}$$

$$= 1 \cdot \det(\mathbf{C}).$$

Factor (x_2-x_1) from the first column of $\mathbf{C}$, (x_3-x_1) from the second column of $\mathbf{C}$, . . . , x_k-x_1 from the k-th column of $\mathbf{C}$ and get $\det(\mathbf{C}) = (x_2-x_1)(x_3-x_1) \dots (x_k-x_1)\det(\mathbf{D})$, where

$$\mathbf{D} = \begin{bmatrix} 1 & 1 & \dots & 1 \\ x_2 & x_3 & \dots & x_k \\ \vdots & \vdots & & \vdots \\ x_2^{k-2} & x_3^{k-2} & \dots & x_k^{k-2} \end{bmatrix}.$$

Clearly $\mathbf{D}$ is a $(k-1) \times (k-1)$ Vandermonde matrix with elements $x_2, x_3, \dots, x_k$. If we apply the same procedure to $\mathbf{D}$ as we did to $\mathbf{X}$, we get

$\det(\mathbf{D}) = (x_3-x_2)(x_4-x_2) \ldots (x_k-x_2)\det(\mathbf{F})$, where

$$\mathbf{F} = \begin{bmatrix} 1 & 1 & \ldots & 1 \\ x_3 & x_4 & \ldots & x_k \\ \vdots & \vdots & & \vdots \\ x_3^{k-3} & x_4^{k-4} & \ldots & x_k^{k-3} \end{bmatrix},$$

which is a $(k-2) \times (k-2)$ Vandermonde matrix. If we continue this procedure, the theorem is proved. ∎

Example 8.12.2. Find the determinant of the matrix $\mathbf{X}$ in Example 8.12.1. We get

$$\begin{aligned}\det(\mathbf{X}) &= (x_2-x_1)(x_3-x_2)(x_3-x_1)(x_4-x_1)(x_4-x_2)(x_4-x_3) \\ &= (4)(-1)(3)(-1)(-5)(-4) = 240.\end{aligned}$$

It is easy to determine the rank of a Vandermonde matrix, since all one has to do is determine the number of distinct numbers in the set $x_1, x_2, \ldots, x_k$. This is the context of the next theorem.

Theorem 8.12.3

Let $\mathbf{X}$ *be a* $k \times k$ *Vandermonde matrix given in Eq.* (8.12.1). *The rank of* $\mathbf{X}$ *is* r, *where* r *is the number of distinct* x_i *values.*

Proof: It is quite general to assume that the first r values are distinct, and this means that the first r columns of $\mathbf{X}$ are distinct, but each of the remaining columns is equal to one of the first r columns. Hence all columns can be obtained as a linear combination of the first r columns, so the rank cannot exceed r. Now consider the $r \times r$ leading principal submatrix that is given by

$$\mathbf{V} = \begin{bmatrix} 1 & 1 & \ldots & 1 \\ x_1 & x_2 & \ldots & x_r \\ \vdots & \vdots & & \vdots \\ x_1^{r-1} & x_2^{r-1} & \ldots & x_r^{r-1} \end{bmatrix}.$$

This is an $r \times r$ Vandermonde matrix, so by Theorem 8.12.2 $\det(\mathbf{V}) = (x_r-x_1) \ldots (x_2-x_1)$ and cannot be zero, since each of the $x_1, x_2, \ldots, x_r$ is distinct. This says that the rank of $\mathbf{V}$ is r and that the rank of $\mathbf{X}$ is greater than or equal to r ($\mathbf{X}$ contains an $r \times r$ nonvanishing determinant). This result, combined with the fact that rank $(\mathbf{X}) \leq r$, implies rank $(\mathbf{X}) = r$ and the theorem is proved. ▮

Corollary 8.12.3

Consider the $n \times p$ (extended Vandermonde) matrix $\mathbf{X}$, where $n \geq p$,

$$\mathbf{X} = \begin{bmatrix} 1 & x_1 & x_1^2 & \cdots & x_1^{p-1} \\ 1 & x_2 & x_2^2 & \cdots & x_2^{p-1} \\ \vdots & \vdots & \vdots & & \vdots \\ 1 & x_n & x_n^2 & \cdots & x_n^{p-1} \end{bmatrix}. \tag{8.12.2}$$

The rank of $\mathbf{X}$ and of $\mathbf{X}'\mathbf{X}$ is equal to $\min(p, d)$*, where d is the number of distinct values of x_i.*

Example 8.12.3. Consider the polynomial regression model

$$y_i = \beta_0 + \beta_1 x_i + \beta_2 x_i^2 + \ldots + \beta_{p-1} x_i^{p-1} + \varepsilon_i$$

for $i = 1, 2, \ldots, n$, where $n \geq p$.

This model can be written as $\mathbf{Y} = \mathbf{X\beta} + \boldsymbol{\varepsilon}$, where $\mathbf{X}$ is given in Eq. (8.12.2), $\mathbf{Y} = [y_1, y_2, \ldots, y_n]'$, and $\boldsymbol{\varepsilon} = [\varepsilon_1, \varepsilon_2, \ldots, \varepsilon_n]'$. The normal equations for this model are $\mathbf{X}'\mathbf{X}\,\hat{\boldsymbol{\beta}} = \mathbf{X}'\mathbf{y}$ and by Corollary 8.12.3 $\mathbf{X}'\mathbf{X}$ is nonsingular if and only if at least p of the x_i values are distinct.

Next we shall derive formulas for finding the inverse of a nonsingular Vandermonde matrix $\mathbf{X}$ given in Eq. (8.12.1), but first we shall exhibit some notation that will be used. Consider k distinct points $x_1, x_2, \ldots, x_k$ and define the $k-1$ degree polynomials $P_i(x)$ by

$$P_i(x) = (x-x_1)(x-x_2) \ldots (x-x_{i-1})(x-x_{i+1}) \ldots (x-x_k)$$

$$= \prod_{\substack{t=1 \\ t \neq i}}^{k} (x-x_t) \text{ for } i = 1, 2, \ldots, k.$$

Note that there are k of these polynomials. If we multiply the factors in $P_i(x)$, we get

$$P_i(x) = \sum_{j=1}^{k} a_{ij} x^{j-1} \text{ for } i = 1, 2, \ldots, k,$$

where a_{ij} are the appropriate products and sums of the numbers $x_1, x_2, \ldots, x_k$.

$$\textit{Note: } P_i(x_t) = 0 \text{ if } t = 1, \ldots, k \text{ and } t \neq i. \qquad (8.12.3)$$
$$P_i(x_i) \neq 0 \text{ for } i = 1, \ldots, k.$$

We now state a theorem for the inverse of **X**.

Theorem 8.12.4

Let **X** *be a* $k \times k$ *Vandermonde matrix given in Eq.* (8.12.1), *where the* x_i *are distinct. Denote* $\mathbf{X}^{-1}$ *by* $\mathbf{B} = [b_{ij}]$. *The elements of* b_{ij} *are*

$$b_{ij} = \frac{a_{ij}}{P_i(x_i)}. \qquad (8.12.4)$$

Proof: Since the x_i are distinct, we know that **X** has an inverse. Consider $\mathbf{BX} = \mathbf{C}$, where $\mathbf{B} = [b_{ij}]$ is given by Eq. (8.12.4). The pq-th element of $\mathbf{BX} = \mathbf{C}$ is given by (let $x_i^0 = 1$)

$$c_{pq} = \sum_{m=1}^{k} b_{pm} x_q^{m-1}.$$

If we substitute for b_{ij}, we get

$$c_{pq} = \sum_{m=1}^{k} \frac{a_{pm} x_q^{m-1}}{P_p(x_p)} = \frac{\sum_{m=1}^{k} a_{pm} x_q^{m-1}}{P_p(x_p)} = \frac{P_p(x_q)}{P_p(x_p)},$$

and by Eq. (8.12.3) we see that $c_{pq} = 0$ if $p \neq q$ and $c_{pp} = 1$, so $\mathbf{C} = \mathbf{I}$ and $\mathbf{B} = \mathbf{X}^{-1}$. ∎

Example 8.12.4. Consider the Vandermonde matrix **X** given in Example 8.12.1. We get $x_1 = -1$, $x_2 = 3$, $x_3 = 2$, $x_4 = -2$, $k = 4$. Also

$$P_1(x) = (x-x_2)(x-x_3)(x-x_4) = (x-3)(x-2)(x+2)$$
$$= x^3 - 3x^2 - 4x + 12,$$
$$P_2(x) = (x-x_1)(x-x_3)(x-x_4) = (x+1)(x-2)(x+2)$$
$$= x^3 + x^2 - 4x - 4,$$

$$P_3(x) = (x-x_1)(x-x_2)(x-x_4) = (x+1)(x-3)(x+2)$$
$$= x^3 - 7x - 6,$$
$$P_4(x) = (x-x_1)(x-x_2)(x-x_3) = (x+1)(x-3)(x-2)$$
$$= x^3 - 4x^2 + x + 6.$$

We then obtain (noting that since x_i are distinct, $P_i(x_i) \neq 0$)

$$P_1(x_1) = 12,\ P_2(x_2) = 20,\ P_3(x_3) = -12,\ P_4(x_4) = -20.$$

Thus

$$\mathbf{X}^{-1} = \mathbf{B} = \begin{bmatrix} \frac{12}{12} & \frac{-4}{12} & \frac{-3}{12} & \frac{1}{12} \\ \frac{-4}{20} & \frac{-4}{20} & \frac{1}{20} & \frac{1}{20} \\ \frac{-6}{-12} & \frac{-7}{-12} & \frac{0}{-12} & \frac{1}{-12} \\ \frac{6}{-20} & \frac{1}{-20} & \frac{-4}{-20} & \frac{1}{-20} \end{bmatrix}.$$

By multiplication it is easily verified that **B** is indeed the inverse of **X**.

A special Vandermonde matrix of importance is the Fourier matrix, and it is useful in time series analysis.

Definition 8.12.2

Fourier Matrix. *Define the symbol* ω *by* $\omega = \cos 2\pi/k - i \sin 2\pi/k$, *where* $i = \sqrt{-1}$. *Then* $1/\sqrt{k}$ *times the* $k \times k$ *Vandermonde matrix* $\mathbf{F}$ *with elements* $1, \omega, \omega^2, \ldots, \omega^{k-1}$ *is defined to be a Fourier matrix.*

Below is a Fourier matrix written in detail.

$$\mathbf{F} = \frac{1}{\sqrt{k}} \begin{bmatrix} 1 & 1 & 1 & \cdots & 1 \\ 1 & \omega & \omega^2 & \cdots & \omega^{k-1} \\ 1 & \omega^2 & \omega^4 & \cdots & \omega^{2k-2} \\ \vdots & \vdots & \vdots & & \vdots \\ 1 & \omega^{k-1} & \omega^{2k-2} & \cdots & \omega^{(k-1)(k-1)} \end{bmatrix}. \tag{8.12.5}$$

Note that the pq-th element of $\mathbf{F}$ is $(1/\sqrt{k})\omega^{(p-1)(q-1)}$.

Before we state and prove some theorems about a Fourier matrix, we shall list some notation that will be used, as well as some elementary properties of complex numbers. Throughout this discussion, the symbol i will always represent $\sqrt{-1}$; hence $i^2 = -1$, $i^3 = -i$, $i^4 = 1$. Let z represent a complex number given by $z = x + iy$, where x and y are any real numbers. We can also write z in polar form as $z = r(\cos\theta + i\sin\theta)$, where $r = \sqrt{x^2 + y^2}$ (called the modulus of z) and $\theta = \arctan(y/x)$, where θ, called the argument of z, is always taken to be such that $0 \le \theta < 2\pi$. Some elementary results are given below. In (1) through (5), the symbol k is any positive integer.

(1) $(\cos\theta + i\sin\theta)^k = \cos k\,\theta + i\sin k\,\theta$.
(2) $(\cos\theta - i\sin\theta)^k = \cos k\,\theta - i\sin k\,\theta$.
(3) $(\cos\theta + i\sin\theta)^{-k} = \cos k\,\theta - i\sin k\,\theta$.
(4) $(\cos\theta - i\sin\theta)^{-k} = \cos k\,\theta + i\sin k\,\theta$. (8.12.6)
(5) $\cos(2\pi j/k) + i\sin(2\pi j/k)$ for $j = 0, 1, \ldots, k-1$ are the k-th roots of unity; that is, $[\cos(2\pi j/k) + i\sin(2\pi j/k)]^k = 1$ for $j = 0, 1, \ldots, k-1$ and each of these k complex numbers is distinct.
(6) $\bar{z}$ will denote the conjugate of z; that is, $\bar{z} = x - iy$. Hence the conjugate of $\cos\theta - i\sin\theta$ is $\cos\theta + i\sin\theta$.
Also $\overline{\mathbf{F}}$ will denote the conjugate of the matrix $\mathbf{F}$, which means that the ij-th element of $\overline{\mathbf{F}}$ is the conjugate of the corresponding ij-th element of $\mathbf{F}$.

From these we obtain some results about ω in the Fourier matrix $\mathbf{F}$ (t used below is any positive integer such that $t < k$).

(1) $\bar{\omega} = \cos 2\pi/k + i\sin 2\pi/k$.
(2) $\omega\bar{\omega} = 1$.
(3) $\omega^{-t} = \bar{\omega}^t$.
(4) $\omega^{-t} = \omega^{k-t} = \bar{\omega}^t$. (8.12.7)
(5) $\omega^k = 1$.
(6) $\omega^{k+t} = \omega^t$ for any positive integer $t < k$.
(7) $\Sigma_{j=0}^{k-1}\,\omega^{tj} = 0$ if $k > 1$ and t is any positive integer $t < k$. (We define $\omega^0 \equiv 1$.)

Now we state some theorems about Fourier matrices.

Theorem 8.12.5

Let $\mathbf{F}$ *be the* $k \times k$ *Fourier matrix given in Def.* 8.12.2 *and Eq.* (8.12.5). *The following results obtain.*

(1) $\mathbf{F} = \mathbf{F}'$; *that is,* $\mathbf{F}$ *is symmetric.*
(2) $\mathbf{F}^{-1} = \overline{\mathbf{F}}$; *that is, the inverse of* $\mathbf{F}$ *is equal to the conjugate of* $\mathbf{F}$.

(3) $\mathbf{F}^2 = \mathbf{P}$, *where* $\mathbf{P}$ *is the* $k \times k$ *permutation matrix*

$$\mathbf{P} = [\mathbf{e}_1, \mathbf{e}_k, \mathbf{e}_{k-1}, \ldots, \mathbf{e}_2],$$

where $\mathbf{e}_j$ *is the j-th column of the* $k \times k$ *identity matrix.*

(4) $\mathbf{F}^4 = \mathbf{I}$.

(5) $\mathbf{F}$ *can be written as* $\sqrt{k}\ \mathbf{F} = \mathbf{C} + i\mathbf{S}$, *where* $\mathbf{C}$ *and* $\mathbf{S}$ *are real matrices, where* $c_{pq} = \cos \dfrac{2\pi}{k}(p-1)(q-1)$ *and* $s_{pq} = \sin \dfrac{2\pi}{k}(p-1)(q-1)$.

(6) *In* (5), $\mathbf{CS} = \mathbf{SC}$.

(7) *In* (5), $\mathbf{C}^2 + \mathbf{S}^2 = \mathbf{I}$.

Proof: The proof of (1) is obvious. To prove (2) we let $\overline{\mathbf{F}} = \mathbf{G}$, $\mathbf{FG} = \mathbf{H}$, and $1/k = a$. Now

$$h_{pq} = \sum_{j=1}^{k} f_{pj}\, g_{jq} = a\sum_{j=1}^{k} \omega^{(p-1)(j-1)}\, \overline{\omega}^{(j-1)(q-1)} = a \sum_{j=0}^{k-1} \omega^{rj}\, \overline{\omega}^{js},$$

where we have substituted $p-1 = r$, $q-1 = s$. Note that rj means r times j (and js means j times s) in the superscript of ω (and $\overline{\omega}$). Thus

$$h_{pq} = a \sum_{j=0}^{k-1} \omega^{rj}\, \omega^{k-js} = a \sum_{j=0}^{k-1} \omega^{k+(r-s)j}.$$

We now get

$$h_{ss} = a\sum_{j=0}^{k-1} \omega^k = ak = 1.$$

Also for $r \neq s$ we get

$$h_{rs} = a\omega^k \sum_{j=0}^{k-1} \omega^{(r-s)j} = a\omega^k \cdot 0 = 0,$$

since by (7) of Eq. (8.12.7) it follows that $\Sigma_{j=0}^{k-1}\ \omega^{tj} = 0$ for any integer t where $0 < t <$ k. This proves (2). A similar proof can be used for (3). To prove (4) we note that $\mathbf{P}^2 = \mathbf{I}$, since $\mathbf{P}$ in (3) is a symmetric permutation matrix. The proof of (5) follows immediately from (2) of Eq. (8.12.6). The proofs for (6) and (7) will be asked for in the problems. This completes the proof of the theorem. ∎

Corollary 8.12.5

The characteristic roots of **F** *are* $1, -1, i, -i$ *(each with appropriate multiplicity).*

8.13 Permutation Matrices

It is sometimes useful to interchange various rows or columns of a matrix **A** so that the resulting matrix is simpler. Sometimes the resulting matrix can be recognized as having a certain pattern or form. If **PA** is the matrix **A** with certain rows interchanged (permuted) or **AP** is the matrix **A** with certain columns interchanged, then **P** is called a *permutation* matrix.

Definition 8.13.1

Permutation Matrix. *Let* **P** *be an* $n \times n$ *matrix that results from permuting the columns of an* $n \times n$ *identity matrix. The resulting matrix is defined to be a permutation matrix (actually a column permutation matrix). The matrix that results from permuting the rows of an identity matrix is a (row) permutation matrix.*

If a matrix **A** is multiplied on the right by a permutation matrix to get $\mathbf{B} = \mathbf{AP}$, then **B** is the result of permuting the *columns* of **A** in accordance as the columns of **I** were permuted to get **P**. Similarly $\mathbf{B} = \mathbf{PA}$ permutes the *rows* of **A**.

Example 8.13.1. Denote $\mathbf{I} = [\mathbf{e}_1, \mathbf{e}_2, \mathbf{e}_3, \mathbf{e}_4]$, where **I** is the 4×4 identity matrix and $\mathbf{e}_i$ is the i-th elementary vector. Then define $\mathbf{P} = [\mathbf{e}_2, \mathbf{e}_4, \mathbf{e}_3, \mathbf{e}_1]$. Thus $\mathbf{AP} = \mathbf{B}$ moves column 1 of **A** to column 4 of **B**, column 2 of **A** to column 1 of **B**, column 3 of **A** to column 3 of **B**, and column 4 of **A** to column 2 of **B**. Also $\mathbf{PA} = \mathbf{C}$ moves row 1 of **A** to row 2 of **C**, row 2 of **A** to row 4 of **C**, row 3 of **A** to row 3 of **C**, and row 4 of **A** to row 1 of **C**. We get

$$\mathbf{P} = \begin{bmatrix} 0 & 0 & 0 & 1 \\ 1 & 0 & 0 & 0 \\ 0 & 0 & 1 & 0 \\ 0 & 1 & 0 & 0 \end{bmatrix};$$

$$\mathbf{B} = \mathbf{AP} = \begin{bmatrix} a_{11} & a_{12} & a_{13} & a_{14} \\ a_{21} & a_{22} & a_{23} & a_{24} \\ a_{31} & a_{32} & a_{33} & a_{34} \\ a_{41} & a_{42} & a_{43} & a_{44} \end{bmatrix} \begin{bmatrix} 0 & 0 & 0 & 1 \\ 1 & 0 & 0 & 0 \\ 0 & 0 & 1 & 0 \\ 0 & 1 & 0 & 0 \end{bmatrix} = \begin{bmatrix} a_{12} & a_{14} & a_{13} & a_{11} \\ a_{22} & a_{24} & a_{23} & a_{21} \\ a_{32} & a_{34} & a_{33} & a_{31} \\ a_{42} & a_{44} & a_{43} & a_{41} \end{bmatrix}; \tag{8.13.1}$$

$$\mathbf{C} = \mathbf{PA} = \begin{bmatrix} 0 & 0 & 0 & 1 \\ 1 & 0 & 0 & 0 \\ 0 & 0 & 1 & 0 \\ 0 & 1 & 0 & 0 \end{bmatrix} \begin{bmatrix} a_{11} & a_{12} & a_{13} & a_{14} \\ a_{21} & a_{22} & a_{23} & a_{24} \\ a_{31} & a_{32} & a_{33} & a_{34} \\ a_{41} & a_{42} & a_{43} & a_{44} \end{bmatrix} = \begin{bmatrix} a_{41} & a_{42} & a_{43} & a_{44} \\ a_{11} & a_{12} & a_{13} & a_{14} \\ a_{31} & a_{32} & a_{33} & a_{34} \\ a_{21} & a_{22} & a_{23} & a_{24} \end{bmatrix}; \tag{8.13.2}$$

$$\mathbf{F} = \mathbf{PAP} = \begin{bmatrix} 0 & 0 & 0 & 1 \\ 1 & 0 & 0 & 0 \\ 0 & 0 & 1 & 0 \\ 0 & 1 & 0 & 0 \end{bmatrix} \begin{bmatrix} a_{11} & a_{12} & a_{13} & a_{14} \\ a_{21} & a_{22} & a_{23} & a_{24} \\ a_{31} & a_{32} & a_{33} & a_{34} \\ a_{41} & a_{42} & a_{43} & a_{44} \end{bmatrix} \begin{bmatrix} 0 & 0 & 0 & 1 \\ 1 & 0 & 0 & 0 \\ 0 & 0 & 1 & 0 \\ 0 & 1 & 0 & 0 \end{bmatrix} = \begin{bmatrix} a_{42} & a_{44} & a_{43} & a_{41} \\ a_{12} & a_{14} & a_{13} & a_{11} \\ a_{32} & a_{34} & a_{33} & a_{31} \\ a_{22} & a_{24} & a_{23} & a_{21} \end{bmatrix}; \tag{8.13.3}$$

$$\mathbf{G} = \mathbf{P'AP} = \begin{bmatrix} 0 & 1 & 0 & 0 \\ 0 & 0 & 0 & 1 \\ 0 & 0 & 1 & 0 \\ 1 & 0 & 0 & 0 \end{bmatrix} \begin{bmatrix} a_{11} & a_{12} & a_{13} & a_{14} \\ a_{21} & a_{22} & a_{23} & a_{24} \\ a_{31} & a_{32} & a_{33} & a_{34} \\ a_{41} & a_{42} & a_{43} & a_{44} \end{bmatrix} \begin{bmatrix} 0 & 0 & 0 & 1 \\ 1 & 0 & 0 & 0 \\ 0 & 0 & 1 & 0 \\ 0 & 1 & 0 & 0 \end{bmatrix} = \begin{bmatrix} a_{22} & a_{24} & a_{23} & a_{21} \\ a_{42} & a_{44} & a_{43} & a_{41} \\ a_{32} & a_{34} & a_{33} & a_{31} \\ a_{12} & a_{14} & a_{13} & a_{11} \end{bmatrix} \tag{8.13.4}$$

More generally, suppose an $n \times n$ permutation matrix $\mathbf{P}$ is given by

$$\mathbf{P} = [\mathbf{e}_{i_1}, \mathbf{e}_{i_2}, \ldots, \mathbf{e}_{i_n}], \tag{8.13.5}$$

where $\mathbf{e}_{i_t}$ is the i_t-th unit vector (i_t-th column of $\mathbf{I}_n$). We write this permutation as Π, where

$$\Pi: \begin{pmatrix} 1 & 2 & \dots & n \\ i_1 & i_2 & \dots & i_n \end{pmatrix}, \tag{8.13.6}$$

and this means that the t-th column of the permutation matrix $\mathbf{P}_\Pi$ is $\mathbf{e}_{i_t}$, the i_t-th unit vector (or the i_t-th column of $\mathbf{I}_n$). This results in the following theorem.

Theorem 8.13.1

Let $\mathbf{P}$ *be a permutation matrix given by Eq.* (8.13.5).

(1) *Column* t *of* $\mathbf{P}$ *is column* i_t *of* $\mathbf{I}_n$ *for* $t = 1, 2, \dots, n$. *Or row* i_t *of* $\mathbf{P}$ *is row* t *of* $\mathbf{I}_n$.

(2) $\mathbf{PA} = \mathbf{C}$ *moves row* t *of* $\mathbf{A}$ *to row* i_t *of* $\mathbf{C}$; *that is,* $c_{i_t, r} = a_{tr}$ *for* $t = 1, 2, \dots, n$ *and* $r = 1, 2, \dots, n$.

(3) $\mathbf{AP} = \mathbf{B}$ *moves column* i_t *of* $\mathbf{A}$ *to column* t *of* $\mathbf{B}$; *that is,* $b_{st} = a_{s, i_t}$ *for* $s = 1, 2, \dots, n$ *and* $t = 1, 2, \dots, n$.

(4) $\mathbf{F} = \mathbf{PAP}$ *moves row* t *of* $\mathbf{A}$ *to row* i_t *of* $\mathbf{F}$ *and column* i_s *of* $\mathbf{A}$ *to column* s *of* $\mathbf{F}$; *that is,* $f_{i_p, q} = a_{p, i_q}$ *for* $p = 1, 2, \dots, n$ *and* $q = 1, 2, \dots, n$.

(5) $\mathbf{P'AP} = \mathbf{G}$ *moves column* i_t *of* $\mathbf{A}$ *to column* t *of* $\mathbf{G}$ *and moves row* i_s *of* $\mathbf{A}$ *to row* s *of* $\mathbf{G}$; *that is,* $g_{st} = a_{i_s, i_t}$.

Example 8.13.2. Consider the permutation Π and the resulting permutation matrix given in Example 8.13.1. We see that

$$\Pi: \begin{pmatrix} 1 & 2 & 3 & 4 \\ i_1 & i_2 & i_3 & i_4 \end{pmatrix}$$

is given by

$$\Pi: \begin{pmatrix} 1 & 2 & 3 & 4 \\ 2 & 4 & 3 & 1 \end{pmatrix}.$$

In $\mathbf{PA} = \mathbf{C}$ we note that rows 1, 2, 3, 4 of $\mathbf{A}$ are respectively rows 2, 4, 3, 1 in $\mathbf{C}$ given in Eq. (8.13.2). Also in $\mathbf{AP} = \mathbf{B}$ in Eq. (8.13.1) we note that columns 2, 4, 3, 1 of $\mathbf{A}$ are respectively columns 1, 2, 3, 4 of $\mathbf{B}$. In Eq. (8.13.2) we note that $c_{42} = c_{i_t, r} = a_{tr} = a_{22}$; $c_{13} = c_{i_t, r} = a_{tr} = a_{43}$, and so forth. In Eq. (8.13.3) we get $f_{24} = f_{i_p, q} = a_{p, i_q} = a_{11}$, since $i_p = 2$

gives $p = 1$ and $q = 4$ gives $i_q = 1$. Finally in Eq. (8.13.4) we get $g_{14} = g_{st} = a_{i_s, i_t} = a_{21}$, since $s = 1$ gives $i_s = 2$ and $t = 4$ gives $i_t = 1$.

Theorem 8.13.2

If $\mathbf{P}$ is an $n \times n$ permutation matrix, then $\mathbf{P}'\mathbf{P} = \mathbf{P}\mathbf{P}' = \mathbf{I}$. Thus $\mathbf{P}' = \mathbf{P}^{-1}$; that is, $\mathbf{P}$ is an orthogonal matrix.

The proof will be asked for in the problems.

Theorem 8.13.3

Let $\mathbf{P}$ be any permutation matrix. The elements in the t-th column of $\mathbf{P}'\mathbf{A}\mathbf{P}$ are the same elements that are in the i_t column of $\mathbf{A}$ (rearranged), where $\mathbf{P}$ is such that the t-th column is $\mathbf{e}_{i_t}$. A similar result holds for rows.

In Eq. (8.13.4) note that the elements in the fourth (t-th) row of $\mathbf{P}'\mathbf{A}\mathbf{P}$ are the same elements that are in the first row of $\mathbf{A}$, and the elements in the first (s-th) column of $\mathbf{P}'\mathbf{A}\mathbf{P}$ are the same elements that are in the second column of $\mathbf{A}$.

Corollary 8.13.3

The diagonal elements of $\mathbf{P}'\mathbf{A}\mathbf{P}$ are the same elements (rearranged) as the diagonal elements of $\mathbf{A}$.

Theorem 8.13.4

Let $\mathbf{P}$ be any $n \times n$ permutation matrix. Then $\mathbf{P}^k$ is also a permutation matrix, where k is any positive integer.

Proof: The proof follows by induction after it is shown by simple multiplication that $\mathbf{P}^2$ is a permutation matrix. ▮

Theorem 8.13.5

Let $\mathbf{P}$ be an $n \times n$ permutation matrix. If we set $\mathbf{B} = \mathbf{A}\mathbf{P}$ and $\mathbf{C} = \mathbf{P}\mathbf{A}$, then $\mathbf{B}$ is a permutation of the columns of $\mathbf{A}$ and $\mathbf{C}$ is the inverse permutation of the rows of $\mathbf{A}$.

In Sec. 9.3 we will exhibit a permutation matrix $\mathbf{P}$ such that $\mathbf{P}'(\mathbf{A} \times \mathbf{B})\mathbf{P} = \mathbf{B} \times \mathbf{A}$; that is, by permuting the rows and columns of $\mathbf{A} \times \mathbf{B}$, we transform a left direct product of two matrices $\mathbf{A}$ and $\mathbf{B}$ to the right direct product.

8.14 Hadamard Matrices

A matrix that is useful in statistics, especially in experimental design, is called a Hadamard matrix. In this section we shall define this matrix and briefly discuss it.

Definition 8.14.1

Hadamard Matrix. *Let* $\mathbf{H}$ *be an* $n \times n$ *matrix with the following properties.*

(1) *The elements of* $\mathbf{H}$ *consist only of* $+1$ *and* -1.
(2) $\mathbf{H}'\mathbf{H} = n\mathbf{I}$.

Then $\mathbf{H}$ *is defined to be a Hadamard matrix.*

Some examples of Hadamard matrices are given below.

$$\mathbf{H}_1 = \begin{bmatrix} 1 & 1 \\ -1 & 1 \end{bmatrix};$$

clearly $\mathbf{H}_1'\mathbf{H}_1 = 2\mathbf{I}$.

$$\mathbf{H}_2 = \begin{bmatrix} 1 & 1 & 1 & 1 \\ 1 & 1 & -1 & -1 \\ 1 & -1 & 1 & -1 \\ 1 & -1 & -1 & 1 \end{bmatrix}; \mathbf{H}_3 = \begin{bmatrix} -1 & 1 & 1 & 1 \\ 1 & -1 & 1 & 1 \\ 1 & 1 & -1 & 1 \\ 1 & 1 & 1 & -1 \end{bmatrix}; \qquad (8.14.1)$$

clearly $\mathbf{H}_2'\mathbf{H}_2 = \mathbf{H}_3'\mathbf{H}_3 = 4\mathbf{I}$.

Hadamard matrices where every element in one row is $+1$ are quite useful in 2^n factorial-treatment structured designs. We shall refer to some applications later, but first we shall state and prove some theorems for general Hadamard matrices. The following theorem is a direct result of the definition.

Theorem 8.14.1

Let $\mathbf{H}$ *be an* $n \times n$ *Hadamard matrix. Then*

(1) $n^{-1/2}\mathbf{H}$ *is an orthogonal matrix,*
(2) $\mathbf{H}\mathbf{H}' = n\mathbf{I}$,
(3) $\mathbf{H}^{-1} = n^{-1}\mathbf{H}'$,
(4) $\mathbf{H}'$ *and* $n\mathbf{H}^{-1}$ *are Hadamard matrices.*

Theorem 8.14.2

Let $\mathbf{H}$ *be an* $n \times n$ *Hadamard matrix and let* $\mathbf{D}_1$ *and* $\mathbf{D}_2$ *be* $n \times n$ *diagonal matrices whose diagonal elements consist only of* $+1$ *and* -1. *Then* $\mathbf{D}_1\mathbf{H}$, $\mathbf{HD}_2$, *and* $\mathbf{D}_1\mathbf{HD}_2$ *are Hadamard matrices.*

Proof: Consider $\mathbf{H}_1 = \mathbf{D}_1\mathbf{H}$. Clearly the elements of $\mathbf{H}_1$ are only $+1$ and -1. Also $\mathbf{H}_1\mathbf{H}_1' = \mathbf{D}_1\mathbf{HH}'\mathbf{D}_1' = \mathbf{D}_1(n\mathbf{I})\mathbf{D}_1' = n\mathbf{I}$. The proof of the other parts is similar. ∎

Theorem 8.14.3

Let $\mathbf{H}$ *be any* $n \times n$ *Hadamard matrix. Then there exists a diagonal matrix* $\mathbf{D}$ *such that* $\mathbf{HD}$ *is a Hadamard matrix with all elements in any specified row equal to* $+1$. *A similar result holds for a specified column of* $\mathbf{DH}$.

Proof: Consider the t-th column of $\mathbf{H}$, denoted by $\mathbf{h}_t$. Let $\mathbf{D}$ be a diagonal matrix whose diagonal is $\mathbf{h}_t$. Then clearly $\mathbf{DH}$ is a Hadamard matrix and each element in the t-th column is $+1$. The proof of the remaining part is similar. ∎

Example 8.14.1. Consider the matrix $\mathbf{H}_3$ in Eq. (8.14.1). Suppose we want the third row of $\mathbf{H}_3\mathbf{D}$ to be all $+1$. We get

$$\mathbf{h}_3 = \begin{bmatrix} 1 \\ 1 \\ -1 \\ 1 \end{bmatrix}; \mathbf{D} = \begin{bmatrix} 1 & 0 & 0 & 0 \\ 0 & 1 & 0 & 0 \\ 0 & 0 & -1 & 0 \\ 0 & 0 & 0 & 1 \end{bmatrix}; \mathbf{H}_3\mathbf{D} = \begin{bmatrix} -1 & 1 & -1 & 1 \\ 1 & -1 & -1 & 1 \\ 1 & 1 & 1 & 1 \\ 1 & 1 & -1 & -1 \end{bmatrix}.$$

Note: A Hadamard matrix with all elements in the *first* row (column) equal to $+1$ is called a normalized (seminormalized) Hadamard matrix.

Theorem 8.14.4

Let $\mathbf{H}_1$ *and* $\mathbf{H}_2$ *be* $n_1 \times n_1$ *and* $n_2 \times n_2$ *Hadamard matrices. Then* $\mathbf{H}_1 \times \mathbf{H}_2$ *is also a Hadamard matrix.*

Proof: Let $\mathbf{H} = \mathbf{H}_1 \times \mathbf{H}_2$. Clearly the elements of $\mathbf{H}$ consist of only $+1$ and -1. Next we see that $\mathbf{H}'\mathbf{H} = (\mathbf{H}_1 \times \mathbf{H}_2)'(\mathbf{H}_1 \times \mathbf{H}_2) = (\mathbf{H}_1' \times \mathbf{H}_2')(\mathbf{H}_1 \times \mathbf{H}_2) = (\mathbf{H}_1'\mathbf{H}_1) \times (\mathbf{H}_2'\mathbf{H}_2) = (n_1\mathbf{I}) \times (n_2\mathbf{I}) = n_1n_2\mathbf{I}$. Thus $\mathbf{H}$ is a Hadamard matrix. ∎

The Hadamard matrices that are useful in 2^n factorial experiments are of size $2^n \times 2^n$ and such that each element in one row (say the first row) is a $+1$. These matrices can be constructed by using the Kronecker product of certain 2×2 Hadamard matrices.

Consider the 2×2 Hadamard matrix denoted by $\mathbf{H}_2$, where

$$\mathbf{H}_2 = \begin{bmatrix} 1 & 1 \\ -1 & 1 \end{bmatrix}. \tag{8.14.2}$$

For any $2^n \times 2^n$ Hadamard matrix $\mathbf{H}_{2^n}$, where n is a positive integer, it follows from Theorem 8.14.4 that $\mathbf{H}_{2^{n+1}}$ is a $2^{n+1} \times 2^{n+1}$ Hadamard matrix, where $\mathbf{H}_{2^{n+1}} = \mathbf{H}_{2^n} \times \mathbf{H}_2$.

Example 8.14.2. If $\mathbf{H}_2$ is as defined in Eq. (8.14.2), then

$$\mathbf{H}_{2^2} = \mathbf{H}_2 \times \mathbf{H}_2 = \begin{bmatrix} 1 & 1 \\ -1 & 1 \end{bmatrix} \times \begin{bmatrix} 1 & 1 \\ -1 & 1 \end{bmatrix} = \begin{bmatrix} 1 & 1 & 1 & 1 \\ -1 & 1 & -1 & 1 \\ -1 & -1 & 1 & 1 \\ 1 & -1 & -1 & 1 \end{bmatrix};$$

$$\mathbf{H}_{2^3} = \mathbf{H}_{2^2} \times \mathbf{H}_2 = \begin{bmatrix} 1 & 1 & 1 & 1 \\ -1 & 1 & -1 & 1 \\ -1 & -1 & 1 & 1 \\ 1 & -1 & -1 & 1 \end{bmatrix} \times \begin{bmatrix} 1 & 1 \\ -1 & 1 \end{bmatrix}$$

$$= \begin{bmatrix} 1 & 1 & 1 & 1 & 1 & 1 & 1 & 1 \\ -1 & 1 & -1 & 1 & -1 & 1 & -1 & 1 \\ -1 & -1 & 1 & 1 & -1 & -1 & 1 & 1 \\ 1 & -1 & -1 & 1 & 1 & -1 & -1 & 1 \\ -1 & -1 & -1 & -1 & 1 & 1 & 1 & 1 \\ 1 & -1 & 1 & -1 & -1 & 1 & -1 & 1 \\ 1 & 1 & -1 & -1 & -1 & -1 & 1 & 1 \\ -1 & 1 & 1 & -1 & 1 & -1 & -1 & 1 \end{bmatrix}.$$

The results of Theorem 8.14.4 can be extended so that *the direct product of any finite number of Hadamard matrices is a Hadamard matrix.* As a result we know that there exist Hadamard matrices of size $2^n \times 2^n$ for every positive integer n, but it is of interest to determine if there is a Hadamard matrix of any size. The following theorem is a partial answer to this problem.

Theorem 8.14.5

Let $\mathbf{H}$ *be an* $n \times n$ *Hadamard matrix. Then* n *must be equal to* 1 *or* 2 *or be a multiple of* 4.

Proof: Clearly for $n = 1$ and 2, respectively,

$$\mathbf{H} = [1],\ \mathbf{H} = \begin{bmatrix} 1 & -1 \\ 1 & 1 \end{bmatrix}$$

are Hadamard matrices. To prove that if $\mathbf{H}$ is an $n \times n$ Hadamard matrix with $n > 2$ then n must be a multiple of 4, we proceed as follows. Let $\mathbf{H}$ be an $n \times n$ Hadamard matrix with each element in the first row equal to $+1$. This is always possible to obtain by Theorem 8.14.3. Let $\mathbf{r}_1'$, $\mathbf{r}_2'$, and $\mathbf{r}_3'$ be the first, second, and third rows of this normalized $\mathbf{H}$. Note each element in $\mathbf{r}_1$ as $+1$. Let $N(+, +)$ be the number of common elements in $\mathbf{r}_2$ and $\mathbf{r}_3$ that are each $+1$; let $N(-, -)$ be the number of common elements in $\mathbf{r}_2$ and $\mathbf{r}_3$ that are each -1; let $N(+, -)$ be the number of common elements that are $+1$ in $\mathbf{r}_2$ and -1 in $\mathbf{r}_3$; let $N(-, +)$ be the number of common elements in $\mathbf{r}_2$ and $\mathbf{r}_3$ that are -1 in $\mathbf{r}_2$ and $+1$ in $\mathbf{r}_3$. Then the following four relations obtain.

(1) $N(+, +) + N(-, -) + N(+, -) + N(-, +) = n =$ total number of elements in any row.

(2) $N(+, +) + N(-, -) = N(+, -) + N(-, +)$; this is due to the fact that $\mathbf{r}_2' \mathbf{r}_3 = 0$.

(3) $N(+, +) + N(+, -) = N(-, -) + N(-, +)$; this says that the number of $+1$ elements in $\mathbf{r}_2$ is equal to the number of -1 elements in $\mathbf{r}_2$. They must be equal since $\mathbf{r}_2' \mathbf{r}_1 = 0$.

(4) Similarly, since $\mathbf{r}_3'\mathbf{r}_1 = 0$, we get $N(+, +) + N(-, +) = N(+, -) + N(-, -)$.

From (1) and (2) we get $2[N(+, +) + N(-, -)] = n$. If we add (3) to (4) and simplify, we get $2[N(+, +) - N(-, -)] = 0$. If we add this to (1) + (2), we get $4N(+, +) = n$ and finally $N(+, +) = N(-, -) = N(+, -) = N(-, +) = n/4$, so n must be a multiple of 4 (if $n > 2$). This proves the theorem. ∎

It is not known if an $n \times n$ Hadamard matrix exists for every n that is a multiple of 4. For further information on this problem see [H–2]. By Theorem 8.14.4 we know a Hadamard matrix exists for size $2^n \times 2^n$ for $n = 0$ and every positive integer.

In [H–2] is a discussion of how Hadamard matrices can be used to construct balanced incomplete block designs, partially balanced incomplete block designs, Youden designs, and fractional factorial designs. It is also explained there how Hadamard matrices can be used in coding theory and in probability and statistics.

We conclude this section with a theorem on the determinant of Hadamard matrices.

Theorem 8.14.6

Let $\mathbf{H}$ *be an* $n \times n$ *Hadamard matrix. Then* $\det(\mathbf{H}) = n^{n/2}$.

Proof: By definition $\mathbf{H}'\mathbf{H} = n\mathbf{I}$, so $\det(\mathbf{H}'\mathbf{H}) = n^n$. But $\det(\mathbf{H}'\mathbf{H}) = \det(\mathbf{H})\det(\mathbf{H}') = \det(\mathbf{H})\det(\mathbf{H}) = [\det(\mathbf{H})]^2$. Hence the result. ■

8.15 Band and Toeplitz Matrices

In this section we give a brief discussion of band and Toeplitz matrices. We concentrate mainly on symmetric Toeplitz matrices, since they arise as covariance matrices in the theory of multivariate analysis and time series.

Definition 8.15.1

Band Matrices of Bandwidth 2K + 1. *Let* $\mathbf{A}$ *be an* $n \times n$ *matrix such that* $a_{ij} = 0$ *if* $|i - j| > K$. *This matrix is defined to be an* $n \times n$ *band matrix of bandwidth* $2K + 1$. *The positive integer* K *can take on values* $0, 1, \ldots, n - 1$.

Example 8.15.1. Consider the matrices below.

$$\mathbf{A}_1 = \begin{bmatrix} 1 & 2 & 4 & 0 & 0 \\ 3 & 4 & 0 & 2 & 0 \\ 3 & -2 & 1 & 6 & 4 \\ 0 & 1 & 4 & 0 & 3 \\ 0 & 0 & 1 & 2 & 5 \end{bmatrix}; \mathbf{A}_2 = \begin{bmatrix} 6 & 3 & 0 & 0 \\ 3 & 6 & 3 & 0 \\ 0 & 3 & 6 & 3 \\ 0 & 0 & 3 & 6 \end{bmatrix}.$$

$\mathbf{A}_1$ is a band matrix with bandwidth 5, and $\mathbf{A}_2$ is a band matrix with bandwidth 3.

The following theorem follows directly from the definition.

Theorem 8.15.1

Let $\mathbf{A}$ *be an* $n \times n$ *band matrix.*

(1) *If* $\mathbf{A}$ *is diagonal, it is a band matrix of bandwidth* 1; *that is,* $K = 0$ *in* $2K + 1$.

(2) *If* $\mathbf{A}$ *is a band matrix with bandwidth* $2K + 1$, *then* $\mathbf{A}'$ *is also a band matrix with bandwidth* $2K + 1$.

(3) *If* $\mathbf{A}_1, \ldots, \mathbf{A}_q$ *are* $n \times n$ *band matrices with bandwidth* $2K + 1$, *then* $\mathbf{A}$ *is also an* $n \times n$ *band matrix with bandwidth* $2K + 1$, where $\mathbf{A} = \Sigma_{i=1}^{q} b_i\mathbf{A}_i$, *where* b_i *are any scalars.*

(4) *If* $\mathbf{A}$ *is a band matrix with bandwidth* $2K + 1$, *then* $\mathbf{A}$ *is also a band matrix with bandwidth* $2M + 1$, *where* $M = K + 1, \ldots, n - 1$.

(5) *An* $n \times n$ *matrix of type 2 (see Def.* 8.3.1) *is a band matrix of bandwidth* 3.

The band matrices that have special significance in statistics are those where all the elements are equal that are on a given diagonal parallel to the main diagonal. For example, the elements a_{ij}, where $i - j = t$, are called the t-th superdiagonal elements (elements on the t-th superband) if t is negative, and they are called the t-th subdiagonal elements (elements on the t-th subband) if t is positive. If $t = 0$, the elements are the diagonal elements. For the matrix $\mathbf{A}_1$ in Example 8.15.1, the elements 2, 0, 6, and 3 make up the first superband (or first superdiagonal); the elements 3, 1, and 1 make up the second subband (or second subdiagonal); and so forth.

The band matrices that we shall discuss are those for which all elements on a given subdiagonal (or superdiagonal) are equal. These are called Toeplitz matrices, and for brevity we shall call them T-matrices.

Definition 8.15.2

Toeplitz Matrices (T-Matrices). *Let* $\mathbf{A}$ *be an* $n \times n$ *matrix such that all elements on each superdiagonal are equal and all elements on each subdiagonal are equal. This matrix is defined to be a Toeplitz matrix.*

Example 8.15.2. The matrices below are T-matrices. $\mathbf{A}_1$ is a band matrix with bandwidth 5 (that is, $K = 2$ in $2K + 1$).

$$\mathbf{A}_1 = \begin{bmatrix} 6 & 2 & 3 & 0 & 0 \\ 4 & 6 & 2 & 3 & 0 \\ 5 & 4 & 6 & 2 & 3 \\ 0 & 5 & 4 & 6 & 2 \\ 0 & 0 & 5 & 4 & 6 \end{bmatrix}; \mathbf{A}_2 = \begin{bmatrix} 3 & 1 & 2 & 4 \\ 4 & 3 & 1 & 2 \\ 2 & 4 & 3 & 1 \\ 1 & 2 & 4 & 3 \end{bmatrix}.$$

Also notice that $\mathbf{A}_2$ is a regular circulant.

The results in the next theorem follow immediately from the definition of T-matrices.

Theorem 8.15.2

Let $\mathbf{A}$ *be an* $n \times n$ *T-matrix.*

(1) *An* $n \times n$ *regular circulant is a T-matrix, but a T-matrix is not necessarily a circulant.*
(2) *A linear combination* $\Sigma_{i=1}^{M} b_i\mathbf{A}_i$ *of* $n \times n$ *T-matrices* $\mathbf{A}_i$ *is a T-matrix.*
(3) *If* a_{ij} *is defined by* $a_{ij} = a_{|i-j|}$, *then* $\mathbf{A}$ *is a symmetric T-matrix.*
(4) $\mathbf{A}'$ *is a T-matrix.*
(5) $\mathbf{A}$ *is symmetric about its secondary diagonal.*

Note: A T-matrix is sometimes referred to as a bandmatrix with constant bands.

There are no reasonably good formulas available for determining the elements of the inverse of a T-matrix, but some work has been done on the problem. For a general $n \times n$ symmetric T-matrix $\mathbf{A}$ with bandwidth $2K + 1$, [M–8] contains some explicit formulas for finding the inverse and some approximations that may be useful in some situations.

In the remainder of this section we state some results that may prove useful in examining T-matrices.

Theorem 8.15.3

Let $\mathbf{A}$ *be an* $n \times n$ *T-matrix with bandwidth* 3, *where* $\mathbf{A}$ *is given by*

$$\mathbf{A} = \begin{bmatrix} a_0 & a_1 & 0 & \dots & 0 & 0 \\ a_2 & a_0 & a_1 & \dots & 0 & 0 \\ \vdots & \vdots & \vdots & & \vdots & \vdots \\ 0 & 0 & 0 & \dots & a_2 & a_0 \end{bmatrix}.$$

The characteristic roots of $\mathbf{A}$ *are*

$$\lambda_m = a_0 + 2\sqrt{a_1 a_2} \cos\left(\frac{m\pi}{n+1}\right) \quad \textit{for } m = 1, 2, \dots, n.$$

Proof: Consider $\mathbf{B} = \mathbf{A} - \lambda\mathbf{I}$ and let $b = a_0 - \lambda$. Then

$$\mathbf{B} = \mathbf{A} - \lambda\mathbf{I} = \begin{bmatrix} b & a_1 & 0 & \dots & 0 & 0 \\ a_2 & b & a_1 & \dots & 0 & 0 \\ \vdots & \vdots & \vdots & & & \vdots \\ 0 & 0 & 0 & \dots & a_2 & b \end{bmatrix}. \tag{8.15.1}$$

We must find values of λ such that $\det(\mathbf{A} - \lambda\mathbf{I}) = 0$, but this is equivalent to finding b such that $\det(\mathbf{B}) = 0$. Denote $\mathbf{B}_n$ as the $n \times n$ matrix $\mathbf{B}$. Note that if we delete the first row and column of $\mathbf{B}$, the remaining $(n - 1) \times (n - 1)$ matrix is of the same form as $\mathbf{B}$ (that is, as $\mathbf{B}_n$), so we denote this $(n - 1) \times (n - 1)$ matrix as $\mathbf{B}_{n-1}$. We denote $\mathbf{B}_{n-2}$ as the resulting matrix when the first two rows and columns are eliminated from $\mathbf{B}_n$. We use similar meanings for $\mathbf{B}_{n-3}, \ldots,$ and so forth. If we expand $\det(\mathbf{B}_n)$ on the first row we get

$$\det(\mathbf{B}_n) = b \det(\mathbf{B}_{n-1}) - a_1a_2 \det(\mathbf{B}_{n-2}). \tag{8.15.2}$$

We know that $\det(\mathbf{B}_m)$ is an m-th degree polynomial in b (and hence in λ). Let $b = 2(a_1a_2)^{1/2}x$ and $\det(\mathbf{B}_n) = (a_1a_2)^{n/2}\, p_n(x)$ where $p_n(x)$ is an n-th degree polynomial in x. Substitute these into Eq. (8.15.2) and get the following recurrence formula for the polynomials $p_m(x)$, $p_{m-1}(x)$, $p_{m-2}(x)$:

$$p_m(x) = 2xp_{m-1}(x) - p_{m-2}(x), \text{ for } m = 3, 4, \ldots, n. \tag{8.15.3}$$

If we can find the values of x for which $p_n(x) = 0$, we can determine values of b and hence of λ for which $\det(\mathbf{B}_n) = 0$, and hence they are the characteristic roots of $\mathbf{B}$. By [A–1] the polynomials that satisfy Eq. (8.15.3) are Chebychev polynomials of the second kind and they give the values of x for which $p_m(x) = 0$ as

$$x = \cos\frac{\pi m}{n + 1} \quad \text{for } m = 1, 2, \ldots, n.$$

Thus $b = 2(a_1a_2)^{1/2} \cos\dfrac{\pi m}{n + 1}$ and so the characteristic roots of $\mathbf{B}$ are $\lambda_m = a_0 + 2(a_1a_2)^{1/2} \cos\dfrac{m\pi}{n + 1}$ for $m = 1, 2, \ldots, n$. This proves the theorem. ∎

Corollary 8.15.3.1

Let $\mathbf{A}$ *be the* $n \times n$ *symmetric T-matrix given by* $a_{ij} = a_{|i-j|}$ *for* $|i - j| \leq 1$ *and* $a_{|i-j|} = 0$ *for* $|i - j| > 1$. *The characteristic roots of* $\mathbf{A}$ *are* $\lambda_m = a_0 + 2a_1\cos\dfrac{m\pi}{n + 1}$ *for* $m = 1, 2, \ldots, n$.

Corollary 8.15.3.2

The matrix $\mathbf{A}$ *in Corollary* 8.15.3.1 *is positive definite if and only if* $a_0 + 2a_1\cos\dfrac{m\pi}{n + 1} > 0$ *for* $m = 1, 2, \ldots, n$.

Corollary 8.15.3.3

The matrix **A** *in Corollary* 8.15.3.1 *is positive definite for all* n *if* $a_0 > 0$ *and* $|a_1/a_0| \leq 1/2$.

Corollary 8.15.3.4

Let **A** *be the* $n \times n$ *matrix given in Theorem* 8.15.3. *Then*

$$\det(\mathbf{A}) = \prod_{m=1}^{n} \left[a_0 + 2\sqrt{a_1 a_2} \cos\left(\frac{m\pi}{n+1}\right) \right]. \tag{8.15.4}$$

If **A** *is symmetric (that is, if* $a_1 = a_2$ *and hence* **A** *is a symmetric T-matrix), then a similar result holds.*

For most of the remainder of this section we will discuss symmetric T-matrices.

Theorem 8.15.4

Let **A** *be an* $n \times n$ *positive definite T-matrix with* $a_{ij} = 0$ *if* $|i - j| > 1$. *This is the matrix exhibited in Theorem* 8.15.3 *with* $a_1 = a_2$ *with the additional assumption that* **A** *is positive definite (also assume* $a_1 \neq 0$*). Then* $\mathbf{A}^{-1}$ *(which we denote by* **B***) is given by*

$$b_{ij} = \begin{cases} \dfrac{(1 - b^{2n-2j+2})(b^{j+i+1} - b^{j-i+1})}{(a_1/a_0)(1 - b^2)(1 - b^{2n+2})} & \text{for } i \leq j \\ b_{ji} & \text{for } i > j \end{cases} \tag{8.15.5}$$

where $b = (1/2)(a_1/a_0)(\sqrt{1 - 4(a_1/a_0)^2} - 1)$.

Proof: The proof can be obtained by multiplication. ∎

In the discussion above we stated that a regular circulant is a T-matrix, but a T-matrix is not always a circulant. It would be useful in some applications if a T-matrix could be viewed as an "approximate" circulant, since properties of circulants are easy to write down as formulas (that is, formulas were given in Sec. 8.10 for characteristic roots, characteristic vectors, and determinants of circulants). Consider the 5 × 5 T-matrix **A** following.

$$\mathbf{A} = \begin{bmatrix} 4 & -7 & 0 & 0 & 0 \\ 2 & 4 & -7 & 0 & 0 \\ 0 & 2 & 4 & -7 & 0 \\ 0 & 0 & 2 & 4 & -7 \\ 0 & 0 & 0 & 2 & 4 \end{bmatrix}. \tag{8.15.6}$$

If the zeros in elements 1, 5 and 5, 1 are replaced by 2 and -7, respectively, we get the matrix **C** below.

$$\mathbf{C} = \begin{bmatrix} 4 & -7 & 0 & 0 & 2 \\ 2 & 4 & -7 & 0 & 0 \\ 0 & 2 & 4 & -7 & 0 \\ 0 & 0 & 2 & 4 & -7 \\ -7 & 0 & 0 & 2 & 4 \end{bmatrix} \tag{8.15.7}$$

and this is a regular circulant with first-row elements 4, -7, 0, 0, 2.

We can use formulas developed in Sec. 8.10 to find the determinant and characteristic roots and vectors of **C**. But the question remains as to whether these are approximately equal to the corresponding determinant, inverse, and characteristic roots and vectors of **A**. For a discussion of this see [G–3].

Another matrix that is related to a T-matrix is sometimes referred to as a *cross-symmetric* or *centrosymmetric* matrix. This type of matrix was discussed in [C–9] and later its application to T-matrices was discussed in [G–2] and [R–5]. We shall briefly define cross-symmetric matrices and show their relationship to T-matrices.

Definition 8.15.3

Cross-Symmetric Matrix. *Consider an* $n \times n$ *matrix* $\mathbf{A} = [a_{ij}]$ *such that* $a_{ij} = a_{n+1-i,\, n+1-j}$ *for all i and j. Then* **A** *is defined to be a cross-symmetric (centrosymmetric) matrix.*

Note: We shall refer to a cross-symmetric matrix as a C-matrix.

Note: A C-matrix **A** can be viewed as follows.

(1) The elements of the first column read downward are the same as the elements read upward in the n-th column; the elements in the second column read downward are the same as the elements read upward in the $(n - 1)$-st column; and so forth.

(2) The elements read from left to right in the first row are the same as the elements read from right to left in the n-th row; the elements read from left to right in the second row are the same as the elements read from right to left in the $(n - 1)$-st row; and so forth.

Below are four C-matrices, **A**, **B**, **C**, and **D**.

$$\mathbf{A} = \begin{bmatrix} 3 & 0 & -1 \\ 1 & 2 & 1 \\ -1 & 0 & 3 \end{bmatrix}; \mathbf{B} = \begin{bmatrix} 1 & 2 & 4 & 0 \\ 2 & 1 & 3 & 6 \\ 6 & 3 & 1 & 2 \\ 0 & 4 & 2 & 1 \end{bmatrix};$$

$$\mathbf{C} = \begin{bmatrix} 1 & 2 & -4 & 3 \\ 2 & 1 & 2 & -4 \\ -4 & 2 & 1 & 2 \\ 3 & -4 & 2 & 1 \end{bmatrix}; \mathbf{D} = \begin{bmatrix} 4 & 10 & 7 & 10 \\ 10 & 8 & 4 & 7 \\ 7 & 4 & 8 & 10 \\ 10 & 7 & 10 & 4 \end{bmatrix}.$$

Notice that **C** and **D** are also symmetric matrices and **C** is a T-matrix.

First we shall state several theorems about C-matrices. Then we shall show how C-matrices are related to T-matrices. The first two theorems are obvious consequences of the definition.

Theorem 8.15.5

The following matrices are C-matrices.

(1) $\mathbf{0}$; $\mathbf{I}$; $\mathbf{J}$; $a\mathbf{I} + b\mathbf{J}$, *where a and b are scalars.*
(2) $\mathbf{A}'$ *if* $\mathbf{A}$ *is a C-matrix.*

Theorem 8.15.6

Let $\mathbf{A}_1, \mathbf{A}_2, \ldots, \mathbf{A}_k$ *be* $n \times n$ *C-matrices. Then* $\mathbf{A}$ *is a C-matrix, where*

$$\mathbf{A} = \sum_{i=1}^{K} a_i \mathbf{A}_i$$

and a_i *are scalars.*

Theorem 8.15.7

Let $\mathbf{A}$ *and* $\mathbf{B}$ *be* $n \times n$ *C-matrices. Then*

(1) $\mathbf{AB}$ *is a C-matrix.*
(2) *If* $\mathbf{A}$ *is nonsingular, then* $\mathbf{A}^{-1}$ *is a C-matrix.*

Proof: Let $\mathbf{AB} = \mathbf{K}$. Then $k_{ij} = \Sigma_{t=1}^{n} a_{it}b_{tj}$

$$= \sum_{t=1}^{n} a_{n+1-i,\, n+1-t} b_{n+1-t,\, n+1-j} = \sum_{t=1}^{n} a_{n+1-i,\, t}\, b_{t,\, n+1-j} = k_{n+1-i,\, n+1-j}$$

and this proves (1). To prove (2) it is clear that if $\mathbf{D}$ is the matrix of cofactors of $\mathbf{A}$, then $\mathbf{D}$ and $\mathbf{D}'$ is a C-matrix, and since $\mathbf{A}^{-1} = \mathbf{D}'/\det(\mathbf{A})$, it follows that $\mathbf{A}^{-1}$ is a C-matrix. ∎

In [R–5] the author shows that the inverse of a nonsingular C-matrix of size $2m \times 2m$ or $(2m + 1) \times (2m + 1)$ can be obtained by inverting two matrices of size $m \times m$.

A symmetric T-matrix is a C-matrix, so the results of C-matrices can sometimes prove useful in examining T-matrices.

Problems

1. If $\mathbf{B} = \begin{bmatrix} a\mathbf{I} & b\mathbf{I} \\ b\mathbf{I} & d\mathbf{I} \end{bmatrix}$, where each identity matrix is of size $m \times m$, find the characteristic roots of $\mathbf{B}$.
2. In Prob. 1, if $ad - b^2 \neq 0$, find $\mathbf{B}^{-1}$.
3. If $\mathbf{B} = \begin{bmatrix} -\mathbf{I} & \mathbf{A}-\mathbf{I} \\ \mathbf{A}-\mathbf{I} & \mathbf{A} \end{bmatrix}$, where $\mathbf{A}$ is an $m \times m$ symmetric matrix such that $\mathbf{A}^2 = \mathbf{A}$, show that $|\mathbf{B}| = (-1)^m$.
4. Let a $k \times k$ matrix $\mathbf{C}$ be defined by Eq. (8.3.13).
   (a) Find the conditions on the constants a, b, and k such that $\mathbf{C}$ is positive definite.
   (b) Find the conditions on the constants a, b, and k such that $\mathbf{C}$ is positive semidefinite.
   (c) Find the conditions on the constants a, b, and k such that $\mathbf{C}^2 = \mathbf{C}$.
5. In Theorem 8.2.1, suppose $n_i = n_j$ and $\mathbf{B}_{11} = \mathbf{B}_{22} = \mathbf{0}$. State a result for the existence of $\mathbf{B}^{-1}$.
6. Find the inverse of the matrix $\mathbf{C}$ where

$$\mathbf{C} = \begin{bmatrix} a_1 & a_2 & a_2 & a_2 \\ a_2 & a_3 & a_4 & a_4 \\ a_2 & a_4 & a_3 & a_4 \\ a_2 & a_4 & a_4 & a_3 \end{bmatrix}$$

   if the a_i are such that the inverse exists.

7. Find the determinant of the matrix.

$$\mathbf{C} = \begin{bmatrix} a\mathbf{I} & \mathbf{J} & \mathbf{J} & \mathbf{J} \\ \mathbf{J} & a\mathbf{I} & \mathbf{J} & \mathbf{J} \\ \mathbf{J} & \mathbf{J} & a\mathbf{I} & \mathbf{J} \\ \mathbf{J} & \mathbf{J} & \mathbf{J} & a\mathbf{I} \end{bmatrix},$$

where $a \neq 0$ and each matrix has dimension $n \times n$. What are the conditions on a and n to insure that the inverse exists?

8. Let

$$\mathbf{A} = \begin{bmatrix} a_1 & a_1 & a_1 & a_1 \\ a_1 & a_2 & a_2 & a_2 \\ a_1 & a_2 & a_3 & a_3 \\ a_1 & a_2 & a_3 & a_4 \end{bmatrix};$$

what are the conditions on the a_i so that $\mathbf{A}$ is nonsingular?

9. If the conditions on the a_i in the matrix $\mathbf{A}$ in Prob. 8 are such that $\mathbf{A}$ is nonsingular, find $\mathbf{A}^{-1}$.

10. Generalize Probs. 8 and 9 to a $k \times k$ matrix.

11. If

$$\mathbf{B} = \begin{bmatrix} 6 & 6 & 6 & 6 & 6 \\ 6 & 8 & 8 & 8 & 8 \\ 6 & 8 & 3 & 3 & 3 \\ 6 & 8 & 3 & 2 & 2 \\ 6 & 8 & 3 & 2 & 4 \end{bmatrix},$$

find $\mathbf{B}^{-1}$.

12. Find the inverse of the triangular matrix $\mathbf{T}$ where

$$\mathbf{T} = \begin{bmatrix} \mathbf{I} & \mathbf{J} & \mathbf{J} \\ \mathbf{0} & \mathbf{I} & \mathbf{J} \\ \mathbf{0} & \mathbf{0} & \mathbf{I} \end{bmatrix}$$

and each submatrix is of order $k \times k$.

13. Extend Prob. 12 to the case in which there are n^2 block matrices and the order of each is $k \times k$.

14. Find the determinant and characteristic roots of the matrix **A** where

$$\mathbf{A} = \begin{bmatrix} \mathbf{0} & \mathbf{I} & \mathbf{I} & \cdots & \mathbf{I} \\ \mathbf{I} & \mathbf{0} & \mathbf{I} & \cdots & \mathbf{I} \\ \mathbf{I} & \mathbf{I} & \mathbf{0} & \cdots & \mathbf{I} \\ \vdots & \vdots & \vdots & & \vdots \\ \mathbf{I} & \mathbf{I} & \mathbf{I} & \cdots & \mathbf{0} \end{bmatrix},$$

each identity has size $k \times k$, and there are n^2 submatrices.

15. Find the inverse of the matrix in Prob. 3.

16. Find $\mathbf{A}^{-1}$ in Prob. 14.

17. Use Theorem 8.2.1 to find the inverse of **A** where

$$\mathbf{A} = \begin{bmatrix} 1 & 0 & 0 & 0 & 3 \\ 0 & 1 & 0 & 0 & 2 \\ 0 & 0 & 1 & 0 & 1 \\ 0 & 0 & 0 & 1 & 2 \\ 3 & 2 & 1 & 2 & 4 \end{bmatrix}.$$

18. Prove Theorem 8.9.5 by using Theorem 8.2.1.

19. Use Theorem 8.9.3 to find the determinant of **B** where

$$\mathbf{B} = \begin{bmatrix} 2 & 2 & 3 \\ 2 & 5 & 6 \\ 3 & 6 & 10 \end{bmatrix}.$$

Note that $\mathbf{B} = \mathbf{I} + \mathbf{b}\mathbf{b}'$ where $\mathbf{b}' = [1, 2, 3]$.

20. Use Theorem 8.9.3 to find the characteristic roots of the matrix in Prob. 19.

21. Use Theorem 8.4.3 to evaluate the determinant of the matrix **V** in Example 8.3.3

22. Evaluate the determinant of the matrix **B** where **B** is defined by

$$\mathbf{B} = \begin{bmatrix} a & b & c & d \\ -b & a & -d & c \\ -c & d & a & b \\ -d & c & -b & a \end{bmatrix}.$$

23. Use Theorem 8.2.1, to find the determinant of the matrix **A** where

$$\mathbf{A} = \begin{bmatrix} 1 & 3 & 1 & 3 \\ 4 & 2 & 2 & 1 \\ 4 & 2 & 2 & 3 \\ 3 & 1 & 4 & 1 \end{bmatrix}.$$

24. Evaluate the determinant of **A** where $x_i = i$ and $k = 4$.

$$\mathbf{A} = \begin{bmatrix} 1 & 1 & 1 & \cdots & 1 \\ x_1 & x_2 & x_3 & \cdots & x_k \\ x_1^2 & x_2^2 & x_3^2 & \cdots & x_k^2 \\ \vdots & \vdots & \vdots & & \vdots \\ x_1^{k-1} & x_2^{k-1} & x_3^{k-1} & \cdots & x_k^{k-1} \end{bmatrix}.$$

25. Let the $2k \times 2k$ matrix **A** be partitioned as follows

$$\mathbf{A} = \begin{bmatrix} \mathbf{A}_{11} & \mathbf{A}_{12} \\ \mathbf{A}_{21} & \mathbf{A}_{22} \end{bmatrix},$$

where $\mathbf{A}_{11}$ is a $k \times k$ matrix; further suppose that $\mathbf{A}_{21}\mathbf{A}_{22} = \mathbf{A}_{22}\mathbf{A}_{21}$, and let $|\mathbf{A}_{22}| \neq 0$. Show that

$$\det(\mathbf{A}) = \det(\mathbf{A}_{11}\mathbf{A}_{22} - \mathbf{A}_{12}\mathbf{A}_{21}).$$

(Use Theorem 8.2.1.)

26. Find the determinant of the matrix

$$\mathbf{A} = \begin{bmatrix} 0 & a_1 & 0 & 0 \\ b_1 & 0 & a_2 & 0 \\ 0 & b_2 & 0 & a_3 \\ 0 & 0 & b_3 & 0 \end{bmatrix},$$

and deduce the conditions on the a_i and b_i such that $\det(\mathbf{A}) \neq 0$.

27. Find the inverse of **A** in Prob. 26 assuming conditions on the a_i and b_i such that the inverse exists.

28. Work Probs. 26 and 27 when **A** is a $k \times k$ matrix.

29. Find the inverse of the 5 × 5 lower triangular matrix **T** where

$$\mathbf{T} = \begin{bmatrix} 1 & 0 & 0 & 0 & 0 \\ 1 & 1 & 0 & 0 & 0 \\ 1 & 1 & 1 & 0 & 0 \\ 1 & 1 & 1 & 1 & 0 \\ 1 & 1 & 1 & 1 & 1 \end{bmatrix}.$$

30. Generalize Prob. 29; that is, find the inverse of the matrix **T**, where **T** is a $k \times k$ lower triangular matrix, where each element on and below the main diagonal is equal to unity.

31. Use the details of the proof of Theorem 8.6.1 to find a lower triangular matrix **R** and an upper triangular matrix **T** such that **A** = **RT** where

$$\mathbf{A} = \begin{bmatrix} 1 & 0 & 0 & 0 & 1 \\ 0 & 1 & 0 & 0 & 1 \\ 0 & 0 & 1 & 0 & 1 \\ 0 & 0 & 0 & 1 & 1 \\ 1 & 1 & 1 & 1 & 1 \end{bmatrix}.$$

32. If **A** and **B** are symmetric matrices, show that **A** × **B** is symmetric.

33. Compute **A** × **B** and **B** × **A** for the matrices below.

$$\mathbf{A} = [1, -1, 0]; \qquad \mathbf{B} = \begin{bmatrix} 3 & 1 \\ 1 & 4 \\ 2 & 0 \end{bmatrix}.$$

34. In Prob. 33 demonstrate that **A** × **B** = (**A** × **I**)(**I** × **B**).

35. Find det (**A** × **B**) and det (**B** × **A**) if

$$\mathbf{A} = \begin{bmatrix} 2 & 1 \\ 3 & 4 \end{bmatrix}, \qquad \mathbf{B} = \begin{bmatrix} 1 & 0 & 2 \\ 2 & 1 & 3 \\ 2 & 4 & 1 \end{bmatrix}.$$

36. For the matrices **A** and **B** in Prob. 33 and **C** defined below, demonstrate that (**A** × **B**) × **C** = **A** × (**B** × **C**).

$$\mathbf{C} = \begin{bmatrix} 1 & 2 \\ -1 & 0 \end{bmatrix}.$$

37. For the matrices defined below demonstrate that $(\mathbf{A} \times \mathbf{B})(\mathbf{F} \times \mathbf{G}) = (\mathbf{AF}) \times (\mathbf{BG})$.

$$\mathbf{A} = \begin{bmatrix} 2 & 1 \\ 1 & 3 \end{bmatrix}; \quad \mathbf{B} = \begin{bmatrix} 1 & 3 & 2 \\ 2 & 0 & -1 \end{bmatrix}; \quad \mathbf{F} = \begin{bmatrix} 3 \\ 1 \end{bmatrix}; \quad \mathbf{G} = \begin{bmatrix} 4 \\ 0 \\ -1 \end{bmatrix}.$$

38. For the matrices defined below demonstrate $(\mathbf{A} \times \mathbf{B})^{-1} = \mathbf{A}^{-1} \times \mathbf{B}^{-1}$.

$$\mathbf{A} = \begin{bmatrix} 3 & 2 \\ 1 & 1 \end{bmatrix}; \qquad \mathbf{B} = \begin{bmatrix} 6 & 0 \\ -1 & 8 \end{bmatrix}.$$

39. Let $\mathbf{A}$ be an $m \times m$ matrix, and $\mathbf{B}$ an $n \times n$ upper triangular matrix; show that $\mathbf{A} \times \mathbf{B}$ is an upper triangular block matrix.
40. In Prob. 39 find $\det(\mathbf{A} \times \mathbf{B})$ in terms of the elements of $\mathbf{A}$ and $\mathbf{B}$.
41. For the matrix $\mathbf{A}$ below, the identity matrices are each 3×3. Find the inverse of $\mathbf{A}$.

$$\mathbf{A} = \begin{bmatrix} 3\mathbf{I} & 2\mathbf{I} \\ -\mathbf{I} & 4\mathbf{I} \end{bmatrix}.$$

42. In Prob. 41, find $\det(\mathbf{A})$.
43. In Prob. 41, find the characteristic roots of $\mathbf{A}$.
44. For the matrices in Prob. 38, demonstrate Theorem 8.8.13.
45. Find the characteristic roots of the matrix $\mathbf{A}$ where

$$\mathbf{A} = \begin{bmatrix} 2 & 1 & -1 & 0 \\ 0 & 2 & 1 & -1 \\ -1 & 0 & 2 & 1 \\ 1 & -1 & 0 & 2 \end{bmatrix}.$$

46. Use Theorem 8.8.7 to find the inverse of the matrix $\mathbf{B}$ in Example 8.2.1. Assume that the inverse exists.
47. Use Theorem 8.8.10 to find the determinant of the matrix in Example 8.2.1.
48. Find the characteristic roots of the matrix in Prob. 31.
49. Find the characteristic vectors of the matrix in Prob. 45.
50. Let $\mathbf{A}$ be an $n \times n$ matrix that is partitioned as follows:

$$\mathbf{A} = \begin{bmatrix} \mathbf{A}_{11} & \mathbf{A}_{12} \\ \mathbf{A}_{21} & \mathbf{A}_{22} \end{bmatrix},$$

where the submatrix has size $n_i \times n_j$, $i, j = 1, 2$, and $n_1 + n_2 = n$. If $\det(\mathbf{A}) \neq 0$

and det $(\mathbf{A}_{11}) \neq 0$, show that the matrix $\mathbf{B}$ is nonsingular where $\mathbf{B}$ is defined by

$$\mathbf{B} = \mathbf{A}_{22} - \mathbf{A}_{21}\mathbf{A}_{11}^{-1}\mathbf{A}_{12}.$$

51. Let $\mathbf{A}$ be partitioned as in Prob. 50. If rank $(\mathbf{A})$ = rank $(\mathbf{A}_{11})$, show that $\mathbf{A}_{22} = \mathbf{A}_{21}\mathbf{A}_{11}^{-1}\mathbf{A}_{12}$.
52. Let the $n \times n$ matrix $\mathbf{A}$ be defined by

$$\mathbf{A} = \begin{bmatrix} \mathbf{I}_1 & \mathbf{0} \\ \mathbf{B} & \mathbf{I}_2 \end{bmatrix},$$

where $\mathbf{B}$ is an $n_1 \times n_2$ matrix and the size of the other submatrices are thus determined. Show that $\mathbf{A}^{-1}$ exists and find it.
53. Find matrices $\mathbf{A}$ and $\mathbf{B}$ such that $\mathbf{AB} = \mathbf{C}$, where $\mathbf{C}$ is defined in Eq. (8.3.1), $\mathbf{A}$ is lower triangular and does not involve the b_i, and $\mathbf{B}$ is a diagonal matrix and does not involve the a_i.
54. In Prob. 53, find $\mathbf{A}^{-1}$ and show that $\mathbf{B}^{-1}\mathbf{A}^{-1} = \mathbf{C}^{-1}$ where $\mathbf{C}^{-1}$ is defined in Eq. (8.3.2).
55. Let $\mathbf{A}$ be partitioned as in Prob. 50, where $\mathbf{A}_{12} = \mathbf{0}$ and det $(\mathbf{A}_{22}) \neq 0$. Find $\mathbf{A}^{-1}$ in terms of $\mathbf{A}_{11}, \mathbf{A}_{21}, \mathbf{A}_{22}$.
56. Find the inverse of the matrix $\mathbf{B}$ defined by

$$\mathbf{B} = \begin{bmatrix} 4 & 1 & 3 & 2 & 1 \\ 1 & 2 & 6 & 4 & 2 \\ 3 & 6 & 12 & 8 & 4 \\ 2 & 4 & 8 & 24 & 12 \\ 1 & 2 & 4 & 12 & 12 \end{bmatrix}.$$

Use Theorem 8.3.7.
57. Find det $(\mathbf{A})$ in Prob. 31.
58. If $\mathbf{AB} = \mathbf{0}$ show that

$$(\mathbf{A} \times \mathbf{F})(\mathbf{B} \times \mathbf{G}) = \mathbf{0}$$

for any matrices $\mathbf{F}$ and $\mathbf{G}$ whose sizes are such that multiplication is defined.
59. If either $\mathbf{A}$ or $\mathbf{B}$ is the null matrix, show that

$$\text{(a)}\ \mathbf{A} \times \mathbf{B} = \mathbf{0}$$

and

$$\text{(b)}\ \mathbf{B} \times \mathbf{A} = \mathbf{0}.$$

60. If **A** is any $k \times k$ matrix, show that there exists a diagonal matrix **D** where $d_{ii} = +1$ or $d_{ii} = -1$ such that $|\mathbf{A} + \mathbf{D}| \neq 0$.
61. Let **C** be defined by

$$\mathbf{C} = \begin{bmatrix} 2\mathbf{B} & -\mathbf{B} & -\mathbf{B} \\ -\mathbf{B} & 2\mathbf{B} & -\mathbf{B} \\ -\mathbf{B} & -\mathbf{B} & 2\mathbf{B} \end{bmatrix}, \quad \text{where} \quad \mathbf{B} = \begin{bmatrix} 1 & -1 \\ -1 & 1 \end{bmatrix};$$

find the characteristic roots of **C**.
62. In the quadratic forms of Eq. (8.8.1), show that each matrix is idempotent and that the product of each pair is equal to the null matrix.
63. Let **R** be an $n \times n$ correlation matrix and let θ^2 be such that $\theta^2 \leq \rho_{ij}^2$ for all $i \neq j$. Show that $|\mathbf{R}| \leq 1 - \theta^2$.
64. Show that the largest characteristic root of a correlation matrix is less than or equal to n, the size of the matrix.
65. Let **V** be an $n \times n$ covariance matrix and **R** the corresponding correlation matrix. Show that $|\mathbf{V}| = v_{11}v_{22} \ldots v_{nn}|\mathbf{R}|$.
66. If **R** is an $n \times n$ correlation matrix, show that $|\mathbf{R}|$ attains its maximum value when $\rho_{ij} = 0$ for all $i \neq j$.
67. If each entry ρ_{ij} of an $n \times n$ correlation matrix **R** satisfies $-1 \leq \rho_{ij} \leq 1$, show that $|\mathbf{R}| = 0$ if and only if at least one ρ_{ij} for $i \neq j$ is equal to plus or minus unity.
68. If **R** and **T** are lower and upper triangular nonsingular matrices, respectively, and if $\mathbf{RT} = \mathbf{D}$ where **D** is diagonal, show that **R** and **T** are also diagonal.
69. Show that any square matrix **A** can be written as the sum of a symmetric and a skew-symmetric matrix.
70. If **T** is an upper (lower) triangular $n \times n$ matrix and **D** is a diagonal $n \times n$ matrix, show that **DT** and **TD** are upper (lower) triangular matrices.
71. If **A** is a symmetric circulant, show that $\mathbf{A}^2$ is a symmetric regular circulant.
72. If **A** and **B** are $k \times k$ symmetric circulants, show that $\mathbf{A}^2\mathbf{B}^2 = \mathbf{B}^2\mathbf{A}^2$, but **AB** may not equal **BA**.
73. If **A** is a $k \times k$ symmetric regular circulant and k is an even integer, exhibit the first row of **A** in terms of $a_0, a_1, \ldots, a_m$, where $m = k/2$.
74. If **A** is a $k \times k$ regular circulant then **B** is a $k \times k$ symmetric circulant, where $\mathbf{P}'\mathbf{A} = \mathbf{B}$ and hence $\mathbf{A} = \mathbf{PB}$, where $\mathbf{P} = [\mathbf{e}_1, \mathbf{e}_k, \mathbf{e}_{k-1}, \ldots, \mathbf{e}_3, \mathbf{e}_2]$, where $\mathbf{e}_i$ is the i-th unit vector (the i-th column of the $k \times k$ identity matrix **I**). Note that **P** is a permutation matrix. Prove this result.
75. In Prob. 74, show that **P** is a symmetric circulant and that $\mathbf{P}' = \mathbf{P}^{-1}$.
76. If **A** is a symmetric circulant, show that $\mathbf{A}^2$ is a T-matrix and a C-matrix.

77. Show that if **A** is a $k \times k$ symmetric circulant, then $\mathbf{P}'\mathbf{A}\mathbf{P} = \mathbf{A}$, where **P** is the permutation matrix in Problem 74.

78. Prove Theorem 8.10.22.

79. Prove Corollary 8.10.22.

80. Prove Theorem 8.10.23 by using the fact that $\mathbf{P}'\mathbf{A}\mathbf{P} = \mathbf{A}$ for an appropriate permutation (and hence orthogonal) matrix **P** and $\mathbf{Q}'\mathbf{B}\mathbf{Q} = \mathbf{B}$ for an appropriate permutation matrix **Q**.

81. If **A** is a regular circulant, show that $\mathbf{A}'\mathbf{A}$ and $\mathbf{A}\mathbf{A}'$ are symmetric regular circulants.

82. If **A** is a regular circulant, show that **A** is also a T-matrix.

Trace and Vector of Matrices: Commutation Matrices

9

9.1 Trace

This section is devoted to the many applications in which the sum of the diagonal elements (trace) of a matrix plays an important role.

Definition 9.1.1

Trace. *The trace of an* $n \times n$ *matrix* $\mathbf{A}$, *which we write as* $\operatorname{tr}(\mathbf{A})$ *is defined to be the sum of the diagonal elements of* $\mathbf{A}$; *that is,*

$$\operatorname{tr}(\mathbf{A}) = \sum_{i=1}^{n} a_{ii}. \tag{9.1.1}$$

Theorem 9.1.1.

Let $\mathbf{A}$ *and* $\mathbf{B}$ *each be* $n \times n$ *matrices; then*

$$\operatorname{tr}(\mathbf{AB}) = \operatorname{tr}(\mathbf{BA}). \tag{9.1.2}$$

Proof: Let $\mathbf{AB} = \mathbf{C}$; then $c_{pq} = \sum_{j=1}^{n} a_{pj} b_{jq}$. Let $\mathbf{G} = \mathbf{BA}$; then

$$g_{rs} = \sum_{i=1}^{n} b_{ri} a_{is}.$$

But

$$\operatorname{tr}(\mathbf{AB}) = \operatorname{tr}(\mathbf{C}) = \sum_{p=1}^{n} c_{pp} = \sum_{p=1}^{n} \sum_{j=1}^{n} a_{pj} b_{jp}.$$

Also

$$\operatorname{tr}(\mathbf{BA}) = \operatorname{tr}(\mathbf{G}) = \sum_{r=1}^{n} g_{rr} = \sum_{r=1}^{n} \sum_{i=1}^{n} b_{ri} a_{ir}.$$

Thus tr (**AB**) = tr (**BA**). ▮

Theorem 9.1.2

Let **A** *be any* $n \times n$ *matrix and let* **P** *be any nonsingular* $n \times n$ *matrix; then*

$$\operatorname{tr}(\mathbf{A}) = \operatorname{tr}(\mathbf{P}^{-1}\mathbf{AP}). \tag{9.1.3}$$

If **P** *is an orthogonal matrix, then*

$$\operatorname{tr}(\mathbf{A}) = \operatorname{tr}(\mathbf{P}'\mathbf{AP}). \tag{9.1.4}$$

Proof: By the previous theorem,

$$\operatorname{tr}[\mathbf{P}(\mathbf{AP}^{-1})] = \operatorname{tr}[(\mathbf{AP}^{-1})\mathbf{P}] = \operatorname{tr}(\mathbf{AI}) = \operatorname{tr}(\mathbf{A}). \quad ▮$$

Theorem 9.1.3

Let **A** *be an* $n \times n$ *matrix with characteristic roots* $\lambda_1, \lambda_2, \ldots, \lambda_n$; *then* $\operatorname{tr}(\mathbf{A}) = \sum_{i=1}^{n} \lambda_i$; *that is, the sum of the diagonal elements of an* $n \times n$ *matrix is equal to the sum of the characteristic roots of the matrix.*

Proof: Let **P** be a nonsingular matrix such that $\mathbf{P}^{-1}\mathbf{AP} = \mathbf{T}$ where **T** is a triangular matrix with characteristic roots λ_i on the diagonal (**P** and **T** may not be real matrices; see Corollary 8.6.9.1). Then

$$\operatorname{tr}(\mathbf{A}) = \operatorname{tr}(\mathbf{P}^{-1}\,\mathbf{AP}) = \operatorname{tr}(\mathbf{T}) = \sum_{i=1}^{n} \lambda_i. \quad ▮ \tag{9.1.5}$$

The proofs of the next fifteen theorems are left for the reader. In most cases the proof involves an application of one or more of the first three theorems in the chapter.

Theorem 9.1.4

If $\mathbf{A}$ *and* $\mathbf{B}$ *are* $n \times n$ *matrices and* a *and* b *are scalars, then*

$$\text{tr}\,(a\mathbf{A} + b\mathbf{B}) = a\,\text{tr}\,(\mathbf{A}) + b\,\text{tr}\,(\mathbf{B}). \tag{9.1.6}$$

Theorem 9.1.5

If $\mathbf{A}$ *is an* $n \times n$ *matrix and* $\mathbf{A}^2 = m\mathbf{A}$, *then*

$$\text{tr}\,(\mathbf{A}) = m\,\text{rank}\,(\mathbf{A}). \tag{9.1.7}$$

Note: If $\mathbf{A}$ is idempotent, then $m = 1$ and $\text{tr}\,(\mathbf{A}) = \text{rank}\,(\mathbf{A})$.

Theorem 9.1.6

Let $\mathbf{A}$ *be an* $m \times n$ *matrix; then* $\text{tr}\,(\mathbf{A}'\mathbf{A}) = 0$ *if and only if* $\mathbf{A} = \mathbf{0}$.

Theorem 9.1.7

If $\mathbf{A}$ *is an* $n \times n$ *matrix, then*

$$\text{tr}\,(\mathbf{A}') = \text{tr}\,(\mathbf{A}). \tag{9.1.8}$$

Theorem 9.1.8

Let $\mathbf{A}$ *be an* $m \times n$ *matrix of rank* r; *then*

$$\text{tr}\,[\mathbf{I} - \mathbf{A}(\mathbf{A}'\mathbf{A})^{-}\mathbf{A}'] = m - r. \tag{9.1.9}$$

Theorem 9.1.9

If $\mathbf{A}$ *is an* $n \times m$ *matrix, then*

$$\text{tr}\,(\mathbf{A}\mathbf{A}') = \text{tr}\,(\mathbf{A}'\mathbf{A}) = \sum_{j=1}^{m} \sum_{i=1}^{n} a_{ij}^2. \tag{9.1.10}$$

Theorem 9.1.10

If $\mathbf{A}$ *is an* $n \times n$ *matrix and* k *is a positive integer, then*

$$\text{tr}\,(\mathbf{A}^k) = \sum_{i=1}^{n} \lambda_i^k, \tag{9.1.11}$$

where $\lambda_1, \lambda_2, \ldots, \lambda_n$ *are the characteristic roots of* $\mathbf{A}$.

Theorem 9.1.11

If $\mathbf{A}$ *is an* $n \times n$ *matrix and* $\mathbf{B}$ *is an* $m \times m$ *matrix and* $\mathbf{A} \times \mathbf{B}$ *is the direct product, then*

$$\operatorname{tr}(\mathbf{A} \times \mathbf{B}) = \operatorname{tr}(\mathbf{A}) \operatorname{tr}(\mathbf{B}). \tag{9.1.12}$$

Theorem 9.1.12

If $\mathbf{A}$ *is an* $n \times m$ *matrix and* $\mathbf{A}^c$ *is any c-inverse of* $\mathbf{A}$ *and* $\mathbf{A}^L$ *is any L-inverse of* $\mathbf{A}$*, then*

$$\operatorname{tr}(\mathbf{A}^c\mathbf{A}) = \operatorname{tr}(\mathbf{A}\mathbf{A}^c) = \operatorname{tr}(\mathbf{A}^L\mathbf{A}) = \operatorname{tr}(\mathbf{A}\mathbf{A}^L) = \operatorname{tr}(\mathbf{A}^-\mathbf{A}) = \operatorname{tr}(\mathbf{A}\mathbf{A}^-) = \operatorname{rank}(\mathbf{A}). \tag{9.1.13}$$

Theorem 9.1.13

If $\mathbf{A}$ *is an* $n \times n$ *symmetric matrix with* r *nonzero characteristic roots* $\lambda_1, \lambda_2, \ldots, \lambda_r$*, then*

$$\operatorname{tr}(\mathbf{A}^-) = \sum_{i=1}^{r} \lambda_i^{-1}. \tag{9.1.14}$$

Theorem 9.1.14

If $\mathbf{A}$ *is an* $n \times n$ *symmetric matrix with characteristic roots* λ_i*, then*

$$\sum_{i=1}^{n} \lambda_i^2 = \sum_i \sum_j a_{ij}^2. \tag{9.1.15}$$

Theorem 9.1.15

If $\mathbf{S}$ *is an* $n \times n$ *skew-symmetric matrix, then*

$$\operatorname{tr}(\mathbf{I} + \mathbf{S}) = n, \quad \textit{and} \quad \operatorname{tr}(\mathbf{S}) = 0. \tag{9.1.16}$$

Theorem 9.1.16

If $\mathbf{A}$ *is an* $n \times n$ *matrix such that* $\mathbf{A}^k = \mathbf{0}$ *for some positive integer* k*, then* $\operatorname{tr}(\mathbf{A}) = 0$.

Theorem 9.1.17

If $\mathbf{A}$ *is a non-negative* $n \times n$ *matrix, then* $\operatorname{tr}(\mathbf{A}) = 0$ *if and only if* $\mathbf{A} = \mathbf{0}$.

Theorem 9.1.18

If $\mathbf{A}$ *and* $\mathbf{B}$ *are* $n \times n$ *matrices, then*

$$\operatorname{tr}(\mathbf{A}^q\mathbf{B}^q) = \operatorname{tr}(\mathbf{B}^q\mathbf{A}^q) \tag{9.1.17}$$

for any positive integer q.

Theorem 9.1.19

If $\mathbf{A}$ *and* $\mathbf{B}$ *are* $n \times n$ *symmetric matrices, then*

$$\operatorname{tr}[(\mathbf{AB})^2] \leq \operatorname{tr}(\mathbf{A}^2\mathbf{B}^2) = \operatorname{tr}(\mathbf{B}^2\mathbf{A}^2). \tag{9.1.18}$$

Proof: We shall prove $\operatorname{tr}[(\mathbf{AB})^2] \leq \operatorname{tr}(\mathbf{A}^2\mathbf{B}^2)$ only, since the result, $\operatorname{tr}(\mathbf{A}^2\mathbf{B}^2) = \operatorname{tr}(\mathbf{B}^2\mathbf{A}^2)$, follows from Theorem 9.1.18. Let $\mathbf{C}$ be defined by

$$\mathbf{C} = \mathbf{AB} - \mathbf{BA}. \tag{9.1.19}$$

Then, since $\mathbf{A} = \mathbf{A}'$ and $\mathbf{B} = \mathbf{B}'$, we get

$$\mathbf{C}' = \mathbf{BA} - \mathbf{AB} = -\mathbf{C}.$$

But

$$\operatorname{tr}(\mathbf{CC}') = \operatorname{tr}[(\mathbf{AB} - \mathbf{BA})(\mathbf{BA} - \mathbf{AB})] = \operatorname{tr}(\mathbf{ABBA}) + \operatorname{tr}(\mathbf{BAAB}) - \operatorname{tr}(\mathbf{BABA}) - \operatorname{tr}(\mathbf{ABAB}),$$

and by Theorem 9.1.1,

$$\operatorname{tr}(\mathbf{ABBA}) = \operatorname{tr}(\mathbf{AABB}) = \operatorname{tr}(\mathbf{A}^2\mathbf{B}^2),$$

$$\operatorname{tr}(\mathbf{BAAB}) = \operatorname{tr}(\mathbf{AABB}) = \operatorname{tr}(\mathbf{A}^2\mathbf{B}^2),$$

and

$$-\operatorname{tr}(\mathbf{BABA}) = -\operatorname{tr}(\mathbf{ABAB}) = -\operatorname{tr}[(\mathbf{AB})^2].$$

Thus we get

$$\operatorname{tr}(\mathbf{CC}') = 2\operatorname{tr}(\mathbf{A}^2\mathbf{B}^2) - 2\operatorname{tr}[(\mathbf{AB})^2];$$

but by Theorem 9.1.9, $\operatorname{tr}(\mathbf{CC}') = \sum\sum c_{ij}^2 \geq 0$; hence,

$$\operatorname{tr}(\mathbf{A}^2\mathbf{B}^2) - \operatorname{tr}[(\mathbf{AB})^2] \geq 0,$$

and the theorem is proved. ▮

The proofs of the next two theorems are left for the reader.

Theorem 9.1.20

If $\mathbf{x}$ *is an* $n \times 1$ *vector and* $\mathbf{A}$ *an* $n \times n$ *matrix, then*

$$\mathbf{x}'\mathbf{A}\mathbf{x} = \operatorname{tr}(\mathbf{A}\mathbf{x}\mathbf{x}'). \tag{9.1.20}$$

Theorem 9.1.21

Let $\mathbf{A}$ *be a* 2×2 *nonsingular matrix. Then*

$$\operatorname{tr}(\mathbf{A}) = [\det(\mathbf{A})][\operatorname{tr}(\mathbf{A}^{-1})]. \tag{9.1.21}$$

Theorem 9.1.22

If $\mathbf{A}$ *is an* $n \times n$ *nonzero symmetric matrix, then*

$$\operatorname{rank}(\mathbf{A}) \geq \frac{[\operatorname{tr}(\mathbf{A})]^2}{\operatorname{tr}(\mathbf{A}^2)}. \tag{9.1.22}$$

Proof: From the fact that for any set of ($n \geq 1$) real numbers $b_1, b_2, \ldots, b_n$, the quantity

$$\sum_{i=1}^{n} \left[b_i - \frac{\sum_{j=1}^{n} b_j}{n} \right]^2$$

is non-negative, we get that

$$\sum_{i=1}^{n} b_i^2 - \frac{\left(\sum_{j=1}^{n} b_j\right)^2}{n} \geq 0;$$

and if at least one $b_i \neq 0$, we have

$$n \geq \frac{\left(\sum_{j=1}^{n} b_j\right)^2}{\sum_{j=1}^{n} b_j^2},$$

where the equality holds if and only if $b_1 = b_2 = \cdots = b_n$. Now suppose that the rank of **A** is $r > 0$. Then **A** has exactly r nonzero characteristic roots (all real); denote them by $\lambda_1, \lambda_2, \ldots, \lambda_r$. Now by Theorems 9.1.3 and 9.1.10, we get

$$\text{tr}(\mathbf{A}) = \sum_{i=1}^{r} \lambda_i, \qquad \text{tr}(\mathbf{A}^2) = \sum_{i=1}^{r} \lambda_i^2.$$

Thus, from the above, if we let $n = r$ and $\lambda_i = b_i$, we get

$$r \geq \frac{\left(\sum_{i=1}^{r} \lambda_i\right)^2}{\sum_{i=1}^{r} \lambda_i^2},$$

and by substitution the theorem is proved. ∎

Theorem 9.1.23

Let **A** *be an* $n \times n$ *matrix with all real characteristic roots and let exactly* t *of them be nonzero; then*

$$[\text{tr}(\mathbf{A})]^2 \leq t\,\text{tr}(\mathbf{A}^2).$$

Proof: If $t = 0$, the theorem is trivial, so assume $t > 0$. Let the characteristic roots of **A** be denoted by $\lambda_1, \ldots \lambda_t, \ldots, \lambda_n$, where we shall assume that the first t characteristic roots are the nonzero roots and that $\lambda_{t+1} = \lambda_{t+2} = \cdots = \lambda_n = 0$. It then follows that the characteristic roots of $\mathbf{A}^2$ are $\lambda_1^2, \ldots, \lambda_t^2, \ldots, \lambda_n^2$ and also the first t are nonzero and the remaining ones are equal to zero. Now we shall examine the following sum of squares

$$S = \sum_{i=1}^{t} (\lambda_i - \bar{\lambda})^2,$$

where

$$\bar{\lambda} = \frac{1}{t}\sum_{i=1}^{t}\lambda_i; \quad \text{but} \quad \sum_{i=1}^{t}\lambda_i = \sum_{i=1}^{n}\lambda_i = \text{tr}(\mathbf{A}) \quad \text{and} \quad \bar{\lambda} = \frac{\text{tr}(\mathbf{A})}{t}.$$

It is clear that $S \geq 0$; also $S = 0$ if and only if $\lambda_1 = \lambda_2 = \cdots = \lambda_t = \bar{\lambda}$, that is, if and only if each and every λ_i is equal to $\bar{\lambda}$ for $i = 1, 2, \ldots, t$. Now we have

$$S = \sum_{i=1}^{t}\lambda_i^2 - t\bar{\lambda}^2$$

or

$$S = \text{tr}(\mathbf{A}^2) - t\left[\frac{\text{tr}(\mathbf{A})}{t}\right]^2,$$

and hence

$$[\text{tr}(\mathbf{A})]^2 \leq t\,\text{tr}(\mathbf{A}^2). \quad \blacksquare$$

Theorem 9.1.24

Let $\mathbf{A}$ *be any* $k \times k$ *matrix with rank* r *and let the number of nonzero characteristic roots be equal to* t*; then* $r \geq t$.

Proof: Let $\mathbf{P}$ be a nonsingular matrix such that $\mathbf{P}^{-1}\mathbf{A}\mathbf{P} = \mathbf{T}$ where $\mathbf{T}$ is an upper triangular matrix with the characteristic roots of $\mathbf{A}$ on the diagonal of $\mathbf{T}$ ($\mathbf{P}$ and $\mathbf{T}$ may not be real matrices). Clearly $\mathbf{T}$ has exactly t nonzero diagonal elements, and the rank of $\mathbf{T}$ is no less than the number of nonzero diagonal elements; that is, rank $(\mathbf{T}) \geq t$. But rank $(\mathbf{A})$ = rank $(\mathbf{P}^{-1}\mathbf{A}\mathbf{P})$ = rank $(\mathbf{T})$ and the theorem is proved. $\blacksquare$

Recall that a sufficient condition (not necessary) for r to be equal to t is that $\mathbf{A}$ be symmetric.

Theorem 9.1.25

Let $\mathbf{A}$ *be an* $n \times n$ *matrix;*

(1) *if* $\mathbf{A}$ *has real characteristic roots, then* $[\text{tr}(\mathbf{A})]^2 \leq \text{rank}(\mathbf{A})\,\text{tr}(\mathbf{A}^2)$,
(2) *if* $\mathbf{A}$ *is symmetric, then* $[\text{tr}(\mathbf{A})]^2 = \text{rank}(\mathbf{A})\,\text{tr}(\mathbf{A}^2)$ *if and only if there is a non-negative integer* m *such that* $\mathbf{A}^2 = m\mathbf{A}$.

Proof: Since the rank of a matrix is greater than or equal to the number of nonzero characteristic roots, the result (1) follows from Theorem 9.1.23. Also by Theorem 9.1.23, if $\mathbf{A}$ is symmetric, we know that rank $(\mathbf{A}) = t$ and

$$[\text{tr}(\mathbf{A})]^2 = \text{rank}(\mathbf{A})\,\text{tr}(\mathbf{A}^2)$$

if and only if $S = 0$, or in other words, if and only if the nonzero characteristic roots are equal. If $m = 0$, the proof to part (2) is trivial, so assume $m > 0$. Part (2) follows if and only if $\mathbf{A}$ has r characteristic roots equal to $\bar{\lambda}$ (say) and the remaining roots equal to zero. The relationship $\mathbf{A}^2 = m\mathbf{A}$ is equivalent to

$$\left(\frac{1}{m}\mathbf{A}\right)^2 = \frac{1}{m}\mathbf{A}$$

and hence $(1/m)\mathbf{A}$ has $n - r$ characteristic roots equal to zero and r roots equal to unity. Therefore $\mathbf{A}$ has $n - r$ characteristic roots equal to zero and r roots equal to m, and the theorem is proved. ▮

Corollary 9.1.25

If $\mathbf{A}$ *is an* $n \times n$ *symmetric matrix, then* $\mathbf{A}^2 = \mathbf{A}$ *if and only if rank* $(\mathbf{A}) = \text{tr}(\mathbf{A}) = \text{tr}(\mathbf{A}^2)$.

The remaining theorems in this chapter play an important role in the theory of quadratic forms and their distribution.

Theorem 9.1.26

Let $\mathbf{A}$ *be an* $n \times n$ *matrix;*

(1) *If* $\mathbf{A}$ *is positive definite, then* $\text{tr}(\mathbf{A}) > 0$.
(2) *If* $\mathbf{A}$ *is positive semidefinite, then* $\text{tr}(\mathbf{A}) \geq 0$.
(3) *If* $\mathbf{A}$ *is non-negative, then* $\text{tr}(\mathbf{A}) \geq 0$.

Proof: The results follow, since a_{ii} is positive for each and every i if $\mathbf{A}$ is positive definite, and a_{ii} is non-negative for each and every i if $\mathbf{A}$ is positive semidefinite. Note that (3) is just a summary of (1) and (2). ▮

Theorem 9.1.27

Let $\mathbf{A}_1, \mathbf{A}_2, \ldots, \mathbf{A}_k$ *be a collection of* $n \times n$ *non-negative matrices; then*

$$\text{tr}\left(\sum_{i=1}^{k} \mathbf{A}_i\right) = \sum_{i=1}^{k} \text{tr}(\mathbf{A}_i) \geq 0,$$

and a strict inequality certainly holds if any one, or more, of the $\mathbf{A}_i$ *is positive definite but may hold even if none of the* $\mathbf{A}_i$ *is positive definite.*

Proof: The results follow from the previous theorem since the sum of non-negative (positive) numbers is a non-negative (positive) number. ▮

Corollary 9.1.27.1

Let $\mathbf{A}_1, \mathbf{A}_2, \ldots, \mathbf{A}_k$ *be a collection of* $n \times n$ *non-negative matrices; then*

$$\sum_{i=1}^{k} \operatorname{tr}(\mathbf{A}_i) = 0,$$

if and only if

$$\mathbf{A}_1 = \mathbf{A}_2 = \cdots = \mathbf{A}_k = \mathbf{0}.$$

Proof: The proof follows from the fact that if the i-th diagonal element of a positive semidefinite matrix is equal to zero, then the entire i-th row and i-th column is equal to zero. ▮

Corollary 9.1.27.2

Let $\mathbf{B}_1, \mathbf{B}_2, \ldots, \mathbf{B}_k$ *be a collection of* $m \times n$ *matrices such that*

$$\sum_{i=1}^{k} \operatorname{tr}(\mathbf{B}_i \mathbf{B}_i') = 0$$

or such that

$$\sum_{i=1}^{k} \operatorname{tr}(\mathbf{B}_i' \mathbf{B}_i) = 0;$$

then $\mathbf{B}_1 = \mathbf{B}_2 = \cdots = \mathbf{B}_k = \mathbf{0}$.

Theorem 9.1.28

Let $\mathbf{A}$ *and* $\mathbf{B}$ *be* $n \times n$ *non-negative matrices; then*

(a) $\operatorname{tr}(\mathbf{AB}) \geq 0$,
(b) $\operatorname{tr}(\mathbf{AB}) = 0$ *if and only if* $\mathbf{AB} = \mathbf{0}$.

Proof: We shall prove part (b) first. Clearly if $\mathbf{AB} = \mathbf{0}$, then $\text{tr}\,(\mathbf{AB}) = 0$. To prove the "only if," we note that since $\mathbf{A}$ and $\mathbf{B}$ are non-negative matrices, there exist matrices $\mathbf{U}$ and $\mathbf{V}$ such that $\mathbf{A} = \mathbf{U}'\mathbf{U}$ and $\mathbf{B} = \mathbf{V}\mathbf{V}'$; thus

$$\text{tr}\,(\mathbf{AB}) = \text{tr}\,(\mathbf{U}'\mathbf{U}\mathbf{V}\mathbf{V}') = \text{tr}\,(\mathbf{V}'\mathbf{U}'\mathbf{U}\mathbf{V}] = \text{tr}\,[(\mathbf{UV})'(\mathbf{UV})].$$

But by the hypothesis of the theorem, $\text{tr}\,(\mathbf{AB}) = 0$; hence $\text{tr}\,[(\mathbf{UV})'(\mathbf{UV})] = 0$, and by Theorem 9.1.6 this implies that $\mathbf{UV} = \mathbf{0}$. If we multiply on the left by $\mathbf{U}'$ and the right by $\mathbf{V}'$, we get $\mathbf{U}'\mathbf{U}\mathbf{V}\mathbf{V}' = \mathbf{0}$, or $\mathbf{AB} = \mathbf{0}$, and part (b) of the theorem is proved. To prove part (a) let $\mathbf{A} = \mathbf{U}'\mathbf{U}$, $\mathbf{B} = \mathbf{V}\mathbf{V}'$ and $\mathbf{UV} = \mathbf{C}$; we obtain

$$\text{tr}\,(\mathbf{AB}) = \text{tr}\,[(\mathbf{UV})'(\mathbf{UV})] = \text{tr}\,(\mathbf{C}'\mathbf{C}) = \sum_i \sum_j c_{ij}^2 \geq 0,$$

and part (a) is proved.

The next three theorems follow directly from Theorems 9.1.27 and 9.1.28.

Theorem 9.1.29

Let $\mathbf{A}_1, \mathbf{A}_2, \ldots, \mathbf{A}_k$ be a collection of $n \times n$ non-negative matrices; then

$$\text{tr}\left[\sum_{j=1}^{k} \sum_{\substack{i=1 \\ i \neq j}}^{k} \mathbf{A}_i \mathbf{A}_j\right] = 0$$

if and only if $\mathbf{A}_i \mathbf{A}_j = \mathbf{0}$ for $i = 1, 2, \ldots, k$; $j = 1, 2, \ldots, k$; $i \neq j$.

Theorem 9.1.30

Let $\mathbf{A}_1, \mathbf{A}_2, \ldots, \mathbf{A}_k$ be a collection of $n \times n$ non-negative matrices; then

$$\text{tr}\left[\sum_{j=1}^{k} \sum_{i=1}^{k} \mathbf{A}_i \mathbf{A}_j\right] = \sum_{j=1}^{k} \sum_{i=1}^{k} \text{tr}\,[\mathbf{A}_i \mathbf{A}_j] \geq 0,$$

and

$$\text{tr}\left[\sum_{j=1}^{k} \sum_{\substack{i=1 \\ i \neq j}}^{k} \mathbf{A}_i \mathbf{A}_j\right] = \sum_{j=1}^{k} \sum_{\substack{i=1 \\ i \neq j}}^{k} \text{tr}\,[\mathbf{A}_i \mathbf{A}_j] \geq 0.$$

Theorem 9.1.31

Let $\mathbf{A}_1, \mathbf{A}_2, \ldots, \mathbf{A}_k$ *be a collection of* $n \times n$ *non-negative matrices, and let* $\lambda_1, \lambda_2, \ldots, \lambda_k$ *be positive numbers; then*

(1) $\operatorname{tr}\left[\sum_{i=1}^{k} \lambda_i \mathbf{A}_i\right] \geq 0.$

(2) $\operatorname{tr}\left[\sum_{i=1}^{k} \lambda_i \mathbf{A}_i\right] = 0$ *if and only if* $\mathbf{A}_1 = \cdots = \mathbf{A}_k = \mathbf{0}$.

9.2 Vector of a Matrix

In the theory of (normal) multivariate analysis, the starting place is generally as follows: an $m \times 1$ vector $\mathbf{x}$ is distributed $\mathbf{N}(\boldsymbol{\mu}, \boldsymbol{\Sigma})$, that is, as an m-variate normal with mean vector $\boldsymbol{\mu}$ and covariance $\boldsymbol{\Sigma}$; also if $\mathbf{x}_1, \mathbf{x}_2, \ldots, \mathbf{x}_n$ is a random sample of n vectors from this normal distribution, then it may be desirable to determine the joint distribution of the scalar random variables x_{ij} for $j = 1, \ldots, n$ and $i = 1, \ldots, m$. This can be viewed as the distribution of the elements of the random matrix $\mathbf{X} = [x_{ij}]$. In distribution theory it is generally easier, notationally, to work with the distribution of a *vector* than of a matrix. For this reason and others we may often want to represent an $m \times n$ matrix, say $\mathbf{A}$, as a vector $\mathbf{a}$. This is the context of the following definition and theorems.

Definition 9.2.1

Vector of a Matrix $\mathbf{A}$. *Let* $\mathbf{A}$ *be an* $m \times n$ *matrix, where* $\mathbf{A} = [\mathbf{a}_1, \mathbf{a}_2, \ldots, \mathbf{a}_n]$; *that is,* $\mathbf{a}_i$ *is the i-th column of* $\mathbf{A}$. *Then vector* ($\mathbf{A}$), *denoted by* vec($\mathbf{A}$), *is defined by*

$$\operatorname{vec}(\mathbf{A}) = \begin{bmatrix} \mathbf{a}_1 \\ \mathbf{a}_2 \\ \vdots \\ \mathbf{a}_n \end{bmatrix}.$$

Note: Some authors refer to the operation vec($\mathbf{A}$) as "rolling out" the columns of $\mathbf{A}$.

The results in the next theorem are a direct result of this definition.

Theorem 9.2.1

(1) *If* $\mathbf{A}$ *is an* $m \times n$ *matrix, then* $\text{vec}(\mathbf{A})$ *is an* $mn \times 1$ *vector.*
(2) *If* $\mathbf{x}$ *is an* $m \times 1$ *vector and* $\mathbf{y}$ *is an* $n \times 1$ *vector, then* $\text{vec}(\mathbf{xy}') = \mathbf{x} \times \mathbf{y}$.
(3) $\text{vec}(\mathbf{y}) = \text{vec}(\mathbf{y}') = \mathbf{y}$ *for any vector* $\mathbf{y}$.
(4) *If* $\mathbf{A}$ *and* $\mathbf{B}$ *are* $n \times n$ *matrices and* a *and* b *are scalars, then* $\text{vec}(a\mathbf{A} + b\mathbf{B}) = a\,\text{vec}(\mathbf{A}) + b\,\text{vec}(\mathbf{B})$.

Example 9.2.1. Let $\mathbf{x}_1, \mathbf{x}_2, \ldots, \mathbf{x}_n$ be a random sample of size n from the m-variable normal distribution with mean vector $\boldsymbol{\mu}$ and covariance matrix $\boldsymbol{\Sigma}$. Define the $m \times n$ matrix $\mathbf{X}$ by $\mathbf{X} = [\mathbf{x}_1, \mathbf{x}_2, \ldots, \mathbf{x}_n]$ and the $mn \times 1$ vector $\mathbf{y}$ by $\mathbf{y} = \text{vec}(\mathbf{X})$. Then $\mathbf{y}$ is distributed as the mn-variable normal with mean $\boldsymbol{\mu}_\mathbf{y} = \boldsymbol{\mu} \times \mathbf{1}_n$ and covariance matrix $\boldsymbol{\Sigma}_\mathbf{y} = \boldsymbol{\Sigma} \times \mathbf{I}_n$.

Theorem 9.2.2

Let $\mathbf{A}$, $\mathbf{B}$, $\mathbf{C}$ *be* $m \times q$, $q \times s$, *and* $n \times s$ *matrices, respectively. Then*

$$\text{vec}(\mathbf{ABC}') = (\mathbf{A} \times \mathbf{C})\text{vec}(\mathbf{B}).$$

Proof: Write

$$\begin{aligned}\text{vec}(\mathbf{ABC}') &= \text{vec}(\mathbf{A}[\mathbf{b}_1, \mathbf{b}_2, \ldots, \mathbf{b}_s]\mathbf{C}') \\ &= \text{vec}([\mathbf{Ab}_1, \mathbf{Ab}_2, \ldots, \mathbf{Ab}_s]\mathbf{C}') \\ &= \text{vec}([\mathbf{A}\sum_j \mathbf{b}_j c_{1j}, \mathbf{A}\sum_j \mathbf{b}_j c_{2j}, \ldots, \mathbf{A}\sum_j \mathbf{b}_j c_{nj}]) \\ &= \begin{bmatrix} \mathbf{A}\sum_j \mathbf{b}_j c_{1j} \\ \mathbf{A}\sum_j \mathbf{b}_j c_{2j} \\ \vdots \\ \mathbf{A}\sum_j \mathbf{b}_j c_{nj} \end{bmatrix}.\end{aligned}$$

Also

$$(\mathbf{A} \times \mathbf{C})\text{vec}(\mathbf{B}) = \begin{bmatrix} \mathbf{A}c_{11} & \mathbf{A}c_{12} & \cdots & \mathbf{A}c_{1s} \\ \vdots & & & \\ \mathbf{A}c_{n1} & \mathbf{A}c_{n2} & \cdots & \mathbf{A}c_{ns} \end{bmatrix} \begin{bmatrix} \mathbf{b}_1 \\ \mathbf{b}_2 \\ \vdots \\ \mathbf{b}_s \end{bmatrix} = \begin{bmatrix} \mathbf{A}\sum_j c_{1j}\mathbf{b}_j \\ \mathbf{A}\sum_j c_{2j}\mathbf{b}_j \\ \vdots \\ \mathbf{A}\sum_j c_{nj}\mathbf{b}_j \end{bmatrix}.$$

So clearly $\text{vec}(\mathbf{ABC}') = (\mathbf{A} \times \mathbf{C})\text{vec}(\mathbf{B})$. ∎

Corollary 9.2.2

Let $\mathbf{A}$ *and* $\mathbf{B}$ *be* $m \times q$ *and* $q \times n$ *matrices, respectively. Write* $\mathbf{B} = [\mathbf{b}_1, \mathbf{b}_2, \ldots, \mathbf{b}_n]$. *Then*

$$\text{vec}(\mathbf{AB}) = (\mathbf{A} \times \mathbf{I})\text{vec}(\mathbf{B})$$

$$= (\mathbf{I} \times \mathbf{B}')\text{vec}(\mathbf{A}) = (\mathbf{A} \times \mathbf{B}')\text{vec}(\mathbf{I}) = \begin{bmatrix} \mathbf{Ab}_1 \\ \mathbf{Ab}_2 \\ \vdots \\ \mathbf{Ab}_n \end{bmatrix}.$$

Theorem 9.2.3

Let $\mathbf{A}$ *be an* $m \times q$ *matrix and* $\mathbf{B}$ *be a* $q \times m$ *matrix. Then* $[\text{vec}(\mathbf{A}')]'\text{vec}(\mathbf{B}) = \text{tr}(\mathbf{AB})$.

Proof: Let $\mathbf{A}' = \mathbf{C} = [\mathbf{c}_1, \mathbf{c}_2, \ldots, \mathbf{c}_m]$ and $\mathbf{B} = [\mathbf{b}_1, \mathbf{b}_2, \ldots, \mathbf{b}_m]$. Then

$$[\text{vec}(\mathbf{A}')]'\text{vec}(\mathbf{B}) = [\mathbf{c}_1', \mathbf{c}_2', \ldots, \mathbf{c}_m'] \begin{bmatrix} \mathbf{b}_1 \\ \mathbf{b}_2 \\ \vdots \\ \mathbf{b}_m \end{bmatrix} = \sum_{j=1}^{m} \mathbf{c}_j'\mathbf{b}_j = \sum_{j=1}^{m}\sum_{i=1}^{q} c_{ij}b_{ij}$$

$$= \sum_{j=1}^{m}\sum_{i=1}^{q} a_{ji}b_{ij} = \text{tr}(\mathbf{AB}). \quad \blacksquare$$

Theorem 9.2.4

Let $\mathbf{A}$, $\mathbf{B}$, $\mathbf{C}$ *be* $m \times n$, $n \times p$, *and* $p \times m$ *matrices, respectively. Then*

$$\text{tr}(\mathbf{ABC}) = [\text{vec}(\mathbf{A})']' \, [\mathbf{B} \times \mathbf{I}]\text{vec}(\mathbf{C}).$$

Proof: Let $\mathbf{BC} = \mathbf{D}$. Then by Theorem 9.2.3 we get

$$\text{tr}(\mathbf{ABC}) = \text{tr}(\mathbf{AD}) = [\text{vec}(\mathbf{A}')]' \, [\text{vec}(\mathbf{D})] = [\text{vec}(\mathbf{A}')]' \, [\text{vec}(\mathbf{BC})].$$

But by Corollary 9.2.2 we get $\text{vec}(\mathbf{BC}) = (\mathbf{B} \times \mathbf{I})\text{vec}(\mathbf{C})$. Hence the result. $\blacksquare$

Corollary 9.2.4

In Theorem 9.2.4 *let* $\mathbf{X}$ *be an* $n \times n$ *matrix and let* $p = n$. *Then*

$$\text{tr}(\mathbf{AX'BXC}) = [\text{vec}(\mathbf{X})]' \, [\mathbf{B} \times \mathbf{A'C'}] \, [\text{vec}(\mathbf{X})].$$

For further information on this subject, see [M–2] and [N–1].

9.3 Commutation Matrices

In Sec. 8.13 we stated that there exists a permutation matrix $\mathbf{P}$ such that $\mathbf{P}'(\mathbf{A} \times \mathbf{B})\mathbf{P} = \mathbf{B} \times \mathbf{A}$, where $\mathbf{P}$ depends only on the sizes of $\mathbf{A}$ and $\mathbf{B}$. This type of matrix $\mathbf{P}$ has other uses. For example, there exists a permutation matrix $\mathbf{P}$ that transforms vec($\mathbf{A}$) into vec($\mathbf{A}'$); that is, vec($\mathbf{A}$) = $\mathbf{P}$vec($\mathbf{A}'$).

Matrices with these characteristics are called *commutation* matrices and they are useful in multivariate analysis, in finding moments of quadratic forms in normal random variables, in variance components, and in several other branches of statistics. In this section we define and discuss commutation matrices. Also see [M–2] and [N–1].

First we list some notation that will be used throughout this section.

Notation:

(1) $\mathbf{d}_i$ is an $m \times 1$ *unit* vector (that is, the i-th element of $\mathbf{d}_i$ is $+1$ and the remaining elements are 0).

(2) $\mathbf{e}_i$ is an $n \times 1$ *unit* vector.

(3) $\mathbf{\Delta}_{ij}$ is an $m \times n$ matrix whose ij-th element is $+1$ and whose remaining elements are 0. We call this an $m \times n$ *unit* matrix.

Note: The reason we use both symbols $\mathbf{d}_i$ and $\mathbf{e}_i$ for unit vectors is that this allows us to denote an $m \times n$ matrix $\mathbf{\Delta}_{ij}$ by $\mathbf{d}_i\mathbf{e}_j'$ without using two subscripts (or a superscript). For example, $\mathbf{e}_i\mathbf{e}_j'$ or $\mathbf{e}_i\mathbf{e}_i'$ are always square matrices, but $\mathbf{d}_i\mathbf{e}_i'$ or $\mathbf{d}_i\mathbf{e}_j'$ may not be square.

Some relationships and results of this notation are given in the next theorem. The proof will be left for the reader.

Theorem 9.3.1

(1) *If* $\mathbf{\Delta}_{ij}$ *is an* $m \times n$ *unit matrix, then* $\mathbf{\Delta}_{ij} = \mathbf{d}_i\mathbf{e}_j' = \mathbf{d}_i \times \mathbf{e}_j'$.

(2) *If* $\mathbf{\Delta}_{ij}$ *is an* $n \times n$ *unit matrix, then* $\mathbf{\Delta}_{ij} = \mathbf{e}_i\mathbf{e}_j' = \mathbf{\Delta}_{ji}'$. *If* $\mathbf{\Delta}_{ij}$ *is an*

$m \times n$ unit matrix, then $\boldsymbol{\Delta}'_{ij} = \boldsymbol{\Delta}_{ji}$, where $\boldsymbol{\Delta}_{ji}$ is an $n \times m$ unit matrix.

(3) *Let $\mathbf{A} = [a_{ij}]$ be an $m \times n$ matrix. Then $\mathbf{A} = \Sigma_{i=1}^{m} \Sigma_{j=1}^{n} a_{ij} \boldsymbol{\Delta}_{ij} = \Sigma_{i=1}^{m} \Sigma_{j=1}^{n} a_{ij}\mathbf{d}_i\mathbf{e}'_j$. Also $a_{ij} = \mathbf{d}'_i\mathbf{A}\mathbf{e}_j = \mathbf{e}'_j\mathbf{A}'\mathbf{d}_i$.*

(4) $\mathbf{I}_n = \Sigma_{i=1}^{n} \boldsymbol{\Delta}_{ii} = \Sigma_{i=1}^{n} \mathbf{e}_i\mathbf{e}'_i$.

(5) $\text{vec}(\mathbf{e}_i\mathbf{e}'_j) = \text{vec}(\boldsymbol{\Delta}_{ij}) = \mathbf{e}_i \times \mathbf{e}_j$.

(6) $\text{vec}(\mathbf{I}_n) = \text{vec}(\Sigma_{i=1}^{n} \boldsymbol{\Delta}_{ii}) = \text{vec}(\Sigma_{i=1}^{n} \mathbf{e}_i\mathbf{e}'_i) = \Sigma_{i=1}^{n} \text{vec}(\boldsymbol{\Delta}_{ii}) = \Sigma_{i=1}^{n} \text{vec}(\mathbf{e}_i\mathbf{e}'_i) = \Sigma_{i=1}^{n} \mathbf{e}_i \times \mathbf{e}_i$.

(7) $\boldsymbol{\Delta}_{ij} \times \boldsymbol{\Delta}_{mn} = \mathbf{e}_i\mathbf{e}'_j \times \mathbf{e}_m\mathbf{e}'_n = (\mathbf{e}_i \times \mathbf{e}_m)(\mathbf{e}'_j \times \mathbf{e}'_n) = [\text{vec}(\mathbf{e}_i\mathbf{e}'_m)]\,[\text{vec}(\mathbf{e}_j\mathbf{e}'_n)]' = [\text{vec}(\boldsymbol{\Delta}_{im})]\,[\text{vec}(\boldsymbol{\Delta}_{jn})]'$,
where each $\boldsymbol{\Delta}$ matrix is $p \times p$.

(8) *Let $\mathbf{A}$ be an $mp \times nq$ block matrix given by*

$$\mathbf{A} = \begin{bmatrix} \mathbf{A}_{11} & \mathbf{A}_{12} & \cdots & \mathbf{A}_{1q} \\ \mathbf{A}_{21} & \mathbf{A}_{22} & \cdots & \mathbf{A}_{2q} \\ \vdots & & & \\ \mathbf{A}_{p1} & \mathbf{A}_{p2} & \cdots & \mathbf{A}_{pq} \end{bmatrix},$$

where $\mathbf{A}_{ij}$ is an $m \times n$ matrix. Then

$$\mathbf{A} = \sum_{i=1}^{p} \sum_{j=1}^{q} (\mathbf{A}_{ij} \times \boldsymbol{\Delta}_{ij}),$$

where $\boldsymbol{\Delta}_{ij}$ is a $p \times q$ unit matrix.

(9) *Let $\mathbf{A} = [a_{ij}]$ be an $m \times n$ matrix, and let $\mathbf{B} = [b_{ij}]$ be a $p \times q$ matrix. Then*

$$\mathbf{A} \times \mathbf{B} = \sum_{i=1}^{m} \sum_{j=1}^{n} [\mathbf{A} \times (b_{ij}\, \boldsymbol{\Delta}_{ij})],$$

where $\boldsymbol{\Delta}_{ij}$ is a $p \times q$ unit matrix.

Theorem 9.3.2

Let $\boldsymbol{\Delta}_{ij}$ be an $m \times n$ unit matrix (that is, $\boldsymbol{\Delta}_{ij} = \mathbf{d}_i\mathbf{e}'_j$). Then

$$\sum_{i=1}^{m} \sum_{j=1}^{n} (\boldsymbol{\Delta}'_{ij} \times \boldsymbol{\Delta}_{ij}) = \mathbf{P}$$

is an $mn \times mn$ permutation matrix.

Proof: First we show that $\mathbf{P'P} = \mathbf{I}$. We get

$$\mathbf{P'P} = \left[\sum_{i=1}^{m}\sum_{j=1}^{n}(\boldsymbol{\Delta}'_{ij} \times \boldsymbol{\Delta}_{ij})\right]'\left[\sum_{p=1}^{m}\sum_{q=1}^{n}(\boldsymbol{\Delta}'_{pq} \times \boldsymbol{\Delta}_{pq})\right]$$

$$= \sum_i\sum_j\sum_p\sum_q [(\boldsymbol{\Delta}_{ij} \times \boldsymbol{\Delta}'_{ij})(\boldsymbol{\Delta}'_{pq} \times \boldsymbol{\Delta}_{pq})]$$

$$= \sum_i\sum_j\sum_p\sum_q [(\boldsymbol{\Delta}_{ij}\boldsymbol{\Delta}'_{pq}) \times (\boldsymbol{\Delta}'_{ij}\boldsymbol{\Delta}_{pq})]$$

$$= \sum_i\sum_j\sum_p\sum_q [(\mathbf{d}_i\mathbf{e}'_j\mathbf{e}_q\mathbf{d}'_p) \times (\mathbf{e}_j\mathbf{d}'_i\mathbf{d}_p\mathbf{e}'_q)]$$

$$= \sum_i\sum_j\sum_p\sum_q [\mathbf{d}_i(\delta_{jq})\mathbf{d}'_p] \times [\mathbf{e}_j(\delta_{ip})\mathbf{e}'_q]$$

$$= \sum_i\sum_j\sum_p\sum_q [(\mathbf{d}_i\mathbf{d}'_p) \times (\mathbf{e}_j\mathbf{e}'_q)\delta_{jq}\delta_{ip}],$$

since $\mathbf{e}'_j\mathbf{e}_q = \delta_{jq}$ and $\mathbf{d}'_i\mathbf{d}_p = \delta_{ip}$, where δ_{ns} are Kronecker deltas and are scalars. Hence

$$\mathbf{P'P} = \sum_i\sum_j [(\mathbf{d}_i\mathbf{d}'_i) \times (\mathbf{e}_j\mathbf{e}'_j)] = \left[\left(\sum_i \boldsymbol{\Delta}_{ii}\right) \times \left(\sum_j \boldsymbol{\Delta}_{jj}\right)\right] = \mathbf{I}_m \times \mathbf{I}_n = \mathbf{I}_{mn}.$$

This establishes that $\mathbf{P'P} = \mathbf{I}$. Next we show that elements of $\mathbf{P}$ consist only of 0 and 1. This is clearly the case, since $\boldsymbol{\Delta}_{ij} \times \boldsymbol{\Delta}'_{ij}$ has only one element equal to 1 and the remaining elements equal to 0. Also for each i and j the 1 occurs in a different place, so all elements in $\mathbf{P}$ are either 0 or 1. Finally we must show that each column of $\mathbf{P}$ contains exactly one value equal to 1 and the remaining values equal to zero, and that no two distinct columns can have the value 1 in the same row. Suppose this is not true; that is, suppose columns t and s have a value 1 in the same row; then $\mathbf{p}'_t\mathbf{p}_s \neq 0$ if $t \neq s$. Also suppose one column of $\mathbf{P}$, say the s-th, does not have exactly one element equal to 1; then $\mathbf{p}'_s\mathbf{p}_s \neq 1$. We have denoted the s-th columns of $\mathbf{P}$ by $\mathbf{p}_s$. Hence each column of $\mathbf{P}$ has exactly one element equal to 1 and each row of $\mathbf{P}$ has exactly one element equal to 1. The remaining elements are 0. This completes the proof of the theorem. ∎

Corollary 9.3.2

Let $\mathbf{\Delta}_{ij} = \mathbf{e}_i\mathbf{e}_j'$ *be an* $n \times n$ *unit matrix. Then*

$$\sum_{i=1}^{n}\sum_{j=1}^{n} (\mathbf{\Delta}_{ij} \times \mathbf{\Delta}_{ij}) = [\text{vec}(\mathbf{I}_n)]\,[\text{vec}(\mathbf{I}_n)]'.$$

The permutation matrix in Theorem 9.3.2 is useful in manipulating direct products and the vector of a matrix. We define and briefly discuss this particular permutation matrix, which we call a commutation matrix.

Definition 9.3.1

Commutation Matrix. *Let* $\mathbf{\Delta}_{ij} = \mathbf{d}_i\mathbf{e}_j'$ *be an* $m \times n$ *unit matrix (*$\mathbf{d}_i$ *is an* $m \times 1$ *unit vector and* $\mathbf{e}_j$ *is an* $n \times 1$ *unit vector). The* $mn \times mn$ *matrix* $\mathbf{K}_{mn}$ *is defined to be an* $mn \times mn$ *commutation matrix, where*

$$\mathbf{K}_{mn} = \sum_{i=1}^{m}\sum_{j=1}^{n} (\mathbf{\Delta}_{ij}' \times \mathbf{\Delta}_{ij}).$$

Some results that follow immediately from the definition are given in the next theorem.

Theorem 9.3.3

Let $\mathbf{K}_{mn}$ *be an* $mn \times mn$ *commutation matrix (*$\mathbf{e}_j$ *is an* $n \times 1$ *vector;* $\mathbf{d}_i$ *is an* $m \times 1$ *vector).*

(1) $\mathbf{K}_{mn}' = \mathbf{K}_{nm}$.

(2) $\mathbf{K}_{1n} = \mathbf{K}_{n1} = \mathbf{I}_n$.

(3) $\mathbf{K}_{mn} = \sum_{j=1}^{n} (\mathbf{e}_j \times \mathbf{I}_m \times \mathbf{e}_j') = \sum_{i=1}^{m} (\mathbf{d}_i' \times \mathbf{I}_n \times \mathbf{d}_i)$.

Proof:

$$(1)\ \mathbf{K}_{mn}' = \left(\sum_{i=1}^{m}\sum_{j=1}^{n} \mathbf{\Delta}_{ij}' \times \mathbf{\Delta}_{ij}\right)' = \sum_{i=1}^{m}\sum_{j=1}^{n} (\mathbf{\Delta}_{ij}' \times \mathbf{\Delta}_{ij})'$$

$$= \sum_{i=1}^{m}\sum_{j=1}^{n} (\mathbf{\Delta}_{ij} \times \mathbf{\Delta}_{ij}') = \sum_{i=1}^{m}\sum_{j=1}^{n} (\mathbf{\Delta}_{ji}' \times \mathbf{\Delta}_{ji}) = \mathbf{K}_{nm}.$$

(2) $\mathbf{K}_{1n} = \sum_{j=1}^{n} (\boldsymbol{\Delta}'_{1j} \times \boldsymbol{\Delta}_{1j}) = \sum_{j=1}^{n} (\mathbf{e}_j \times \mathbf{e}'_j) = \mathbf{I}_n$

and by (1) we get $\mathbf{K}'_{n1} = \mathbf{K}_{1n} = \mathbf{I}_n$, so $\mathbf{K}_{n1} = \mathbf{I}_n$.

(3)
$$\begin{aligned}\mathbf{K}_{mn} &= \sum_{i=1}^{m} \sum_{j=1}^{n} [(\mathbf{d}_i\mathbf{e}'_j)' \times (\mathbf{d}_i\mathbf{e}'_j)] \\ &= \sum_{i=1}^{m} \sum_{j=1}^{n} [(\mathbf{e}_j\mathbf{d}'_i) \times (\mathbf{d}_i\mathbf{e}'_j)] \\ &= \sum_{i=1}^{m} \sum_{j=1}^{n} [(\mathbf{e}_j \times \mathbf{d}'_i) \times (\mathbf{d}_i \times \mathbf{e}'_j)] \\ &= \sum_{j=1}^{n} \left(\mathbf{e}_j \times \left[\sum_{i=1}^{m} \mathbf{d}'_i \times \mathbf{d}_i \right] \times \mathbf{e}'_j \right) \\ &= \sum_{j=1}^{n} (\mathbf{e}_j \times \mathbf{I}_m \times \mathbf{e}'_j).\end{aligned}$$

The proof of the other result is similar. ∎

Note: By Theorem 9.3.2 it follows that $\mathbf{K}_{mn}$ is a *special* $mn \times mn$ permutation matrix.

First we shall show how $\mathbf{K}_{mn}$ can transform a vector of a matrix $\mathbf{A}$ into the vector of its transpose, $\mathbf{A}'$. Then we shall discuss various properties of $\mathbf{K}_{mn}$.

Theorem 9.3.4

Let $\mathbf{A}$ *be an* $m \times n$ *matrix with (ij)-th element* a_{ij}. *Then*

$$\text{vec}(\mathbf{A}) = \sum_{i=1}^{m} \sum_{j=1}^{n} (\boldsymbol{\Delta}_{ij} \times \boldsymbol{\Delta}'_{ij})\text{vec}(\mathbf{A}') = \mathbf{K}'_{mn}\, \text{vec}(\mathbf{A}').$$

Proof: By (3) of Theorem 9.3.1 we can write

$$\begin{aligned}\mathbf{A} &= \sum_j \sum_i a_{ij}\boldsymbol{\Delta}_{ij} = \sum_j \sum_i (\mathbf{d}'_i\mathbf{A}\,\mathbf{e}_j)\mathbf{d}_i\mathbf{e}'_j = \sum_j \sum_i \mathbf{d}_i(\mathbf{d}'_i\mathbf{A}\,\mathbf{e}_j)\mathbf{e}'_j \\ &= \sum_j \sum_i \mathbf{d}_i(\mathbf{e}'_j\mathbf{A}'\mathbf{d}_i)\mathbf{e}'_j.\end{aligned}$$

Note that $a_{ij} = \mathbf{d}_i'\mathbf{A}\mathbf{e}_j = \mathbf{e}_j'\mathbf{A}'\mathbf{d}_i$ is a scalar and hence can be inserted in $\mathbf{d}_i\mathbf{e}_j'$ as $\mathbf{d}_i(\mathbf{d}_i'\mathbf{A}\mathbf{e}_j)\mathbf{e}_j' = (\mathbf{d}_i'\mathbf{A}\mathbf{e}_j)\mathbf{d}_i\mathbf{e}_j' = (\mathbf{e}_j'\mathbf{A}'\mathbf{d}_i)\mathbf{d}_i\mathbf{e}_j' = \mathbf{d}_i(\mathbf{e}_j'\mathbf{A}'\mathbf{d}_i)\mathbf{e}_j'$. Thus

$$\begin{aligned}
\text{vec}(\mathbf{A}) &= \text{vec}\left(\sum_j \sum_i a_{ij}\, \boldsymbol{\Delta}_{ij}\right) = \text{vec}\left(\sum_j \sum_i [\mathbf{d}_i'\, \mathbf{A}\mathbf{e}_j]\mathbf{d}_i\, \mathbf{e}_j'\right) \\
&= \text{vec}\left(\sum_j \sum_i \mathbf{d}_i[\mathbf{d}_i'\, \mathbf{A}\mathbf{e}_j]\mathbf{e}_j'\right) = \text{vec}\left(\sum_j \sum_i \mathbf{d}_i[\mathbf{e}_j'\, \mathbf{A}'\mathbf{d}_i]\mathbf{e}_j'\right) \\
&= \text{vec}\left(\sum_j \sum_i (\mathbf{d}_i\mathbf{e}_j')\mathbf{A}'(\mathbf{d}_i\mathbf{e}_j')\right) = \text{vec}\left(\sum_j \sum_i [\boldsymbol{\Delta}_{ij}\, \mathbf{A}'\, \boldsymbol{\Delta}_{ij}]\right) \\
&= \sum_j \sum_i \text{vec}(\boldsymbol{\Delta}_{ij}\, \mathbf{A}'\, \boldsymbol{\Delta}_{ij}) = \sum_j \sum_i (\boldsymbol{\Delta}_{ij} \times \boldsymbol{\Delta}_{ij}')\text{vec}(\mathbf{A}') \\
&= \mathbf{K}_{mn}'\, \text{vec}(\mathbf{A}')
\end{aligned}$$

and this completes the proof. The next to last equality is a result of Theorem 9.2.2. ▮

Corollary 9.3.4

For any $m \times n$ matrix $\mathbf{A}$ there exists a permutation matrix $\mathbf{K}_{mn}$, which depends only on the size of $\mathbf{A}$ (that is, on m and n) such that $\text{vec}(\mathbf{A}) = \mathbf{K}_{mn}'\, \text{vec}(\mathbf{A}') = \mathbf{K}_{nm}\, \text{vec}(\mathbf{A}')$. *Also* $\mathbf{K}_{mn}\, \text{vec}(\mathbf{A}) = \text{vec}(\mathbf{A}')$.

The definition of $\mathbf{K}_{mn}$ may appear to be not very revealing in its structure, so we state a theorem that is actually for the purpose of helping exhibit a $\mathbf{K}_{mn}$ matrix for any value of m and n.

Theorem 9.3.5

An $mn \times mn$ commutation matrix $\mathbf{K}_{mn}$ can be written as

$$\mathbf{K}_{mn} = \begin{bmatrix} \mathbf{C}_{11} & \mathbf{C}_{12} & \cdots & \mathbf{C}_{1n} \\ \mathbf{C}_{21} & \mathbf{C}_{22} & \cdots & \mathbf{C}_{2n} \\ \vdots & \vdots & & \vdots \\ \mathbf{C}_{m1} & \mathbf{C}_{m2} & \cdots & \mathbf{C}_{mn} \end{bmatrix},$$

where $\mathbf{C}_{pq}$ is an $n \times m$ matrix with every element 0 except the (qp)-th element, which is $+1$.

Proof: The proof follows immediately from the fact that

$$\mathbf{K}_{mn} = \sum_{i=1}^{m} \sum_{j=1}^{n} (\mathbf{\Delta}'_{ij} \times \mathbf{\Delta}_{ij}). \quad \blacksquare$$

Theorem 9.3.6

Let **A** *be an* $m \times p$ *matrix and* **B** *be an* $n \times q$ *matrix. Then*

$$\mathbf{K}_{mn}(\mathbf{A} \times \mathbf{B})\mathbf{K}'_{pq} = \mathbf{B} \times \mathbf{A}.$$

Proof: Let **C** be any $q \times p$ matrix. Then

$\mathbf{K}_{mn}(\mathbf{A} \times \mathbf{B})\mathbf{K}_{qp}[\text{vec}(\mathbf{C})] = \mathbf{K}_{mn}(\mathbf{A} \times \mathbf{B})\text{vec}(\mathbf{C}') = \mathbf{Q}$ (say) by Corollary 9.3.4.

Use Theorem 9.2.2 and Theorem 9.3.4 to get

$$\mathbf{Q} = \mathbf{K}_{mn}\,\text{vec}(\mathbf{AC}'\mathbf{B}') = \text{vec}[(\mathbf{AC}'\mathbf{B}')'] = \text{vec}(\mathbf{BCA}') = (\mathbf{B} \times \mathbf{A})[\text{vec}(\mathbf{C})].$$

So for any **C** we get

$$\mathbf{K}_{mn}(\mathbf{A} \times \mathbf{B})\mathbf{K}_{qp}[\text{vec}(\mathbf{C})] = (\mathbf{B} \times \mathbf{A})[\text{vec}(\mathbf{C})],$$

so the result follows. $\blacksquare$

Note: We can remember which particular commutation matrices to use in Theorem 9.3.6 by the following: **A** has m rows and **B** has n rows, so the matrix on the left is $\mathbf{K}_{mn}$; **A** has p columns and **B** has q columns, so the matrix on the right is $\mathbf{K}'_{pq}$.

Corollary 9.3.6.1

In Theorem 9.3.6 *we can write*

$$\mathbf{K}_{mn}(\mathbf{A} \times \mathbf{B}) = (\mathbf{B} \times \mathbf{A})\mathbf{K}'_{qp}.$$

Corollary 9.3.6.2

Let **A** *be an* $m \times m$ *(square) matrix and let* **B** *be an* $n \times n$ *(square) matrix. Then*

$$\mathbf{K}_{mn}(\mathbf{A} \times \mathbf{B})\mathbf{K}'_{mn} = \mathbf{B} \times \mathbf{A}.$$

Corollary 9.3.6.3

Let $\mathbf{A}$ *be an* $m \times n$ *matrix, let* $\mathbf{x}$ *be a* $p \times 1$ *vector, and let* $\mathbf{y}$ *be a* $q \times 1$ *vector. Then*

(1) $\mathbf{K}_{mp}(\mathbf{A} \times \mathbf{x}) = \mathbf{x} \times \mathbf{A}$,
(2) $(\mathbf{x}' \times \mathbf{A})\mathbf{K}'_{pn} = \mathbf{A} \times \mathbf{x}'$,
(3) $\mathbf{K}_{pq}(\mathbf{x} \times \mathbf{y}) = \mathbf{y} \times \mathbf{x}$.

Corollary 9.3.6.4

Let $\mathbf{A}$ *be an* $m \times p$ *matrix, let* $\mathbf{B}$ *be an* $n \times q$ *matrix, let* $\mathbf{x}$ *be an* $m \times 1$ *vector, let* $\mathbf{y}$ *be an* $n \times 1$ *vector, and let* $\mathbf{z}$ *be a* $k \times 1$ *vector. Then*

(1) $\mathbf{y} \times \mathbf{A} \times \mathbf{z}' = \mathbf{K}_{mn}(\mathbf{A} \times \mathbf{yz}')$,
(2) $\mathbf{z}' \times \mathbf{B} \times \mathbf{x} = \mathbf{K}_{mn}(\mathbf{xz}' \times \mathbf{B})$.

Proof: The proof follows from (2) of Theorem 9.3.3. ∎

In the next theorem we will be interested in commutation matrices $\mathbf{K}_{mn}$, where $m = pq$ (that is, m is equal to p times q), where p and q (and hence m) are positive integers. To avoid confusion, we shall write these as $\mathbf{K}_{mn} = \mathbf{K}_{pq,\,n}$; that is, when only two letters (say m and n) occur in the subscript of $\mathbf{K}$, it will mean an $mn \times mn$ commutation matrix. When it is written as $\mathbf{K}_{pq,\,n}$, it will mean that the first subscript is p times q and the second subscript is n.

Theorem 9.3.7

Let m, p, *and* q *be any positive integers. Then*

$$\mathbf{K}_{mp,\,q}\,\mathbf{K}_{pq,\,m}\,\mathbf{K}_{qm,\,p} = \mathbf{I}_{mpq}.$$

Proof: By (3) of Theorem 9.3.3 we get ($\mathbf{e}_j$ is a $q \times 1$ vector)

$$\mathbf{K}_{mp,\,q} = \sum_{j=1}^{q} \mathbf{e}_j \times \mathbf{I}_{mp} \times \mathbf{e}'_j = \sum_{j=1}^{q} \mathbf{e}_j \times (\mathbf{I}_m \times \mathbf{I}_p) \times \mathbf{e}'_j$$

$$= \sum_{j=1}^{q} [(\mathbf{e}_j \times \mathbf{I}_m) \times (\mathbf{I}_p \times \mathbf{e}'_j)] = \sum_{j=1}^{q} (\mathbf{A}_j \times \mathbf{B}_j),$$

where we let $\mathbf{A}_j = \mathbf{e}_j \times \mathbf{I}_m$ ($\mathbf{A}_j$ is an $mq \times m$ matrix) and $\mathbf{B}_j = \mathbf{I}_p \times \mathbf{e}'_j$ ($\mathbf{B}_j$ is a $p \times pq$ matrix). By Theorem 9.3.6 we get $\mathbf{A}_j \times \mathbf{B}_j = \mathbf{K}_{p,\,mq} \times$

$(\mathbf{B}_j \times \mathbf{A}_j)\mathbf{K}'_{pq,\,m}$. Thus

$$\mathbf{K}_{mp,\,q} = \sum_{j=1}^{q} \mathbf{K}_{p,\,mq}\,[(\mathbf{I}_p \times \mathbf{e}'_j) \times (\mathbf{e}_j \times \mathbf{I}_m)]\mathbf{K}'_{pq,\,m}$$

$$= \mathbf{K}_{p,\,mq}\left[\mathbf{I}_p \times \sum_{j=1}^{q} (\mathbf{e}'_j \times \mathbf{e}_j) \times \mathbf{I}_m\right]\mathbf{K}'_{pq,\,m}$$

$$= \mathbf{K}_{p,\,mq}[\mathbf{I}_p \times (\mathbf{I}_q) \times \mathbf{I}_m]\mathbf{K}'_{pq,\,m}.$$

So

$$\mathbf{K}_{mp,\,q} = \mathbf{K}_{p,\,mq}\,\mathbf{I}_{pqm}\,\mathbf{K}'_{pq,\,m} = \mathbf{K}_{p,\,mq}\,\mathbf{K}'_{pq,\,m}.$$

Multiply both sides on the right by $\mathbf{K}_{pq,\,m}\,\mathbf{K}_{mq,\,p}$ and the theorem is proved. Note that $\mathbf{K}_{pq,\,mn} = \mathbf{K}_{qp,\,mn} = \mathbf{K}_{pq,\,nm}$, and so forth. ▮

Corollary 9.3.7

Let m, p, and q be any positive integers. Then

(1) $\mathbf{K}_{mp,\,q}\,\mathbf{K}_{pq,\,m} = \mathbf{K}_{p,\,mq}$,
(2) $\mathbf{K}_{m,\,qp}\,\mathbf{K}_{q,\,mp} = \mathbf{K}_{mq,\,p}$,
(3) $\mathbf{K}_{m,\,pq}\,\mathbf{K}_{q,\,mp} = \mathbf{K}_{q,\,mp}\,\mathbf{K}_{m,\,pq}$,
(4) $\mathbf{K}_{mp,\,q}\,\mathbf{K}_{pq,\,m} = \mathbf{K}_{pq,\,m}\,\mathbf{K}_{mp,\,q}$,
(5) $\mathbf{K}_{m,\,pq}\,\mathbf{K}_{mp,\,q} = \mathbf{K}_{mp,\,q}\,\mathbf{K}_{m,\,pq}$.

Note: The results of (3), (4), and (5) extend to any permutation of the letters *mpq*. An easy way to remember (3) is as follows: Write down any two permutations of the three letters (integers) *mpq* (for example, *mpq*; *pqm*). In each permutation put a comma after the first letter (for example, *m*, *pq*; *p*, *qm*). Use these as subscripts on **K** and the resulting two matrices commute (for example, $\mathbf{K}_{m,\,pq}\,\mathbf{K}_{p,\,qm} = \mathbf{K}_{p,\,qm} \times \mathbf{K}_{m,\,pq}$). For (4) follow the same procedure, except place the comma after the first two letters (between the second and third letters) in each permutation. In (5) write down the same permutation of the letters (integers) *mpq* two times (for example, *pqm*; *pqm*). In one permutation place a comma after the first letter; in the second permutation place a comma after the second letter (for example, *p*, *qm*; *pq*, *m*). Then use these as subscripts on **K** and the resulting two matrices commute (for example, $\mathbf{K}_{p,\,qm}\,\mathbf{K}_{pq,\,m} = \mathbf{K}_{pq,\,m}\,\mathbf{K}_{p,\,qm}$).

We now state a theorem about commutation matrices and omit the proof. The proof can be found in [M–2].

Theorem 9.3.8

Consider the $mn \times mn$ commutation matrix $\mathbf{K}_{mn}$.

(1) $\text{tr}(\mathbf{K}_{mn}) = 1 +$ *greatest common divisor of* $(m - 1, n - 1)$, *where* $m > 1, n > 1$.
(2) $\text{tr}(\mathbf{K}_{m1}) = \text{tr}(\mathbf{K}_{1m}) = m$.
(3) $\text{tr}(\mathbf{K}_{nn}) = n$.
(4) *The characteristic roots of* $\mathbf{K}_{nn}$ *are* $+1$ *with multiplicity* $n(n + 1)/2$ *and* -1 *with multiplicity* $n(n - 1)/2$.
(5) $\det(\mathbf{K}_{mn}) = (-1)^{m(m-1)n(n-1)/4}$.
(6) *Let* $\mathbf{A}$ *and* $\mathbf{B}$ *be* $m \times n$ *matrices. Then*

$$\text{tr}[\mathbf{K}_{mn}(\mathbf{B} \times \mathbf{A}')] = \text{tr}(\mathbf{A}'\mathbf{B}) = [\text{vec}(\mathbf{A}')]'\mathbf{K}_{mn}[\text{vec}(\mathbf{B})].$$

Next we state a theorem that gives permutation matrices that will cyclic permute the Kronecker product of three matrices. This is an extension of Theorem 9.3.6.

Theorem 9.3.9

Let $\mathbf{A}$, $\mathbf{B}$, *and* $\mathbf{C}$ *be* $m \times n$, $p \times q$, *and* $r \times s$ *matrices, respectively. Then*

$$\begin{aligned}(\mathbf{A} \times \mathbf{B} \times \mathbf{C}) &= \mathbf{K}_{r,\,mp}(\mathbf{C} \times \mathbf{A} \times \mathbf{B})\mathbf{K}'_{s,\,nq}\\ &= \mathbf{K}_{rp,\,m}(\mathbf{B} \times \mathbf{C} \times \mathbf{A})\mathbf{K}'_{qs,\,n}.\end{aligned}$$

Proof: Let $\mathbf{D}$ denote $\mathbf{A} \times \mathbf{B}$. Then $\mathbf{D}$ is an $mp \times nq$ matrix. Hence by Theorem 9.3.6 we get

$$\mathbf{A} \times \mathbf{B} \times \mathbf{C} = \mathbf{D} \times \mathbf{C} = \mathbf{K}_{r,\,mp}(\mathbf{C} \times \mathbf{D})\mathbf{K}'_{s,\,nq} = \mathbf{K}_{r,\,mp}(\mathbf{C} \times \mathbf{A} \times \mathbf{B})\mathbf{K}'_{s,nq}.$$

The remaining part is proved similarly by letting $\mathbf{B} \times \mathbf{C} = \mathbf{F}$. ∎

Theorem 9.3.10

Let $\mathbf{A}$, $\mathbf{B}$, $\mathbf{C}$, *and* $\mathbf{D}$ *be* $m \times n$, $p \times q$, $n \times r$, *and* $q \times s$ *matrices. Then*

$$\begin{aligned}\mathbf{K}_{mp}(\mathbf{AC} \times \mathbf{BD}) &= \mathbf{K}_{mp}(\mathbf{A} \times \mathbf{B})(\mathbf{C} \times \mathbf{D}) = (\mathbf{B} \times \mathbf{A})\mathbf{K}'_{qn}(\mathbf{C} \times \mathbf{D})\\ &= (\mathbf{B} \times \mathbf{A})(\mathbf{D} \times \mathbf{C})\mathbf{K}'_{sr} = (\mathbf{BD} \times \mathbf{AC})\mathbf{K}'_{sr}\\ &= (\mathbf{BD} \times \mathbf{I}_m)(\mathbf{I}_s \times \mathbf{AC})\mathbf{K}'_{sr} = (\mathbf{BD} \times \mathbf{I}_m)\mathbf{K}'_{sm}(\mathbf{AC} \times \mathbf{I}_s)\\ &= (\mathbf{I}_p \times \mathbf{AC})\mathbf{K}_{rp}(\mathbf{I}_r \times \mathbf{BD}).\end{aligned}$$

Proof: The proof is obtained by repeated applications of Theorem 9.3.6. ∎

In Chapter 10 we shall show how some of the theorems in this section can be used to obtain expectations of Wishart matrices and of products of quadratic forms of normal random variables.

Problems

1. Prove Theorem 9.1.5.
2. Show that tr $(a\mathbf{I}) = na$ where $\mathbf{I}$ is the $n \times n$ identity matrix.
3. Prove Theorem 9.1.9.
4. Prove Theorem 9.1.10.
5. If $\mathbf{A}$, $\mathbf{B}$, and $\mathbf{AB}$ are symmetric $n \times n$ matrices and the characteristic roots of $\mathbf{A}$ are $a_1, a_2, \ldots, a_n$ and of $\mathbf{B}$ are $b_1, b_2, \ldots, b_n$, show that

$$\text{tr}\,(\mathbf{AB}) = \sum_{i=1}^{n} a_{j_i} b_i,$$

where $a_{j1}, a_{j_2}, \ldots, a_{j_n}$ is some ordering of $a_1, a_2, \ldots, a_n$.

6. Prove Theorem 9.1.29.
7. Prove Theorem 9.1.13.
8. If $\mathbf{x}_i$ is an $n \times 1$ vector for each $i = 1, 2, \ldots, k$, and $\mathbf{A}$ is an $n \times n$ symmetric matrix, show that

$$\text{tr}\left[\mathbf{A}\sum_{i=1}^{k} \mathbf{x}_i\mathbf{x}_i'\right] = \sum_{i=1}^{k} \mathbf{x}_i'\mathbf{A}\mathbf{x}_i.$$

9. If $\mathbf{A}$ is an $n \times n$ symmetric idempotent matrix and $\mathbf{V}$ is an $n \times n$ positive definite matrix, show that

$$\text{rank}\,(\mathbf{AV}^{-1}\mathbf{A}) = \text{tr}(\mathbf{A}).$$

10. Prove Theorem 9.1.14.
11. Prove Theorem 9.1.16.

12. If $\mathbf{A}$ is defined below, find a 4×4 matrix $\mathbf{B}$ such that $\operatorname{tr}(\mathbf{AB}) = \operatorname{rank}(\mathbf{A})$.

$$\mathbf{A} = \begin{bmatrix} 3 & 1 & -2 & 0 \\ 1 & 2 & 3 & -1 \\ -2 & 1 & 3 & 4 \\ 6 & 2 & -2 & -2 \end{bmatrix}.$$

13. Let $\mathbf{A}$ be an $n \times n$ (real) matrix with characteristic roots $\lambda_1, \lambda_2, \ldots, \lambda_n$, where any λ_t may not be a real number. Denote λ_t by $x_t + iy_t$ where x_t and y_t are real numbers and where $i = \sqrt{-1}$. Show that:

 (a) $\sum_{t=1}^{n} y_t = 0$.

 (b) $\sum_{t=1}^{n} x_t y_t = 0$.

 (c) $\operatorname{tr}(\mathbf{A}^2) = \sum_{t=1}^{n} x_t^2 - \sum_{t=1}^{n} y_t^2$.

14. Let $\mathbf{A}$ and $\mathbf{B}$ be two $n \times m$ matrices such that $\mathbf{AB}' = \mathbf{0}$. Is $\mathbf{B}'\mathbf{A}$ necessarily equal to zero? Show that $\operatorname{tr}(\mathbf{B}'\mathbf{A}) = 0$.
15. If $\mathbf{A}$ and $\mathbf{B}$ are $n \times n$ matrices such that $\mathbf{AB} = \mathbf{0}$, show that

$$\operatorname{tr}[(\mathbf{A} + \mathbf{B})^3] = \operatorname{tr}(\mathbf{A}^3) + \operatorname{tr}(\mathbf{B}^3).$$

16. Let $\mathbf{X}$ be an $n \times p$ matrix of rank p. Partition $\mathbf{X}$ such that $\mathbf{X} = [\mathbf{X}_1, \mathbf{X}_2]$, where $\mathbf{X}_1$ has size $n \times p_1$ and $\mathbf{X}_2$ has size $n \times p_2$ where $p_1 + p_2 = p$. Show that the rank of $\mathbf{B}$ is p_2 where $\mathbf{B}$ is defined by

$$\mathbf{B} = \mathbf{X}(\mathbf{X}'\mathbf{X})^{-1}\mathbf{X}' - \mathbf{X}_1'(\mathbf{X}_1\mathbf{X}_1)^{-1}\mathbf{X}_1'.$$

17. If $\mathbf{A}$ and $\mathbf{B}$ are $n \times n$ matrices, show that $\operatorname{tr}[(\mathbf{AB} - \mathbf{BA})(\mathbf{AB} + \mathbf{BA})] = 0$.
18. Let $\mathbf{A}$ be an orthogonal $n \times n$ matrix such that $\det(\mathbf{A} + \mathbf{I}) \neq 0$. Show that $\operatorname{tr}[2(\mathbf{A} + \mathbf{I})^{-1} - \mathbf{I}] = 0$.
19. If $\mathbf{A}$ and $\mathbf{A} + \mathbf{I}$ are nonsingular $n \times n$ matrices, show that

$$\operatorname{tr}[(\mathbf{A} + \mathbf{I})^{-1}] + \operatorname{tr}[(\mathbf{A}^{-1} + \mathbf{I})^{-1}] = n.$$

20. Let $h(x) = \sum_{i=0}^{m} a_i x^i$ be a polynomial. Define $\mathbf{h}(\mathbf{A})$ by

$$\mathbf{h}(\mathbf{A}) = \sum_{i=0}^{m} a_i \mathbf{A}^i,$$

where $\mathbf{A}^0 = \mathbf{I}$ and a_i are scalars. If $\lambda_1, \lambda_2, \ldots, \lambda_n$ are the characteristic roots of the $n \times n$ matrix $\mathbf{A}$, show that

$$\text{tr}\,[\mathbf{h}(\mathbf{A})] = \sum_{t=1}^{n} h(\lambda_t).$$

21. Let $\mathbf{A}$ be any $n \times n$ matrix of rank k. Show that there exists a nonsingular $n \times n$ matrix $\mathbf{B}$ such that $\text{tr}\,(\mathbf{BA}) = k$.
22. Let $\mathbf{A}$ be an $n \times n$ symmetric matrix. Show that $\mathbf{A}$ is a positive definite matrix if and only if $\text{tr}\,(\mathbf{AB}) > 0$ for every non-negative matrix $\mathbf{B}$ of rank 1.
23. If $\mathbf{A}$ is an $n \times n$ matrix and $\mathbf{A}'\mathbf{A} = \mathbf{A}^2$, show that $\text{tr}\,[(\mathbf{A}' - \mathbf{A})(\mathbf{A} - \mathbf{A}')] = 0$.
24. Use Prob. 23 to show that $\mathbf{A}'\mathbf{A} = \mathbf{A}^2$, if and only if $\mathbf{A}$ is symmetric.
25. Let $\mathbf{V}$, $\mathbf{A}$, $\mathbf{B}$ be non-negative $n \times n$ matrices. Show that $\mathbf{AVB} = \mathbf{0}$ if and only if $\text{tr}\,(\mathbf{VAVB}) = 0$ but that $\text{tr}\,(\mathbf{AVB}) = 0$ does not imply that $\mathbf{AVB} = \mathbf{0}$.
26. Let $\mathbf{A}$ and $\mathbf{B}'$ be $m \times n$ matrices such that $\mathbf{AB} = \mathbf{0}$. Show that $\text{tr}\,(\mathbf{BCA}) = 0$ for any $m \times m$ matrix $\mathbf{C}$.
27. If $\mathbf{A}$ is an $n \times n$ matrix, show that $\text{tr}\,(\mathbf{A}^k) = 0$ for $k = 1, 2, 3, \ldots,$ if and only if $\mathbf{A}^t = \mathbf{0}$ for some positive integer t.
28. If $\mathbf{A}$ is a symmetric $n \times n$ matrix and $\mathbf{B}$ is an $n \times n$ skew-symmetric matrix, show that $\text{tr}\,(\mathbf{AB}) = 0$.
29. If $\mathbf{A}$ is an $n \times n$ matrix show that $\text{tr}\,(\mathbf{A}^2) \leq \text{tr}\,(\mathbf{AA}')$.
30. Let $\mathbf{A}$ and $\mathbf{B}$ be $m \times n$ matrices. Show that $\text{tr}\,(\mathbf{A}'\mathbf{B}) = \text{tr}\,(\mathbf{AB}')$.
31. If $\mathbf{A}$ is an $n \times n$ matrix, show that $\text{tr}[\{\text{vec}(\mathbf{A})\}\{\text{vec}(\mathbf{I})\}'] = \text{tr}(\mathbf{A})$.
32. If $\mathbf{A}$ is an $m \times n$ matrix with columns $\mathbf{a}_i$, show that $\text{vec}(\mathbf{A}) = \Sigma_{i=1}^{n} (\mathbf{a}_i \times \mathbf{e}_i) = \Sigma_{i=1}^{n} \text{vec}(\mathbf{a}_i\mathbf{e}_i') = \text{vec}[\Sigma \mathbf{a}_i\mathbf{e}_i']$, where $\mathbf{e}_i$ is the i-th column of the $n \times n$ identity matrix.
33. In Theorem 9.2.2, if $\mathbf{A} = \mathbf{a}'$ and $\mathbf{C} = \mathbf{c}'$, where $\mathbf{a}$ and $\mathbf{c}$ are $q \times 1$ and $s \times 1$ vectors, respectively, and if $\mathbf{B}$ is a $q \times s$ matrix, show that $(\mathbf{a}' \times \mathbf{c}')\text{vec}(\mathbf{B}) = \mathbf{a}'\mathbf{Bc}$.
34. Prove Theorem 9.2.1.

35. If **A** and **B** are given by

$$\mathbf{A} = \begin{bmatrix} 3 & 2 \\ 0 & 1 \\ 1 & -1 \end{bmatrix}; \mathbf{B} = \begin{bmatrix} 0 & 1 & -2 \\ 1 & 3 & 1 \end{bmatrix},$$

show that $[\mathrm{vec}(\mathbf{A}')]'[\mathrm{vec}(\mathbf{B})] = \mathrm{tr}(\mathbf{AB})$.

36. Prove Corollary 9.2.4.

37. If

$$\mathbf{A} = \begin{bmatrix} 1 & 0 \\ 2 & 3 \end{bmatrix}; \mathbf{B} = \begin{bmatrix} 1 \\ 1 \end{bmatrix},$$

find the commutation matrices $\mathbf{K}_{22}$ and $\mathbf{K}_{21}$ such that $\mathbf{K}_{22}(\mathbf{A} \times \mathbf{B})\mathbf{K}'_{21} = \mathbf{B} \times \mathbf{A}$.

38. Prove Corollary 9.3.6.3.

39. If $\mathbf{P}$ is an $n \times n$ orthogonal matrix, show that $[\mathrm{vec}(\mathbf{P})]'[\mathrm{vec}(\mathbf{P})] = n$.

40. If $\mathbf{A}$ is an $n \times n$ idempotent matrix of rank k, show that $[\mathrm{vec}(\mathbf{A}')]'[\mathrm{vec}(\mathbf{A})] = k$.

41. If **A** is given by

$$\mathbf{A} = \begin{bmatrix} 2 & 1 & 0 \\ 1 & 3 & 1 \end{bmatrix},$$

find the commutation matrix $\mathbf{K}_{23}$ such that $\mathbf{K}_{23}\ \mathrm{vec}(\mathbf{A}) = \mathrm{vec}(\mathbf{A}')$.

Integration and Differentiation

10

10.1 Introduction

This chapter demonstrates how matrices, vectors, and determinants can be used in transforming random variables, in evaluating multiple integrals, and in differentiation. It also shows how matrices and vectors are used in the multivariate normal density—one of the most frequently used densities in statistics.

10.2 Transformation of Random Variables

One of the basic quantities in mathematical statistics is a *joint density function* of n continuous *random variables* $x_1, x_2, \ldots, x_n$. Any function f can serve as a density function if it satisfies the following two conditions.

$$\begin{aligned} &(1)\quad f(x_1, x_2, \ldots, x_n) \geq 0; \quad -\infty < x_i < \infty; \quad i = 1, 2, \ldots, n, \\ &(2)\quad \int_{-\infty}^{\infty} \int_{-\infty}^{\infty} \cdots \int_{-\infty}^{\infty} f(x_1, x_2, \ldots, x_n)\, dx_1\, dx_2 \cdots dx_n = 1. \end{aligned} \tag{10.2.1}$$

For example, suppose $n = 2$ and f is defined by

$$f(x_1, x_2) = \begin{cases} e^{-(x_1+x_2)}, & \text{for } x_1 > 0, x_2 > 0, \\ 0, & \text{elsewhere.} \end{cases}$$

Then, clearly, conditions (1) and (2) in Eq. (10.2.1) are satisfied, and f is a density function.

Many of the density functions that are important for applications in statistics are defined only in a portion of the n-dimensional space. When this is the case, they can be defined to be zero at the remaining points. For example, sometimes the density above would be written

$$f(x_1, x_2) = e^{-(x_1+x_2)} \qquad x_1 > 0; \quad x_2 > 0.$$

The fact that f has been defined only for a portion of the x_1x_2 space will imply that it is zero for all remaining points.

Let

$$f(x_1, x_2, \dots, x_n); \qquad a_i < x_i < b_i; \quad i = 1, 2, \dots, n,$$

(where any a_i may be $-\infty$ and any b_i may be $+\infty$) be a density function of n continuous random variables $x_1, x_2, \dots, x_n$, and suppose that f is bounded (and positive) in its domain of definition D and continuous except at most for a finite number of points; that is, the domain D is given by

$$D = \{(x_1, x_2, \dots, x_n)\colon \ a_i < x_i < b_i; \quad i = 1, 2, \dots, n\}.$$

We can state this in an alternate way. Let f be a density function of n continuous variables where f is defined in the set E_n where

$$E_n = \{(x_1, \dots, x_n) : -\infty < x_i < \infty; i = 1, 2, \dots, n\}.$$

Let D, which is defined above, be the set of points in E_n such that $f(x_1, \dots, x_n) > 0$. The complement of D with respect to E_n is the set of points such that $f(x_1, \dots, x_n) = 0$. Suppose that we want to find the density function g of n different random variables $y_1, y_2, \dots, y_n$ defined by

$$y_1 = t_1(x_1, \dots, x_n); \qquad y_2 = t_2(x_1, \dots, x_n); \qquad \dots; \qquad y_n = t_n(x_1, \dots, x_n). \quad (10.2.2)$$

To find g we shall assume the following conditions on the t_i:

(1) Each t_i has continuous first partial derivatives with respect to each x_j at each point in D.
(2) The determinant J, called the *Jacobian*, vanishes for at most a finite number of points in D. J is given by

$$J = \begin{vmatrix} \frac{\partial t_1}{\partial x_1} & \frac{\partial t_1}{\partial x_2} & \cdots & \frac{\partial t_1}{\partial x_n} \\ \frac{\partial t_2}{\partial x_1} & \frac{\partial t_2}{\partial x_2} & \cdots & \frac{\partial t_2}{\partial x_n} \\ \vdots & \vdots & & \vdots \\ \frac{\partial t_n}{\partial x_1} & \frac{\partial t_n}{\partial x_2} & \cdots & \frac{\partial t_n}{\partial x_n} \end{vmatrix}. \tag{10.2.3}$$

(3) The transformation from the x's to the y's is one to one, and suppose D^* is the set of points $(y_1, y_2, \ldots, y_n)$ such that Eq. (10.2.2) has a solution for the x_i, $i = 1, 2, \ldots, n$. Denote the solution by

$$x_1 = s_1(y_1, \ldots, y_n); \ldots; x_n = s_n(y_1, \ldots, y_n).$$

When conditions (1), (2), and (3) are satisfied, the density function g of the random variables $y_1, y_2, \ldots, y_n$ is given by

$$g(y_1, y_2, \ldots, y_n) = \begin{cases} f[s_1(y_1, \ldots, y_n), \ldots, s_n(y_1, \ldots, y_n)]|J|^{-1} \text{ for} \\ \qquad \text{all points}(y_1, y_2, \ldots, y_n) \text{ in } D^*, \\ 0 \quad \text{for all points } (y_1, y_2, \ldots, y_n) \text{ not in } D^*, \end{cases} \tag{10.2.4}$$

where $|J|^{-1}$ is the inverse of the absolute value of the Jacobian if $J \neq 0$ and $|J|^{-1} = 0$ at the points where $J = 0$. Conditions (1), (2), and (3) in Eq. (10.2.3) are also sufficient conditions for changing variables in *multiple integration*. That is, if these conditions are satisfied,

$$\iint_D \cdots \int f(x_1, x_2, \ldots, x_n)\, dx_1\, dx_2 \cdots dx_n$$

is equal to

$$\iint_{D^*} \cdots \int g(y_1, y_2, \ldots, y_n)\, dy_1\, dy_2 \cdots dy_n.$$

For example, suppose that the joint density of two random variables x_1, x_2 is given by

$$f(x_1, x_2) = e^{-(x_1+x_2)}, \qquad 0 < x_1 < \infty, \quad 0 < x_2 < \infty,$$

and suppose that we want the joint density of the random variables y_1, y_2, where the transformation equations are

$$y_1 = 6x_1 + x_2 - 4 = t_1(x_1, x_2),$$

$$y_2 = 3x_1 + 4x_2 = t_2(x_1, x_2).$$

The solution equations are

$$x_1 = \frac{4y_1}{21} - \frac{y_2}{21} + \frac{16}{21} = s_1(y_1, y_2),$$

$$x_2 = -\frac{y_1}{7} + \frac{2y_2}{7} - \frac{4}{7} = s_2(y_1, y_2).$$

The domain D^* is

$$D^* = \left\{(y_1, y_2): \frac{1}{2} y_1 + 2 < y_2 < 4y_1 + 16;\ -4 < y_1 < \infty\right\}.$$

The Jacobian J is given by

$$J = \begin{vmatrix} 6 & 1 \\ 3 & 4 \end{vmatrix} = 21.$$

It is clear that conditions (1), (2), and (3) of Eq. (10.2.3) are satisfied, so the joint density function g of the random variables y_1, y_2 is given by Eq. (10.2.4) and is

$$g(y_1, y_2) = \begin{cases} e^{-(4y_1/21 - y_2/21 + 16/21 + y_1/7 + 2y_2/7 - 4/7)} \dfrac{1}{21}, \\ \qquad \qquad \text{for } \left\{\begin{matrix} \frac{1}{2} y_1 + 2 < y_2 < 4y_1 + 16 \\ -4 < y_1 < \infty \end{matrix}\right\}, \\ 0, \quad \text{elsewhere.} \end{cases}$$

Upon simplification, we get

$$g(y_1, y_2) = \begin{cases} \dfrac{1}{21} e^{-1/21(y_1+5y_2+4)}, & \text{for } \left\{\begin{matrix} \dfrac{1}{2} y_1 + 2 < y_2 < 4y_1 + 16 \\ -4 < y_1 < \infty \end{matrix}\right\}, \\ 0, & \text{elsewhere.} \end{cases}$$

It is easy to verify that g satisfies the two conditions in Eq. (10.2.1) that qualify it to be a density function.

10.3 Multivariate Normal Density

One of the most important density functions in statistics is the *n-variate normal* that is defined by

$$N(x_1, x_2, \ldots, x_n) = K \exp\left(-\frac{1}{2}\sum_{i=1}^{n}\sum_{j=1}^{n}(x_i - c_i)(x_j - c_j)r_{ij}\right),$$
$$-\infty < x_i < \infty; \quad i = 1, 2, \ldots, n, \qquad (10.3.1)$$

where K, the c_i; $i = 1, 2, \ldots, n$; and the r_{ij}; $i = 1, 2, \ldots, n$; $j = 1, 2, \ldots, n$; are constants, and the matrix $\mathbf{R} = [r_{ij}]$ is positive definite. Eq. (10.3.1), of course, can also be written in matrix notation as

$$N(x_1, x_2, \ldots, x_n) = K \exp[-\tfrac{1}{2}(\mathbf{x} - \mathbf{c})'\mathbf{R}(\mathbf{x} - \mathbf{c})], \qquad (10.3.1a)$$

where $\mathbf{x} = [x_i]$, $\mathbf{c} = [c_i]$. For N to be a density, it is clear that K must be a positive constant and

$$\int_{-\infty}^{\infty}\int_{-\infty}^{\infty}\cdots\int_{-\infty}^{\infty} N(x_1, x_2, \ldots, x_n)\, dx_1\, dx_2 \cdots dx_n = 1. \qquad (10.3.2)$$

To show that Eq. (10.3.2) is satisfied, we prove the following theorem.

Theorem 10.3.1

Let $\mathbf{x}' = [x_1, x_2, \ldots, x_n]$ *be any point in* E_n, *and define the vector* $\mathbf{y}$ *by* $\mathbf{y}' = [y_1, y_2, \ldots, y_n]$, *where*

$$y_i = \sum_{j=1}^{n} a_{ij}(x_j - c_j); \quad i = 1, 2, \ldots, n, \qquad (10.3.3)$$

or, in matrix notation,

$$\mathbf{y} = \mathbf{A}(\mathbf{x} - \mathbf{c}), \tag{10.3.3a}$$

which can also be written $\mathbf{x} - \mathbf{c} = \mathbf{A}^{-1}\mathbf{y}$, *where the* c_i *are constants and where* $\mathbf{A} = [a_{ij}]$ *is nonsingular. Then*

(1) *All the first partial derivatives* $\partial y_i/\partial x_j$ *are continuous* (*for all* $i = 1, 2, \ldots, n; j = 1, 2, \ldots, n$).
(2) *The Jacobian of the transformation equation in Eq.* (10.3.3) *is*

$$J = |\mathbf{A}|.$$

(3) *The transformation from* $\mathbf{x}$ *to* $\mathbf{y}$ *is one-to-one.*
(4) *If* $D = \{\mathbf{x} : -\infty < x_i < \infty;\ i = 1, 2, \ldots, n\}$, *then* $D^* = \{\mathbf{y} : -\infty < y_i < \infty;\ i = 1, 2, \ldots, n\}$; that is, *if* $\mathbf{x}$ *can take on any value in* E_n, *then* $\mathbf{y}$ *can also take on any value in* E_n.

This theorem actually states that the transformation given in Eq. (10.3.3) satisfies the conditions in Eq. (10.2.3).

Proof: Clearly $\partial y_i/\partial x_j = a_{ij}$ for $i = 1, 2, \ldots, n; j = 1, 2, \ldots, n$, and $J = |\mathbf{A}|$, hence (1) and (2) in Eq. (10.2.3) follow immediately. Conditions (3) and (4) of the theorem follow from the fact that $\mathbf{A}$ is nonsingular, and we can write $\mathbf{x} = \mathbf{A}^{-1}\mathbf{y} + \mathbf{c}$, and a unique value of $\mathbf{y}$ gives a unique value of $\mathbf{x}$. From the equation $\mathbf{x} = \mathbf{A}^{-1}\mathbf{y} + \mathbf{c}$, it is clear that $\mathbf{y}$ takes on any value in D^*. ▮

Corollary 10.3.1.1

In Theorem 10.3.1 *let* $\mathbf{A}$ *be an orthogonal matrix such that* $\mathbf{ARA}' = \mathbf{D}$ *where* $\mathbf{D} = [d_{ij}]$ *is a diagonal matrix with characteristic roots of* $\mathbf{R}$ *displayed on the diagonal. Then*

(1) $|J| = 1$,
(2) *Eq.* (10.3.2) *can be written*

$$\int_{-\infty}^{\infty}\int_{-\infty}^{\infty}\cdots\int_{-\infty}^{\infty} K \exp\left[-\frac{1}{2}(\mathbf{x} - \mathbf{c})'\mathbf{R}(\mathbf{x} - \mathbf{c})\right] dx_1\, dx_2 \cdots dx_n$$

$$= K \prod_{i=1}^{n}\left[\int_{-\infty}^{\infty} \exp\left(-\frac{1}{2} d_{ii} y_i^2\right) dy_i\right] = 1.$$

Proof: By Theorem 10.3.1, $J = |\mathbf{A}|$ and, since the determinant of an orthogonal matrix is equal to ± 1, the result (1) follows. To prove (2), substitute $\mathbf{A}'\mathbf{y}$ for $\mathbf{x} - \mathbf{c}$, and (since the absolute value of the Jacobian is $+1$) the multiple integral becomes

$$K \int_{-\infty}^{\infty} \int_{-\infty}^{\infty} \cdots \int_{-\infty}^{\infty} e^{-(1/2)\mathbf{y}'(\mathbf{ARA}')\mathbf{y}} \, dy_1 \cdots dy_n$$

$$= K \int_{-\infty}^{\infty} \int_{-\infty}^{\infty} \cdots \int_{-\infty}^{\infty} e^{-(1/2)\mathbf{y}'\mathbf{D}\mathbf{y}} \, dy_1 \cdots dy_n$$

$$= K \int_{-\infty}^{\infty} \cdots \int_{-\infty}^{\infty} \exp\left(-\frac{1}{2} \sum_{i=1}^{n} y_i^2 \, d_{ii}\right) dy_1 \cdots dy_n$$

$$= K \left[\int_{-\infty}^{\infty} e^{(-d_{11}y_1^2)/2} \, dy_1\right] \cdots \left[\int_{-\infty}^{\infty} e^{(-d_{nn}y_n^2)/2} \, dy_n\right].$$

The result then follows. ▮

Corollary 10.3.1.2

In Eq. (10.3.1), $K = |\mathbf{R}|^{1/2}(2\pi)^{-n/2}$ *and*

$$\int_{-\infty}^{\infty} \cdots \int_{-\infty}^{\infty} e^{-(1/2)(\mathbf{x}-\mathbf{c})'\mathbf{R}(\mathbf{x}-\mathbf{c})} \, dx_1 \cdots dx_n = (2\pi)^{n/2} |\mathbf{R}|^{-1/2}.$$

Proof: It is generally shown in a course in calculus that for all $a > 0$,

$$\begin{aligned} &(1) \quad \int_{-\infty}^{\infty} e^{-ay^2/2} \, dy = \sqrt{\frac{2\pi}{a}}; \\ &(2) \quad \int_{-\infty}^{\infty} y \, e^{-ay^2/2} \, dy = 0; \quad \text{and} \\ &(3) \quad \int_{-\infty}^{\infty} y^2 \, e^{-ay^2/2} \, dy = \sqrt{\frac{2\pi}{a^3}}. \end{aligned} \tag{10.3.4}$$

If we substitute into (2) of Corollary 10.3.1.1, we get

$$1 = K \prod_{i=1}^{n} \left[\sqrt{\frac{2\pi}{d_{ii}}}\right] = K(2\pi)^{n/2} \prod_{i=1}^{n} d_{ii}^{-1/2} = K(2\pi)^{n/2} |\mathbf{D}|^{-1/2},$$

or

$$K = (2\pi)^{-n/2}|\mathbf{D}|^{1/2}.$$

But $\mathbf{ARA}' = \mathbf{D}$; $|\mathbf{D}| = |\mathbf{ARA}'| = |\mathbf{A}||\mathbf{R}||\mathbf{A}'| = |\mathbf{R}|$, so $K = (2\pi)^{-n/2} \times |\mathbf{R}|^{1/2}$. Therefore the *n-variate normal density*, defined by Eq. (10.3.1), can be written

$$N(x_1, x_2, \ldots, x_n) = \frac{|\mathbf{R}|^{1/2}}{(2\pi)^{n/2}} \exp\left[-\frac{1}{2}(\mathbf{x}-\mathbf{c})'\mathbf{R}(\mathbf{x}-\mathbf{c})\right],$$
$$-\infty < x_i < \infty; \quad i = 1, 2, \ldots, n. \quad \blacksquare \tag{10.3.5}$$

Since a large amount of the material in mathematical statistics involves the multivariate normal density given in Eq. (10.3.5), it is often necessary to evaluate integrals involving the density function defined in this equation. Here we shall study some of these integrals, discuss some of their uses in statistics, and show that often these definite integrals can be evaluated by simply manipulating matrices.

Theorem 10.3.2

If $\mathbf{R}$ *is a positive definite* $n \times n$ *matrix of constants,* $\mathbf{A}$ *is an* $n \times n$ *matrix of constants, and* $\mathbf{c}$ *is an* $n \times 1$ *vector of constants, then* $S = \operatorname{tr}(\mathbf{AR}^{-1})$, *where*

$$S = \frac{|\mathbf{R}|^{1/2}}{(2\pi)^{n/2}} \int_{-\infty}^{\infty} \cdots \int_{-\infty}^{\infty} (\mathbf{x}-\mathbf{c})'\mathbf{A}(\mathbf{x}-\mathbf{c})\, e^{-(1/2)(\mathbf{x}-\mathbf{c})'\mathbf{R}(\mathbf{x}-\mathbf{c})}\, dx_1 \cdots dx_n. \tag{10.3.6}$$

Note: S does not depend on the vector $\mathbf{c}$.

Proof: Let $\mathbf{P}$ be an orthogonal matrix such that $\mathbf{P}'\mathbf{RP} = \mathbf{D}$, where $\mathbf{D}$ is a diagonal matrix, and make the following transformation:

$$\mathbf{P}'(\mathbf{x}-\mathbf{c}) = \mathbf{y}, \quad \text{or} \quad \mathbf{x}-\mathbf{c} = \mathbf{Py}.$$

The integral in Eq. (10.3.6) becomes

$$S = \frac{|\mathbf{R}|^{1/2}}{(2\pi)^{n/2}} \int_{-\infty}^{\infty} \cdots \int_{-\infty}^{\infty} (\mathbf{x}-\mathbf{c})'\mathbf{A}(\mathbf{x}-\mathbf{c})\, e^{-(1/2)(\mathbf{x}-\mathbf{c})'\mathbf{R}(\mathbf{x}-\mathbf{c})}\, dx_1 \cdots dx_n$$
$$= \frac{|\mathbf{R}|^{1/2}}{(2\pi)^{n/2}} \int_{-\infty}^{\infty} \cdots \int_{-\infty}^{\infty} \mathbf{y}'(\mathbf{P}'\mathbf{AP})\mathbf{y}\, e^{-(1/2)\mathbf{y}'\mathbf{Dy}}\, dy_1 \cdots dy_n,$$

or, if we let $\mathbf{P}'\mathbf{A}\mathbf{P} = \mathbf{B}$, we get

$$S = \frac{|\mathbf{R}|^{1/2}}{(2\pi)^{n/2}} \int_{-\infty}^{\infty} \cdots \int_{-\infty}^{\infty} \sum_u \sum_v y_u y_v b_{uv} \exp\left[-\frac{1}{2}\sum_t y_t^2 d_{tt}\right] dy_1 \cdots dy_n$$

$$= \frac{|\mathbf{R}|^{1/2}}{(2\pi)^{n/2}} \sum_u \sum_v b_{uv} \int_{-\infty}^{\infty} \cdots \int_{-\infty}^{\infty} y_u y_v \exp\left[-\frac{1}{2}\sum_t y_t^2 d_{tt}\right] dy_1 \cdots dy_n.$$

But, by (2) and (3) of Eq. (10.3.4), the value of the integral

$$\int_{-\infty}^{\infty} \cdots \int_{-\infty}^{\infty} y_u y_v \exp\left[-\frac{1}{2}\sum_t y_t^2 d_{tt}\right] dy_1 \cdots dy_n \tag{10.3.7}$$

is equal to zero if $u \neq v$, and if $u = v$ the value of the integral in Eq. (10.3.7) is (see the proof to Corollary 10.3.1.2)

$$\frac{1}{d_{uu}} \frac{(2\pi)^{n/2}}{\prod_{i=1}^{n} d_{ii}^{1/2}} = \frac{(2\pi)^{n/2}}{d_{uu}|\mathbf{R}|^{1/2}}.$$

So

$$S = \sum_{u=1}^{n} \frac{b_{uu}}{d_{uu}} = \sum_{u=1}^{n} b_{uu} d_{uu}^{-1}.$$

But since $\mathbf{P}'\mathbf{R}\mathbf{P} = \mathbf{D}$, we get

$$\mathbf{D}^{-1} = (\mathbf{P}'\mathbf{R}\mathbf{P})^{-1} = \mathbf{P}'\mathbf{R}^{-1}\mathbf{P},$$

and

$$\sum_{u=1}^{n} b_{uu} d_{uu}^{-1} = \operatorname{tr}(\mathbf{B}\mathbf{D}^{-1}) = \operatorname{tr}(\mathbf{B}\mathbf{P}'\mathbf{R}^{-1}\mathbf{P}).$$

Using $\mathbf{P}'\mathbf{A}\mathbf{P} = \mathbf{B}$, we get

$$S = \sum_{u=1}^{n} b_{uu} d_{uu}^{-1} = \operatorname{tr}(\mathbf{B}\mathbf{P}'\mathbf{R}^{-1}\mathbf{P}) = \operatorname{tr}[(\mathbf{P}'\mathbf{A}\mathbf{P})(\mathbf{P}'\mathbf{R}^{-1}\mathbf{P})]$$
$$= \operatorname{tr}(\mathbf{P}'\mathbf{A}\mathbf{R}^{-1}\mathbf{P}) = \operatorname{tr}(\mathbf{A}\mathbf{R}^{-1}\mathbf{P}\mathbf{P}') = \operatorname{tr}(\mathbf{A}\mathbf{R}^{-1}),$$

and the theorem is proved. ▮

Note: This integral is of fundamental importance in the study of the multivariate normal density.

10.4 Moments of Density Functions and Expected Values of Random Matrices

Another important subject in mathematical statistics is that of moments.

Definition 10.4.1

Mean, Variance, Covariance. *Let the joint density of the random variables* $x_1, x_2, \ldots, x_n$ *be* $f(x_1, x_2, \ldots, x_n)$; *then*

(1) *The first moment of* x_p *(called the mean of* x_p*) is denoted by* μ_p, *where*

$$\mu_p = \int_{-\infty}^{\infty} \cdots \int_{-\infty}^{\infty} x_p f(x_1, x_2, \ldots, x_n)\, dx_1\, dx_2 \cdots dx_n\, ; \; p = 1, 2, \ldots, n. \tag{10.4.1}$$

(2) *The covariance of* x_p *and* x_q *is defined by*

$$v_{pq} = \mu'_{pq} - \mu_p \mu_q, \tag{10.4.2}$$

where

$$\mu'_{pq} = \int_{-\infty}^{\infty} \cdots \int_{-\infty}^{\infty} x_p x_q f(x_1, x_2, \ldots, x_n)\, dx_1\, dx_2 \cdots dx_n,$$
$$p = 1, 2, \ldots, n, \quad q = 1, 2, \ldots, n; \tag{10.4.3}$$

v_{pp} *is called the variance of* x_p.

(3) *The* $n \times 1$ *vector* $\boldsymbol{\mu}$, *whose* p-th *element is* μ_p, *is called the vector mean (or vector of first moments) of the density* $f(x_1, \ldots, x_n)$; that is, $\boldsymbol{\mu} = [\mu_p]$.

(4) *The* $n \times n$ *matrix* $\mathbf{V}$, *whose* pq-th *element is* v_{pq}, *is called the covariance matrix of the density* $f(x_1, x_2, \ldots, x_n)$, *and the* pq-th *element is the covariance of the random variable* x_p *with the random variable* x_q.

Note: $\mathbf{V}$ is symmetric, and we can also write

$$v_{pq} = \int_{-\infty}^{\infty} \cdots \int_{-\infty}^{\infty} (x_p - \mu_p)(x_q - \mu_q) f(x_1, x_2, \ldots, x_n)\, dx_1\, dx_2 \cdots dx_n.$$

Next we shall prove a theorem for moments of the n-variate normal density defined in Eq. (10.3.1).

Theorem 10.4.1

In the n-variate normal density defined in Eq. (10.3.1), *the mean vector* $\boldsymbol{\mu}$ *is equal to* $\mathbf{c}$ *and the covariance matrix* $\mathbf{V}$ *is equal to* $\mathbf{R}^{-1}$.

Proof: To prove that $\boldsymbol{\mu} = \mathbf{c}$ we must show that

$$\frac{|\mathbf{R}|^{1/2}}{(2\pi)^{n/2}} \int_{-\infty}^{\infty} \cdots \int_{-\infty}^{\infty} x_p \exp\left[-\frac{1}{2}(\mathbf{x}-\mathbf{c})'\mathbf{R}(\mathbf{x}-\mathbf{c})\right] dx_1\, dx_2 \cdots dx_n = c_p. \tag{10.4.4}$$

If we use Corollary 10.3.1.1, we get (since $\mathbf{x} = \mathbf{c} + \mathbf{A}^{-1}\mathbf{y}$)

$$x_p = c_p + \sum_{j=1}^{n} a_{pj}^{(-1)} y_j,$$

where $a_{pj}^{(-1)}$ is the pj-th element of $\mathbf{A}^{-1}$, and the integral becomes

$$\frac{|\mathbf{R}|^{1/2}}{(2\pi)^{n/2}} \int_{-\infty}^{\infty} \cdots \int_{-\infty}^{\infty} \left(c_p + \sum_{j=1}^{n} a_{pj}^{(-1)} y_j\right)$$
$$\times\, e^{-(1/2)d_{11}y_1^2}\, e^{-(1/2)d_{22}y_2^2} \cdots e^{-(1/2)d_{nn}y_n^2}\, dy_1\, dy_2 \cdots dy_n$$
$$= c_p \frac{|\mathbf{R}|^{1/2}}{(2\pi)^{n/2}} \int_{-\infty}^{\infty} \cdots \int_{-\infty}^{\infty} \exp\left[-\frac{1}{2}\sum_{i=1}^{n} d_{ii}y_i^2\right] dy_1 \cdots dy_n$$
$$+ \frac{|\mathbf{R}|^{1/2}}{(2\pi)^{n/2}} \sum_{j=1}^{n} a_{pj}^{(-1)} \prod_{i=1}^{n} \left[\int_{-\infty}^{\infty} y_j^{\delta_{ij}}\, e^{-(1/2)d_{ii}y_i^2}\, dy_i\right]$$
$$= c_p,$$

since

$$\int_{-\infty}^{\infty} y_j\, e^{-(1/2)d_{jj}y_j^2}\, dy_j = 0.$$

In the above, δ_{ij} is the Kronecker delta.

Thus we have shown that $\mu_p = c_p$ for $p = 1, 2, \ldots, n$, and therefore $\boldsymbol{\mu} = \mathbf{c}$. To show that $\mathbf{V} = \mathbf{R}^{-1}$ (or $\mathbf{V}^{-1} = \mathbf{R}$), we must show that v_{pq}, defined in (2) of Def. 10.4.1, is the pq-th element of $\mathbf{R}^{-1}$, which we write as $r_{pq}^{(-1)}$.

Clearly, we can write v_{pq} as

$$v_{pq} = \frac{|\mathbf{R}|^{1/2}}{(2\pi)^{n/2}} \int_{-\infty}^{\infty} \cdots \int_{-\infty}^{\infty} (x_p - \mu_p)(x_q - \mu_q)$$

$$\exp\left[-\frac{1}{2}(\mathbf{x} - \boldsymbol{\mu})'\mathbf{R}(\mathbf{x} - \boldsymbol{\mu})\right] dx_1\, dx_2 \cdots dx_n. \quad (10.4.5)$$

We now show how Theorem 10.3.2 can be used to evaluate Eq. (10.4.5). The quantity $(x_p - \mu_p)(x_q - \mu_q)$ in the integrand of the integral of Eq. (10.4.5) can be written as

$$(x_p - \mu_p)(x_q - \mu_q) = (\mathbf{x} - \boldsymbol{\mu})'\mathbf{A}(\mathbf{x} - \boldsymbol{\mu}),$$

where $\mathbf{A}$ is an $n \times n$ matrix with every element equal to zero except

$$a_{pq} = a_{qp} = \frac{1}{2}.$$

Thus by Theorem 10.3.2 $v_{pq} = S = \operatorname{tr}(\mathbf{AR}^{-1})$, but, by the structure of $\mathbf{A}$ described above, we get $\operatorname{tr}(\mathbf{AR}^{-1}) = r_{pq}^{(-1)}$, and Theorem 10.4.1 is proved. ∎

Example 10.4.1. Evaluate the following integral.

$$I = \int_{-\infty}^{\infty} \int_{-\infty}^{\infty} \int_{-\infty}^{\infty} (x_1^2 + x_2^2)\, e^{-(x_1^2 + x_2^2 + 2x_3^2)}\, dx_1\, dx_2\, dx_3.$$

We can use Theorem 10.3.2. The exponent of the integrand of the integral to be evaluated can be written

$$-\frac{1}{2}(2x_1^2 + 2x_2^2 + 4x_3^2);$$

hence

$$\mathbf{x} = \begin{bmatrix} x_1 \\ x_2 \\ x_3 \end{bmatrix}; \quad \mathbf{R} = \begin{bmatrix} 2 & 0 & 0 \\ 0 & 2 & 0 \\ 0 & 0 & 4 \end{bmatrix}; \quad \mathbf{c} = \begin{bmatrix} 0 \\ 0 \\ 0 \end{bmatrix}; \quad \text{and} \quad \mathbf{A} = \begin{bmatrix} 1 & 0 & 0 \\ 0 & 1 & 0 \\ 0 & 0 & 0 \end{bmatrix}.$$

Since $\mathbf{R}$ is positive definite, the hypothesis of Theorem 10.3.2 is satisfied. We get

$$S = \frac{|\mathbf{R}|^{1/2}}{(2\pi)^{3/2}} I,$$

but $S = \text{tr}\,(\mathbf{AR}^{-1}) = 1$. So

$$I = \frac{(2\pi)^{3/2}}{4}.$$

Since we shall have occasion to evaluate multiple integrals involving density functions quite often, we now define an operator (called expected value) that will help to shorten our notation.

Definition 10.4.2

Expected Value of $t(x_1, x_2, \ldots, x_n)$. *Let the* $n \times 1$ *random vector* $\mathbf{x}$ *have the density defined by* $f(x_1, x_2, \ldots, x_n)$. *Then the expected value of* $t(x_1, x_2, \ldots, x_n)$ *is denoted by*

$$\mathscr{E}[t(x_1, x_2, \ldots, x_n)]$$

and is defined by

$$\mathscr{E}[t(x_1, x_2, \ldots, x_n)] = \int_{-\infty}^{\infty} \int_{-\infty}^{\infty} \cdots \int_{-\infty}^{\infty} t(x_1, x_2, \ldots, x_n) f(x_1, x_2, \ldots, x_n)\, dx_1\, dx_2 \cdots dx_n$$

if the integral exists.

For example, by Def. (10.4.1), we get $\mu_1 = \mathscr{E}(x_1)$; the variance of x_p is equal to

$$v_{pp} = \mathscr{E}(x_p^2) - [\mathscr{E}(x_p)]^2;$$

the covariance of x_p and x_q equals

$$v_{pq} = \mathscr{E}(x_p x_q) - [\mathscr{E}(x_p)][\mathscr{E}(x_q)].$$

Next we extend this by defining the expected value of a matrix.

Definition 10.4.3

Expected Value of a Random Matrix. *Let* $\mathbf{W}$ *be a* $k_1 \times k_2$ *random matrix (a matrix of functions of the* $n \times 1$ *random vector* $\mathbf{x}$*); that is, let* $w_{ij} = t_{ij}(x_1, x_2, \ldots, x_n)$*; then the expected value of the matrix* $\mathbf{W}$ *is denoted by* $\mathscr{E}(\mathbf{W})$ *and is defined by the* $k_1 \times k_2$ *matrix* $\mathbf{A}$ *where*

$$a_{ij} = \mathscr{E}[t_{ij}(x_1, x_2, \ldots, x_n)].$$

For example, if

$$\mathbf{W} = \begin{bmatrix} x_1 & x_2 \\ x_2 & x_3 \end{bmatrix},$$

then

$$\mathscr{E}(\mathbf{W}) = \begin{bmatrix} \mathscr{E}(x_1) & \mathscr{E}(x_2) \\ \mathscr{E}(x_2) & \mathscr{E}(x_3) \end{bmatrix}.$$

Theorem 10.4.2

Let $\mathbf{x}$ *be a random vector and* $t_1, t_2, \ldots, t_m$ be *m functions of the elements in* $\mathbf{x}$*. If* $b_1, b_2, \ldots, b_m$ *are constants, then*

$$\mathscr{E}[b_1 t_1(x_1, \ldots, x_n) + b_2 t_2(x_1, \ldots, x_n) + \cdots + b_m t_m(x_1, \ldots, x_n)]$$
$$= b_1 \mathscr{E}[t_1(x_1, \ldots, x_n)] + b_2 \mathscr{E}[t_2(x_1, \ldots, x_n)] + \cdots + b_m \mathscr{E}[t_m(x_1, \ldots, x_n)]$$

if all integrals exist.

The proof follows from a property of multiple integrals.

Theorem 10.4.3

If $\mathbf{W}$ *is a* $k_1 \times k_2$ *random matrix,* $\mathbf{T}$ *is a* $k_1 \times k_2$ *random matrix,* $\mathbf{A}_1$ *is an* $m_1 \times k_1$ *matrix of constants, and* $\mathbf{A}_2$ *is a* $k_2 \times m_2$ *matrix of constants, then the following relationships hold:*

(1) $\mathscr{E}(\mathbf{A}_1) = \mathbf{A}_1$,
(2) $\mathscr{E}(\mathbf{A}_1\mathbf{W}) = \mathbf{A}_1[\mathscr{E}(\mathbf{W})]$,
(3) $\mathscr{E}(\mathbf{W}\mathbf{A}_2) = [\mathscr{E}(\mathbf{W})]\mathbf{A}_2$,
(4) $\mathscr{E}[\mathbf{A}_1\mathbf{W}\mathbf{A}_2] = \mathbf{A}_1[\mathscr{E}(\mathbf{W})]\mathbf{A}_2$, *and*
(5) $\mathscr{E}[\mathbf{T} + \mathbf{W}] = \mathscr{E}(\mathbf{T}) + \mathscr{E}(\mathbf{W})$,

if all integrals involved exist.

Proof: We shall prove relationship (4), and relationships (1) through (3) will follow by setting $\mathbf{A}_1$ and $\mathbf{A}_2$ equal to the proper identity matrix.

Let $\mathscr{E}(w_{pq}) = c_{pq}$ and hence $\mathscr{E}(\mathbf{W}) = \mathbf{C}$. Now if we set $\mathbf{A}_1\mathbf{W}\mathbf{A}_2 = \mathbf{U}$, then

$$u_{ij} = \sum_{p=1}^{k_1} \sum_{q=1}^{k_2} a_{ip}^{(1)} w_{pq} a_{qj}^{(2)},$$

where $\mathbf{A}_i = [a_{pq}^{(i)}]$. By Theorem 10.4.2 we get

$$\mathscr{E}(u_{ij}) = \sum_{p=1}^{k_1} \sum_{q=1}^{k_2} a_{ip}^{(1)} c_{pq} a_{qj}^{(2)},$$

and hence

$$\mathscr{E}(\mathbf{W}) = \mathbf{A}_1\mathbf{C}\mathbf{A}_2 = \mathbf{A}_1[\mathscr{E}(\mathbf{W})]\mathbf{A}_2 .$$

Relationship (5) follows from the fact that

$$\mathscr{E}(t_{ij} + w_{ij}) = \mathscr{E}(t_{ij}) + \mathscr{E}(w_{ij}). \quad \blacksquare$$

Example 10.4.2. Notice that Theorem 10.4.1 gives us

$$\boldsymbol{\mu} = \mathscr{E}(\mathbf{x})$$

and

$$\mathbf{V} = \mathscr{E}[(\mathbf{x} - \boldsymbol{\mu})(\mathbf{x} - \boldsymbol{\mu})'] = \mathscr{E}\{[\mathbf{x} - \mathscr{E}(\mathbf{x})][\mathbf{x} - \mathscr{E}(\mathbf{x})]'\}.$$

Example 10.4.3. Suppose that $\mathbf{x}$ is a 2×1 random vector and $\mathbf{a}$ is a 2×1 constant vector; then

$$\mathscr{E}(\mathbf{a}'\mathbf{x}) = a_1\mathscr{E}(x_1) + a_2\mathscr{E}(x_2).$$

Theorem 10.4.4

Let $\mathbf{x}$ be an $n \times 1$ random vector and let $\mathbf{A}$ be an $n \times n$ symmetric matrix of constants; then the expected value of Q, the quadratic form $\mathbf{x}'\mathbf{A}\mathbf{x}$, is given by

$$\mathscr{E}(Q) = \sum_{i=1}^{n} \sum_{j=1}^{n} a_{ij}\mathscr{E}(x_i x_j).$$

Proof: To evaluate $\mathscr{E}(Q)$, we write

$$Q = \sum_{i=1}^{n} \sum_{j=1}^{n} a_{ij} x_i x_j$$

and use Theorem 10.4.2. ∎

Example 10.4.4. Note that Theorem 10.3.2 can be written in terms of expected values, since

$$S = \mathscr{E}[(\mathbf{x} - \mathbf{c})'\mathbf{A}(\mathbf{x} - \mathbf{c})].$$

By expanding the quadratic form, we get

$$\begin{aligned} S &= \mathscr{E}[\mathbf{x}'\mathbf{A}\mathbf{x} - \mathbf{x}'\mathbf{A}\mathbf{c} - \mathbf{c}'\mathbf{A}\mathbf{x} + \mathbf{c}'\mathbf{A}\mathbf{c}] \\ &= \mathscr{E}[\mathbf{x}'\mathbf{A}\mathbf{x} - 2\mathbf{x}'\mathbf{A}\mathbf{c} + \mathbf{c}'\mathbf{A}\mathbf{c}], \end{aligned}$$

since $\mathbf{x}'\mathbf{A}\mathbf{c}$ is a scalar and hence is equal to its transpose, $\mathbf{c}'\mathbf{A}\mathbf{x}$. By Theorem 10.4.2, we get

$$S = \mathscr{E}(\mathbf{x}'\mathbf{A}\mathbf{x}) - 2\mathscr{E}(\mathbf{x}'\mathbf{A}\mathbf{c}) + \mathscr{E}(\mathbf{c}'\mathbf{A}\mathbf{c}).$$

By Theorem 10.4.4, the first term becomes

$$\mathscr{E}(\mathbf{x}'\mathbf{A}\mathbf{x}) = \sum_{j=1}^{n} \sum_{i=1}^{n} a_{ij} \mathscr{E}(x_i x_j).$$

But by Eqs. (10.4.2) and (10.4.3), we get

$$\mathscr{E}(x_i x_j) = v_{ij} + [\mathscr{E}(x_i)][\mathscr{E}(x_j)] = v_{ij} + \mu_i \mu_j;$$

so we get

$$\mathscr{E}(\mathbf{x}'\mathbf{A}\mathbf{x}) = \sum_{i=1}^{n} \sum_{j=1}^{n} a_{ij}(v_{ij} + \mu_i \mu_j) = \operatorname{tr}(\mathbf{A}\mathbf{V}) + \boldsymbol{\mu}'\mathbf{A}\boldsymbol{\mu}.$$

Also by Theorem 10.4.3, the other terms of S can be evaluated, and we have

$$\begin{aligned} S &= \operatorname{tr}(\mathbf{A}\mathbf{V}) + \boldsymbol{\mu}'\mathbf{A}\boldsymbol{\mu} - 2\boldsymbol{\mu}'\mathbf{A}\mathbf{c} + \mathbf{c}'\mathbf{A}\mathbf{c} \\ &= \operatorname{tr}(\mathbf{A}\mathbf{V}) = \operatorname{tr}(\mathbf{A}\mathbf{R}^{-1}), \end{aligned}$$

by using the result of Theorem 10.4.1 that $\boldsymbol{\mu} = \mathbf{c}$.

10.5 Evaluation of a General Multiple Integral

There are many areas of mathematics and statistics where the integrals given in Eqs. (10.3.2), (10.3.4), (10.4.4), and (10.3.6) are very important. These integrals are special cases of the integral given in Theorem (10.5.1), which follows.

Theorem 10.5.1

*Let a_0 and b_0 be scalar constants; let **a** be an $n \times 1$ vector of constants; let **b** be an $n \times 1$ vector of constants; let **A** be an $n \times n$ symmetric matrix of constants; let **B** be a positive definite matrix of constants. The value of the multiple integral in Eq.* (10.5.1) *is given in Eq.* (10.5.2); *that is,*

$$I = \int_{-\infty}^{\infty} \int_{-\infty}^{\infty} \cdots \int_{-\infty}^{\infty} (\mathbf{x}'\mathbf{A}\mathbf{x} + \mathbf{x}'\mathbf{a} + a_0)\, e^{-(\mathbf{x}'\mathbf{B}\mathbf{x} + \mathbf{x}'\mathbf{b} + b_0)}\, dx_1\, dx_2 \cdots dx_n, \tag{10.5.1}$$

and

$$I = \frac{1}{2}\pi^{n/2} |\mathbf{B}|^{-1/2} e^{(1/4)\mathbf{b}'\mathbf{B}^{-1}\mathbf{b} - b_0} \left[\operatorname{tr}(\mathbf{A}\mathbf{B}^{-1}) - \mathbf{b}'\mathbf{B}^{-1}\mathbf{a} + \frac{1}{2}\mathbf{b}'\mathbf{B}^{-1}\mathbf{A}\mathbf{B}^{-1}\mathbf{b} + 2a_0\right], \tag{10.5.2}$$

*where the $n \times 1$ vector **x** has components $x_1, x_2, \ldots, x_n$.*

Proof: First we shall examine the exponent of Eq. (10.5.1). It is easily shown that the exponent can be written as

$$\mathbf{x}'\mathbf{B}\mathbf{x} + \mathbf{x}'\mathbf{b} + b_0 = \frac{1}{2}\left(\mathbf{x} + \frac{1}{2}\mathbf{B}^{-1}\mathbf{b}\right)'(2\mathbf{B})\left(\mathbf{x} + \frac{1}{2}\mathbf{B}^{-1}\mathbf{b}\right) - \frac{1}{4}\mathbf{b}'\mathbf{B}^{-1}\mathbf{b} + b_0 \tag{10.5.3}$$

Also the terms in the integrand that are not part of the exponential term can be written as

$$\mathbf{x}'\mathbf{A}\mathbf{x} + \mathbf{x}'\mathbf{a} + a_0$$

$$= \left(\mathbf{x} + \frac{1}{2}\mathbf{B}^{-1}\mathbf{b}\right)'\mathbf{A}\left(\mathbf{x} + \frac{1}{2}\mathbf{B}^{-1}\mathbf{b}\right) + \mathbf{x}'(\mathbf{a} - \mathbf{A}\mathbf{B}^{-1}\mathbf{b}) - \frac{1}{4}\mathbf{b}'\mathbf{B}^{-1}\mathbf{A}\mathbf{B}^{-1}\mathbf{b} + a_0. \tag{10.5.4}$$

Equations (10.5.3) and (10.5.4) can be verified by simply expanding the right-hand member in each case and showing that it reduces to the left-hand member. If we use Eq. (10.5.4) and substitute into the integrand, we get

$$I =$$

$$e^{(1/4)\mathbf{b}'\mathbf{B}^{-1}\mathbf{b}-b_0}\left[\int_{-\infty}^{\infty}\cdots\int_{-\infty}^{\infty}(\mathbf{x}-\mathbf{c})'\mathbf{A}(\mathbf{x}-\mathbf{c})\,e^{-(1/2)(\mathbf{x}-\mathbf{c})'\mathbf{R}(\mathbf{x}-\mathbf{c})}\,dx_1\,dx_2\cdots dx_n\right.$$

$$+\int_{-\infty}^{\infty}\cdots\int_{-\infty}^{\infty}\mathbf{x}'\mathbf{d}\,e^{-(1/2)(\mathbf{x}-\mathbf{c})'\mathbf{R}(\mathbf{x}-\mathbf{c})}\,dx_1\,dx_2\cdots dx_n$$

$$\left.+\int_{-\infty}^{\infty}\cdots\int_{-\infty}^{\infty}\left(-\frac{1}{4}\mathbf{b}'\mathbf{B}^{-1}\mathbf{A}\mathbf{B}^{-1}\mathbf{b}+a_0\right)e^{-(1/2)(\mathbf{x}-\mathbf{c})'\mathbf{R}(\mathbf{x}-\mathbf{c})}\,dx_1\,dx_2\cdots dx_n\right]$$

$$= e^{(1/4)\mathbf{b}'\mathbf{B}^{-1}\mathbf{b}-b_0}[I_1 + I_2 + I_3], \tag{10.5.5}$$

written as a sum of three integrals for brevity.

In Eq. (10.5.5) we used the following notation:

$$\mathbf{c} = -\frac{1}{2}\mathbf{B}^{-1}\mathbf{b}, \quad \mathbf{d} = \mathbf{a} - \mathbf{A}\mathbf{B}^{-1}\mathbf{b}, \quad \text{and} \quad \mathbf{R} = 2\mathbf{B}. \tag{10.5.6}$$

Notice that $\mathbf{R}$ is positive definite, since we assumed in the statement of the theorem that $\mathbf{B}$ is positive definite. Notice also that $\mathbf{c}$, $\mathbf{d}$, and $\mathbf{R}$ are constant vectors and a constant matrix, respectively. Also, the quantity $(1/4)\mathbf{b}'\mathbf{B}^{-1}\mathbf{b} - b_0$ in the exponent is a constant; hence it is factored out, and

$$e^{(1/4)\mathbf{b}'\mathbf{B}^{-1}\mathbf{b}-b_0} \tag{10.5.7}$$

appears as a coefficient. Now by using Theorem 10.3.2, we get

$$I_1 = (2\pi)^{n/2}|\mathbf{R}|^{-1/2}\,\mathrm{tr}\,(\mathbf{A}\mathbf{R}^{-1}),$$

and by substituting for $\mathbf{R}$ in Eq. (10.5.6), we get

$$I_1 = \frac{1}{2}\pi^{n/2}|\mathbf{B}|^{-1/2}\,\mathrm{tr}\,(\mathbf{A}\mathbf{B}^{-1}). \tag{10.5.8}$$

Since we can write $\mathbf{x}'\mathbf{d}$ as $\sum_{i=1}^{n} x_i d_i$, the integral represented by I_2 is a sum of n integrals; thus by Theorem 10.4.1, and specifically Eq. (10.4.4), we get

$$(2\pi)^{n/2}\,|\mathbf{R}|^{-1/2}c_p\,d_p$$

for the p-th integral. Therefore,

$$I_2 = \sum_{i=1}^{n} (2\pi)^{n/2} |\mathbf{R}|^{-1/2} c_i d_i = (2\pi)^{n/2} |\mathbf{R}|^{-1/2} \mathbf{c}'\mathbf{d}.$$

If we substitute the pertinent quantities from Eq. (10.5.6), we get

$$I_2 = \frac{1}{2} \pi^{n/2} |\mathbf{B}|^{-1/2} \mathbf{b}'\mathbf{B}^{-1}(\mathbf{A}\mathbf{B}^{-1}\mathbf{b} - \mathbf{a}). \tag{10.5.9}$$

The integral denoted by I_3 in Eq. (10.5.5) can be written as

$$I_3 = \left(-\frac{1}{4}\mathbf{b}'\mathbf{B}^{-1}\mathbf{A}\mathbf{B}^{-1}\mathbf{b} + a_0\right) \int_{-\infty}^{\infty} \cdots \int_{-\infty}^{\infty} e^{-(1/2)(\mathbf{x}-\mathbf{c})'\mathbf{R}(\mathbf{x}-\mathbf{c})} \, dx_1 \cdots dx_n,$$

and, by Corollaries (10.3.1.1) and (10.3.1.2), we get

$$I_3 = \left(-\frac{1}{4}\mathbf{b}'\mathbf{B}^{-1}\mathbf{A}\mathbf{B}^{-1}\mathbf{b} + a_0\right)(2\pi)^{n/2} |\mathbf{R}|^{-1/2}.$$

If we substitute the pertinent quantities from Eq. (10.5.6), we get

$$I_3 = \pi^{n/2} |\mathbf{B}|^{-1/2} \left(a_0 - \frac{1}{4}\mathbf{b}'\mathbf{B}^{-1}\mathbf{A}\mathbf{B}^{-1}\mathbf{b}\right). \tag{10.5.10}$$

If we now substitute the quantities for I_1, I_2, and I_3 of Eqs. (10.5.8), (10.5.9), and (10.5.10) into Eq. (10.5.5), we obtain the result in Eq. (10.5.2). ∎

10.6 Marginal Density Function

If $f(x_1, \ldots, x_n)$ is the joint density of n random variables $x_1, x_2, \ldots, x_n$, it is often desirable to find the density (sometimes called the marginal density) of a subset of p of these random variables. It is perfectly general if we consider the first p of these variables, so we state the following definition.

Definition 10.6.1

Marginal Density. *Let $f(x_1, x_2, \ldots, x_n)$ be the joint density of n continuous* *random variables* $x_1, x_2, \ldots, x_n$. *The marginal density of a subset of p of these random variables (that is, of* $x_1, x_2, \ldots, x_p$; $p < n$) *is defined by*

$$g(x_1, x_2, \ldots, x_p) = \int_{-\infty}^{\infty} \int_{-\infty}^{\infty} \cdots \int_{-\infty}^{\infty} f(x_1, x_2, \ldots, x_n)\, dx_{p+1}\, dx_{p+2} \cdots dx_n\,; \qquad -\infty < x_i < \infty; \quad i = 1, 2, \ldots, p. \tag{10.6.1}$$

For example, if the joint density of the random variables x_1, x_2 is given by

$$f(x_1, x_2) = \begin{cases} e^{-(x_1+x_2)}, & 0 < x_1 < \infty; \quad 0 < x_2 < \infty, \\ 0, & \text{elsewhere,} \end{cases}$$

then the marginal density of x_1 is defined by

$$g(x_1) = \int_{-\infty}^{\infty} f(x_1, x_2)\, dx_2\,,$$

and we get

$$g(x_1) = \begin{cases} 0, & \text{for} \quad -\infty < x_1 \le 0, \\ e^{-x_1}, & \text{for} \quad 0 < x_1 < \infty. \end{cases}$$

When a set of random variables $x_1, x_2, \ldots, x_n$ has an n-variate density given by Eq. (10.3.1), we shall state "the random vector $\mathbf{x}$ has a density $N(\mathbf{x}; \boldsymbol{\mu}, \mathbf{V})$," to mean that the components of the vector $\mathbf{x}$, that is, $x_1, \ldots, x_n$, have an n-variate normal density with mean vector $\boldsymbol{\mu}$ and covariance matrix $\mathbf{V}$. The functional form can be written as in Eq. (10.3.1), but more often it is written as

$$N(\mathbf{x}\,; \boldsymbol{\mu}, \mathbf{V}) = \frac{e^{-(1/2)(\mathbf{x}-\boldsymbol{\mu})'\mathbf{V}^{-1}(\mathbf{x}-\boldsymbol{\mu})}}{(2\pi)^{n/2}\,|\mathbf{V}|^{1/2}}, \qquad -\infty < x_i < \infty; \quad i = 1, 2, \ldots, n. \tag{10.6.2}$$

If the n random variables $x_1, x_2, \ldots, x_n$ have a density given by Eq. (10.6.2), that is, an n-variate normal density, then the density of any subset consisting of p of these random variables ($0 < p < n$) is a p-variate normal, and the mean vector and covariance matrix of this p-variate normal density can be obtained from the original n-variate normal density simply by operations on matrices and vectors. This is the context of the next theorem.

Theorem 10.6.1

Let the $n \times 1$ random vector $\mathbf{x}$ have a normal density with mean vector $\boldsymbol{\mu}$ and covariance matrix $\mathbf{V}$; that is, $N(\mathbf{x}; \boldsymbol{\mu}, \mathbf{V})$ as given in Eq. (10.6.2). Then the marginal density of $x_1, x_2, \ldots, x_p$ is normal with mean vector $\boldsymbol{\mu}_1$ and covariance matrix $\mathbf{V}_{11}$ where $\boldsymbol{\mu}_1$ and $\mathbf{V}_{11}$ are defined in Eq. (10.6.5).

This theorem states that

$$g(x_1, \ldots, x_p) = \frac{e^{-(1/2)[(\mathbf{x}_1-\boldsymbol{\mu}_1)'\mathbf{V}_{11}^{-1}(\mathbf{x}_1-\boldsymbol{\mu}_1)]}}{(2\pi)^{p/2}|\mathbf{V}_{11}|^{1/2}}, \tag{10.6.3}$$

which is of the same form as Eq. (10.6.2), so we can write

$$g(x_1, x_2, \ldots, x_p) = N(\mathbf{x}_1; \boldsymbol{\mu}_1, \mathbf{V}_{11}).$$

This theorem can also be stated as an integration formula in the following form

$$\int_{-\infty}^{\infty}\int_{-\infty}^{\infty}\cdots\int_{-\infty}^{\infty} \frac{e^{-(1/2)[(\mathbf{x}-\boldsymbol{\mu})'\mathbf{V}^{-1}(\mathbf{x}-\boldsymbol{\mu})]}}{(2\pi)^{n/2}|\mathbf{V}|^{1/2}}\, dx_{p+1}\, dx_{p+2} \cdots dx_n = \frac{e^{-(1/2)[(\mathbf{x}_1-\boldsymbol{\mu}_1)'\mathbf{V}_{11}^{-1}(\mathbf{x}_1-\boldsymbol{\mu}_1)]}}{(2\pi)^{p/2}|\mathbf{V}_{11}|^{1/2}}. \tag{10.6.4}$$

We have used the following notation:

$$\mathbf{x}_1 = \begin{bmatrix} x_1 \\ \vdots \\ x_p \end{bmatrix}, \quad \mathbf{x}_2 = \begin{bmatrix} x_{p+1} \\ \vdots \\ x_n \end{bmatrix}, \quad \mathbf{x} = \begin{bmatrix} \mathbf{x}_1 \\ \mathbf{x}_2 \end{bmatrix},$$

$$\boldsymbol{\mu}_1 = \begin{bmatrix} \mu_1 \\ \vdots \\ \mu_p \end{bmatrix}, \quad \boldsymbol{\mu}_2 = \begin{bmatrix} \mu_{p+1} \\ \vdots \\ \mu_n \end{bmatrix}, \quad \boldsymbol{\mu} = \begin{bmatrix} \boldsymbol{\mu}_1 \\ \boldsymbol{\mu}_2 \end{bmatrix}, \tag{10.6.5}$$

$$\mathbf{V} = \begin{bmatrix} \mathbf{V}_{11} & \mathbf{V}_{12} \\ \mathbf{V}_{21} & \mathbf{V}_{22} \end{bmatrix}, \quad \mathbf{V}^{-1} = \mathbf{R} = \begin{bmatrix} \mathbf{R}_{11} & \mathbf{R}_{12} \\ \mathbf{R}_{21} & \mathbf{R}_{22} \end{bmatrix},$$

where $\mathbf{V}_{11}$ (and $\mathbf{R}_{11}$) is a $p \times p$ matrix.

Note that $\mathbf{x}_1$ contains the first p components of $\mathbf{x}$, and $\boldsymbol{\mu}$ and $\mathbf{V}$ have been partitioned so that $\boldsymbol{\mu}_1$ is a $p \times 1$ vector and $\mathbf{V}_{11}$ is a $p \times p$ matrix. Similarly for $\mathbf{R}$.

Proof: The exponent in the numerator of Eq. (10.6.2) can be written as

$$-\frac{1}{2}[(\mathbf{x}-\boldsymbol{\mu})'\mathbf{V}^{-1}(\mathbf{x}-\boldsymbol{\mu})] = -\frac{1}{2}\begin{bmatrix}\mathbf{x}_1-\boldsymbol{\mu}_1\\ \mathbf{x}_2-\boldsymbol{\mu}_2\end{bmatrix}'\begin{bmatrix}\mathbf{R}_{11} & \mathbf{R}_{12}\\ \mathbf{R}_{21} & \mathbf{R}_{22}\end{bmatrix}\begin{bmatrix}\mathbf{x}_1-\boldsymbol{\mu}_1\\ \mathbf{x}_2-\boldsymbol{\mu}_2\end{bmatrix}$$

$$= -\frac{1}{2}\{[(\mathbf{x}_1-\boldsymbol{\mu}_1)'(\mathbf{R}_{11}-\mathbf{R}_{12}\mathbf{R}_{22}^{-1}\mathbf{R}_{21})(\mathbf{x}_1-\boldsymbol{\mu}_1)]$$

$$+ [(\mathbf{x}_2-\boldsymbol{\mu}_2)+\mathbf{R}_{22}^{-1}\mathbf{R}_{21}(\mathbf{x}_1-\boldsymbol{\mu}_1)]'$$

$$\times \mathbf{R}_{22}[(\mathbf{x}_2-\boldsymbol{\mu}_2)+\mathbf{R}_{22}^{-1}\mathbf{R}_{21}(\mathbf{x}_1-\boldsymbol{\mu}_1)]\}$$

$$= -\frac{1}{2}[(\mathbf{x}_1-\boldsymbol{\mu}_1)'\mathbf{V}_{11}^{-1}(\mathbf{x}_1-\boldsymbol{\mu}_1)+(\mathbf{x}_2-\mathbf{h})'\mathbf{R}_{22}(\mathbf{x}_2-\mathbf{h})], \tag{10.6.6}$$

where we have defined $\mathbf{h}$ by $\mathbf{h} = \boldsymbol{\mu}_2 - \mathbf{R}_{22}^{-1}\mathbf{R}_{21}(\mathbf{x}_1-\boldsymbol{\mu}_1)$. Notice we have replaced $\mathbf{R}_{11}-\mathbf{R}_{12}\mathbf{R}_{22}^{-1}\mathbf{R}_{21}$ with $\mathbf{V}_{11}^{-1}$ by using (1) of Theorem 8.2.1. Notice that the elements in $\mathbf{x}_1$ are constants with respect to the variables of integration. So, if we denote the expression on the left of Eq. (10.6.4) by S, we get

$$S = \frac{e^{-(1/2)[(\mathbf{x}_1-\boldsymbol{\mu}_1)'\mathbf{V}_{11}^{-1}(\mathbf{x}_1-\boldsymbol{\mu}_1)]}}{(2\pi)^{n/2}\,|\mathbf{V}|^{1/2}}$$

$$\times \int_{-\infty}^{\infty}\cdots\int_{-\infty}^{\infty} e^{-(1/2)(\mathbf{x}_2-\mathbf{h})'\mathbf{R}_{22}(\mathbf{x}_2-\mathbf{h})}\,dx_{p+1}\cdots dx_n. \tag{10.6.7}$$

By Corollary 10.3.1.2, the value of the multiple integral in Eq. (10.6.7) is $(2\pi)^{(n-p)/2}|\mathbf{R}_{22}|^{-1/2}$. Therefore,

$$S = \frac{e^{-(1/2)[(\mathbf{x}_1-\boldsymbol{\mu}_1)'\mathbf{V}_{11}^{-1}(\mathbf{x}_1-\boldsymbol{\mu}_1)]}}{(2\pi)^{p/2}\,|\mathbf{V}|^{1/2}\,|\mathbf{R}_{22}|^{1/2}}.$$

But by Theorem 8.2.1, $|\mathbf{V}| = |\mathbf{V}_{11}|/|\mathbf{R}_{22}|$, so we get

$$S = \frac{e^{-(1/2)[(\mathbf{x}_1-\boldsymbol{\mu}_1)'\mathbf{V}_{11}^{-1}(\mathbf{x}_1-\boldsymbol{\mu}_1)]}}{(2\pi)^{p/2}\,|\mathbf{V}_{11}|^{1/2}}, \tag{10.6.8}$$

which proves the theorem. ▌

10.7 Examples

We now illustrate some of the previous theorems with examples.

Example 10.7.1. Evaluate the integral

$$\int_{-\infty}^{\infty}\int_{-\infty}^{\infty} e^{-(3x_1^2+4x_1x_2+2x_2^2)}\,dx_1\,dx_2 .$$

Clearly the exponent can be written

$$-\frac{1}{2}Q = -\frac{1}{2}\mathbf{x}'\mathbf{R}\mathbf{x}, \quad \text{where} \quad \mathbf{R} = \begin{bmatrix} 6 & 4 \\ 4 & 4 \end{bmatrix};$$

observe that $\mathbf{R}$ is positive definite. So, by Corollary 10.3.1.2, the value of the integral is $1/K$ or $|\mathbf{R}|^{1/2}2\pi = \pi/\sqrt{2}$.

Example 10.7.2. Evaluate the integral

$$I = \int_{-\infty}^{\infty}\int_{-\infty}^{\infty} x_2(x_1 - 2)e^{-(3x_1^2-4x_1x_2+2x_2^2)}\,dx_1\,dx_2 . \tag{10.7.1}$$

It can be written as

$$I = \int_{-\infty}^{\infty}\int_{-\infty}^{\infty} (x_1x_2 - 2x_2)\,e^{-(3x_1^2-4x_1x_2+2x_2^2)}\,dx_1\,dx_2 ,$$

where we identify the quantities in Theorem 10.5.1 (actually in Eq. (10.5.1)), with the quantities here as follows:

$$\mathbf{x}'\mathbf{A}\mathbf{x} = x_1x_2 , \quad \text{so} \quad \mathbf{A} = \begin{bmatrix} 0 & \frac{1}{2} \\ \frac{1}{2} & 0 \end{bmatrix},$$

$$\mathbf{x}'\mathbf{a} = -2x_2 , \quad \text{so} \quad \mathbf{a} = \begin{bmatrix} 0 \\ -2 \end{bmatrix},$$

$$a_0 = 0,$$

$$\mathbf{x}'\mathbf{B}\mathbf{x} = 3x_1^2 - 4x_1x_2 + 2x_2^2, \quad \text{so} \quad \mathbf{B} = \begin{bmatrix} 3 & -2 \\ -2 & 2 \end{bmatrix},$$

$\mathbf{x}'\mathbf{b} = 0, \quad \text{so} \quad \mathbf{b} = \mathbf{0},$

$b_0 = 0, \quad \text{and}$

$n = 2.$

Now $\mathbf{B}$ is a positive definite matrix, so, by Eq. (10.5.2), the value of the integral is

$$I = \frac{\pi}{2\sqrt{2}}.$$

Example 10.7.3. Let the 2×1 random vector $\mathbf{x} = \begin{bmatrix} x_1 \\ x_2 \end{bmatrix}$ have a normal density with mean $\boldsymbol{\mu} = \begin{bmatrix} 6 \\ 3 \end{bmatrix}$ and the covariance matrix $\mathbf{V} = \begin{bmatrix} 3 & 1 \\ 1 & 2 \end{bmatrix}$. Find the density of the random variable x_1. By Theorem 10.6.1 the random variable x_1 has a normal density with mean $\mu_1 = 6$ and the covariance matrix $\mathbf{V}_{11} = v_{11} = 3$ (notice that x_1, v_{11} and μ_1 are scalars in this example). So we can write

$$N(x_1; \mu_1, V_{11}) = \frac{1}{\sqrt{6\pi}} e^{-(1/6)(x_1 - 6)^2}, \qquad -\infty < x_1 < \infty$$

for the density of the random variable x_1. Another way to solve this example of course is to evaluate the integral

$$\int_{-\infty}^{\infty} f(x_1, x_2)\, dx_2,$$

where $f(x_1, x_2)$ is the 2-variate normal density with mean $\boldsymbol{\mu}$ and covariance matrix $\mathbf{V}$ given above. The result is given in Eq. (10.6.4).

Example 10.7.4. Assume that the multiple integral

$$M(\theta_1, \theta_2, \ldots, \theta_n) = \int_{-\infty}^{\infty} \cdots \int_{-\infty}^{\infty} e^{\mathbf{x}'\boldsymbol{\theta}} f(x_1, x_2, \ldots, x_n)\, dx_1 \cdots dx_n \qquad (10.7.2)$$

exists for all values of θ_i such that $|\theta_i| < a$ for some $a > 0$, $i = 1, 2, \ldots, n$. M is called the *moment generating function* of the random vector $\mathbf{x}$ where the the density of the elements of $\mathbf{x}$ is $f(x_1, \ldots, x_n)$.

If the random vector $\mathbf{x}$ has a multivariate normal density with mean vector $\boldsymbol{\mu}$ and covariance matrix $\mathbf{V}$, find the moment generating function of $\mathbf{x}$.

By referring to Eq. (10.7.2), we have

$$M(\theta_1, \ldots, \theta_n) = \int_{-\infty}^{\infty} \cdots \int_{-\infty}^{\infty} e^{\mathbf{x}'\boldsymbol{\theta}} \frac{1}{(2\pi)^{n/2}|\mathbf{V}|^{1/2}} e^{-(1/2)(\mathbf{x}-\boldsymbol{\mu})'\mathbf{V}^{-1}(\mathbf{x}-\boldsymbol{\mu})}\, dx_1 \cdots dx_n$$

$$= \frac{1}{(2\pi)^{n/2}|\mathbf{V}|^{1/2}} \int_{-\infty}^{\infty} \cdots \int_{-\infty}^{\infty} e^{-[(1/2)\mathbf{x}'\mathbf{V}^{-1}\mathbf{x} - \mathbf{x}'(\mathbf{V}^{-1}\boldsymbol{\mu}+\boldsymbol{\theta}) + (1/2)\boldsymbol{\mu}'\mathbf{V}^{-1}\boldsymbol{\mu}]}\, dx_1 \cdots dx_n.$$

By using Theorem 10.5.1, we get

$$M(\theta_1, \theta_2, \ldots, \theta_n) = e^{\boldsymbol{\mu}'\boldsymbol{\theta} + (1/2)\boldsymbol{\theta}'\mathbf{V}\boldsymbol{\theta}},$$

and this is the moment generating function of the random vector $\mathbf{x}$ that has a normal density with mean vector $\boldsymbol{\mu}$ and covariance matrix $\mathbf{V}$.

10.8 Derivatives

In many situations, it is necessary to obtain the partial derivatives of a function with respect to a number of variables. For example, consider the function f of the real variables x_1, x_2, and x_3 given by

$$f(x_1, x_2, x_3) = 6x_1^2 - 2x_1x_2 + 2x_3^2; \qquad -\infty < x_i < \infty, \quad i = 1, 2, 3, \tag{10.8.1}$$

and suppose that it is necessary to obtain the three partial derivatives

$$\frac{\partial f}{\partial x_1}, \quad \frac{\partial f}{\partial x_2}, \quad \text{and} \quad \frac{\partial f}{\partial x_3}. \tag{10.8.2}$$

We recognize that f can be written as a function of the vector, $\mathbf{x}$, where

$$\mathbf{x} = \begin{bmatrix} x_1 \\ x_2 \\ x_3 \end{bmatrix}, \tag{10.8.3}$$

and it may be desirable to express the three partial derivatives as a vector. We define this by

$$\frac{\partial f}{\partial \mathbf{x}} = \begin{bmatrix} \dfrac{\partial f}{\partial x_1} \\ \dfrac{\partial f}{\partial x_2} \\ \dfrac{\partial f}{\partial x_3} \end{bmatrix} \tag{10.8.4}$$

and obtain

$$\frac{\partial f}{\partial \mathbf{x}} = \begin{bmatrix} 12x_1 \quad - 2x_2 \\ -2x_1 \\ 4x_3 \end{bmatrix} \tag{10.8.5}$$

from Eq. (10.8.1). This leads to the next definition.

Definition 10.8.1

Derivative of a Function with Respect to a Vector. *Let f be a function of k independent real variables $x_1, x_2, \ldots, x_k$. The derivative of the function f with respect to the vector* $\mathbf{x}$, *where*

$$\mathbf{x} = \begin{bmatrix} x_1 \\ x_2 \\ \vdots \\ x_k \end{bmatrix}, \tag{10.8.6}$$

is denoted by $\partial f/\partial \mathbf{x}$ and is defined by

$$\frac{\partial f}{\partial \mathbf{x}} = \begin{bmatrix} \dfrac{\partial f}{\partial x_1} \\ \dfrac{\partial f}{\partial x_2} \\ \vdots \\ \dfrac{\partial f}{\partial x_k} \end{bmatrix}. \tag{10.8.7}$$

We now state and prove some theorems that are useful in statistical applications.

Theorem 10.8.1

Let ℓ be a linear function of k independent real variables defined by $\ell(\mathbf{x}) = \sum_{i=1}^{k} a_i x_i = \mathbf{a}'\mathbf{x} = \mathbf{x}'\mathbf{a}$, where

$$\mathbf{a} = \begin{bmatrix} a_1 \\ a_2 \\ \vdots \\ a_k \end{bmatrix} \tag{10.8.8}$$

and the a_i are any constants. Then

$$\frac{\partial \ell}{\partial \mathbf{x}} = \mathbf{a}. \tag{10.8.9}$$

Proof: The t-th element of $\partial\ell/\partial\mathbf{x}$ is, by definition, equal to $\partial\ell/\partial x_t$ and it is clearly a_t. ∎

Theorem 10.8.2

Let q be a quadratic form in the k independent real variables $x_1, x_2, \ldots, x_k$ defined by

$$q(\mathbf{x}) = \mathbf{x}'\mathbf{A}\mathbf{x}, \tag{10.8.10}$$

where $\mathbf{A} = [a_{ij}]$ is a $k \times k$ symmetric matrix of constants. Then

$$\frac{\partial \mathbf{q}}{\partial \mathbf{x}} = 2\mathbf{A}\mathbf{x}. \tag{10.8.11}$$

Proof: We can write

$$q(\mathbf{x}) = \sum_{j=1}^{k} \sum_{i=1}^{k} x_i x_j a_{ij}.$$

The t-th element of $\partial q/\partial \mathbf{x}$ is $\partial q/\partial x_t$, and clearly

$$\left[\frac{\partial q}{\partial x_t}\right] = \left[\sum_{j=1}^{k} x_j a_{tj} + \sum_{i=1}^{k} x_i a_{it}\right] = 2\left[\sum_{j=1}^{k} x_j a_{tj}\right] = 2\mathbf{Ax}, \qquad (10.8.12)$$

since $\mathbf{A}$ is symmetric. ∎

Definition 10.8.2

Derivative of a Function with Respect to a Matrix. *Let f be a function of the mn independent real variables $x_{11}, x_{12}, \ldots, x_{mn}$ or, in other words, a function of the $m \times n$ matrix $\mathbf{X}$ defined by*

$$\mathbf{X} = \begin{bmatrix} x_{11} & x_{12} & \cdots & x_{1n} \\ x_{21} & x_{22} & \cdots & x_{2n} \\ \vdots & \vdots & & \vdots \\ x_{m1} & x_{m2} & \cdots & x_{mn} \end{bmatrix}, \qquad (10.8.13)$$

and assume that each partial derivative $\partial f/\partial x_{ij}$ exists. Then the derivative of f with respect to the matrix $\mathbf{X}$ is denoted by $\partial f/\partial \mathbf{X}$ and defined by

$$\frac{\partial f}{\partial \mathbf{X}} = \left[\frac{\partial f}{\partial x_{ij}}\right]. \qquad (10.8.14)$$

The definition states that $\partial f/\partial \mathbf{X}$ is an $m \times n$ matrix and that the ij-th element of this matrix is $\partial f/\partial x_{ij}$.

Theorem 10.8.3

Let f be defined by

$$f(\mathbf{X}) = \mathbf{a}'\mathbf{Xb}, \qquad (10.8.15)$$

where $\mathbf{a}$ is an $m \times 1$ vector of constants, $\mathbf{b}$ is an $n \times 1$ vector of constants, and $\mathbf{X}$ is an $m \times n$ matrix of independent real variables. Then

$$\frac{\partial f}{\partial \mathbf{X}} = \mathbf{ab}'. \qquad (10.8.16)$$

Proof: We can write $f(\mathbf{X}) = \sum_{p=1}^{m} \sum_{q=1}^{n} a_p b_q x_{pq}$, and clearly $\partial f / \partial x_{ij} = a_i b_j$, so

$$\frac{\partial f}{\partial \mathbf{X}} = [a_i b_j] = \mathbf{ab}'. \quad \blacksquare \tag{10.8.17}$$

Theorem 10.8.4

Let **a** *be a* $k \times 1$ *vector of constants and let* **X** *be a* $k \times k$ *symmetric matrix of independent real variables (except that* $x_{ij} = x_{ji}$*) and define the function* u *by*

$$u(\mathbf{X}) = \mathbf{a}'\mathbf{X}\mathbf{a}; \tag{10.8.18}$$

then

$$\frac{\partial u}{\partial \mathbf{X}} = 2\mathbf{aa}' - \mathbf{D}_{\mathbf{aa}'} \tag{10.8.19}$$

where $\mathbf{D}_{\mathbf{aa}'}$ *is defined to be a* $k \times k$ *diagonal matrix whose* i-th *diagonal element is equal to the diagonal element of the matrix* $\mathbf{aa}'$.

Proof: Since $u(\mathbf{X}) = \sum_{q=1}^{k} \sum_{p=1}^{k} a_p a_q x_{pq}$, then

$$\frac{\partial u}{\partial x_{ij}} = a_i a_j + a_j a_i, \quad \text{if} \quad i \neq j,$$

and (10.8.20)

$$\frac{\partial u}{\partial x_{ii}} = a_i^2.$$

Hence

$$\frac{\partial u}{\partial x_{ij}} = 2a_i a_j - \delta_{ij} a_i a_j,$$

where δ_{ij} is the Kronecker delta, and hence

$$\frac{\partial u}{\partial \mathbf{X}} = 2\mathbf{aa}' - \mathbf{D}_{\mathbf{aa}'}. \quad \blacksquare$$

In the discussion above we have considered the partial derivative of a scalar function of independent variables with respect to each variable. There are also situations when each element in a matrix is a function of other real variables and we want to obtain the partial derivatives of each function with respect to one of these variables. For example, suppose

$$\mathbf{Y} = \begin{bmatrix} y_{11} & y_{12} \\ y_{21} & y_{22} \end{bmatrix},$$

and $y_{11} = x_{11} + x_{21}, y_{12} = x_{22}^2, y_{21} = x_{12}^{1/2}, y_{22} = x_{21} + x_{22}^2$ (that is, each y_{pq} is a function of independent variables x_{ij}). Then we may want to determine $\partial y_{pq}/\partial x_{ij}$ for any i, j and p, q. We shall do this by defining and proving some theorems on the derivative of a matrix with respect to a scalar.

Definition 10.8.3

Derivative of a Matrix with Respect to a Scalar. *Let* $\mathbf{Y}$ *be a* $k \times k$ *matrix with elements denoted by* y_{pq} *where* $y_{pq} = f_{pq}(x_{11}, x_{12}, \ldots, x_{mn})$; *that is, each* y_{pq} *is a function of an* $m \times n$ *matrix of independent real variables* x_{ij}. *The partial derivative of the matrix* $\mathbf{Y}$ *with respect to the scalar* x_{ij}, *denoted by* $\partial \mathbf{Y}/\partial x_{ij}$, *is defined to be*

$$\frac{\partial \mathbf{Y}}{\partial x_{ij}} = \left[\frac{\partial f_{pq}(x_{11}, \ldots, x_{mn})}{\partial x_{ij}}\right].$$

Since we sometimes need to take the derivative of a determinant with respect to the elements in the determinant, we now consider two theorems on this subject.

Theorem 10.8.5

Let $\mathbf{X}$ *be a* $k \times k$ *matrix of independent real variables* x_{ij}; *then*

$$\frac{\partial |\mathbf{X}|}{\partial \mathbf{X}} = [X_{ij}], \tag{10.8.21}$$

where X_{ij} *is the cofactor of* x_{ij}.

Proof: By Theorem 1.5.10, we can write

$$|\mathbf{X}| = \sum_{p=1}^{k} x_{pj} X_{pj}, \qquad j = 1, 2, \ldots, k,$$

and X_{ij} does not involve x_{ij} and hence $\partial |\mathbf{X}|/\partial x_{ij} = X_{ij}$. ∎

Theorem 10.8.6

Let $\mathbf{Y}$ *be a* $k \times k$ *matrix such that each element* y_{pq} *of* $\mathbf{Y}$ *is a real function of mn independent real variables* $x_{11}, \ldots, x_{mn}$; *that is,* $y_{pq} = f_{pq}(x_{11}, \ldots, x_{mn})$. *Then*

$$\frac{\partial|\mathbf{Y}|}{\partial x_{ij}} = \operatorname{tr}\left[\mathbf{Y}^* \frac{\partial \mathbf{Y}'}{\partial x_{ij}}\right] \qquad (10.8.22)$$

for any fixed i and j where $i = 1, 2, \ldots, m$; $j = 1, 2, \ldots, n$ *and where* $\mathbf{Y}^*$ *is the matrix of cofactors of the matrix* $\mathbf{Y}$.

Proof: From the theory of partial derivatives, we obtain

$$\frac{\partial|\mathbf{Y}|}{\partial x_{ij}} = \sum_{q=1}^{k}\sum_{p=1}^{k} \frac{\partial|\mathbf{Y}|}{\partial y_{pq}} \frac{\partial y_{pq}}{\partial x_{ij}},$$

where in the expression $\partial|\mathbf{Y}|/\partial y_{pq}$ the quantities y_{pq} are considered to be independent real variables and hence by Theorem 10.8.5 we obtain $[\partial|\mathbf{Y}|/\partial y_{pq}] = [Y_{pq}] = \mathbf{Y}^*$, where Y_{pq} is the cofactor of y_{pq} in $\mathbf{Y}$.
Thus we get

$$\frac{\partial|\mathbf{Y}|}{\partial x_{ij}} = \sum_{q=1}^{k}\sum_{p=1}^{k} Y_{pq} \frac{\partial y_{pq}}{\partial x_{ij}}.$$

Now $\partial\mathbf{Y}/\partial x_{ij} = [\partial y_{pq}/\partial x_{ij}]$ by Definition 10.8.3 and hence

$$\sum_{q=1}^{k}\sum_{p=1}^{k} Y_{pq} \frac{\partial y_{pq}}{\partial x_{ij}} = \operatorname{tr}\left[\mathbf{Y}^* \frac{\partial \mathbf{Y}'}{\partial x_{ij}}\right],$$

and the theorem is proved. ▮

Theorem 10.8.7

Let $\mathbf{X}$ *be a* $k \times k$ *symmetric matrix of independent real variables (except* $x_{ij} = x_{ji}$); *then*

$$\frac{\partial|\mathbf{X}|}{\partial \mathbf{X}} = 2[X_{ij}] - \mathbf{D}_{[X_{ij}]}, \qquad (10.8.23)$$

where X_{ij} *is the cofactor of* x_{ij} *and* $\mathbf{D}_{[X_{ij}]}$ *is a diagonal matrix with i-*th *diagonal element equal to* X_{ii}, *the cofactor of* x_{ii}.

Proof: The proof of this theorem is obtained by using Theorem 10.8.6. ▮

Theorem 10.8.8

Let $\mathbf{X}$ *be a* $k \times k$ *symmetric nonsingular matrix of independent real variables (except* $x_{ij} = x_{ji}$*); then*

$$\frac{\partial(\log |\mathbf{X}|)}{\partial \mathbf{X}} = 2\mathbf{X}^{-1} - \mathbf{D}_{\mathbf{X}^{-1}}, \tag{10.8.24}$$

where $\mathbf{D}_{\mathbf{X}^{-1}}$ *is a diagonal matrix with* i-th *diagonal element equal to the* i-th *diagonal element of* $\mathbf{X}^{-1}$.

Proof: Clearly $\partial(\log |\mathbf{X}|)/\partial \mathbf{X} = (1/|\mathbf{X}|)\partial|\mathbf{X}|/\partial \mathbf{X}$, and by Theorem 10.8.7 the result follows. ∎

Theorem 10.8.9

Let $\mathbf{X}$ *be a* $k \times k$ *matrix of independent real variables; then*

$$\frac{\partial[\operatorname{tr}(\mathbf{X})]}{\partial \mathbf{X}} = \mathbf{I}. \tag{10.8.25}$$

Proof: $\operatorname{tr}(\mathbf{X}) = \sum_{t=1}^{k} x_{tt}$ and clearly

$$\frac{\partial\left(\sum_{t=1}^{k} x_{tt}\right)}{\partial x_{ij}} = 0 \quad \text{if } i \neq j$$

and

$$\frac{\partial\left(\sum_{t=1}^{k} x_{tt}\right)}{\partial x_{ii}} = 1 \quad \text{for } i = 1, 2, \ldots, k. \quad ∎$$

The next theorem and corollary are useful in situations in which one wants to take the derivative of the inverse of a matrix $\mathbf{X}$ with respect to the elements of $\mathbf{X}$.

Theorem 10.8.10

Let $\mathbf{X}$ *be a* $k \times k$ *nonsingular matrix of independent real variables. Then*

$$\frac{\partial \mathbf{X}^{-1}}{\partial x_{pq}} = -\mathbf{X}^{-1}\mathbf{\Delta}_{pq}\mathbf{X}^{-1},$$

where $\mathbf{\Delta}_{pq}$ *is a* $k \times k$ *matrix with* pq-th *element equal to plus one and the remaining elements equal to zero.*

Proof: To simplify the notation we let **A** denote $\mathbf{X}^{-1}$. Consider the *ij*-th element of $\mathbf{I} = \mathbf{AX}$, which is $\delta_{ij} = \sum_{t=1}^{k} a_{it}x_{tj}$. If we take the derivative of both sides with respect to x_{pq}, we obtain

$$0 = \frac{\partial}{\partial x_{pq}}\left(\sum_{t=1}^{k} a_{it}x_{tj}\right) = \sum_{t=1}^{k} \frac{\partial}{\partial x_{pq}}(a_{it}x_{tj}) = \sum_{t=1}^{k}\left[a_{it}\frac{\partial x_{tj}}{\partial x_{pq}} + \frac{\partial a_{it}}{\partial x_{pq}}x_{tj}\right].$$

But $\partial x_{tj}/\partial x_{pq} = 0$, unless $t = p$ and $j = q$, in which case it is equal to 1. By using the notation

$$\frac{\partial x_{tj}}{\partial x_{pq}} = \delta_{tj}^{pq}$$

where $\delta_{tj}^{pq} = 0$, unless $p = t$ and $j = q$, and where $\delta_{pq}^{pq} = 1$, we get

$$0 = \sum_{t=1}^{k} a_{it}\delta_{tj}^{pq} + \sum_{t=1}^{k}\frac{\partial a_{it}}{\partial x_{pq}}x_{tj} \quad \text{or} \quad a_{ip}\delta_{pj}^{pq} = -\sum_{t=1}^{k}\frac{\partial a_{it}}{\partial x_{pq}}x_{tj}.$$

But this is the *ij*-th element of

$$\mathbf{A}\boldsymbol{\Delta}_{pq} = -\frac{\partial\mathbf{A}}{\partial x_{pq}}\mathbf{X},$$

where δ_{ij}^{pq} is the *ij*-th element of $\boldsymbol{\Delta}_{pq}$, where $\boldsymbol{\Delta}_{pq}$ is defined in the theorem. Hence we have

$$\frac{\partial\mathbf{X}^{-1}}{\partial x_{pq}} = -\mathbf{X}^{-1}\boldsymbol{\Delta}_{pq}\mathbf{X}^{-1}. \quad \blacksquare$$

Corollary 10.8.10

If **X** *is a* $k \times k$ *nonsingular symmetric matrix of independent real variables (except* $x_{ij} = x_{ji}$*),*

$$\frac{\partial\mathbf{X}^{-1}}{\partial x_{pq}} = -\mathbf{X}^{-1}\boldsymbol{\Delta}_{pq}^{*}\mathbf{X}^{-1}$$

where $\boldsymbol{\Delta}_{pq}^{*}$ *is an* $n \times n$ *matrix whose every element is zero except the* pq-th *and the* qp-th *elements, and these are equal to plus one.*

In applications of derivatives of matrices in statistics the multivariate normal density is of particular importance. The next theorem contains a number of important results related to derivatives of variables that occur in the multivariate normal. The proofs are immediate applications of the theorems above. Additional material can be found in [D–6], [D–7].

Theorem 10.8.11

Let $\mathbf{x}$ *and* $\mathbf{y}$ *be* $k \times 1$ *vectors, let* $\mathbf{V}$ *be a* $k \times k$ *positive definite matrix, and let all variables be independent except* $v_{ij} = v_{ji}$, *all* $i \neq j$. *Then*

$$\frac{\partial}{\partial \mathbf{y}}[e^{-(1/2)(\mathbf{x}-\mathbf{y})'\mathbf{V}^{-1}(\mathbf{x}-\mathbf{y})}] = e^{-(1/2)(\mathbf{x}-\mathbf{y})'\mathbf{V}^{-1}(\mathbf{x}-\mathbf{y})}[\mathbf{V}^{-1}(\mathbf{x}-\mathbf{y})], \tag{10.8.26}$$

$$\frac{\partial}{\partial \mathbf{V}^{-1}}[e^{-(1/2)(\mathbf{x}-\mathbf{y})'\mathbf{V}^{-1}(\mathbf{x}-\mathbf{y})}] = e^{-(1/2)(\mathbf{x}-\mathbf{y})'\mathbf{V}^{-1}(\mathbf{x}-\mathbf{y})}\left[\frac{1}{2}\mathbf{D}_{(\mathbf{x}-\mathbf{y})(\mathbf{x}-\mathbf{y})'} - (\mathbf{x}-\mathbf{y})(\mathbf{x}-\mathbf{y})'\right], \tag{10.8.27}$$

$$\frac{\partial}{\partial \mathbf{V}}[(\mathbf{x}-\mathbf{y})'\mathbf{V}^{-1}(\mathbf{x}-\mathbf{y})] = -2\mathbf{V}^{-1}(\mathbf{x}-\mathbf{y})(\mathbf{x}-\mathbf{y})'\mathbf{V}^{-1} + \mathbf{D}_{\mathbf{V}^{-1}(\mathbf{x}-\mathbf{y})(\mathbf{x}-\mathbf{y})'\mathbf{V}^{-1}}. \tag{10.8.28}$$

In making transformations in multiple integrals, it is necessary to evaluate Jacobians (Sec. 10.2). The transformation from a vector $\mathbf{x}$ to a vector $\mathbf{y}$ can be written

$$\mathbf{y} = \begin{bmatrix} f_1(\mathbf{x}) \\ f_2(\mathbf{x}) \\ \vdots \\ f_k(\mathbf{x}) \end{bmatrix}.$$

The matrix of the Jacobian, which we denote $\partial\mathbf{y}/\partial\mathbf{x}$, is defined by

$$\frac{\partial \mathbf{y}}{\partial \mathbf{x}} = \left[\frac{\partial y_i}{\partial x_j}\right].$$

We shall prove one theorem that we have used in finding the Jacobian of a transformation in the multivariate normal distribution.

Theorem 10.8.12

Let the transformation from the $k \times 1$ vector $\mathbf{x}$ to the $k \times 1$ vector $\mathbf{y}$ be given by $\mathbf{y} = \mathbf{A}\mathbf{x} + \mathbf{c}$, where $\mathbf{A}$ is a $k \times k$ matrix of constants and $\mathbf{c}$ is a $k \times 1$ vector of constants. The matrix of the Jacobian of the transformation is $\mathbf{A}$.

Proof: Clearly $y_p = \sum_{q=1}^{k} a_{pq}x_q + c_p$, $p = 1, 2, \ldots, k$ and $\partial y_i/\partial x_j = a_{ij}$, so

$$\frac{\partial \mathbf{y}}{\partial \mathbf{x}} = [a_{ij}] = \mathbf{A}. \quad \blacksquare$$

Example 10.8.1 Consider the linear model $\mathbf{y} = \mathbf{X}\boldsymbol{\beta} + \mathbf{e}$ defined in the Introduction. To find the least squares estimator of $\boldsymbol{\beta}$, we find a value of $\boldsymbol{\beta}$, denoted by $\hat{\boldsymbol{\beta}}$, that minimizes $\mathbf{e}'\mathbf{e}$. To do this we consider $\mathbf{e}'\mathbf{e}$ as a function of $\boldsymbol{\beta}$ and write

$$f(\boldsymbol{\beta}) = \mathbf{e}'\mathbf{e} = (\mathbf{y} - \mathbf{X}\boldsymbol{\beta})'(\mathbf{y} - \mathbf{X}\boldsymbol{\beta})$$

for $\boldsymbol{\beta} \in E_p$. One method of finding the value of $\boldsymbol{\beta}$ that minimizes $f(\boldsymbol{\beta})$ is to solve the system of equations $\partial f(\boldsymbol{\beta})/\partial \beta_i = 0$, $i = 1, 2, \ldots, p$, or, in other words, to solve

$$\frac{\partial[f(\boldsymbol{\beta})]}{\partial \boldsymbol{\beta}} = \mathbf{0}.$$

We get

$$\frac{\partial[f(\boldsymbol{\beta})]}{\partial \boldsymbol{\beta}} = \frac{\partial}{\partial \boldsymbol{\beta}}[\mathbf{y}'\mathbf{y} - 2\mathbf{y}'\mathbf{X}\boldsymbol{\beta} + \boldsymbol{\beta}'\mathbf{X}'\mathbf{X}\boldsymbol{\beta}] = -2\mathbf{X}'\mathbf{y} + 2\mathbf{X}'\mathbf{X}\boldsymbol{\beta}$$

by using Theorems 10.8.1 and 10.8.2. If $\hat{\boldsymbol{\beta}}$ is a vector such that $\partial[f(\boldsymbol{\beta})]/\partial \boldsymbol{\beta} = \mathbf{0}$, we obtain the normal equations $\mathbf{X}'\mathbf{X}\hat{\boldsymbol{\beta}} = \mathbf{X}'\mathbf{y}$, and clearly these have a solution and any solution minimizes $\mathbf{e}'\mathbf{e}$ and hence is a least squares solution of $\mathbf{y} = \mathbf{X}\boldsymbol{\beta} + \mathbf{e}$.

10.9 Expected Values of Quadratic Forms

In linear model theory in statistics it is often necessary to find the expected value and the variance of quadratic forms of normal random variables. It is also often required to obtain the covariance of two quadratic forms and the expectation of the product of two or more quadratic forms in normal random variables. Theorem 10.5.1 can be used to find the expected value of a quadratic form in normal random variables. In this section we show how some of the theorems in Secs. 9.2 and 9.3 can help with the problem. First we state a theorem that will be used quite often throughout this section. We assume throughout this section that $\mathbf{V}$ is p.d.

Theorem 10.9.1

Let $\mathbf{z} = [z_i]$ *be an* $n \times 1$ *random vector that is distributed* $N(\mathbf{z}\colon \mathbf{0}, \mathbf{I})$; *see Eq.* (10.6.2). *Then*

(1) $\mathscr{E}[z_p\mathbf{z}\,\mathbf{z}'] = \mathbf{0}$,
(2) $\mathscr{E}[z_p z_q \mathbf{z}\mathbf{z}'] = \mathbf{\Delta}_{pq} + \mathbf{\Delta}_{qp} + \delta_{pq}\,\mathbf{I}$,

where p *and* q *can assume any values* $1, 2, \ldots, n$; *where* δ_{pq} *is the Kronecker delta and* $\mathbf{\Delta}_{pq}$ *is defined in Sec.* 9.3.

Proof: It is easily shown that $\mathscr{E}[z_i^k] = 0$ if k is an odd integer, and from this result (1) follows. Also $\mathscr{E}[z_i^2] = 1$, $\mathscr{E}[z_i^4] = 3$, $\mathscr{E}[z_i^6] = 15$. The (rs)-th element of $\mathbf{z}\mathbf{z}'$ is $z_r z_s$, so the (rs)-th element of the $n \times n$ matrix $z_p z_q \mathbf{z}\mathbf{z}'$ is $z_p z_q z_r z_s$ and clearly

$$\mathscr{E}[z_p z_q z_r z_s] = \begin{cases} 1 & \text{if } p = q \neq r = s \\ 1 & \text{if } p = r \neq q = s \\ 1 & \text{if } p = s \neq q = r \\ 3 & \text{if } p = q = r = s \\ 0 & \text{otherwise} \end{cases}.$$

The result (2) is the matrix representation of this expectation. ∎

Theorem 10.9.2

Let $\mathbf{z}$ *be an* $n \times 1$ *vector that is distributed* $N(\mathbf{z}\colon \mathbf{0}, \mathbf{I})$. *Then*

(1) $\mathscr{E}(\mathbf{z}\mathbf{z}' \times \mathbf{z}\mathbf{z}') = \mathbf{I}_{n^2} + \mathbf{K}_{nn} + [\text{vec}(\mathbf{I}_n)]\,[\text{vec}(\mathbf{I}_n)]'$,
(2) $\mathscr{E}[\mathbf{z} \times \mathbf{z}\mathbf{z}'] = \mathbf{0}$.

Proof: Clearly $\mathbf{zz}' \times \mathbf{zz}' = \Sigma_{p=1}^{n} \Sigma_{q=1}^{n} [\mathbf{zz}' \times (z_p z_q \mathbf{\Delta}_{pq})]$ by (9) of Theorem 9.3.1. But $\mathbf{zz}' \times (z_p z_q \mathbf{\Delta}_{pq}) = \mathbf{zz}' z_p z_q \times \mathbf{\Delta}_{pq}$ and $\mathscr{E}[\mathbf{zz}' z_p z_q \times \mathbf{\Delta}_{pq}] = (\mathbf{\Delta}_{pq} + \mathbf{\Delta}'_{pq} + \delta_{pq}\mathbf{I}) \times \mathbf{\Delta}_{pq}$, where $\mathbf{\Delta}_{pq}$ is an $n \times n$ unit matrix. Hence

$$\begin{aligned}\mathscr{E}(\mathbf{zz}' \times \mathbf{zz}') &= \sum_{p=1}^{n} \sum_{q=1}^{n} [(\delta_{pq}\mathbf{I} + \mathbf{\Delta}'_{pq} + \mathbf{\Delta}_{pq}) \times \mathbf{\Delta}_{pq}] \\ &= \mathbf{I}_{n^2} + \mathbf{K}_{nn} + [\mathrm{vec}(\mathbf{I}_n)]\,[\mathrm{vec}(\mathbf{I}_n)]'.\end{aligned}$$

This proves (1). The proof of (2) follows from $\mathscr{E}[z_i^k] = 0$ if k is an odd integer. ▮

Corollary 10.9.2

If the $n \times 1$ random vector $\mathbf{z}$ has density $N(\mathbf{z}\colon \mathbf{0}, \mathbf{I})$, and $\mathbf{c}$ is an $n \times 1$ constant vector, and $\mathbf{A}$ and $\mathbf{B}$ are $n \times n$ matrices of constants, then

(1) $\mathscr{E}[\mathbf{zc}' \times \mathbf{zz}'] = \mathscr{E}[\mathbf{zz}' \times \mathbf{zc}'] = \mathscr{E}[\mathbf{cz}' \times \mathbf{zz}'] = \mathscr{E}[(\mathbf{z} \times \mathbf{c})\,(\mathbf{z}' \times \mathbf{z}')] = \mathbf{0}$,
(2) $\mathscr{E}[(\mathbf{z}'\mathbf{Az})(\mathbf{z}'\mathbf{Bc})] = \mathbf{0}$,
(3) *If $\mathbf{z}$ has a density $N(\mathbf{z}\colon \mathbf{0}, \mathbf{V})$, then the results* (1) *and* (2) *are also correct.*

Theorem 10.9.3

Let $\mathbf{z}$ be an $n \times 1$ random vector defined in Theorem 10.9.2. *Then*

(1) $\mathscr{E}[z_i z_j z_h z_k \mathbf{zz}'] = \mathbf{0}$ *if i, j, h, and k are distinct,*
(2) $\mathscr{E}[z_i^2 z_h z_k \mathbf{zz}'] = \mathbf{\Delta}_{hk}^*$ *if i, h, and k are distinct,*
(3) $\mathscr{E}[z_i^3 z_j \mathbf{zz}'] = 3\mathbf{\Delta}_{ij}^*$ *if i and j are distinct,*
(4) $\mathscr{E}[z_i^2 z_h^2 \mathbf{zz}'] = \mathbf{I} + 2\mathbf{\Delta}_{ii} + 2\mathbf{\Delta}_{hh}$ *if i and h are distinct,*
(5) $\mathscr{E}[z_i^4 \mathbf{zz}'] = 3\mathbf{I} + 12\mathbf{\Delta}_{ii}$,

where $\mathbf{\Delta}_{ij}^*$ is defined in Corollary 10.8.10

Proof: The proof is a matrix representation of straightforward evaluations of each expectation. ▮

Theorem 10.9.4

Let $\mathbf{z}$ be defined as in Theorem 10.9.2. *Then*

(1) $\mathscr{E}[z_i^2(\mathbf{zz}' \times \mathbf{zz}')] = \tfrac{1}{2} \Sigma_{s=1}^{n} \Sigma_{t=1}^{n} (\overline{\mathbf{\Delta}}_{st} \times \overline{\mathbf{\Delta}}_{st}) + (\overline{\mathbf{\Delta}}_{ii} \times \mathbf{I}) + (\mathbf{I} \times \mathbf{I}) + (\mathbf{I} \times \overline{\mathbf{\Delta}}_{ii}) + 2\,\Sigma_{s=1}^{n} (\overline{\mathbf{\Delta}}_{is} \times \overline{\mathbf{\Delta}}_{is})$,

(2) $\mathscr{E}[z_i z_j(\mathbf{zz}' \times \mathbf{zz}')] = (\overline{\mathbf{\Delta}}_{ij} \times \mathbf{I}) + \sum_{s=1}^{n} (\overline{\mathbf{\Delta}}_{is} \times \overline{\mathbf{\Delta}}_{js}) + \sum_{s=1}^{n} (\overline{\mathbf{\Delta}}_{js} \times \overline{\mathbf{\Delta}}_{is}) + (\mathbf{I} \times \overline{\mathbf{\Delta}}_{ij})$ *if* $i \neq j$,

(3) $\mathscr{E}[\mathbf{zz}' \times \mathbf{zz}' \times \mathbf{zz}'] = \mathbf{I} \times \mathbf{I} \times \mathbf{I} + \tfrac{1}{2}\{\sum_{j=1}^{n} \sum_{i=1}^{n} (\overline{\mathbf{\Delta}}_{ij} \times \overline{\mathbf{\Delta}}_{ij} \times \mathbf{I}) + \sum_{j=1}^{n} \sum_{i=1}^{n} (\overline{\mathbf{\Delta}}_{ij} \times \mathbf{I} \times \overline{\mathbf{\Delta}}_{ij}) + \sum_{j=1}^{n} \sum_{i=1}^{n} (\mathbf{I} \times \overline{\mathbf{\Delta}}_{ij} \times \overline{\mathbf{\Delta}}_{ij})\} + \sum_{k=1}^{n} \sum_{j=1}^{n} \sum_{i=1}^{n} (\overline{\mathbf{\Delta}}_{jk} \times \overline{\mathbf{\Delta}}_{ik} \times \overline{\mathbf{\Delta}}_{ij})$.

Note: We use $\overline{\mathbf{\Delta}}_{st} = \overline{\mathbf{\Delta}}_{ts} = \mathbf{\Delta}_{st} + \mathbf{\Delta}_{ts} = \mathbf{\Delta}^*_{st} + \delta_{st}\mathbf{I}$.

Proof: We give the proof for (1). The proofs for (2) and (3) are left for the reader. Details can be found in [M–2].

Let $\mathbf{zz}' = \mathbf{Z} = [z_{pq}]$. Thus $z_{pq} = z_p z_q$. Define $\mathbf{W}$ by

$$\mathbf{W} = z_i^2(\mathbf{zz}' \times \mathbf{zz}') = \begin{bmatrix} \mathbf{W}_{11} & \mathbf{W}_{12} & \cdots & \mathbf{W}_{1n} \\ \mathbf{W}_{21} & \mathbf{W}_{22} & \cdots & \mathbf{W}_{2n} \\ \vdots & \vdots & & \vdots \\ \mathbf{W}_{n1} & \mathbf{W}_{n2} & \cdots & \mathbf{W}_{nn} \end{bmatrix},$$

where $\mathbf{W}_{st} = z_i^2 \mathbf{Z} z_s z_t = z_i^2 z_s z_t \mathbf{zz}'$. We get

$$\mathscr{E}[\mathbf{W}_{ss}] = \mathscr{E}[z_i^2 z_s^2 \mathbf{zz}'] = (1 + 2\delta_{is})(\mathbf{I} + 2\mathbf{\Delta}_{ii} + 2\mathbf{\Delta}_{ss})$$

by using (4) and (5) of Theorem 10.9.3. Thus we get

$$\mathscr{E}\begin{bmatrix} \mathbf{W}_{11} & \mathbf{0} & \cdots & \mathbf{0} \\ \mathbf{0} & \mathbf{W}_{22} & \cdots & \mathbf{0} \\ \vdots & \vdots & & \vdots \\ \mathbf{0} & \mathbf{0} & \cdots & \mathbf{W}_{nn} \end{bmatrix} = \sum_{s=1}^{n} \{[(1 + 2\delta_{is})(\mathbf{I} + 2\mathbf{\Delta}_{ii} + 2\mathbf{\Delta}_{ss})] \times \mathbf{\Delta}_{ss}\}.$$

Next we examine $\mathscr{E}[\mathbf{W}_{st}]$ for $s \neq t$. We get for $s \neq t$,

$$\mathscr{E}[z_i^2 z_s z_t \mathbf{zz}'] = \begin{cases} 3\mathbf{\Delta}^*_{is} & \text{if } i = t \\ 3\mathbf{\Delta}^*_{it} & \text{if } i = s \\ \mathbf{\Delta}^*_{st} & \text{if } i, s, \text{ and } t \text{ are distinct} \end{cases}$$

Thus $\mathscr{E}[\mathbf{W}_{st}] = (1 + 2\delta_{is})(1 + 2\delta_{it})\mathbf{\Delta}^*_{st}$ for $s \neq t$ and

$$\mathscr{E}\begin{bmatrix} \mathbf{0} & \mathbf{W}_{12} & \cdots & \mathbf{W}_{1n} \\ \mathbf{W}_{21} & \mathbf{0} & \cdots & \mathbf{W}_{2n} \\ \vdots & \vdots & & \vdots \\ \mathbf{W}_{n1} & \mathbf{W}_{n2} & \cdots & \mathbf{0} \end{bmatrix} = \sum_{\substack{t=1 \\ t \neq s}}^{n} \sum_{s=1}^{n} [(1 + 2\delta_{is})(1 + 2\delta_{it})\mathbf{\Delta}^*_{st}] \times \mathbf{\Delta}_{st}.$$

Thus

$$\mathscr{E}[\mathbf{W}] = \sum_{s=1}^{n} [(1 + 2\delta_{is})(\mathbf{I} + 2\boldsymbol{\Delta}_{ii} + 2\boldsymbol{\Delta}_{ss})] \times \boldsymbol{\Delta}_{ss}$$

$$+ \sum_{t=1}^{n} \sum_{\substack{s=1 \\ s \neq t}}^{n} [(1 + 2\delta_{is})(1 + 2\delta_{it})\boldsymbol{\Delta}_{st}^{*}] \times \boldsymbol{\Delta}_{st}$$

$$= \tfrac{1}{2} \sum_{s=1}^{n} \sum_{t=1}^{n} (\overline{\boldsymbol{\Delta}}_{st} \times \overline{\boldsymbol{\Delta}}_{st}) + (\mathbf{I} \times \overline{\boldsymbol{\Delta}}_{ii}) + (\mathbf{I} \times \mathbf{I}) + (\overline{\boldsymbol{\Delta}}_{ii} \times \mathbf{I})$$

$$+ 2 \sum_{s=1}^{n} (\overline{\boldsymbol{\Delta}}_{is} \times \overline{\boldsymbol{\Delta}}_{is}).$$

This proves (1); the result (2) is proved similarly, and (3) is a matrix representation of (1) and (2). ∎

Note: $\overline{\boldsymbol{\Delta}}_{st} = \boldsymbol{\Delta}_{st}^{*}$ if $s \neq t$. But $\overline{\boldsymbol{\Delta}}_{ss} = 2\boldsymbol{\Delta}_{ss}$ and $\boldsymbol{\Delta}_{ss}^{*} = \boldsymbol{\Delta}_{ss}$.

We demonstrate some cases where these theorems can be used to find expectations of quadratic forms in normal variables and variances of Wishart matrices.

Theorem 10.9.5

Let $\mathbf{z}$ *be an* $n \times 1$ *random vector with density* $N(\mathbf{z}: \mathbf{0}, \mathbf{I})$. *Then*

(1) $\mathscr{E}[\mathbf{z} \times \mathbf{z}] = \text{vec}(\mathbf{I}_n)$,
(2) $\text{Cov}(\mathbf{z} \times \mathbf{z}) = \mathbf{I} + \mathbf{K}_{nn}$.

Proof: To prove (1) we note that $\mathscr{E}[\mathbf{z} \times \mathbf{z}] = \mathscr{E}[\text{vec}(\mathbf{zz}')] = \text{vec}(\mathscr{E}[\mathbf{zz}']) = \text{vec}(\mathbf{I})$. To prove (2) we get $\text{Cov}[\mathbf{z} \times \mathbf{z}] = \mathscr{E}[(\mathbf{z} \times \mathbf{z})(\mathbf{z} \times \mathbf{z})'] - \mathscr{E}[\mathbf{z} \times \mathbf{z}] \times \mathscr{E}[\mathbf{z} \times \mathbf{z}]' = \mathbf{I} + \mathbf{K}_{nn} + [\text{vec}(\mathbf{I})]\,[\text{vec}(\mathbf{I})]' - [\text{vec}(\mathbf{I})]\,[\text{vec}(\mathbf{I})]' = \mathbf{I} + \mathbf{K}_{nn}$. ∎

Theorem 10.9.6

In Theorem 10.9.5, *if* $\mathbf{x}$ *is an* $n \times 1$ *normal random vector with density* $N(\mathbf{x}: \mathbf{0}, \mathbf{V})$, *then*

(1) $\mathscr{E}[\mathbf{x} \times \mathbf{x}] = \text{vec}(\mathbf{V})$,
(2) $\text{Cov}(\mathbf{x} \times \mathbf{x}) = (\mathbf{V} \times \mathbf{V})(\mathbf{I} + \mathbf{K}_{nn}) = (\mathbf{I} + \mathbf{K}_{nn})(\mathbf{V} \times \mathbf{V})$.

Proof: To prove (1) we let $\mathbf{V} = \mathbf{T}'\mathbf{T}$ and let $\mathbf{z} = \mathbf{T}'^{-1}\mathbf{x}$, so $\mathbf{z}$ has density $N(\mathbf{z}: \mathbf{0}, \mathbf{I})$. From Theorem 10.9.5 we get $\mathscr{E}[\mathbf{x} \times \mathbf{x}] = \mathscr{E}[(\mathbf{T}'\mathbf{z}) \times (\mathbf{T}'\mathbf{z})] =$

$\mathscr{E}[(\mathbf{T}' \times \mathbf{T}')(\mathbf{z} \times \mathbf{z})] = (\mathbf{T}' \times \mathbf{T}')\text{vec}(\mathbf{I})$ and from Corollary 9.2.2 this gives us $\mathscr{E}[\mathbf{x} \times \mathbf{x}] = \text{vec}(\mathbf{T}'\mathbf{T}) = \text{vec}(\mathbf{V})$. To prove (2) we get $\text{Cov}(\mathbf{x} \times \mathbf{x}) = \mathscr{E}[(\mathbf{x} \times \mathbf{x})(\mathbf{x} \times \mathbf{x})'] - \mathscr{E}[\mathbf{x} \times \mathbf{x}]\, \mathscr{E}[(\mathbf{x} \times \mathbf{x})']$, and $\mathscr{E}[(\mathbf{x} \times \mathbf{x})(\mathbf{x} \times \mathbf{x})'] = \mathscr{E}[(\mathbf{T}'\mathbf{z} \times \mathbf{T}'\mathbf{z})(\mathbf{T}'\mathbf{z} \times \mathbf{T}'\mathbf{z})' = \mathscr{E}(\mathbf{T}' \times \mathbf{T}')(\mathbf{z} \times \mathbf{z})(\mathbf{z} \times \mathbf{z})'(\mathbf{T} \times \mathbf{T})]$. Use the proof of Theorem 10.9.5 and the result follows. ∎

Theorem 10.9.7

Let $\mathbf{x}$ *be an* $n \times 1$ *random vector with density* $N(\mathbf{x}: \boldsymbol{\mu}, \mathbf{V})$. *Then*

(1) $\mathscr{E}[\mathbf{x} \times \mathbf{x}] = \text{vec}(\mathbf{V}) + (\boldsymbol{\mu} \times \boldsymbol{\mu})$,
(2) $\text{Cov}(\mathbf{x} \times \mathbf{x}) = (\mathbf{I} + \mathbf{K}_{nn})(\mathbf{V} \times \mathbf{V} + \mathbf{V} \times \boldsymbol{\mu}\boldsymbol{\mu}' + \boldsymbol{\mu}\boldsymbol{\mu}' \times \mathbf{V})$.
(3) *rank of* $\text{Cov}(\mathbf{x} \times \mathbf{x}) = (\tfrac{1}{2})n(n + 1)$.

The proof will be omitted. It can be found in [M–2].

Next we state a theorem and a corollary that will be useful when considering covariance of quadratic forms.

Theorem 10.9.8

Let $\mathbf{w}$, $\mathbf{x}$, $\mathbf{y}$, *and* $\mathbf{z}$ *be* $n \times 1$ *vectors and let* $\mathbf{A}$ *and* $\mathbf{B}$ *be* $n \times n$ *matrices. Then*

$$(\mathbf{w}'\mathbf{A}\mathbf{x})(\mathbf{y}'\mathbf{B}\mathbf{z}) = (\mathbf{w}' \times \mathbf{y}')(\mathbf{A} \times \mathbf{B})(\mathbf{x} \times \mathbf{z})$$
$$= \text{tr}[(\mathbf{A} \times \mathbf{B})(\mathbf{x} \times \mathbf{z})(\mathbf{w}' \times \mathbf{y}')]$$
$$= \text{tr}[(\mathbf{A} \times \mathbf{B})(\mathbf{x}\mathbf{w}' \times \mathbf{z}\mathbf{y}')]$$
$$= \text{tr}[(\mathbf{x}\mathbf{w}' \times \mathbf{z}\mathbf{y}')(\mathbf{A} \times \mathbf{B})].$$

Proof: The proof is straightforward by using Theorem 8.8.6 and the fact that $(\mathbf{w}'\mathbf{A}\mathbf{x})(\mathbf{y}'\mathbf{B}\mathbf{z})$ is a scalar and hence equal to its trace. ∎

Corollary 10.9.8

In Theorem 10.9.8 *let* $\mathbf{w}$ *be a constant vector (say* $\boldsymbol{\mu}$*) and let* $\mathbf{z}$ *have a density* $N(\mathbf{z}: \mathbf{0}, \mathbf{I})$. *Then (let* $\mathbf{A}$ *and* $\mathbf{B}$ *be symmetric)*

$$\mathscr{E}[(\boldsymbol{\mu}'\mathbf{A}\mathbf{z})(\mathbf{z}'\mathbf{B}\mathbf{z})] = \mathscr{E}[(\mathbf{z}'\mathbf{A}\boldsymbol{\mu})(\mathbf{z}'\mathbf{B}\mathbf{z})]$$
$$= \mathscr{E}[(\mathbf{z}'\mathbf{B}\mathbf{z})(\mathbf{z}'\mathbf{A}\boldsymbol{\mu})] = \mathscr{E}[(\mathbf{z}'\mathbf{B}\mathbf{z})(\boldsymbol{\mu}'\mathbf{A}\mathbf{z})]$$
$$= \mathscr{E}[\text{tr}\{(\boldsymbol{\mu}' \times \mathbf{z}')(\mathbf{A} \times \mathbf{B})(\mathbf{z} \times \mathbf{z})\}] = \mathscr{E}[\text{tr}\{(\mathbf{A} \times \mathbf{B})(\mathbf{z}\boldsymbol{\mu}' \times \mathbf{z}\mathbf{z}')\}]$$
$$= \text{tr}[(\mathbf{A} \times \mathbf{B})\, \mathscr{E}[\mathbf{z}\boldsymbol{\mu}' \times \mathbf{z}\mathbf{z}'] = 0.$$

Theorem 10.9.9

Let $\mathbf{x}_i' \mathbf{A}_i \mathbf{y}_i$ *for* $i = 1, 2, 3$ *be bilinear forms, where* $\mathbf{A}_i$ *are* $n \times n$ *matrices. Then*

$$(\mathbf{x}_1' \mathbf{A}_1 \mathbf{y}_1)(\mathbf{x}_2' \mathbf{A}_2 \mathbf{y}_2)(\mathbf{x}_3' \mathbf{A}_3 \mathbf{y}_3) = \text{tr}[(\mathbf{A}_1 \times \mathbf{A}_2 \times \mathbf{A}_3)(\mathbf{y}_1 \mathbf{x}_1' \times \mathbf{y}_2 \mathbf{x}_2' \times \mathbf{y}_3 \mathbf{x}_3')].$$

Proof: Let $\mathbf{A}_1 \times \mathbf{A}_2 = \mathbf{A}$ and $\mathbf{y}_1 \mathbf{x}_1' \times \mathbf{y}_2 \mathbf{x}_2' = \mathbf{x}$. Then

$$\text{tr}\{[(\mathbf{A}_1 \times \mathbf{A}_2) \times \mathbf{A}_3]\,[(\mathbf{y}_1 \mathbf{x}_1' \times \mathbf{y}_2 \mathbf{x}_2') \times \mathbf{y}_3 \mathbf{x}_3']\} = \text{tr}[(\mathbf{A} \times \mathbf{A}_3)(\mathbf{x} \times \mathbf{y}_3 \mathbf{x}_3')]$$
$$= \text{tr}[\mathbf{A}\mathbf{x} \times \mathbf{A}_3 \mathbf{y}_3 \mathbf{x}_3'] = \text{tr}(\mathbf{A}\mathbf{x})\text{tr}(\mathbf{A}_3 \mathbf{y}_3 \mathbf{x}_3') = K \text{ (say)}.$$

Also $\text{tr}\{(\mathbf{A}_1 \times \mathbf{A}_2)(\mathbf{y}_1\mathbf{x}_1' \times \mathbf{y}_2\mathbf{x}_2')\} = (\mathbf{x}_1' \mathbf{A}_1\mathbf{y}_1)(\mathbf{x}_2' \mathbf{A}_2 \mathbf{y}_2)$ by Theorem 10.9.8, and $\text{tr}(\mathbf{A}_3\mathbf{y}_3\mathbf{x}_3') = \mathbf{x}_3'\mathbf{A}_3\mathbf{y}_3$. Substituting these into the expressions for K gives the result. ∎

Corollary 10.9.9

In Theorem 10.9.9 *let* $\mathbf{x}_i = \mathbf{y}_i = \mathbf{z}$ *for* $i = 1, 2, 3$. *Then*

(1) $(\mathbf{z}'\mathbf{A}_1\mathbf{z})(\mathbf{z}'\mathbf{A}_2\mathbf{z})(\mathbf{z}'\mathbf{A}_3\mathbf{z}) = \text{tr}[(\mathbf{A}_1 \times \mathbf{A}_2 \times \mathbf{A}_3)(\mathbf{z}\mathbf{z}' \times \mathbf{z}\mathbf{z}' \times \mathbf{z}\mathbf{z}')]$,
(2) *If* $\mathbf{A}_1 = \mathbf{A}_2$, *then* $(\mathbf{z}'\mathbf{A}_1\mathbf{z})^2(\mathbf{z}'\mathbf{A}_3\mathbf{z})$
$= \text{tr}[(\mathbf{A}_1 \times \mathbf{A}_1 \times \mathbf{A}_3)(\mathbf{z}\mathbf{z}' \times \mathbf{z}\mathbf{z}' \times \mathbf{z}\mathbf{z}')]$,
(3) *If* $\mathbf{A}_1 = \mathbf{A}_2 = \mathbf{A}_3 = \mathbf{A}$, *then* $(\mathbf{z}' \mathbf{A}\mathbf{z})^3$
$= \text{tr}[(\mathbf{A} \times \mathbf{A} \times \mathbf{A})(\mathbf{z}\mathbf{z}' \times \mathbf{z}\mathbf{z}' \times \mathbf{z}\mathbf{z}')]$.

Note: The results of Theorem 10.9.9 and its corollary can be extended to any finite number of bilinear or quadratic forms in $n \times 1$ vectors.

Next we show how the above results can be used to find expectations and covariance of quadratic forms of multivariate random variables.

Theorem 10.9.10

Let $\mathbf{x}$ *be an* $n \times 1$ *vector with distribution* $N(\mathbf{x}\colon \mathbf{0}, \mathbf{V})$; *let* $\mathbf{A}$, $\mathbf{B}$, *and* $\mathbf{C}$ *be symmetric matrices of constants. Then*

(1) $\mathscr{E}[(\mathbf{x}' \mathbf{A}\mathbf{x})(\mathbf{x}' \mathbf{B}\mathbf{x})] = [\text{tr}(\mathbf{A}\mathbf{V})]\,[\text{tr}(\mathbf{B}\mathbf{V})] + 2\,\text{tr}(\mathbf{A}\mathbf{V}\mathbf{B}\mathbf{V})$,
(2) $\text{Cov}[\mathbf{x}'\mathbf{A}\mathbf{x}, \mathbf{x}'\mathbf{B}\mathbf{x}] = 2\,\text{tr}(\mathbf{A}\mathbf{V}\mathbf{B}\mathbf{V})$,
(3) $\text{var}[\mathbf{x}'\mathbf{A}\mathbf{x}] = 2\,\text{tr}(\mathbf{A}\mathbf{V})^2$.

Proof: To prove (1) we use $\mathbf{z} = \mathbf{T}'^{-1}\mathbf{x}$, where $\mathbf{T}'\mathbf{T} = \mathbf{V}$. Then $\mathbf{z}$ has a distribution $N(\mathbf{z}\colon \mathbf{0}, \mathbf{I})$; also $\mathbf{x}'\mathbf{A}\mathbf{x} = \mathbf{z}'\mathbf{T}\mathbf{A}\mathbf{T}'\mathbf{z} = \mathbf{z}'\mathbf{D}\mathbf{z}$ (say) and $\mathbf{x}'\mathbf{B}\mathbf{x} = \mathbf{z}'\mathbf{T}\mathbf{B}\mathbf{T}'\mathbf{z} = \mathbf{z}'\mathbf{F}\mathbf{z}$ (say). Let

$$K = \mathscr{E}[(\mathbf{x}'\mathbf{A}\mathbf{x})(\mathbf{x}'\mathbf{B}\mathbf{x})] = \mathscr{E}[(\mathbf{z}'\mathbf{D}\mathbf{z})(\mathbf{z}'\mathbf{F}\mathbf{z})] = \mathscr{E}[(\mathbf{z}' \times \mathbf{z}')(\mathbf{D} \times \mathbf{F})(\mathbf{z} \times \mathbf{z})] = \mathscr{E}\{\mathrm{tr}[(\mathbf{z}' \times \mathbf{z}')(\mathbf{D} \times \mathbf{F})(\mathbf{z} \times \mathbf{z})]\} = \mathscr{E}\{\mathrm{tr}[(\mathbf{D} \times \mathbf{F})(\mathbf{z} \times \mathbf{z})(\mathbf{z}' \times \mathbf{z}')]\} = \mathrm{tr}[(\mathbf{D} \times \mathbf{F})\mathscr{E}(\mathbf{z}\mathbf{z}' \times \mathbf{z}\mathbf{z}')] = \mathrm{tr}[(\mathbf{D} \times \mathbf{F})\,[\mathbf{I} + \mathbf{K}_{nn} + \{\mathrm{vec}(\mathbf{I})\}\{\mathrm{vec}(\mathbf{I})\}']] = \mathrm{tr}[\mathbf{D} \times \mathbf{F}] + \mathrm{tr}[(\mathbf{D} \times \mathbf{F})\mathbf{K}_{nn}] + \mathrm{tr}[(\mathbf{D} \times \mathbf{F})\{\mathrm{vec}(\mathbf{I})\}\{\mathrm{vec}(\mathbf{I})\}'] = \mathrm{tr}(\mathbf{D})\mathrm{tr}(\mathbf{F}) + \mathrm{tr}(\mathbf{F}\mathbf{D}) + \mathrm{tr}[\{\mathrm{vec}(\mathbf{D}\mathbf{F}')\}\{\mathrm{vec}(\mathbf{I})\}'] = \mathrm{tr}(\mathbf{D})\mathrm{tr}(\mathbf{F}) + \mathrm{tr}(\mathbf{F}\mathbf{D}) + \mathrm{tr}(\mathbf{D}\mathbf{F}').$$

If we substitute for $\mathbf{F}$ and $\mathbf{D}$ we get the result (1). In the last four equalities we use Theorems 10.9.2, (6) of Theorem 9.3.8, Theorem 9.2.2, and Theorem 9.2.3. We leave the proof of (2) and (3) for the reader; they follow immediately from (1). ▮

Theorem 10.9.11

Let $\mathbf{x}$ *be an* $n \times 1$ *random vector with density* $N(\mathbf{x}\colon \boldsymbol{\mu}, \mathbf{V})$. *Then*

(1) $\mathscr{E}[(\mathbf{x}'\mathbf{A}\mathbf{x})(\mathbf{x}'\mathbf{B}\mathbf{x})] = \mathrm{tr}(\mathbf{A}\mathbf{V})\mathrm{tr}(\mathbf{B}\mathbf{V}) + 2\mathrm{tr}(\mathbf{A}\mathbf{V}\mathbf{B}\mathbf{V}) + (\boldsymbol{\mu}'\mathbf{A}\boldsymbol{\mu})\mathrm{tr}(\mathbf{B}\mathbf{V}) + (\boldsymbol{\mu}'\mathbf{B}\boldsymbol{\mu})\mathrm{tr}(\mathbf{A}\mathbf{V}) + 4\boldsymbol{\mu}'(\mathbf{A}\mathbf{V}\mathbf{B})\boldsymbol{\mu} + (\boldsymbol{\mu}'\mathbf{A}\boldsymbol{\mu})(\boldsymbol{\mu}'\mathbf{B}\boldsymbol{\mu})$,
(2) $\mathrm{Cov}[\mathbf{x}'\mathbf{A}\mathbf{x}, \mathbf{x}'\mathbf{B}\mathbf{x}] = 2\mathrm{tr}(\mathbf{A}\mathbf{V}\mathbf{B}\mathbf{V}) + 4\boldsymbol{\mu}'\mathbf{A}\mathbf{V}\mathbf{B}\boldsymbol{\mu}$,
(3) $\mathrm{var}[\mathbf{x}'\mathbf{A}\mathbf{x}] = 2\mathrm{tr}[(\mathbf{A}\mathbf{V})^2] + 4\boldsymbol{\mu}'\mathbf{A}\mathbf{V}\mathbf{A}\boldsymbol{\mu}$.

Proof: Let $\mathbf{x} - \boldsymbol{\mu} = \mathbf{z}$ and then $\mathbf{z}$ has a density $N(\mathbf{z}\colon \mathbf{0}, \mathbf{V})$. We get

$$\begin{aligned}
&\mathscr{E}[(\mathbf{x}'\mathbf{A}\mathbf{x}) \cdot (\mathbf{x}'\mathbf{B}\mathbf{x})] = \mathscr{E}[(\mathbf{z} + \boldsymbol{\mu})'\mathbf{A}(\mathbf{z} + \boldsymbol{\mu}) \cdot (\mathbf{z} + \boldsymbol{\mu})'\mathbf{B}(\mathbf{z} + \boldsymbol{\mu})] \\
&= \mathscr{E}[(\mathbf{z}'\mathbf{A}\mathbf{z} + 2\boldsymbol{\mu}'\mathbf{A}\mathbf{z} + \boldsymbol{\mu}'\mathbf{A}\boldsymbol{\mu})(\mathbf{z}'\mathbf{B}\mathbf{z} + 2\boldsymbol{\mu}'\mathbf{B}\mathbf{z} + \boldsymbol{\mu}'\mathbf{B}\boldsymbol{\mu})] \\
&= \mathscr{E}\{(\mathbf{z}'\mathbf{A}\mathbf{z})(\mathbf{z}'\mathbf{B}\mathbf{z})\} + 2\mathscr{E}[(\mathbf{z}'\mathbf{A}\mathbf{z})(\boldsymbol{\mu}'\mathbf{B}\mathbf{z})] \\
&\quad + (\boldsymbol{\mu}'\mathbf{B}\boldsymbol{\mu})\mathscr{E}[\mathbf{z}'\mathbf{A}\mathbf{z}] + 2\mathscr{E}[(\boldsymbol{\mu}'\mathbf{A}\mathbf{z})(\mathbf{z}'\mathbf{B}\mathbf{z})] + 4\mathscr{E}[(\boldsymbol{\mu}'\mathbf{A}\mathbf{z})(\boldsymbol{\mu}'\mathbf{B}\mathbf{z})] \\
&\quad + 2(\boldsymbol{\mu}'\mathbf{B}\boldsymbol{\mu})\mathscr{E}[\boldsymbol{\mu}'\mathbf{A}\mathbf{z}] + (\boldsymbol{\mu}'\mathbf{A}\boldsymbol{\mu})\mathscr{E}[\mathbf{z}'\mathbf{B}\mathbf{z}] + 2(\boldsymbol{\mu}'\mathbf{A}\boldsymbol{\mu})\mathscr{E}[\boldsymbol{\mu}'\mathbf{B}\mathbf{z}] \\
&\quad + (\boldsymbol{\mu}'\mathbf{A}\boldsymbol{\mu})(\boldsymbol{\mu}'\mathbf{B}\boldsymbol{\mu}) = \{\mathrm{tr}(\mathbf{A}\mathbf{V})\mathrm{tr}(\mathbf{B}\mathbf{V}) + 2\mathrm{tr}(\mathbf{A}\mathbf{V}\mathbf{B}\mathbf{V})\} + 2 \cdot 0 \\
&\quad + (\boldsymbol{\mu}'\mathbf{B}\boldsymbol{\mu}) \cdot \mathrm{tr}(\mathbf{A}\mathbf{V}) + 2 \cdot 0 + 4\boldsymbol{\mu}'(\mathbf{A}\mathbf{V}\mathbf{B})\boldsymbol{\mu} + 2 \cdot 0 \\
&\quad + (\boldsymbol{\mu}'\mathbf{A}\boldsymbol{\mu})\mathrm{tr}(\mathbf{B}\mathbf{V}) + 2 \cdot 0 + (\boldsymbol{\mu}'\mathbf{A}\boldsymbol{\mu})(\boldsymbol{\mu}'\mathbf{B}\boldsymbol{\mu}).
\end{aligned}$$

This completes the proof of (1). The proofs for (2) and (3) follow directly from this. ▮

Theorem 10.9.12

Let the $n \times 1$ *random vector* $\mathbf{x}$ *have a density* $N(\mathbf{x}: \mathbf{0}, \mathbf{V})$. *Then*

(1) $\mathscr{E}[(\mathbf{x}'\mathbf{A}\mathbf{x})(\mathbf{x}'\mathbf{B}\mathbf{x})(\mathbf{x}'\mathbf{C}\mathbf{x})] = [\mathrm{tr}(\mathbf{AV})][\mathrm{tr}(\mathbf{BV})][\mathrm{tr}(\mathbf{CV})]$
$+ 2[\mathrm{tr}(\mathbf{AV})][\mathrm{tr}(\mathbf{BVCV})] + 2[\mathrm{tr}(\mathbf{BV})][\mathrm{tr}(\mathbf{AVCV})]$
$+ 2[\mathrm{tr}(\mathbf{CV})][\mathrm{tr}(\mathbf{AVBV})] + 8\mathrm{tr}(\mathbf{AVBVCV})$,

(2) $\mathscr{E}[(\mathbf{x}'\mathbf{A}\mathbf{x})^2(\mathbf{x}'\mathbf{B}\mathbf{x})] = [\mathrm{tr}(\mathbf{AV})]^2[\mathrm{tr}(\mathbf{BV})]$
$+ 4[\mathrm{tr}(\mathbf{AV})][\mathrm{tr}(\mathbf{AVBV})] + 2[\mathrm{tr}(\mathbf{BV})][\mathrm{tr}(\mathbf{AV})^2] + 8\mathrm{tr}[(\mathbf{AV})^2\mathbf{BV}]$,

(3) $\mathrm{Cov}[(\mathbf{x}'\mathbf{A}\mathbf{x})^2, (\mathbf{x}'\mathbf{B}\mathbf{x})] = 4[\mathrm{tr}(\mathbf{AV})][\mathrm{tr}(\mathbf{AVBV})] + 8\mathrm{tr}[(\mathbf{AV})^2(\mathbf{BV})]$,

(4) $\mathscr{E}[(\mathbf{x}'\mathbf{A}\mathbf{x})^3] = [\mathrm{tr}(\mathbf{AV})]^3 + 6[\mathrm{tr}(\mathbf{AV})][\mathrm{tr}(\mathbf{AV})^2] + 8\mathrm{tr}[(\mathbf{AV})^3]$.

Proof: The proof of (1) follows from the fact that $\mathbf{V} = \mathbf{T}'\mathbf{T}$ and letting $\mathbf{z} = \mathbf{T}'^{-1}\mathbf{x}$ and $\mathbf{x} = \mathbf{T}'\mathbf{z}$. Thus $\mathbf{z}$ has a density $N(\mathbf{z}: \mathbf{0}, \mathbf{I})$. Hence

$\mathscr{E}[(\mathbf{x}'\mathbf{A}\mathbf{x})(\mathbf{x}'\mathbf{B}\mathbf{x})(\mathbf{x}'\mathbf{C}\mathbf{x})] = \mathscr{E}[(\mathbf{z}'\mathbf{TAT}'\mathbf{z})(\mathbf{z}'\mathbf{TBT}'\mathbf{z})(\mathbf{z}'\mathbf{TCT}'\mathbf{z})] = \mathscr{E}\{\mathrm{tr}[(\mathbf{TAT}' \times \mathbf{TBT}' \times \mathbf{TCT}') \cdot (\mathbf{zz}' \times \mathbf{zz}' \times \mathbf{zz}')]\}$.

Use (3) of Theorem 10.9.4 to evaluate $\mathscr{E}[\mathbf{zz}' \times \mathbf{zz}' \times \mathbf{zz}']$ and simplify. The results in (2), (3), and (4) follow immediately from this. Details can be found in [M–2]. ∎

We leave the proof of the next theorem for the reader. It is similar to the proof for Theorem 10.9.6.

Theorem 10.9.13

Let the $n \times 1$ *random vector* $\mathbf{x}$ *have a density* $N(\mathbf{x}: \boldsymbol{\mu}, \mathbf{V})$ *and let*

$$\mathbf{x} = \begin{bmatrix} \mathbf{x}_1 \\ \mathbf{x}_2 \end{bmatrix}, \quad \boldsymbol{\mu} = \begin{bmatrix} \boldsymbol{\mu}_1 \\ \boldsymbol{\mu}_2 \end{bmatrix}, \quad \mathbf{V} = \begin{bmatrix} \mathbf{V}_{11} & \mathbf{V}_{12} \\ \mathbf{V}_{21} & \mathbf{V}_{22} \end{bmatrix},$$

where $\mathbf{x}_i$ *and* $\boldsymbol{\mu}_i$ *are* $n_i \times 1$, $\mathbf{V}_{ij}$ *is* $n_i \times n_j$ *for* $i = 1, 2$; $j = 1, 2$, *and* $n_1 + n_2 = n$. *Then*

(1) $\mathscr{E}[\mathbf{x}_1 \times \mathbf{x}_2] = \mathrm{vec}(\mathbf{V}_{12}) + (\boldsymbol{\mu}_1 \times \boldsymbol{\mu}_2)$,

(2) $\mathrm{Cov}(\mathbf{x}_1 \times \mathbf{x}_2) = \mathbf{V}_{11} \times \mathbf{V}_{22} + \mathbf{V}_{11} \times \boldsymbol{\mu}_2\boldsymbol{\mu}_2' + \boldsymbol{\mu}_1\boldsymbol{\mu}_1' \times \mathbf{V}_{22}$
$+ \mathbf{K}_{n_2n_1}(\mathbf{V}_{21} \times \mathbf{V}_{12} + \mathbf{V}_{21} \times \boldsymbol{\mu}_1\boldsymbol{\mu}_2'$
$+ \boldsymbol{\mu}_2\boldsymbol{\mu}_1' \times \mathbf{V}_{12})$,

(3) $\text{tr}[\text{Cov}(\mathbf{x}_1 \times \mathbf{x}_2)] = \text{tr}(\mathbf{V}_{11})\text{tr}(\mathbf{V}_{22}) + \boldsymbol{\mu}_1' \boldsymbol{\mu}_1 \text{tr}(\mathbf{V}_{22}) + \boldsymbol{\mu}_2' \boldsymbol{\mu}_2 \text{tr}(\mathbf{V}_{11}) + \text{tr}(\mathbf{V}_{12} \mathbf{V}_{21}) + \boldsymbol{\mu}_2' \mathbf{V}_{21} \boldsymbol{\mu}_1 + \boldsymbol{\mu}_1' \mathbf{V}_{12} \boldsymbol{\mu}_2.$

10.10 Expectation of the Elements of a Wishart Matrix

In this section we illustrate briefly how the previous theory can be used to determine the mean vector and covariance matrix of a Wishart matrix.

Definition 10.10.1

Wishart Matrix. *Let $\mathbf{x}_i$ for $i = 1, 2, \ldots, K$ be $n \times 1$ vectors that are jointly independent and let $\mathbf{x}_i$ have the density $N(\mathbf{x}_i\text{: } \boldsymbol{\mu}_i, \mathbf{V})$. Define $\mathbf{S}$ and $\mathbf{M}$ by $\mathbf{S} = [s_{pq}] = \Sigma_{i=1}^K \mathbf{x}_i \mathbf{x}_i'$ and $\mathbf{M} = [\boldsymbol{\mu}_1, \boldsymbol{\mu}_2, \ldots, \boldsymbol{\mu}_K]$. Then $\mathbf{S}$ is defined to be a Wishart matrix.*

Note: The density of $s_{11}, s_{12}, \ldots, s_{KK}$ (referred to as the distribution of $\mathbf{S}$) is defined as a noncentral Wishart density; see [R–4] for a derivation and the functional form of this density. This density is denoted by $W_n(\mathbf{S}; K; \mathbf{V}, \mathbf{M})$.

We now state and prove a theorem about the expectation and covariance of the elements of $\mathbf{S}$.

Theorem 10.10.1

Let $\mathbf{x}_i$ for $i = 1, \ldots, K$ be independent $n \times 1$ vectors with density $N(\mathbf{x}_i\text{: } \boldsymbol{\mu}_i, \mathbf{V})$ and thus $\mathbf{S} = \Sigma_{i=1}^K \mathbf{x}_i \mathbf{x}_i'$ is a Wishart matrix with density $W_n(\mathbf{S}\text{: } K; \mathbf{V}, \mathbf{M})$. Then

(1) $\mathscr{E}[\mathbf{S}] = K\mathbf{V} + \mathbf{M}\mathbf{M}'$,

(2) $\text{Cov}[\text{vec}(\mathbf{S})] = (\mathbf{I} + \mathbf{K}_{nn})[K(\mathbf{V} \times \mathbf{V}) + (\mathbf{V} \times \mathbf{M}\mathbf{M}') + (\mathbf{M}\mathbf{M}' \times \mathbf{V})]$.

Proof: $\mathscr{E}[\mathbf{S}] = \mathscr{E}[\Sigma\, \mathbf{x}_i \mathbf{x}_i'] = \Sigma\, \mathscr{E}[\mathbf{x}_i \mathbf{x}_i'] = \Sigma(\boldsymbol{\mu}_i \boldsymbol{\mu}_i' + \mathbf{V}) = \mathbf{M}\mathbf{M}' + K\mathbf{V}$. This proves (1). To prove (2) we get $\text{Cov}[\text{vec}(\mathbf{S})] = \text{Cov}[\text{vec}(\Sigma\, \mathbf{x}_i \mathbf{x}_i')] = \text{Cov}[\Sigma\, \text{vec}(\mathbf{x}_i \mathbf{x}_i')] = \Sigma\, \text{Cov}[\text{vec}(\mathbf{x}_i \mathbf{x}_i')]$. The last equality follows from the fact that $\mathbf{x}_1, \mathbf{x}_2, \ldots, \mathbf{x}_K$ are jointly independent. We use (2) of Theorem 9.2.1 to get $\text{vec}(\mathbf{x}_i \mathbf{x}_i') = \mathbf{x}_i \times \mathbf{x}_i$. Thus $\text{Cov}[\text{vec}(\mathbf{S})] = \Sigma\, \text{Cov}[\mathbf{x}_i \times \mathbf{x}_i] = \Sigma(\mathbf{I} + \mathbf{K}_{nn})[(\mathbf{V} \times \mathbf{V}) + (\mathbf{V} \times \boldsymbol{\mu}_i \boldsymbol{\mu}_i') + (\boldsymbol{\mu}_i \boldsymbol{\mu}_i' \times \mathbf{V})]$ by (2) of Theorem 10.9.7. If we sum this expression and simplify, the result is obtained. ∎

Problems

1. The bivariate normal density can be written as

$$N(x, y) = \frac{1}{2\pi\sigma_x\sigma_y\sqrt{1-\rho^2}}$$

$$\times \exp\left\{-\frac{1}{2(1-\rho^2)}\left[\left(\frac{x-\mu_x}{\sigma_x}\right)^2 - 2\rho\left(\frac{x-\mu_x}{\sigma_x}\right)\left(\frac{y-\mu_y}{\sigma_y}\right) + \left(\frac{y-\mu_y}{\sigma_y}\right)^2\right]\right\},$$

where $|\rho| < 1$, $\sigma_x > 0$, $\sigma_y > 0$.
Put this in the form of Eq. (10.3.1a) by identifying $\mathbf{R}$, $\mathbf{c}$, $\mathbf{x}$, and K.

2. In Prob. 1 find $\mathbf{V}$ and hence show that the covariance of x and y is equal to $\rho\sigma_x\sigma_y$.
3. Find the characteristic roots and characteristic vectors of $\mathbf{R}$ in Prob. 1.
4. Use the results of Prob. 3 to find an orthogonal matrix $\mathbf{P}$ such that $\mathbf{P}'\mathbf{R}\mathbf{P}$ is a diagonal matrix (see Corollary 10.3.1.1).
5. Use Theorem 10.5.1 to evaluate $\mathscr{E}(xy)$ for the random vector with density given in Prob. 1.
6. Define a vector $\mathbf{z}$ as $\mathbf{z} = \mathbf{P}'(\mathbf{x} - \boldsymbol{\mu})$, where $\mathbf{x}$ is a random $n \times 1$ vector with a normal density given by Eq. (10.6.2) and $\mathbf{P}$ is an orthogonal matrix of constants such that $\mathbf{P}'\mathbf{V}\mathbf{P} = \mathbf{D}$, a diagonal matrix. Show that

$$\mathscr{E}(\mathbf{z}'\mathbf{z}) = \sum_{i=1}^{n} d_{ii},$$

where d_{ii} is the i-th diagonal element of $\mathbf{D}$.

7. Find the constant K such that the following function is a normal density

$$f(x_1, x_2) = K\, e^{-(2x_1{}^2 + 4x_2{}^2 - 2x_1x_2 - 6x_1 - 4x_2 + 8)}.$$

8. In the normal density given by Eq. (10.6.2) show that $\mathscr{E}(\mathbf{x})$ is the vector $\boldsymbol{\mu}$ that satisfies $\frac{\partial}{\partial \mathbf{x}} N(\mathbf{x}; \boldsymbol{\mu}, \mathbf{V}) = \mathbf{0}$.
9. If the $k \times 1$ vector $\mathbf{x}$ has a normal density and $\mathbf{B}$ is an $m \times k$ matrix of rank m, then use Theorem 10.3.1 to show that the density of the m $\times$ 1 random vector $\mathbf{y}$ has a normal density, where $\mathbf{y} = \mathbf{B}\mathbf{x}$.
10. In Prob. 9 show that the mean vector of the density of the random vector $\mathbf{y}$ is $\mathbf{B}\boldsymbol{\mu}$ and that the covariance matrix is $\mathbf{B}\mathbf{V}\mathbf{B}'$, where $\boldsymbol{\mu}$ and $\mathbf{V}$ are the mean vector and and covariance matrix, respectively, of the random $k \times 1$ vector $\mathbf{x}$.

11. Evaluate the integral

$$\int_{-\infty}^{\infty}\int_{-\infty}^{\infty}\int_{-\infty}^{\infty}\int_{-\infty}^{\infty}(x_1^2 - 2x_1x_4)\, e^{-(1/2)Q}\, dx_1\, dx_2\, dx_3\, dx_4$$

where

$$Q = 3x_1^2 + 2x_2^2 + 2x_3^2 + x_4^2 + 2x_1x_2 + 2x_3x_4 - 6x_1 - 2x_2 - 6x_3 - 2x_4 + 8.$$

12. Find the number K such that the function defined by

$$Ke^{-(1/2)Q}$$

is a normal density function where Q is defined in Prob. 11.

13. Let the $n \times 1$ random vector $\mathbf{x}$ have a normal density defined by Eq. (10.6.2). Find $\mathscr{E}(Q)$ where

$$Q = (\mathbf{x} - \boldsymbol{\mu})'\mathbf{V}(\mathbf{x} - \boldsymbol{\mu}).$$

14. In Prob. 13 let $Q = (\mathbf{x} - \boldsymbol{\mu})'\mathbf{A}(\mathbf{x} - \boldsymbol{\mu})$ and show that

$$\mathscr{E}\{[Q - \mathscr{E}(Q)]^2\} = \mathscr{E}(Q^2) - [\mathscr{E}(Q)]^2 = 2\,\mathrm{tr}\,[(\mathbf{AV})^2].$$

15. Let the $n \times 1$ random vector $\mathbf{x}$ have a normal density with mean vector equal to $\boldsymbol{\mu}$ and covariance matrix equal to $\mathbf{D}$, where $\mathbf{D}$ is a diagonal matrix. Show that

$$\mathscr{E}[(\mathbf{x} - \boldsymbol{\mu})'\mathbf{A}(\mathbf{x} - \boldsymbol{\mu})] = \sum_{i=1}^{n} a_{ii}\, d_{ii}$$

where d_{ii} is the i-th diagonal element of $\mathbf{D}$.

16. Let the $n \times 1$ random vector $\mathbf{x}$ have a normal density with mean vector equal to zero and with covariance matrix $\mathbf{I}$; that is, by Eq. (10.6.2) the density is denoted by $N(\mathbf{x}; \mathbf{0}, \mathbf{I})$. Show that

$$\mathscr{E}[\mathbf{x}'\mathbf{A}(\mathbf{A}'\mathbf{A})^{-1}\mathbf{A}'\mathbf{x}] = m,$$

where $\mathbf{A}$ is an $n \times m$ matrix of rank m.

17. Let the $n \times 1$ random vector $\mathbf{x}$ have a normal density given by Eq. (10.6.2) with $\boldsymbol{\mu} = \mathbf{0}$ and $\mathbf{V} = \mathbf{I}$. Find the matrix $\mathbf{C}$, where $\mathbf{C} = \mathscr{E}(\mathbf{xx}')$.

18. In Prob. 17 let the two scalar random variables y and z be defined by

$$y = \mathbf{a}'\mathbf{x}, \qquad z = \mathbf{b}'\mathbf{x},$$

where $\mathbf{a}$ and $\mathbf{b}$ are vectors of constants. Find $\mathscr{E}(y)$, $\mathscr{E}(z)$, $\mathscr{E}(yz)$.

19. In Prob. 18 show that $\mathscr{E}(yz) = 0$ if and only if $\mathbf{a}'\mathbf{b} = 0$.

20. Let $\mathbf{R}$ be a positive definite $n \times n$ matrix and $\mathbf{A}$ be a symmetric $n \times n$ matrix. Show that, for some positive value of the real number λ, the matrix $\mathbf{B}$ is a positive definite matrix, where $\mathbf{B} = \mathbf{R} - \lambda\mathbf{A}$.

21. Let the $n \times 1$ random vector $\mathbf{x}$ have a normal density given by Eq. (10.6.2) and let $\mathbf{A}$ be a symmetric $n \times n$ matrix of constants. Find the set of numbers (values of λ) such that the following expected value exists: $\mathscr{E}[e^{\lambda(\mathbf{x}'\mathbf{A}\mathbf{x})}]$.

22. If $\mathbf{x}$, $\mathbf{a}$, and $\mathbf{b}$ are $n \times 1$ vectors and $\mathbf{A}$ and $\mathbf{B}$ are $n \times n$ matrices such that $\mathbf{A} + \mathbf{B}$ is nonsingular, show that

$$\begin{aligned}(\mathbf{x} - \mathbf{a})'\mathbf{A}(\mathbf{x} - \mathbf{a}) &+ (\mathbf{x} - \mathbf{b})'\mathbf{B}(\mathbf{x} - \mathbf{b}) \\ &= (\mathbf{x} - \mathbf{c})'(\mathbf{A} + \mathbf{B})(\mathbf{x} - \mathbf{c}) + (\mathbf{a} - \mathbf{b})'\mathbf{A}(\mathbf{A} + \mathbf{B})^{-1}\mathbf{B}(\mathbf{a} - \mathbf{b}),\end{aligned}$$

where $\mathbf{c} = (\mathbf{A} + \mathbf{B})^{-1}(\mathbf{A}\mathbf{a} + \mathbf{B}\mathbf{b})$.

23. If $\mathbf{A}$ is an $n \times n$ positive definite matrix and $\mathbf{B}$ is a symmetric matrix, show that the integral

$$\int_{-\infty}^{\infty} \cdots \int_{-\infty}^{\infty} e^{-(\mathbf{x}'\mathbf{A}\mathbf{x} + \theta\mathbf{x}'\mathbf{B}\mathbf{x})}\, dx_1 \cdots dx_n$$

exists for all θ such that $|\theta| < \theta_0$ for a suitable positive number θ_0.

24. In Prob. 23 find θ_0 as a function of the characteristic roots of $\mathbf{A}$ and $\mathbf{B}$.

25. In Prob. 23 show that the value of the integral is $\pi^{n/2}|\mathbf{A} + \theta\mathbf{B}|^{-1/2}$ for all $|\theta| < \theta_0$.

26. If $\mathbf{A}$ and $\mathbf{B}$ are n $\times$ n positive definite matrices and $\mathbf{A} = \mathbf{C}\mathbf{C}'$, show that

$$\int_{-\infty}^{\infty} \cdots \int_{-\infty}^{\infty} e^{-\mathbf{x}'\mathbf{C}'\mathbf{B}\mathbf{C}\mathbf{x}}\, dx_1 \cdots dx_n = \pi^{n/2}\,|\mathbf{A}\mathbf{B}|^{-1/2}.$$

27. Let $\mathbf{A}$ and $\mathbf{B}$ be $n \times n$ symmetric matrices such that $|\mathbf{I} - x\mathbf{A}||\mathbf{I} - y\mathbf{B}| = |\mathbf{I} - x\mathbf{A} - y\mathbf{B}|$ for all x and y such that $|x| < a$, $|y| < a$ for some positive number a. Show that $\mathbf{A}\mathbf{B} = \mathbf{B}\mathbf{A} = \mathbf{0}$.

28. Let $\mathbf{A}$, $\mathbf{B}$, and $\mathbf{A}\mathbf{B}$ be symmetric $n \times n$ matrices and let $\mathbf{A}$ and $\mathbf{B}$ be positive definite. Show that

$$\int_{-\infty}^{\infty} \cdots \int_{-\infty}^{\infty} e^{-\mathbf{x}'\mathbf{A}\mathbf{B}\mathbf{x}}\, dx_1 \cdots dx_n = \pi^{n/2}\,|\mathbf{A}\mathbf{B}|^{-1/2}.$$

Inverse Positive Matrices and Matrices with Nonpositive Off-Diagonal Elements

11

11.1 Introduction and Definitions

In this chapter we briefly discuss two general types of matrices: matrices that have nonpositive off-diagonal elements, and matrices whose inverses have all elements non-negative. The types of matrices that satisfy these conditions are Z-matrices, M-matrices, and other special types. They have applications in economics, in the physical sciences, in probability, in biology, in the study of the convergence of some iterative procedures, and in other areas.

Terminology and Notation:

(1) Let $\mathbf{A}$ be an $m \times n$ matrix. We call $\mathbf{A}$ a matrix with positive (non-negative) elements if $a_{ij} > 0$ ($a_{ij} \geq 0$) for $i = 1, \ldots, m$ and $j = 1, \ldots, n$.

(2) In (1) we write $\mathbf{A} > \mathbf{0}$ ($\mathbf{A} \geq \mathbf{0}$) if $\mathbf{A}$ is a matrix with positive (non-negative) elements.

(3) We write $\mathbf{A} > \mathbf{B}$ ($\mathbf{A} \geq \mathbf{B}$) if $\mathbf{A} - \mathbf{B} > \mathbf{0}$ ($\mathbf{A} - \mathbf{B} \geq \mathbf{0}$). This means that $a_{ij} > b_{ij}$ ($a_{ij} \geq b_{ij}$) for $i = 1, \ldots, m$ and $j = 1, \ldots, n$.

(4) We use $|\lambda|$ for the modulus of a complex number; that is, if $\lambda = x + iy$, where $i = \sqrt{-1}$, then $|\lambda| = (x^2 + y^2)^{1/2}$.

(5) If $\lambda_1, \lambda_2, \ldots, \lambda_n$ are the characteristic roots of an $n \times n$ matrix $\mathbf{A}$ and if $|\lambda_1| \geq |\lambda_2| \geq \ldots \geq |\lambda_n|$, then λ_1 is called the *spectral radius* of $\mathbf{A}$ and is denoted by $\rho(\mathbf{A})$.

Note: We remind the reader that all elements of matrices are real numbers, but the characteristic roots may be complex numbers that are not real.

We conclude this section by stating the famous Perron-Frobenius theorem, which will be used throughout this chapter. We omit the proof, which can be found in [V–1].

Theorem 11.1.1

Let $\mathbf{A}$ *be an* $n \times n$ *matrix with non-negative elements; that is,* $\mathbf{A} \geq \mathbf{0}$, *which means* $a_{ij} \geq 0$ *for* $i = 1, \ldots, n$ *and* $j = 1, 2, \ldots, n$.

(1) *There exists at least one characteristic root (say* $\lambda_{\mathbf{A}}$*) of* $\mathbf{A}$ *that is real and non-negative, and* $|\lambda| \leq \lambda_{\mathbf{A}}$ *for all characteristic roots of* $\mathbf{A}$*; that is, the moduli of all roots of* $\mathbf{A}$ *are less than or equal to* $\lambda_{\mathbf{A}}$. *Hence* $\lambda_{\mathbf{A}}$ *is the spectral radius of* $\mathbf{A}$ *and is often denoted by* $\rho(\mathbf{A})$*; also a characteristic vector corresponding to this root can be chosen to have non-negative elements.*

(2) *If* $\mathbf{B}$ *is an* $n \times n$ *matrix such that* $\mathbf{0} \leq \mathbf{A} \leq \mathbf{B}$, *then* (1) *holds for the roots of* $\mathbf{B}$ *(since* $\mathbf{B} \geq \mathbf{0}$*).*

(3) *If* $\mathbf{A} \geq \mathbf{0}$ *is an irreducible matrix (see Def.* 8.11.2*), there exists a characteristic root of* $\mathbf{A}$, *denoted by* $\lambda_{\mathbf{A}}$ *or* $\rho(\mathbf{A})$, *such that* $\rho(\mathbf{A})$ *is real and positive and* $|\lambda| \leq \rho(\mathbf{A})$ *for all roots* λ *of* $\mathbf{A}$*; in addition* $\rho(\mathbf{A})$ *is distinct (that is, only one root has this value); also a characteristic vector corresponding to the root can be chosen to have all positive elements.*

(4) *If* $\mathbf{0} \leq \mathbf{A} \leq \mathbf{B}$, *then* $\rho(\mathbf{A}) \leq \rho(\mathbf{B})$ in (1) and (3).

We illustrate this theorem with two simple examples.

Example 11.1.1. Consider the following 3×3 matrices $\mathbf{A}_1$ and $\mathbf{A}_2$ given by

$$\mathbf{A}_1 = \begin{bmatrix} 1 & 0 & 2 \\ 1 & 2 & 4 \\ 0 & 1 & 1 \end{bmatrix}; \mathbf{A}_2 = \begin{bmatrix} 1 & 0 & 2 \\ 3 & 2 & 4 \\ 0 & 1 & 1 \end{bmatrix}.$$

Clearly $\mathbf{A}_1 \geq \mathbf{0}$, $\mathbf{A}_2 \geq \mathbf{0}$; also $\mathbf{A}_1$ and $\mathbf{A}_2$ are irreducible. Thus by Theorem 11.1.1 each matrix must have a real root that is positive, and no root can have a larger modulus. The roots of $\mathbf{A}_1$ are $2 - \sqrt{3}$, 0, and $2 + \sqrt{3}$. The roots of $\mathbf{A}_2$ are $\sqrt{-1}$, $-\sqrt{-1}$, and 4. A characteristic vector $\mathbf{x}$ corresponding to $\rho(\mathbf{A}_2) = 4$ is $\mathbf{x}' = [2, 9, 3] > \mathbf{0}$. Also $\mathbf{A}_1 \leq \mathbf{A}_2$, so $\rho(\mathbf{A}_1) \leq \rho(\mathbf{A}_2)$; that is, $2 + \sqrt{3} \leq 4$.

Example 11.1.2. Clearly the matrix $\mathbf{A}_3 \geq \mathbf{0}$ is reducible, where

$$\mathbf{A}_3 = \begin{bmatrix} 4 & 0 & 3 \\ 0 & 4 & 0 \\ 0 & 1 & 0 \end{bmatrix}.$$

The roots are 0, 4, and 4, so $\rho(\mathbf{A}_3) = 4$, and this is not a simple (distinct) root.

11.2 Matrices with Positive Principal Minors

In this section we state and prove some theorems about square matrices whose principal *minors* are positive. First we state a theorem concerning matrices whose *leading* principal minors are positive.

Theorem 11.2.1

Let $\mathbf{A}$ *be an* $n \times n$ *matrix. There exists a lower triangular matrix* $\mathbf{R}$ *and an upper triangular matrix* $\mathbf{S}$ *each with positive diagonal elements and such that* $\mathbf{A} = \mathbf{RS}$ *if and only if the leading principal minors of* $\mathbf{A}$ *are positive.*

Proof: Assume that the leading principal minors are positive. By Theorem 8.6.1 there exist upper and lower triangular matrices $\mathbf{R}$ and $\mathbf{S}$ such that $\mathbf{A} = \mathbf{RS}$. To prove that the diagonal elements of $\mathbf{R}$ and $\mathbf{S}$ are positive, we use induction. Clearly if $n = 1$, we can choose $\mathbf{R} = r$ and $\mathbf{S} = s$ to be positive, since $\mathbf{A} = a > 0$. Assume that the theorem is true if $n = k > 1$. We must show that it is true if $n = k + 1$. Partition $\mathbf{A}$, which is $(k + 1) \times (k + 1)$, as

$$\mathbf{A} = \begin{bmatrix} \mathbf{A}_{11} & \mathbf{a}_{12} \\ \mathbf{a}_{21} & a \end{bmatrix},$$

where $\mathbf{A}_{11}$ is $k \times k$, and by the induction hypothesis $\mathbf{A}_{11} = \mathbf{R}_{11}\mathbf{S}_{11}$, where $\mathbf{R}_{11}$ and $\mathbf{S}_{11}$ are lower and upper triangular matrices, respectively, and each has positive diagonal elements; hence $\mathbf{S}_{11}^{-1}$ and $\mathbf{R}_{11}^{-1}$ exist, which imply $\mathbf{A}_{11}^{-1}$ exists. Also since $\mathbf{A}$ has positive leading principal minors, this implies $\det(\mathbf{A}) > 0$ and hence $\mathbf{A}^{-1}$ exists. We can write

$$\mathbf{A} = \begin{bmatrix} \mathbf{A}_{11} & \mathbf{a}_{12} \\ \mathbf{a}_{21} & a \end{bmatrix} = \begin{bmatrix} \mathbf{R}_{11} & \mathbf{0} \\ \mathbf{a}_{21}\mathbf{S}_{11}^{-1} & 1 \end{bmatrix} \begin{bmatrix} \mathbf{S}_{11} & \mathbf{R}_{11}^{-1}\mathbf{a}_{12} \\ \mathbf{0} & b \end{bmatrix} = \mathbf{RS},$$

where $b = a - \mathbf{a}_{21}\mathbf{A}_{11}^{-1}\mathbf{a}_{12}$. But by Theorem 8.1.1, $\det(\mathbf{A}) = \det(\mathbf{A}_{11})\det(a - \mathbf{a}_{21}\mathbf{A}_{11}^{-1}\mathbf{a}_{12})$ and from this we get (note $a - \mathbf{a}_{21}\mathbf{A}_{11}^{-1}\mathbf{a}_{12}$ is a scalar) $b = a - \mathbf{a}_{21}\mathbf{A}_{11}^{-1}\mathbf{a}_{12} = \det(\mathbf{A})/\det(\mathbf{A}_{11})$. But $\det(\mathbf{A}) > 0$ and $\det(\mathbf{A}_{11}) > 0$, so $b > 0$. So $\mathbf{R}$ and $\mathbf{S}$ are $(k + 1) \times (k + 1)$ upper and lower triangular matrices, respectively, with positive diagonal elements. The induction hypothesis proves this part of the theorem. To prove that $\mathbf{A}$ has positive leading principal minors if $\mathbf{A} = \mathbf{RS}$, where $\mathbf{R}$ and $\mathbf{S}$ are upper and lower triangular matrices, respectively, with positive diagonal elements, note that a $k \times k$ leading principal minor of an $n \times n$ matrix $\mathbf{A} = \mathbf{RS}$ is equal to the product of the first k diagonal elements of $\mathbf{R}$ and the first k diagonal elements of $\mathbf{S}$. This completes the proof. ∎

Next we define a matrix with all principal minors positive and prove some results that can be used for this type of matrix.

Definition 11.2.1

P-Matrices. *Let* $\mathbf{A}$ *be an* $n \times n$ *matrix such that all principal minors are positive. Then* $\mathbf{A}$ *is defined to be a P-matrix.*

Theorem 11.2.2

Let $\mathbf{A}$ *be a P-matrix of size* $n \times n$. *Then*

(1) $\mathbf{DA}$ *and* $\mathbf{AD}$ *are also P-matrices, where* $\mathbf{D}$ *is an* $n \times n$ *diagonal matrix with positive diagonal elements,*
(2) *Every principal submatrix of* $\mathbf{A}$ *is also a P-matrix,*
(3) $a_{ii} > 0$ *for each* $i = 1, 2, \ldots, n$,
(4) $\mathbf{R}'\mathbf{A}\mathbf{R}$ *is a P-matrix, where* $\mathbf{R}$ *is any* $n \times n$ *permutation matrix,*
(5) $\mathbf{A} + \mathbf{D}$ *is a P-matrix, where* $\mathbf{D}$ *is a diagonal matrix with* $d_{ii} \geq 0$ *for all* i,
(6) $\mathbf{A}'$ *is a P-matrix.*

The proof follows directly from the definition and will be left for the reader.

Example 11.2.1. Consider the 3×3 matrix $\mathbf{A}$ given below.

$$\mathbf{A} = \begin{bmatrix} 3 & 2 & 1 \\ 3 & 7 & 2 \\ 1 & 4 & 2 \end{bmatrix}.$$

The 3×3 principal minor is $\det(\mathbf{A}) = 15$. The 2×2 principal minors are

$$\det\begin{bmatrix} 3 & 2 \\ 3 & 7 \end{bmatrix} = 15; \det\begin{bmatrix} 3 & 1 \\ 1 & 2 \end{bmatrix} = 5; \det\begin{bmatrix} 7 & 2 \\ 4 & 2 \end{bmatrix} = 6.$$

The 1×1 principal minors are 3, 7, and 2. Hence all principal minors are positive, so $\mathbf{A}$ is a P-matrix.

Note: Theorem 8.11.7 and its corollaries state some results on dominant diagonal matrices and P-matrices.

Theorem 11.2.3

Let $\mathbf{A}$ *be an* $n \times n$ *matrix. Each of the four conditions below is necessary and sufficient for* $\mathbf{A}$ *to be a P-matrix.*

(1) *Define* $\mathbf{y}$ *by* $\mathbf{y} = \mathbf{Ax}$. *For every* $n \times 1$ *vector* $\mathbf{x} \neq \mathbf{0}$, *there is an element in* $\mathbf{x}$ *(say the q-th element) and the corresponding element in* $\mathbf{y}$ *such that* $x_q y_q > 0$.
(2) *For every* $n \times 1$ *vector* $\mathbf{x} \neq \mathbf{0}$, *there exists a diagonal matrix* $\mathbf{D}$ *with positive diagonal elements such that* $\mathbf{x}'\mathbf{DAx} > 0$.
(3) *For every* $n \times 1$ *vector* $\mathbf{x} \neq \mathbf{0}$, *there exists a diagonal matrix* $\mathbf{D}$ *with non-negative diagonal elements such that* $\mathbf{x}'\mathbf{DAx} > 0$.
(4) *Every real characteristic root of* $\mathbf{A}$ *and of each principal submatrix of* $\mathbf{A}$ *is positive.*

Proof: To prove this theorem, we shall prove that if $\mathbf{A}$ is a P-matrix, this implies (1), which implies (2), which implies (3), which implies (4), and (4) implies $\mathbf{A}$ is a P-matrix. To prove that if $\mathbf{A}$ is a P-matrix this implies (1), we assume the contrary; that is, for a vector $\mathbf{x} \neq \mathbf{0}$ and $\mathbf{y} = \mathbf{Ax}$ we assume $x_i y_i \leq 0$ for $i = 1, 2, \ldots, n$ and show that this leads to a contradiction.

To do this let K be the set of subscripts of $\mathbf{x}$ and $\mathbf{y}$ such that $x_i \neq 0$ and $x_i y_i \leq 0$. By hypothesis K is not the empty set. Let $\mathbf{A}_1$ be the submatrix of $\mathbf{A}$ that results from deleting from $\mathbf{A}$ the rows and columns corresponding to subscripts *not* in K. Let $\mathbf{x}_1$ be the vector resulting from eliminating the zero elements of $\mathbf{x}$ and let $\mathbf{y}_1 = \mathbf{A}_1\mathbf{x}_1$. Clearly $\mathbf{y}_1$ is the vector resulting from $\mathbf{y}$ with the elements with subscripts not in K deleted. Note that $\mathbf{A}_1$ is a principal submatrix of $\mathbf{A}$ and hence $\mathbf{A}_1$ is a P-matrix. Clearly there exists a diagonal matrix $\mathbf{D}$ with $d_{ii} \geq 0$ such that $\mathbf{y}_1 = -\mathbf{Dx}_1$, which gives $\mathbf{A}_1\mathbf{x}_1 = -\mathbf{Dx}_1$ or $(\mathbf{A}_1 + \mathbf{D})\mathbf{x}_1 = \mathbf{0}$. But by hypothesis $\mathbf{x}_1 \neq \mathbf{0}$, so $\mathbf{A}_1 + \mathbf{D}$ must be singular.

But by (5) of Theorem 11.2.2 this matrix is *not* singular, so this contradiction proves that if **A** is a P-matrix it implies (1).

To prove that (1) implies (2), let $\mathbf{x} \neq \mathbf{0}$ be an arbitrary vector and let $\mathbf{y} = \mathbf{A}\mathbf{x}$. So (1) implies $x_q y_q > 0$ for some q in $1, 2, \ldots, n$. Then there exists a $c > 0$ such that

$$x_q y_q + c \sum_{\substack{i=1 \\ i \neq q}}^{n} x_i y_i > 0.$$

Then take $d_{qq} = 1$, $d_{ii} = c$ for $i = 1, 2, \ldots, n$ $(i \neq q)$ as the diagonal elements of **D** and clearly $\mathbf{x}'\mathbf{D}\mathbf{A}\mathbf{x} = \mathbf{x}'\mathbf{D}\mathbf{y} > 0$, which proves that (1) implies (2). The proof that (2) implies (3) is obvious, since if $d_{ii} > 0$, then $d_{ii} \geq 0$. To prove that (3) implies (4), consider any principal submatrix of **A**, denoted by $\mathbf{A}_1$; let λ be a *real* characteristic root of $\mathbf{A}_1$, and let $\mathbf{x}_1$ be a characteristic vector of $\mathbf{A}_1$ corresponding to the root λ. Clearly $\mathbf{x}_1 \neq \mathbf{0}$. We must show that (3) implies $\lambda > 0$. We can obtain $\mathbf{A}_1$ from **A** by deleting certain rows (and the same columns) of **A**. Define an $n \times 1$ vector **x** to have zero elements corresponding to these deleted rows of **A** and let the remaining elements of **x** correspond to the elements of $\mathbf{x}_1$, the characteristic vector of $\mathbf{A}_1$ corresponding to λ. Let **D** be the diagonal matrix in (3) and $\mathbf{D}_1$ the matrix resulting from elimination of the rows and columns of **D** that were eliminated from **A** to get $\mathbf{A}_1$. We get

$$\mathbf{x}'\mathbf{D}\mathbf{A}\mathbf{x} = \mathbf{x}_1'\mathbf{D}_1\mathbf{A}_1\mathbf{x}_1 = \mathbf{x}_1'\mathbf{D}_1(\lambda\, \mathbf{x}_1) = \lambda\, \mathbf{x}_1'\mathbf{D}_1\mathbf{x}_1 \text{ and } \mathbf{x}_1'\mathbf{D}_1\mathbf{x}_1 > 0.$$

But by (3) we have $\mathbf{x}'\mathbf{D}\mathbf{A}\mathbf{x} > 0$, so $\lambda > 0$ also. This proves that (3) implies (4). We must now show that (4) implies that **A** is a P-matrix. Note that

$$\det(\mathbf{B}) = \prod_{i=1}^{k} \lambda_i,$$

where λ_i are the roots of the $k \times k$ matrix **B**. Also all complex, nonreal roots of **B** must occur in conjugate pairs; thus their product is positive. By (4) every real root of every principal submatrix of **A** is positive. Hence the product of the roots of every principal submatrix of **A** is positive, which says that every principal minor of **A** is positive. So by definition **A** is a P-matrix. This completes the proof of the theorem. ▮

Note: Condition (4) of Theorem 11.2.3 does not imply that a P-matrix must have a real root; but if it does have real roots, they must be positive. For example, consider the matrix **A** given below.

$$\mathbf{A} = \begin{bmatrix} 2 & -1 \\ 3 & 2 \end{bmatrix}.$$

Clearly **A** is a P-matrix, since all principal minors are positive. The characteristic roots of **A** are $2 + i\sqrt{3}$ and $2 - i\sqrt{3}$.

11.3 Matrices with Nonpositive Off-Diagonal Elements

In this section we discuss a special kind of matrix, often referred to as a Z-matrix, which is a matrix whose off-diagonal elements are nonpositive.

Definition 11.3.1

Z-Matrix. *Let* **A** *be an* $n \times n$ *matrix such that* $a_{ij} \leq 0$ *for all* $i \neq j$. *Then* **A** *is defined to be a Z-matrix.*

Theorem 11.3.1

The following $n \times n$ *matrices are Z-matrices.*

(1) *Any diagonal matrix (for example,* **I** *and* **0***).*
(2) *An upper (lower) triangular matrix whose upper (lower) elements are nonpositive.*
(3) *The sum of a finite number of Z-matrices.*
(4) *All leading principal submatrices of a Z-matrix.*
(5) **AD** *and* **DA**, *where* **A** *is a Z-matrix and* **D** *is a diagonal matrix with all diagonal elements non-negative.*
(6) *The transpose of a Z-matrix.*
(7) $a\mathbf{A}$, *where* $a \geq 0$ *and* **A** *is a Z-matrix.*
(8) $\sum_{i=1}^{k} a_i\mathbf{A}_i$, *where each* $\mathbf{A}_i$ *is a Z-matrix and each* $a_i \geq 0$.

The proof is a direct result of the definition and will be left for the reader.

Theorem 11.3.2

Let $\mathbf{A}$ *be a Z-matrix of size* $n \times n$ *and suppose there exist lower and upper triangular matrices* $\mathbf{R}$ *and* $\mathbf{S}$*, each with positive diagonal elements, such that* $\mathbf{A} = \mathbf{RS}$. *Then*

(1) $\mathbf{A}$ *has positive leading principal minors,*
(2) $\mathbf{R}^{-1}$ *and* $\mathbf{S}^{-1}$ *each exist,*
(3) *The off-diagonal elements of both* $\mathbf{R}$ *and* $\mathbf{S}$ *are nonpositive,*
(4) *No element of* $\mathbf{R}^{-1}$ *or* $\mathbf{S}^{-1}$ *is negative and the diagonal elements of* $\mathbf{R}^{-1}$ *and* $\mathbf{S}^{-1}$ *are all positive.*

Proof: Clearly (1) is true, since the value of the $k \times k$ leading principal minor of $\mathbf{A}$ is $\Pi_{i=1}^{k} r_{ii} s_{ii}$, but $r_{ii} > 0$ and $s_{ii} > 0$ for $i = 1, 2, \ldots, n$ by the hypothesis of the theorem. Result (2) follows from $\det(\mathbf{R}) = \Pi_{i=1}^{n} r_{ii} > 0$ and $\det(\mathbf{S}) = \Pi_{i=1}^{n} s_{ii} > 0$. The proof of (3) is by direct multiplication of the matrices displayed in Eq. (8.6.1). The proof of (4) will be by induction. We prove it for $\mathbf{R}^{-1}$ only, since the proof for $\mathbf{S}^{-1}$ is similar. Clearly the theorem is true for $n = 1$ (and $n = 2$). Assume that the theorem is true for any $(k - 1) \times (k - 1)$ upper triangular matrix $\mathbf{U}$ such that $u_{ii} > 0$, $u_{ij} = 0$ for $i > j$, $u_{ij} \leq 0$ for $i < j$; this means that $\mathbf{U}^{-1} = \mathbf{V}$ (say) is such that $v_{ii} > 0$, $v_{ij} = 0$ for $i > j$, $v_{ij} \geq 0$ for $i < j$. Now let $\mathbf{R}$ be any upper $k \times k$ triangular matrix with $r_{ii} > 0$, $r_{ij} = 0$ for $i > j$, $r_{ij} \leq 0$ for $i < j$. We write $\mathbf{R}$ and $\mathbf{R}^{-1}$ in partitioned form as

$$\mathbf{R} = \begin{bmatrix} \mathbf{U} & \mathbf{r}_1 \\ \mathbf{0} & r_{kk} \end{bmatrix}; \mathbf{R}^{-1} = \begin{bmatrix} \mathbf{U}^{-1} & -\mathbf{U}^{-1}\mathbf{r}_1 r_{kk}^{-1} \\ \mathbf{0} & r_{kk}^{-1} \end{bmatrix}.$$

By the hypothesis of the theorem, r_{kk} and hence r_{kk}^{-1} are positive, and all elements in $\mathbf{r}_1$ are nonpositive. By the induction hypothesis the upper diagonal elements of $\mathbf{U}^{-1}$ are non-negative, and by Theorem 8.6.12 the diagonal elements of $\mathbf{R}^{-1}$ are positive and the elements of the vector $-\mathbf{U}^{-1}\mathbf{r}_1 r_{kk}^{-1}$ are non-negative. So the $k \times k$ matrix $\mathbf{R}^{-1}$ is upper triangular with non-negative elements above the diagonal and positive diagonal elements. With the use of the induction hypothesis this proves the theorem. ∎

Theorem 11.3.3

Let $\mathbf{A}$ *be a Z-matrix such that each real characteristic root of* $\mathbf{A}$ *is positive. Let* $\mathbf{B}$ *be a Z-matrix such that* $\mathbf{A} \leq \mathbf{B}$. *Then*

(1) *Both* $\mathbf{A}^{-1}$ *and* $\mathbf{B}^{-1}$ *exist,*
(2) $\mathbf{0} \leq \mathbf{B}^{-1} \leq \mathbf{A}^{-1}$,

(3) *Each real characteristic root of* **B** *is positive,*
(4) $\det(\mathbf{B}) \geq \det(\mathbf{A}) > 0$.

Proof: To prove (1) and (2), consider the matrices **F** and **G** defined by $\mathbf{F} = \mathbf{I} - \delta\mathbf{A}$ and $\mathbf{G} = \mathbf{I} - \delta\mathbf{B}$, where δ is a positive number such that $\mathbf{F} \geq \mathbf{0}$ and $\mathbf{G} \geq \mathbf{0}$; thus $\mathbf{F} \geq \mathbf{G} \geq \mathbf{0}$. Let $\rho = \rho(\mathbf{F})$ be the spectral radius of **F** (hence $\rho \geq 0$). By Theorem 11.1.1 we get $0 = \det(\mathbf{F} - \rho\mathbf{I}) = \det(\mathbf{I} - \delta\mathbf{A} - \rho\mathbf{I}) = \det([1 - \rho]\mathbf{I} - \delta\mathbf{A})$, so $(1 - \rho)/\delta$ is a real characteristic root of **A**, which is positive by the hypothesis of the theorem. Hence $1 - \rho > 0$, which implies $0 \leq \rho < 1$, and by Corollary 5.6.10.1 the series $\mathbf{I} + \mathbf{F} + \mathbf{F}^2 + \ldots$ converges to $(\mathbf{I} - \mathbf{F})^{-1}$. But $\mathbf{I} - \mathbf{F} = \delta\mathbf{A}$, so $(\mathbf{I} - \mathbf{F})^{-1} = (\delta\mathbf{A})^{-1}$ and this implies that $\mathbf{A}^{-1}$ exists. But since $\mathbf{F} \geq \mathbf{0}$, clearly $\mathbf{F}^k \geq \mathbf{0}$ for $k = 1, 2, \ldots$, so $\mathbf{A}^{-1} \geq \mathbf{0}$. Also since $\mathbf{0} \leq \mathbf{G} \leq \mathbf{F}$, it follows that $\mathbf{0} \leq \mathbf{G}^k \leq \mathbf{F}^k$ for $k = 1, 2, \ldots$, and hence $\mathbf{I} + \mathbf{G} + \mathbf{G}^2 + \ldots$ converges to $(\mathbf{I} - \mathbf{G})^{-1}$, which equals $(\delta\mathbf{B})^{-1}$. Also $(\mathbf{I} - \mathbf{G})^{-1} \geq \mathbf{0}$, so $\mathbf{0} \leq \mathbf{B}^{-1} \leq \mathbf{A}^{-1}$ and this proves (1) and (2). To prove (3), consider $\mathbf{B} - \lambda\mathbf{I} = \mathbf{M}$, where $\lambda \leq 0$. Clearly $\mathbf{A} \leq \mathbf{M}$, and by (1) it follows that $\mathbf{M}^{-1}$ exists, so $\det(\mathbf{M}) \neq 0$ and λ cannot be a characteristic root of **B**; this proves (3). To prove (4), one can use induction. We leave the details for the reader. ∎

Example 11.3.1. Consider the 3×3 matrix **A** given below.

$$\mathbf{A} = \begin{bmatrix} 6 & -2 & -4 \\ -3 & 7 & -1 \\ 0 & -6 & 5 \end{bmatrix}.$$

Clearly **A** is a Z-matrix, since $a_{ij} \leq 0$ for all $i \neq j$. Also by simple multiplication it follows that $\mathbf{A} = \mathbf{RS}$, where

$$\mathbf{R} = \begin{bmatrix} 2 & 0 & 0 \\ -1 & 3 & 0 \\ 0 & -3 & 2 \end{bmatrix}; \mathbf{S} = \begin{bmatrix} 3 & -1 & -2 \\ 0 & 2 & -1 \\ 0 & 0 & 1 \end{bmatrix}.$$

The conditions of Theorem 11.3.2 are satisfied. If we let δ_i be the i-th leading principal minor of **A**, we get $\delta_1 = 6$; $\delta_2 = 36$; $\delta_3 = 72$; so (1) is satisfied. Clearly **R** and **S** are nonsingular, so (2) is satisfied, and by inspection we note that (3) is satisfied. Also

$$\mathbf{R}^{-1} = \frac{1}{12}\begin{bmatrix} 6 & 0 & 0 \\ 2 & 4 & 0 \\ 3 & 6 & 6 \end{bmatrix}; \mathbf{S}^{-1} = \frac{1}{6}\begin{bmatrix} 2 & 1 & 5 \\ 0 & 3 & 3 \\ 0 & 0 & 6 \end{bmatrix},$$

and (4) is satisfied.

11.4 M-Matrices (Z-Matrices with Positive Principal Minors)

In this section we discuss M-matrices, which are especially useful in the field of economics.

Definition 11.4.1

M-Matrix. *If an $n \times n$ matrix $\mathbf{A}$ satisfies the following two conditions, it is defined to be an M-matrix.*

(1) *$\mathbf{A}$ is a Z-matrix (that is, $a_{ij} \leq 0$ for all $i \neq j$).*
(2) *All principal minors of $\mathbf{A}$ are positive.*

Example 11.4.1. The matrices below are examples of M-matrices.

$$\mathbf{A}_1 = \begin{bmatrix} 1 & -2 \\ -2 & 5 \end{bmatrix}; \mathbf{A}_2 = \begin{bmatrix} 2 & 0 & 0 \\ -1 & 1 & 0 \\ -3 & -5 & 1 \end{bmatrix}; \mathbf{A}_3 = \begin{bmatrix} 2 & -2 & 0 \\ -1 & 6 & -2 \\ -1 & -2 & 8 \end{bmatrix}.$$

Clearly each matrix is a Z-matrix, and by direct calculation it is straightforward to show that all principal minors of each matrix are positive.

The following two theorems are immediate consequences of the definition. The proofs will be left for the reader.

Theorem 11.4.1

If a matrix $\mathbf{A}$ is an M-matrix, it is also a Z-matrix and a P-matrix.

Theorem 11.4.2

The following are M-matrices.

(1) *A diagonal matrix with positive diagonal elements.*
(2) *A triangular matrix whose diagonal elements are positive and whose off-diagonal elements are nonpositive.*
(3) *The transpose of an M-matrix.*
(4) *$c\mathbf{A}$, where $\mathbf{A}$ is an M-matrix and $c > 0$.*
(5) *Each principal submatrix of an M-matrix.*
(6) *$\mathbf{AD}$ and $\mathbf{DA}$, where $\mathbf{D}$ is a diagonal matrix with positive diagonal elements and $\mathbf{A}$ is an M-matrix.*

(7) *A dominant diagonal matrix with positive diagonal elements and non-positive off-diagonal elements (Corollary* 8.11.7.1*).*

Theorem 11.4.3

Let $\mathbf{A}$ *be a Z-matrix of size* $n \times n$. *Each of the following conditions is necessary and sufficient for* $\mathbf{A}$ *to be an M-matrix.*

(1) *All principal minors of* $\mathbf{A}$ *are positive (that is,* $\mathbf{A}$ *is a P-matrix).*
(2) *There exist lower and upper triangular matrices* $\mathbf{R}$ *and* $\mathbf{S}$ *such that* $\mathbf{A} = \mathbf{RS}$, $\mathbf{R}$ *is a Z-matrix with positive diagonal elements, and* $\mathbf{S}$ *is a Z-matrix with positive diagonal elements.*
(3) $\det(\mathbf{A}) \neq 0$ *and* $\mathbf{A}^{-1} \geq \mathbf{0}$.

Proof: Since $\mathbf{A}$ is a Z-matrix and $\mathbf{A}$ also satisfies (1), it is an M-matrix by Def. 11.4.1. We need to prove that conditions (1), (2), and (3) are equivalent. To do this we shall prove that (1) implies (2), then (2) implies (3), and finally (3) implies (1). To prove that (1) implies (2), we note that since $\mathbf{A}$ has positive principal minors, it has positive *leading* principal minors. So by Theorem 11.2.1 there exist lower and upper triangular matrices $\mathbf{R}$ and $\mathbf{S}$ with positive diagonal elements such that $\mathbf{A} = \mathbf{RS}$. By (3) of Theorem 11.3.2 it follows that $\mathbf{R}$ and $\mathbf{S}$ are Z-matrices. Hence (1) implies (2). To show that (2) implies (3), we note that $\det(\mathbf{A}) > 0$, since $\det(\mathbf{A}) = \det(\mathbf{RS}) = \det(\mathbf{R})\det(\mathbf{S})$. $\mathbf{R}$ and $\mathbf{S}$ in (2) are nonsingular and by (4) of Theorem 11.3.2, all elements of $\mathbf{R}^{-1}$ and $\mathbf{S}^{-1}$ are non-negative. Thus by simple multiplication, $\mathbf{A}^{-1} = \mathbf{S}^{-1}\mathbf{R}^{-1}$ gives $\mathbf{A}^{-1} \geq \mathbf{0}$. Thus (2) implies (3). To prove that (3) implies (1), assume that $\mathbf{A}$ is a Z-matrix, $\mathbf{A}^{-1}$ exists, and $\mathbf{A}^{-1} \geq \mathbf{0}$. Consider $\mathbf{A}^{-1}\mathbf{1} = \mathbf{x}$, so $\mathbf{x} > \mathbf{0}$ and $\mathbf{Ax} = \mathbf{1} > \mathbf{0}$. Since $a_{ij} \leq 0$ for all $i \neq j$ and $x_i > 0$ for all i, this implies $a_{ii} > 0$ for all i. Let $\mathbf{D}$ be a diagonal matrix with i-th diagonal element x_i and this gives $\mathbf{D1} = \mathbf{x}$. Let $\mathbf{B} = \mathbf{AD}$, and since $\mathbf{AD1} = \mathbf{Ax} > \mathbf{0}$, this results in $\mathbf{B1} > \mathbf{0}$. Since $\mathbf{A}$ is a Z-matrix and $\mathbf{D}$ is diagonal with positive diagonal elements, it follows that $\mathbf{B} = \mathbf{AD}$ is a Z-matrix and $b_{ii} > 0$. Thus $\mathbf{B}$ has positive dominant diagonals, so by Corollary 8.11.7.1, $\mathbf{B}$ has positive principal minors and thus $\mathbf{A}$ has positive principal minors ($\mathbf{A}$ is a P-matrix). This completes the proof of the theorem. ∎

The next theorem gives some conditions on the characteristic roots of a Z-matrix $\mathbf{A}$ so that $\mathbf{A}$ is an M-matrix.

Theorem 11.4.4

Let $\mathbf{A}$ *be an* $n \times n$ *Z-matrix. Conditions* (1) *and* (2) *are necessary and sufficient for* $\mathbf{A}$ *to be an M-matrix.*

(1) *Each real characteristic root of* $\mathbf{A}$ *is positive.*
(2) *The real part of each characteristic root of* $\mathbf{A}$ *is positive.*

The proof of this theorem is omitted. It can be found in [F–3].

In a system of equations $\mathbf{y} = \mathbf{Ax}$, it is often of interest to determine what the conditions on the matrix $\mathbf{A}$ are so that for some vector $\mathbf{x} > \mathbf{0}$ it follows that $\mathbf{y} > \mathbf{0}$. Clearly if $\mathbf{A} > \mathbf{0}$, then $\mathbf{y} > \mathbf{0}$ if $\mathbf{x} > \mathbf{0}$. Also the following theorem gives some results on this problem for Z- and M-matrices.

Theorem 11.4.5

Let the $n \times n$ *matrix* $\mathbf{A}$ *be a Z-matrix. Then each condition* (1), (2), *and* (3) *below is necessary and sufficient for* $\mathbf{A}$ *to be an M-matrix.*

(1) *There exists a vector* $\mathbf{x} \geq \mathbf{0}$ *such that* $\mathbf{Ax} > \mathbf{0}$.
(2) *There exists a vector* $\mathbf{x} > \mathbf{0}$ *such that* $\mathbf{Ax} > \mathbf{0}$.
(3) *There exists a diagonal matrix* $\mathbf{D}$ *with positive diagonal elements such that the sum of each row of* $\mathbf{AD}$ *is positive (that is, such that* $\mathbf{AD1} > \mathbf{0}$*).*

Proof: We shall prove that conditions (1), (2), and (3) are equivalent and then prove that condition (3) is a necessary and sufficient condition for $\mathbf{A}$ to be an M-matrix. Assume (1); that is, $\mathbf{x} \geq \mathbf{0}$ and $\mathbf{Ax} > \mathbf{0}$. Consider the vector $\mathbf{z} = \mathbf{x} + \varepsilon\mathbf{1}$, where $\varepsilon > 0$. Then $\mathbf{Az} = \mathbf{A}(\mathbf{x} + \varepsilon\mathbf{1}) = \mathbf{Ax} + \varepsilon\mathbf{A1}$. So $\mathbf{Az} > \mathbf{0}$ for a small enough ε, and (1) implies (2). Next assume (2); that is, let $\mathbf{x} > \mathbf{0}$ be a vector so that $\mathbf{Ax} > \mathbf{0}$. Let $\mathbf{D}$ be a diagonal matrix with diagonal elements equal to the elements of $\mathbf{x}$. Then $\mathbf{x} = \mathbf{D1}$ and $\mathbf{D}$ has positive diagonal elements. Thus $\mathbf{AD1} = \mathbf{Ax} > \mathbf{0}$, so (2) implies (3). To show that (3) implies (1), let $\mathbf{D}$ be a diagonal matrix with positive diagonal elements, so that by (3) we get $\mathbf{AD1} > \mathbf{0}$. Let $\mathbf{x} = \mathbf{D1} \geq \mathbf{0}$; then $\mathbf{Ax} > \mathbf{0}$; thus (3) implies (1). The proof of the theorem is complete if we show that condition (3) is a necessary and sufficient condition for $\mathbf{A}$ to be a P-matrix. Theorems 8.11.9 and 11.4.4 can be used. We leave the details for the reader. ∎

The next theorem gives some rather unrelated conditions, each of which is necessary and sufficient for a Z-matrix to be an M-matrix. We leave the proof for the reader.

Theorem 11.4.6

Let the $n \times n$ *matrix* $\mathbf{A}$ *be a Z-matrix. Each condition* (1), (2), (3), and (4) *below is a necessary and sufficient condition for* $\mathbf{A}$ *to be an M-matrix.*

(1) *There exists a diagonal matrix* **D** *with positive diagonal elements such that* **AD** *is d.d. with positive diagonal elements.*
(2) *For each diagonal matrix* **D** *such that* $\mathbf{D} \geq \mathbf{A}$, $\mathbf{D}^{-1}$ *exists and the spectral radius of* **B** *is less than* 1 *(that is,* $\rho(\mathbf{B}) < 1$*), where* $\mathbf{B} = \mathbf{D}^{-1}(\mathbf{D_A} - \mathbf{A})$*, and* $\mathbf{D_A}$ *is a diagonal matrix such that the diagonal of* **A** *is the same as the diagonal of* $\mathbf{D_A}$.
(3) *If* **B** *is a Z-matrix and* $\mathbf{B} \geq \mathbf{A}$*, then* $\mathbf{B}^{-1}$ *exists.*
(4) *For each* $n \times 1$ *vector* $\mathbf{x} \neq \mathbf{0}$*, there is a subscript* q *such that* $x_q y_q > 0$*, where* $\mathbf{y} = \mathbf{Ax}$.

Note: Each of the conditions in Theorems 11.4.3, 11.4.4, 11.4.5, and 11.4.6 is a necessary and sufficient condition for a Z-matrix **A** to be an M-matrix. Hence all twelve of these conditions are equivalent.

Example 11.4.2. Consider the 3×3 matrix **A** given by

$$\mathbf{A} = \begin{bmatrix} 6 & -3 & -4 \\ -3 & 2 & 0 \\ -1 & -1 & 20 \end{bmatrix}.$$

Clearly **A** is a Z-matrix. By calculating the principal minors, it is easy to show that **A** is an M-matrix. By using **D** below, it follows that $\mathbf{B} = \mathbf{AD}$ is d.d.

$$\mathbf{B} = \begin{bmatrix} 6 & -3 & -4 \\ -3 & 2 & 0 \\ -1 & -1 & 20 \end{bmatrix} \begin{bmatrix} 1.2 & 0 & 0 \\ 0 & 2 & 0 \\ 0 & 0 & .2 \end{bmatrix} = \begin{bmatrix} 7.2 & -6 & -.8 \\ -3.6 & 4 & 0 \\ -1.2 & -2 & 4 \end{bmatrix}.$$

Since **B** is d.d., we can use Theorem 8.11.13 to state that for any 3×1 vector $\mathbf{g} \geq \mathbf{0}$, there is a unique vector $\mathbf{x} \geq \mathbf{0}$ such that $\mathbf{Ax} = \mathbf{g}$. Another way to view this is that $\mathbf{B} = \mathbf{AD}$ is a Z-matrix, and since **B** is d.d. with positive diagonal elements and nonpositive off-diagonal elements, it is an M-matrix (see (7) of Theorem 11.4.2). Hence $\mathbf{B}^{-1} \geq \mathbf{0}$ by (3) of Theorem 11.4.3. But $\mathbf{B}^{-1} = \mathbf{D}^{-1}\mathbf{A}^{-1} \geq \mathbf{0}$, so $\mathbf{A}^{-1} \geq \mathbf{0}$ and $\mathbf{A}^{-1}\mathbf{g} \geq \mathbf{0}$ if $\mathbf{g} \geq \mathbf{0}$.

The next two theorems contain some interesting conditions that an M-matrix satisfies.

Theorem 11.4.7

Let **A** *be an M-matrix. There exists a positive characteristic root of* **A***, which we denote by* $\ell(\mathbf{A})$*, such that the real part of any characteristic root of* **A** *is greater than or equal to* $\ell(\mathbf{A})$.

The proof of this theorem is left for the reader.

Theorem 11.4.8

Let $\mathbf{A}$ *be an M-matrix and let* $\mathbf{B}$ *be a Z-matrix (both* $n \times n$*). Also let* $\mathbf{B} \geq \mathbf{A}$. *Then*

(1) $\mathbf{B}$ *is also an M-matrix,*
(2) $\mathbf{0} \leq \mathbf{B}^{-1} \leq \mathbf{A}^{-1}$,
(3) $\det(\mathbf{B}) \geq \det(\mathbf{A}) > 0$,
(4) $\mathbf{A}^{-1}\mathbf{B} \geq \mathbf{I}$ *and* $\mathbf{B}\mathbf{A}^{-1} \geq \mathbf{I}$,
(5) $\mathbf{B}^{-1}\mathbf{A} \leq \mathbf{I}$ *and* $\mathbf{A}\mathbf{B}^{-1} \leq \mathbf{I}$,
(6) $\mathbf{A}\mathbf{B}^{-1}$ *and* $\mathbf{B}^{-1}\mathbf{A}$ *are M-matrices.*

Proof: Both $\mathbf{A}$ and $\mathbf{B}$ are Z-matrices, and since $\mathbf{A}$ is an M-matrix, it has positive real characteristics roots. Hence by Theorem 11.3.13, $\mathbf{A}^{-1}$ and $\mathbf{B}^{-1}$ exist and (1), (2), and (3) follow. To prove (4), we know that $\mathbf{B} \geq \mathbf{A}$, which is $\mathbf{B} - \mathbf{A} \geq \mathbf{0}$, and also $\mathbf{A}^{-1} \geq \mathbf{0}$; hence $\mathbf{A}^{-1}(\mathbf{B} - \mathbf{A}) \geq \mathbf{0}$, which gives $\mathbf{A}^{-1}\mathbf{B} \geq \mathbf{I}$. We leave the proof of the remaining parts for the reader. ∎

Corollary 11.4.8.1

If $\mathbf{A}$ *and* $\mathbf{B}$ *are as defined in Theorem* 11.4.8, *then*

(1) $\rho(\mathbf{I} - \mathbf{B}^{-1}\mathbf{A}) < 1$ *and* $\rho(\mathbf{I} - \mathbf{A}\mathbf{B}^{-1}) < 1$,
(2) $\ell(\mathbf{B}) \geq \ell(\mathbf{A})$,

where $\rho(\mathbf{K})$ *is the spectral radius of* $\mathbf{K}$ *and* $\ell(\mathbf{A})$ *is defined in Theorem* 11.4.7.

Corollary 11.4.8.2

Let $\mathbf{A}$ *be an M-matrix then* $\ell(\mathbf{A}) \leq a_{ii}$ *for* $i = 1, 2, \ldots, n$.

Corollary 11.4.8.3

Let $\mathbf{A}_{11}$ *be any principal submatrix of* $\mathbf{A}$. *Then* $\ell(\mathbf{A}) \leq \ell(\mathbf{A}_{11})$.

Example 11.4.3. An easy example is given to illustrate some of the preceding theorems. Let $\mathbf{A}$ be defined by

$$\mathbf{A} = \begin{bmatrix} 6 & -3 & -2 \\ -3 & 2 & 0 \\ -1 & -1 & 5 \end{bmatrix}.$$

Clearly **A** is a Z-matrix, and by calculating the principal minors, it is easy to show that **A** is also an M-matrix. The characteristic roots are $4 + \sqrt{15}$, 5, and $4 - \sqrt{15}$, which are all positive, as they must be, since they are all real (the real part of each root of an M-matrix is positive); also $\ell(\mathbf{A}) = 4 - \sqrt{15}$. A principal submatrix of **A** is

$$\mathbf{A}_{11} = \begin{bmatrix} 6 & -3 \\ -3 & 2 \end{bmatrix}$$

and the roots of $\mathbf{A}_{11}$ are $4 + \sqrt{13}$ and $4 - \sqrt{13}$. Clearly $\ell(\mathbf{A}_{11}) = 4 - \sqrt{13}$ and $\ell(\mathbf{A}) < \ell(\mathbf{A}_{11})$.

Let **B** be defined by

$$\mathbf{B} = \begin{bmatrix} 7 & -2 & 0 \\ -1 & 2 & 0 \\ -1 & -1 & 6 \end{bmatrix}.$$

Clearly **B** is a Z-matrix and $\mathbf{B} \geq \mathbf{A}$, so **B** is also an M-matrix; $\mathbf{B}^{-1}$ is given by

$$\mathbf{B}^{-1} = \frac{1}{72}\begin{bmatrix} 12 & 12 & 0 \\ 6 & 42 & 0 \\ 3 & 9 & 12 \end{bmatrix}.$$

Also

$$\mathbf{A}^{-1} = \frac{1}{5}\begin{bmatrix} 10 & 17 & 4 \\ 15 & 28 & 6 \\ 5 & 9 & 3 \end{bmatrix}$$

and $\mathbf{0} \leq \mathbf{B}^{-1} \leq \mathbf{A}^{-1}$. By direct computation, $\det(\mathbf{B}) = 72$, $\det(\mathbf{A}) = 5$, so this illustrates that $\det(\mathbf{B}) \geq \det(\mathbf{A})$. We get

$$\mathbf{AB}^{-1} = \frac{1}{72}\begin{bmatrix} 48 & -72 & -24 \\ -24 & 48 & 0 \\ -3 & -9 & 60 \end{bmatrix}$$

and $\mathbf{AB}^{-1}$ is an M-matrix with $\mathbf{AB}^{-1} \leq \mathbf{I}$.

The roots of **B** are $(9 + \sqrt{33})/2$, 6, and $(9 - \sqrt{33})/2$, so $\ell(\mathbf{B}) = (9 - \sqrt{33})/2 > 4 - \sqrt{15} = \ell(\mathbf{A})$. Finally

$$\mathbf{BA}^{-1} = \frac{1}{5}\begin{bmatrix} 40 & 63 & 16 \\ 20 & 39 & 8 \\ 5 & 9 & 8 \end{bmatrix}$$

and $\mathbf{BA}^{-1} \geq \mathbf{I}$.

11.5 Z-Matrices with Non-Negative Principal Minors

In this section we discuss a slightly more general matrix than an M-matrix, which we denote by M_0.

Definition 11.5.1

M_0-Matrix. *Let an $n \times n$ matrix **A** be a Z-matrix. If all principal minors of **A** are non-negative, then **A** is defined to be an M_0-matrix.*

We shall state some theorems about M_0-matrices.

Theorem 11.5.1

*Let **A** be a Z-matrix. Each of the three conditions below is necessary and sufficient for **A** to be an M_0-matrix.*

(1) *All real characteristic roots of **A** and of all principal submatrices of **A** are non-negative.*
(2) $\mathbf{A} + \varepsilon\mathbf{I}$ *is an M-matrix for all real numbers ε such that $\varepsilon > 0$.*
(3) *All real characteristic roots of **A** are non-negative.*

We leave the proof of this theorem and the next for the reader.

Theorem 11.5.2

*If **A** is an M-matrix, it is also an M_0-matrix.*

Theorem 11.5.3

*If **A** is an M_0-matrix and if **A** is nonsingular, then **A** is an M-matrix.*

Proof: If λ is a real characteristic root of **A**, then by (3) of Theorem 11.5.1 it follows that $\lambda \geq 0$. But since $\det(\mathbf{A}) \neq 0$, no characteristic roots of **A** can be zero. Thus all real characteristic roots of **A** must be positive, and hence by Theorem 11.4.4 **A** is an M-matrix. ▮

The proof of the next four theorems will be left for the reader.

Theorem 11.5.4

*Let **A** be a Z-matrix. If there exists a vector $\mathbf{x} > \mathbf{0}$ such that $\mathbf{Ax} \geq \mathbf{0}$, then **A** is an M_0-matrix.*

Theorem 11.5.5

If **A** *is an* M_0*-matrix, there exists a nonzero vector* $\mathbf{x} \geq \mathbf{0}$ *such that* $\mathbf{Ax} \geq \mathbf{0}$.

Theorem 11.5.6

Let **A** *be an* $n \times n$ *irreducible* M_0*-matrix.*

(1) *If* **A** *is singular, then* **A** *has rank* $n - 1$.
(2) *If* **A** *is singular, then there exists a vector* $\mathbf{x} > \mathbf{0}$ *such that* $\mathbf{Ax} = \mathbf{0}$.

Theorem 11.5.7

Let **A** *be an* M_0*-matrix and let* **B** *be a Z-matrix such that* $\mathbf{B} \geq \mathbf{A}$. *Then* **B** *is also an* M_0*-matrix.*

For further information on P-, Z-, M-, and M_0-matrices, see [B–7], [B–8], [F–3], [F–4], [H–1], [M–3], [P–3], [P–4], and [V–1].

Problems

1. Find the spectral radius $\rho(\mathbf{A})$ of the matrix **A**, where

$$\mathbf{A} = \begin{bmatrix} 1 & 3 & 2 \\ 2 & 1 & 3 \\ 3 & 2 & 1 \end{bmatrix}$$

   and illustrate Theorem 11.1.1.
2. Find a characteristic vector **x** of the matrix **A** in Prob. 1 corresponding to the root $\rho(\mathbf{A})$ such that $\mathbf{x} \geq \mathbf{0}$.
3. For the matrix **B**, find $\rho(\mathbf{B})$ and note that $\rho(\mathbf{B}) \leq \rho(\mathbf{A})$ since $\mathbf{B} \leq \mathbf{A}$, where **A** is defined in Prob. 1 and **B** is given below.

$$\mathbf{B} = \begin{bmatrix} 1 & 3 & 1 \\ 1 & 1 & 3 \\ 3 & 1 & 1 \end{bmatrix}.$$

   This illustrates (4) of Theorem 11.1.1.

4. Show that the matrix **A** has positive leading principal minors, where

$$\mathbf{A} = \begin{bmatrix} 2 & 4 & -6 \\ 1 & 8 & 0 \\ 1 & 4 & 1 \end{bmatrix}.$$

5. Find upper and lower triangular matrices **R** and **S**, respectively, such that **A** = **RS** (where **A** is defined in Prob. 4) and such that $r_{ii} > 0$, $s_{ii} > 0$. This illustrates Theorem 11.2.1.

6. Show that the matrix **A** below is a P-matrix.

$$\mathbf{A} = \begin{bmatrix} 3 & 10 & -5 \\ 1 & 4 & -2 \\ -1 & 0 & 7 \end{bmatrix}.$$

7. If **D** is given below, show that **DA**, **AD**, and **A** + **D** are P-matrices, where **A** is given in Prob. 6. This illustrates (1) and (5) of Theorem 11.2.2.

$$\mathbf{D} = \begin{bmatrix} 1 & 0 & 0 \\ 0 & 3 & 0 \\ 0 & 0 & 5 \end{bmatrix}.$$

8. If the permutation matrix **R** is given by

$$\mathbf{R} = \begin{bmatrix} 0 & 1 & 0 \\ 1 & 0 & 0 \\ 0 & 0 & 1 \end{bmatrix},$$

show that **R'AR** is a P-matrix, where **A** is given in Prob. 6. This illustrates (4) of Theorem 11.2.2.

9. Prove Theorem 11.2.2.

10. Consider the P-matrix **A** below.

$$\mathbf{A} = \begin{bmatrix} 3 & -2 \\ 3 & 4 \end{bmatrix}.$$

To illustrate (1) and (2) of Theorem 11.2.3, do the following:
(1) If $\mathbf{x}' = [-1, 2]$, show there exists a subscript q such that $x_q y_q > 0$, where $\mathbf{y} = \mathbf{Ax}$.
(2) If $\mathbf{x}' = [-1, 3]$, exhibit a diagonal matrix **D** with $d_{ii} > 0$ such that $\mathbf{x}'\mathbf{DAx} > 0$.

11. Prove Theorem 11.3.1.
12. For the Z-matrix $\mathbf{A}$

$$\mathbf{A} = \begin{bmatrix} 2 & -1 & 0 \\ -1 & 2 & 0 \\ -2 & -1 & 1 \end{bmatrix},$$

find lower and upper triangular matrices $\mathbf{R}$ and $\mathbf{S}$, respectively, each with positive diagonal elements such that $\mathbf{A} = \mathbf{RS}$.
13. In Prob. 12, show that $\mathbf{R}^{-1}$ and $\mathbf{S}^{-1}$ exist, and find them. This illustrates (3) of Theorem 11.3.2.
14. In Prob. 13, demonstrate that all off-diagonal elements of $\mathbf{R}^{-1}$ and $\mathbf{S}^{-1}$ are nonnegative and that the diagonal elements are positive. This illustrates (4) of Theorem 11.3.2.
15. Probs. 15, 16, 17, and 18 will illustrate Theorem 11.3.3. Define the Z-matrices $\mathbf{A}$ and $\mathbf{B}$ by (note $\mathbf{A} \le \mathbf{B}$)

$$\mathbf{A} = \begin{bmatrix} 2 & -1 & 0 \\ -1 & 2 & 0 \\ -2 & -1 & 1 \end{bmatrix}; \mathbf{B} = \begin{bmatrix} 3 & -1 & 0 \\ 0 & 2 & 0 \\ -1 & -1 & 1 \end{bmatrix}.$$

Show that each real characteristic of $\mathbf{A}$ is positive.
16. Show that $\mathbf{A}^{-1}$ and $\mathbf{B}^{-1}$ exist, find them, and show $\mathbf{0} \le \mathbf{B}^{-1} \le \mathbf{A}^{-1}$, where $\mathbf{A}$ and $\mathbf{B}$ are as defined in Prob. 15.
17. In Prob. 15, show that each real characteristic root of $\mathbf{B}$ is positive.
18. In Prob. 15, show $\det(\mathbf{B}) \ge \det(\mathbf{A}) > 0$.
19. Prove (4) of Theorem 11.3.3.
20. Prove Theorem 11.4.2.
21. Prove that (3) of Theorem 11.4.5 is necessary and sufficient for $\mathbf{A}$ to be an M-matrix.
22. Use (3) of Theorem 11.4.3 to determine if $\mathbf{A}$ is an M-matrix, where

$$\mathbf{A} = \begin{bmatrix} 6 & -1 & -2 \\ 0 & 2 & -1 \\ -1 & -1 & 1 \end{bmatrix}.$$

23. Show that there exists a unique vector $\mathbf{x} \ge \mathbf{0}$ such that $\mathbf{Ax} = \mathbf{g}$, where

$$\mathbf{A} = \begin{bmatrix} 3 & -1 & 0 \\ -2 & 4 & -1 \\ -1 & -2 & 4 \end{bmatrix}; \mathbf{g} = \begin{bmatrix} 1 \\ 2 \\ 1 \end{bmatrix}.$$

24. In Prob. 23, find $\mathbf{x}$.
25. Find the spectral radius of $\mathbf{B}$, where $\mathbf{B} = \mathbf{D}^{-1} (\mathbf{D_A} - \mathbf{A})$, where $\mathbf{A}$ is as defined in Prob. 23, and where

$$\mathbf{D_A} = \begin{bmatrix} 3 & 0 & 0 \\ 0 & 4 & 0 \\ 0 & 0 & 4 \end{bmatrix}; \mathbf{D} = \begin{bmatrix} 3 & 0 & 0 \\ 0 & 5 & 0 \\ 0 & 0 & 4 \end{bmatrix}.$$

This illustrates (2) of Theorem 11.4.6.
26. In Example 11.4.2, find the spectral radius of $\delta\mathbf{I} - \mathbf{A}$, where $\delta = \max a_{ii}$.
27. Use the matrix in Prob. 22 to illustrate Corollary 11.4.8.3.
28. Show that the matrix $\mathbf{A}$ given below is an $\mathrm{M_0}$-matrix but not an M-matrix.

$$\mathbf{A} = \begin{bmatrix} 6 & -1 & -3 \\ -2 & 1 & 0 \\ -2 & -1 & 3 \end{bmatrix}.$$

29. In Prob. 28, let $\varepsilon = 0.001$ and show that $\mathbf{A} + \varepsilon\mathbf{I}$ is an M-matrix. This illustrates (2) of Theorem 11.5.1.
30. Show that $\mathbf{A}$ in Prob. 28 is irreducible.
31. Show that the rank of $\mathbf{A}$ in Prob. 28 is 2. This illustrates (1) of Theorem 11.5.6.
32. Exhibit a vector $\mathbf{x} > \mathbf{0}$ such that $\mathbf{Ax} = \mathbf{0}$, where $\mathbf{A}$ is given in Prob. 28. This illustrates (2) of Theorem 11.5.6.
33. If $\mathbf{A} \geq \mathbf{0}$ and $\mathbf{x} \geq \mathbf{y}$, show that $\mathbf{Ax} \geq \mathbf{Ay}$.
34. If $\mathbf{Ax} \geq \mathbf{0}$ for all $\mathbf{x} \geq \mathbf{0}$, show that $\mathbf{A} \geq \mathbf{0}$.
35. If $\mathbf{A} \geq \mathbf{0}$ is an $n \times n$ irreducible matrix, show that $(\mathbf{I} + \mathbf{A})^{n-1} > \mathbf{0}$.
36. Demonstrate Prob. 35 by showing that $(\mathbf{I} + \mathbf{A})^2 > \mathbf{0}$, where $\mathbf{A}$ is given in Prob. 28.
37. If $\mathbf{A}$ in Prob. 35 is such that $a_{ii} > 0$ for all $i = 1, 2, \ldots, n$ show that $\mathbf{A}^{n-1} > \mathbf{0}$.
38. Consider $\mathbf{A}$ in Prob. 28. Show that $\mathbf{A}^{n-1} > \mathbf{0}$, and thus demonstrate the result in Prob. 37.
39. In Prob. 35, show that $(\mathbf{I} + \mathbf{A})^k > \mathbf{0}$ may not be true if $\mathbf{A}$ is reducible or if $k < n - 1$.

Each problem 40–46 is stated as a theorem. Prove each.

40. Let $\mathbf{A}$ be a Z-matrix of size $n \times n$ with $n \leq 3$ and with positive diagonals. Then $\mathbf{A}$ is an M-matrix if and only if $\det(\mathbf{A}) > 0$.

41. If $\mathbf{A}$ is an M-matrix, then $\mathbf{A}^k$ is an M-matrix if and only if $\mathbf{A}^k$ is a Z-matrix with positive diagonal elements (k is any positive integer).
42. If

$$\mathbf{A} = \begin{bmatrix} \mathbf{A}_{11} & \mathbf{A}_{12} \\ \mathbf{0} & \mathbf{A}_{22} \end{bmatrix}$$

is a Z-matrix with positive diagonal elements, then $\mathbf{A}$ is an M-matrix if and only if $\mathbf{A}_{11}$ and $\mathbf{A}_{22}$ are M-matrices.
43. All square matrices $\mathbf{A}$ with $a_{ij} \leq 0$ for all $i \neq j$ such that $\mathbf{A}^k = \mathbf{0}$ for some positive integer k are M_0-matrices.
44. If $\mathbf{A}$ can be written as $\mathbf{A} = \delta\mathbf{I} - \mathbf{B}$, where $\mathbf{B} \geq \mathbf{0}$ and $\delta > \rho(\mathbf{B})$, then $\mathbf{A}$ is an M-matrix.
45. If $\mathbf{A}$ is a Z-matrix and all row sums are positive, then $\det(\mathbf{A}) > 0$.
46. If $\mathbf{A}$ is an $n \times n$ matrix, $n \geq 2$, then $\mathbf{A}$ is an M-matrix if and only if any of the following hold.
    (1) $(\mathbf{A} + \mathbf{D})^{-1} > \mathbf{0}$ for each diagonal nonsingular matrix $\mathbf{D}$.
    (2) $(\mathbf{A} + \delta\mathbf{I})^{-1} > \mathbf{0}$ for each scalar $\delta \geq 0$.
    (3) Each principal matrix of order 1, 2, and n has a positive inverse.

Non-Negative Matrices; Idempotent and Tripotent Matrices; Projections

12

12.1 Introduction

Quadratic forms, particularly those with non-negative and idempotent matrices, play a central role in the theories of regression, correlation, experimental design, and analysis of variance. We cannot give the details of the statistical theory here, but we shall outline some of the procedures and indicate briefly the role that non-negative matrices and idempotent matrices play before considering some theorems about these special matrices. The reader who wishes to obtain further information about the statistical theory can consult [G–7].

The various statistical areas just mentioned have one common mathematical procedure. An observer makes n observations denoted by $y_1, y_2, \ldots, y_n$. These observations can be considered as the elements of a vector $\mathbf{y}$. The sum of squares of these elements is computed and denoted by $\mathbf{y}'\mathbf{y}$ (sometimes a quantity $\mathbf{y}'\mathbf{A}_0\mathbf{y}$ is computed instead of $\mathbf{y}'\mathbf{y}$ when $\mathbf{A}_0$ is a positive semidefinite or positive definite matrix). Then the quadratic form $\mathbf{y}'\mathbf{y}$ is partitioned into the sum of k quadratic forms; that is,

$$\mathbf{y}'\mathbf{y} = \sum_{i=1}^{k} \mathbf{y}'\mathbf{A}_i\mathbf{y} \tag{12.1.1}$$

or

$$\mathbf{y}'\mathbf{A}_0\mathbf{y} = \sum_{i=1}^{k} \mathbf{y}'\mathbf{A}_i\mathbf{y}, \tag{12.1.2}$$

and of course this gives a matrix equation to study, namely,

$$\mathbf{I} = \sum_{i=1}^{k} \mathbf{A}_i \tag{12.1.3}$$

or

$$\mathbf{A}_0 = \sum_{i=1}^{k} \mathbf{A}_i . \tag{12.1.4}$$

The matrices involved are symmetric, and often it is known that they are also non-negative; for example if any quadratic form, say $\mathbf{y}'\mathbf{A}_j\mathbf{y}$, is obtained by squaring numbers and adding them or by adding numbers and squaring the result, then it follows that $\mathbf{y}'\mathbf{A}_j\mathbf{y}$ is either positive semidefinite or positive definite.

In order to discuss various aspects of the theory, it may be necessary to determine whether some or all of the $\mathbf{A}_i$ are idempotent and whether the product $\mathbf{A}_i\mathbf{A}_j$ is the zero matrix if $i \neq j$.

For a simple example, consider the identity

$$\sum_{i=1}^{n} y_i^2 = n\bar{y}^2 + \sum_{i=1}^{n} (y_i - \bar{y})^2 \quad \text{where } \bar{y} = \frac{1}{n} \sum_{i=1}^{n} y_i , \tag{12.1.5}$$

which we can write as

$$\mathbf{y}'\mathbf{y} = \mathbf{y}'\mathbf{A}_1\mathbf{y} + \mathbf{y}'\mathbf{A}_2\mathbf{y},$$

where

$$\mathbf{A}_1 = \frac{1}{n}\mathbf{J} \quad \text{and} \quad \mathbf{A}_2 = \mathbf{I} - \frac{1}{n}\mathbf{J}. \tag{12.1.6}$$

We recognize that

(1) $\mathbf{A}_1$ is a symmetric and idempotent matrix,
(2) $\mathbf{A}_2$ is a symmetric and idempotent matrix,
(3) $\mathbf{A}_1\mathbf{A}_2 = \mathbf{0}$;

and in statistical theory these facts are important to note, because they lead to important results.

12.2 Non-Negative Matrices

In Chapter 1 we defined positive definite and semidefinite matrices and stated some theorems about them. This section states a number of additional theorems about these types of matrices. Since Theorems 12.2.1 through 12.2.4 are generally proved in a first course in matrix algebra, we omit the proofs here.

Definition 12.2.1

Positive Semidefinite Matrix. *An $n \times n$ matrix $\mathbf{A}$ is defined to be positive semidefinite if and only if*

(1) $\mathbf{A} = \mathbf{A}'$,

(2) $\mathbf{y}'\mathbf{A}\mathbf{y} \geq 0$ *for each and every vector* $\mathbf{y}$ *in* E_n *and the equality holds for at least one vector* $\mathbf{y}$ *such that* $\mathbf{y} \neq \mathbf{0}$.

Definition 12.2.2

Positive Definite Matrix. *An $n \times n$ matrix $\mathbf{A}$ is defined to be positive definite if and only if*

(1) $\mathbf{A} = \mathbf{A}'$,

(2) $\mathbf{y}'\mathbf{A}\mathbf{y} > 0$ *for each and every vector* $\mathbf{y}$ *in* E_n *such that* $\mathbf{y} \neq \mathbf{0}$.

Definition 12.2.3

Non-negative Matrix. *A matrix is defined to be non-negative if and only if it is either positive definite or positive semidefinite.*

Theorem 12.2.1

The results (1a), (2a), (3a) *follow if* $\mathbf{A}$ *is an $n \times n$ positive semidefinite matrix, and the results* (1b), (2b), (3b) *follow if* $\mathbf{A}$ *is an $n \times n$ positive definite matrix.*

(1a) *The rank of* $\mathbf{A}$ *is less than* n.

(2a) $a_{ii} \geq 0$ *for all* $i = 1, 2, \ldots, n$; *if* $a_{tt} = 0$, *then each element in the t-th row and the t-th column of* $\mathbf{A}$ *is equal to zero.*

(3a) $\mathbf{P}'\mathbf{A}\mathbf{P}$ *is a non-negative matrix for any $n \times n$ matrix* $\mathbf{P}$.

(1b) *The rank of* $\mathbf{A}$ *is equal to* n.

(2b) $a_{ii} > 0$ *for all* $i = 1, 2, \ldots, n$.

(3b) $\mathbf{P}'\mathbf{A}\mathbf{P}$ *is a positive definite matrix for any nonsingular $n \times n$ matrix* $\mathbf{P}$ (*in particular,* $\mathbf{A}^{-1}$ *is positive definite*).

The next theorem states some necessary and sufficient conditions for a matrix to be positive semidefinite or positive definite.

Theorem 12.2.2

Let $\mathbf{A}$ *be an $n \times n$ symmetric matrix. Conditions* (1a), (2a) *are necessary and* *sufficient for* **A** *to be a positive semidefinite matrix. Conditions* (1b), (2b), (3b) *are necessary and sufficient for* **A** *to be a positive definite matrix.*

(1a) *There exists an* $n \times n$ *matrix* **B** *of rank less than* n *such that* $\mathbf{B}'\mathbf{B} = \mathbf{A}$.

(2a) *The characteristic roots of* **A** *are non-negative and at least one root is equal to zero.*

(1b) *There exists an* $n \times n$ *matrix* **B** *of rank* n *such that* $\mathbf{B}'\mathbf{B} = \mathbf{A}$.

(2b) *The characteristic roots of* **A** *are all positive.*

(3b) $a_{11} > 0; \begin{vmatrix} a_{11} & a_{12} \\ a_{21} & a_{22} \end{vmatrix} > 0; \ldots ; |\mathbf{A}| > 0.$

Corollary 12.2.2

If **B** *is a* $p \times n$ *matrix of rank* r, *then*

(1) $\mathbf{B}'\mathbf{B}$ *and* $\mathbf{B}\mathbf{B}'$ *are non-negative matrices.*

(2) $\mathbf{B}'\mathbf{B}$ *is a positive semidefinite matrix if* $r < n$.

(3) $\mathbf{B}'\mathbf{B}$ *is a positive definite matrix if* $r = n$.

The next two theorems state some useful results concerning the trace of non-negative matrices. These results were proved in Chapter 9.

Theorem 12.2.3

If **A** *and* **B** *are* $n \times n$ *non-negative matrices, then*

(1) $\text{tr}\,(\mathbf{A}) \geq 0$,

(2) $\text{tr}\,(\mathbf{A}) = 0$ *if and only if* $\mathbf{A} = \mathbf{0}$,

(3) $\text{tr}\,(\mathbf{AB}) \geq 0$,

(4) $\text{tr}\,(\mathbf{AB}) = 0$ *if and only if* $\mathbf{AB} = \mathbf{0}$.

If **A** *and* **B** *are* $n \times n$ *positive definite matrices, then*

(5) $\text{tr}\,(\mathbf{A}) > 0$,

(6) $\text{tr}\,(\mathbf{AB}) > 0$.

Theorem 12.2.4

Let $\{\mathbf{A}_1, \mathbf{A}_2, \ldots, \mathbf{A}_k\}$ *be a collection of* $n \times n$ *non-negative matrices. Then*

(1) $\text{tr}\left(\sum_{i=1}^{k} \mathbf{A}_i\right) = \sum_{i=1}^{k} \text{tr}\,(\mathbf{A}_i) \geq 0,$

(2) $\sum_{i=1}^{k} \text{tr}\,(\mathbf{A}_i) = 0$ *if and only if* $\mathbf{A}_1 = \mathbf{A}_2 = \cdots = \mathbf{A}_k = \mathbf{0}$,

(3) $\sum_{j=1}^{k} \sum_{i=1}^{k} \operatorname{tr}(\mathbf{A}_i \mathbf{A}_j) \geq 0$ *and* $\sum_{j=1}^{k} \sum_{\substack{i=1 \\ i \neq j}}^{k} \operatorname{tr}(\mathbf{A}_i \mathbf{A}_j) \geq 0,$

(4) $\sum_{j=1}^{k} \sum_{\substack{i=1 \\ i \neq j}}^{k} \operatorname{tr}(\mathbf{A}_i \mathbf{A}_j) = 0$ *if and only if* $\mathbf{A}_i \mathbf{A}_j = \mathbf{0}$ *for all* $i \neq j$.

If $\mathbf{A}_1, \mathbf{A}_2, \ldots, \mathbf{A}_k$ *are positive definite matrices, then*

(5) $\sum_{i=1}^{k} \operatorname{tr}(\mathbf{A}_i) > 0,$

(6) $\sum_{i} \sum_{j} \operatorname{tr}(\mathbf{A}_i \mathbf{A}_j) > 0$ *and* $\sum_{i} \sum_{\substack{j \\ i \neq j}} \operatorname{tr}(\mathbf{A}_i \mathbf{A}_j) > 0.$

Note: In Theorem 12.2.3 the matrix **AB** may be neither positive definite nor positive semidefinite. In fact, some of the diagonal elements may be negative.

Now we shall state and prove some additional theorems about non-negative matrices.

Theorem 12.2.5

Let $\{\mathbf{A}_1, \mathbf{A}_2, \ldots, \mathbf{A}_k\}$ *be a collection of* $n \times n$ *non-negative matrices. Then*

(1) $\sum_{i=1}^{k} \mathbf{A}_i = \mathbf{0}$ *if and only if* $\mathbf{A}_i = \mathbf{0}$ *for* $i = 1, 2, \ldots, k$,

(2) $\sum\sum_S \mathbf{A}_i \mathbf{A}_j = \mathbf{0}$ *if and only if* $\mathbf{A}_p \mathbf{A}_q = \mathbf{0}$ *for each and every* p, q *in* S, *where the summation over* S *for* i *and* j *is over any subset of* $1, 2, \ldots, k$ *and* $\mathbf{A}_p \mathbf{A}_q$ *is a term of the summation.*

Proof: Since (1) is a special case of (2), we shall prove (2) only. Clearly if $\mathbf{A}_p \mathbf{A}_q = \mathbf{0}$ for each term of the summation, then the summation is the null matrix. On the other hand, if $\sum\sum_S \mathbf{A}_i \mathbf{A}_j = \mathbf{0}$, then $\sum\sum_S \operatorname{tr}(\mathbf{A}_i \mathbf{A}_j) = 0$, and, by (3) of Theorem 12.2.3, each term is non-negative, so each term is equal to zero; that is $\operatorname{tr}(\mathbf{A}_p \mathbf{A}_q) = 0$. But, by (4) of Theorem 12.2.3, this implies that $\mathbf{A}_p \mathbf{A}_q = \mathbf{0}$, and the proof is complete. ∎

Theorem 12.2.6

Let $\mathbf{A}$ *be an* $n \times n$ *symmetric idempotent matrix, let* $\{\mathbf{B}_1, \mathbf{B}_2, \ldots, \mathbf{B}_k\}$ *be a collection of* $n \times n$ *non-negative matrices, and suppose*

$$\mathbf{I} = \mathbf{A} + \sum_{i=1}^{k} \mathbf{B}_i ;$$

then $\mathbf{A}\mathbf{B}_i = \mathbf{B}_i \mathbf{A} = \mathbf{0}$ *for* $i = 1, 2, \ldots, k$.

Proof: Multiply both sides of the equation by $\mathbf{A}$ and obtain

$$\mathbf{A} = \mathbf{A}^2 + \sum_{i=1}^{k} \mathbf{AB}_i ;$$

since $\mathbf{A}$ is idempotent, this reduces to (note that $\mathbf{A}$ is also a non-negative matrix)

$$\sum_{i=1}^{k} \mathbf{AB}_i = \mathbf{0},$$

and by Theorem 12.2.5 we have $\mathbf{AB}_i = \mathbf{0}$ for $i = 1, 2, \ldots, k$. But $(\mathbf{AB}_i)' = \mathbf{B}_i'\mathbf{A}' = \mathbf{B}_i\mathbf{A}$, so $\mathbf{B}_i\mathbf{A} = \mathbf{0}$ for $i = 1, 2, \ldots, k$, and the theorem is proved. ▮

Theorem 12.2.6 has a number of important applications in statistics.

Example 12.2.1. Consider Eq. (12.1.5) and partition the second term into $k - 1$ quadratic forms. We obtain

$$\sum_{i=1}^{n} y_i^2 = n\bar{y}^2 + \mathbf{y}'\mathbf{A}_2\mathbf{y} + \cdots + \mathbf{y}'\mathbf{A}_k\mathbf{y},$$

where $n\bar{y}^2$ is called the "sum of squares due to the mean." Generally it is known that each quadratic form $\mathbf{y}'\mathbf{A}_i\mathbf{y}$ is a sum of squares, and hence each $\mathbf{A}_i$ is a non-negative matrix. Also by Eq. (12.1.6), we see that $n\bar{y}^2 = \mathbf{y}'\mathbf{A}_1\mathbf{y}$, where $\mathbf{A}_1 = (1/n)\mathbf{J}$ and hence is an idempotent matrix. Thus the hypothesis of Theorem 12.2.6 is satisfied, and we get the fact that

$$\mathbf{A}_1\mathbf{A}_2 = \mathbf{0}, \quad \mathbf{A}_1\mathbf{A}_3 = \mathbf{0}, \quad \ldots, \quad \mathbf{A}_1\mathbf{A}_k = \mathbf{0}.$$

Under usual circumstances this result implies that the sum of squares due to the mean, $n\bar{y}^2$, is independent of each of the quadratic forms $\mathbf{y}'\mathbf{A}_2\mathbf{y}$, $\mathbf{y}'\mathbf{A}_3\mathbf{y}, \ldots, \mathbf{y}'\mathbf{A}_k\mathbf{y}$.

Since quadratic forms play such an important role in statistics and since we assume without loss of generality that the matrix of a quadratic form is symmetric, we go now to a theorem that characterizes a symmetric matrix.

Theorem 12.2.7

Let $\mathbf{C}$ be any $n \times n$ symmetric matrix. There exist two unique, non-negative, *disjoint (that is,* $\mathbf{AB} = \mathbf{0}$*)* $n \times n$ *matrices* **A** *and* **B** *such that*

$$\mathbf{C} = \mathbf{A} - \mathbf{B}.$$

Proof: Let **P** be an orthogonal matrix such that $\mathbf{P}'\mathbf{C}\mathbf{P}$ is a diagonal matrix. We shall write

$$\mathbf{P}'\mathbf{C}\mathbf{P} = \begin{bmatrix} \mathbf{D}_1 & \mathbf{0} \\ \mathbf{0} & -\mathbf{D}_2 \end{bmatrix},$$

where the elements on the diagonal of $\mathbf{D}_1$ are the positive characteristic roots of **C** and the elements on the diagonal of $-\mathbf{D}_2$ are the negative and zero characteristic roots of **C**. Thus the diagonal elements of $\mathbf{D}_2$ are non-negative. We define **A** and **B** by

$$\mathbf{A} = \mathbf{P}\begin{bmatrix} \mathbf{D}_1 & \mathbf{0} \\ \mathbf{0} & \mathbf{0} \end{bmatrix}\mathbf{P}'; \qquad \mathbf{B} = \mathbf{P}\begin{bmatrix} \mathbf{0} & \mathbf{0} \\ \mathbf{0} & \mathbf{D}_2 \end{bmatrix}\mathbf{P}',$$

and, by Theorem 12.2.1, **A** and **B** are non-negative. Also it is clear that

$$\mathbf{C} = \mathbf{A} - \mathbf{B} \quad \text{and} \quad \mathbf{AB} = \mathbf{BA} = \mathbf{0}.$$

To show that **A** and **B** are unique, let **F** and **G** be two matrices that are disjoint and non-negative (hence also symmetric) such that $\mathbf{F} - \mathbf{G} = \mathbf{C}$. Let **U** and **V** be matrices such that $\mathbf{F} - \mathbf{A} = \mathbf{U}$ and $\mathbf{G} - \mathbf{B} = \mathbf{V}$; clearly $\mathbf{U} = \mathbf{U}'$ and $\mathbf{V} = \mathbf{V}'$. Since $\mathbf{F} - \mathbf{G} = \mathbf{C}$ and $\mathbf{A} - \mathbf{B} = \mathbf{C}$, we obtain $\mathbf{U} = \mathbf{V}$. Also from $\mathbf{FG} = \mathbf{0}$, we obtain

$$(\mathbf{A} + \mathbf{U})(\mathbf{B} + \mathbf{U}) = \mathbf{0} \tag{12.2.1}$$

or

$$\mathbf{UB} + \mathbf{AU} + \mathbf{U}^2 = \mathbf{0}, \text{ since } \mathbf{AB} = \mathbf{BA} = \mathbf{0}. \tag{12.2.2}$$

Multiply this equation by **B** and obtain

$$\mathbf{BUB} + \mathbf{BU}^2 = \mathbf{0}. \tag{12.2.3}$$

But since **B** and **F**, and hence $\mathbf{A} + \mathbf{U}$, are non-negative, it follows that $\mathbf{B}(\mathbf{A} + \mathbf{U})\mathbf{B}$ and $(\mathbf{A} + \mathbf{U})\mathbf{B}(\mathbf{A} + \mathbf{U})$ are also non-negative. But

$$\mathbf{B}(\mathbf{A} + \mathbf{U})\mathbf{B} = \mathbf{BUB} \quad \text{and} \quad (\mathbf{A} + \mathbf{U})\mathbf{B}(\mathbf{A} + \mathbf{U}) = \mathbf{UBU},$$

and hence **BUB** and **UBU** are non-negative. But **B** and $\mathbf{U}^2$ are also non-negative. Thus, by Theorem 12.2.5 and Eq. (12.2.3), it follows that

$$\mathbf{BUB} = \mathbf{0}, \quad \text{and} \quad \mathbf{BU}^2 = \mathbf{0}.$$

But $\mathbf{BU}^2 = \mathbf{0}$ implies that $\mathbf{UBU} = \mathbf{0}$. By a similar procedure we obtain $\mathbf{AUA} = \mathbf{0}$, $\mathbf{UAU} = \mathbf{0}$, $\mathbf{AU}^2 = \mathbf{0}$.
If we multiply Eq. (12.2.2) on the left by **U**, we obtain

$$\mathbf{UUB} + \mathbf{UAU} + \mathbf{U}^3 = \mathbf{0},$$

which gives $\mathbf{U}^3 = \mathbf{0}$ or $\mathbf{U} = \mathbf{0}$, and hence

$$\mathbf{A} = \mathbf{F} \quad \text{and} \quad \mathbf{B} = \mathbf{G},$$

and the uniqueness of **A** and **B** is obtained. ▌

Proofs of some of the additional theorems about non-negative matrices that follow will be asked for in the problems section.

Theorem 12.2.8

Let **A** *and* **B** *be positive definite (semidefinite)* $n \times n$ *matrices. The matrix* **C** *is a positive definite (semidefinite) matrix where* **C** *is defined by*

$$\mathbf{C} = [c_{ij}] = [a_{ij} b_{ij}].$$

Example 12.2.2. Consider the 2×2 matrices

$$\mathbf{A} = \begin{bmatrix} 1 & 2 \\ 2 & 5 \end{bmatrix}, \qquad \mathbf{B} = \begin{bmatrix} 6 & -1 \\ -1 & 1 \end{bmatrix}.$$

Clearly **A** and **B** are positive definite matrices. The matrix **C**, defined by

$$\mathbf{C} = [c_{ij}] = [a_{ij} b_{ij}],$$

is given by

$$\mathbf{C} = \begin{bmatrix} 6 & -2 \\ -2 & 5 \end{bmatrix},$$

and **C** is clearly a positive definite matrix.

Theorem 12.2.9

Let $\mathbf{A}$ *be a* $k \times k$ *positive definite matrix. If* $a_{ij} < 0$ *for all* $i \neq j$*, then every element in* $\mathbf{A}^{-1}$ *is positive.*

Proof: We shall use induction. If $k = 2$ and $a_{12} < 0$, we get

$$\mathbf{A}_2 = \begin{bmatrix} a_{11} & a_{12} \\ a_{21} & a_{22} \end{bmatrix};$$

but if $\mathbf{A}_2$ is positive definite, we have $a_{11} > 0$, $a_{22} > 0$, and $d = a_{11}a_{22} - a_{12}^2 > 0$. The inverse is

$$\mathbf{A}_2^{-1} = \frac{1}{d}\begin{bmatrix} a_{22} & -a_{12} \\ -a_{21} & a_{11} \end{bmatrix},$$

so clearly, since $d > 0$, if a_{12} is negative, then every element in $\mathbf{A}_2^{-1}$ is positive. Now assume that the theorem is true for every positive definite $(k-1) \times (k-1)$ matrix $\mathbf{B}$. That is, let $\mathbf{B}$ be a positive definite $(k-1) \times (k-1)$ matrix with $b_{ij} < 0$ for all $i \neq j$, and we assume that every element in $\mathbf{B}^{-1}$ is positive. We want to show that if $\mathbf{C}$ is a $k \times k$ positive definite matrix such that $c_{ij} < 0$ for all $i \neq j$, then every element in $\mathbf{C}^{-1}$ is positive. We write

$$\mathbf{C} = \begin{bmatrix} \mathbf{B} & \mathbf{c}_{12} \\ \mathbf{c}_{21} & c_{22} \end{bmatrix}; \qquad \mathbf{C}^{-1} = \mathbf{G} = \begin{bmatrix} \mathbf{G}_{11} & \mathbf{g}_{12} \\ \mathbf{g}_{21} & g_{22} \end{bmatrix}.$$

We obtain $\mathbf{G}_{11}^{-1} = \mathbf{B} - \mathbf{c}_{12}c_{22}^{-1}\mathbf{c}_{12}'$. By hypothesis, $b_{ij} < 0$, $i \neq j$, and each element in $\mathbf{c}_{12}$ is negative (also c_{22} is positive, since $\mathbf{C}$ is positive definite). Hence, each off-diagonal element in $\mathbf{G}_{11}^{-1}$ is negative, and by the induction hypothesis the inverse of this matrix (since it is $(k-1) \times (k-1)$) has all positive off-diagonal elements. Next we consider the vector $\mathbf{g}_{12}$. We obtain

$$\mathbf{g}_{12} = -\mathbf{B}^{-1}\mathbf{c}_{12}g_{22} = \mathbf{B}^{-1}(-\mathbf{c}_{12})g_{22};$$

but, by the induction hypothesis, each element in $\mathbf{B}^{-1}$ is positive, and since g_{22} is positive and each element in $-\mathbf{c}_{12}$ is positive, it follows that each element in $\mathbf{g}_{12}$ is positive. Thus we have proved that if the theorem is true for every $(k-1) \times (k-1)$ positive definite matrix, it is true for every $k \times k$ positive definite matrix. Induction completes the proof. ∎

Example 12.2.3. Consider the 3×3 positive definite matrix

$$\mathbf{A} = \begin{bmatrix} 1 & -2 & -1 \\ -2 & 8 & -3 \\ -1 & -3 & 8 \end{bmatrix}.$$

The inverse is given by

$$\mathbf{A}^{-1} = \frac{1}{3}\begin{bmatrix} 55 & 19 & 14 \\ 19 & 7 & 5 \\ 14 & 5 & 4 \end{bmatrix},$$

and each element is positive.

There is a relationship between the diagonal elements of a positive definite matrix and the corresponding elements of the inverse matrix. For example, in the 3×3 matrix $\mathbf{A}$ above, $a_{11} = 1$ and the corresponding element in the inverse is $a_{11}^{(-1)} = 55/3$. The product is greater than unity. This result is given in the next theorem.

Theorem 12.2.10

Let $\mathbf{A}$ be a positive definite $k \times k$ matrix and let $\mathbf{B} = \mathbf{A}^{-1}$; then

(1) $a_{ii}b_{ii} \geq 1$ for $i = 1, 2, \ldots, k$,

(2) *if* $\mathbf{A} = \begin{bmatrix} \mathbf{A}_{11} & \mathbf{A}_{12} \\ \mathbf{A}_{21} & \mathbf{A}_{22} \end{bmatrix}, \quad \mathbf{B} = \begin{bmatrix} \mathbf{B}_{11} & \mathbf{B}_{12} \\ \mathbf{B}_{21} & \mathbf{B}_{22} \end{bmatrix},$

where $\mathbf{A}_{11}$ and $\mathbf{B}_{11}$ are $k_1 \times k_1$ matrices, then the i-th diagonal element of $\mathbf{A}_{11}$ is greater than or equal to the i-th diagonal element of $\mathbf{B}_{11}^{-1}$.

Proof: Using Theorem 8.2.1, we shall prove part (1) for a_{11}. If we permute rows and columns of $\mathbf{A}$ to bring a_{ii} into the first diagonal element, the proof will work for any i. We write

$$\mathbf{A} = \begin{bmatrix} a_{11} & \mathbf{a}_{12} \\ \mathbf{a}_{21} & \mathbf{A}_{22} \end{bmatrix}, \qquad \mathbf{A}^{-1} = \mathbf{B} = \begin{bmatrix} b_{11} & \mathbf{b}_{12} \\ \mathbf{b}_{21} & \mathbf{B}_{22} \end{bmatrix},$$

and

$$b_{11}^{-1} = a_{11} - \mathbf{a}_{12}\mathbf{A}_{22}^{-1}\mathbf{a}_{21};$$

but since $\mathbf{A}_{22}$, and hence $\mathbf{A}_{22}^{-1}$, is positive definite, it follows that

$$\mathbf{a}_{12}\mathbf{A}_{22}^{-1}\mathbf{a}_{21} \geq 0 \quad \text{and} \quad b_{11}^{-1} \leq a_{11},$$

or $b_{11}a_{11} \geq 1$.

To prove part (2), partition **A** and **B** as it is in the theorem. We obtain

$$\mathbf{B}_{11}^{-1} = \mathbf{A}_{11} - \mathbf{A}_{12}\mathbf{A}_{22}^{-1}\mathbf{A}_{21}.$$

The matrix $\mathbf{A}_{12}\mathbf{A}_{22}^{-1}\mathbf{A}_{21}$ is non-negative, and hence the diagonal elements are non-negative. Therefore the i-th diagonal element of $\mathbf{A}_{11}$ is greater than or equal to the i-th diagonal element of $\mathbf{B}_{11}^{-1}$. ∎

Example 12.2.4. To illustrate this theorem, suppose that **A** is defined by

$$\mathbf{A} = \begin{bmatrix} 1 & 2 & 2 \\ 2 & 5 & 3 \\ 2 & 3 & 6 \end{bmatrix}.$$

We partition **A** such that $\mathbf{A}_{11} = \begin{bmatrix} 1 & 2 \\ 2 & 5 \end{bmatrix}$,

and, by Theorem 8.2.1, we get

$$\mathbf{B}_{11}^{-1} = \begin{bmatrix} 1/3 & 1 \\ 1 & 7/2 \end{bmatrix}.$$

We notice that $1 > 1/3$ and $5 > 7/2$.

From Theorem 3.4.1 we know that the roots of a symmetric matrix are real. The next theorem on this subject is more general.

Theorem 12.2.11

Let **A** *and* **B** *be symmetric* $k \times k$ *matrices.*

(1) *The characteristic roots of* **AB** *are real if either* **A** *or* **B** *is non-negative.*

(2) *If* **B** *is positive definite, the values of* λ *that satisfy* $|\mathbf{A} - \lambda\mathbf{B}| = 0$ *are real.*

Proof: To prove (1), assume that **A** is non-negative of rank $m \leq k$. Let **P** be a nonsingular matrix such that

$$\mathbf{P}'\mathbf{A}\mathbf{P} = \begin{bmatrix} \mathbf{I} & \mathbf{0} \\ \mathbf{0} & \mathbf{0} \end{bmatrix},$$

where $\mathbf{I}$ is the $m \times m$ identity matrix. The characteristic roots of $\mathbf{AB}$ are the same as the characteristic roots of $\mathbf{P'ABP'^{-1}}$, and (let $\mathbf{P^{-1}BP'^{-1}} = \mathbf{C}$)

$$\mathbf{P'ABP'^{-1}} = \mathbf{P'APP^{-1}BP'^{-1}} = \begin{bmatrix} \mathbf{I} & \mathbf{0} \\ \mathbf{0} & \mathbf{0} \end{bmatrix} \begin{bmatrix} \mathbf{C}_{11} & \mathbf{C}_{12} \\ \mathbf{C}_{21} & \mathbf{C}_{22} \end{bmatrix} = \begin{bmatrix} \mathbf{C}_{11} & \mathbf{C}_{12} \\ \mathbf{0} & \mathbf{0} \end{bmatrix}.$$

So the characteristic roots of $\mathbf{P'ABP'^{-1}}$ are the values of λ that satisfy

$$\begin{vmatrix} \mathbf{C}_{11} - \lambda\mathbf{I} & \mathbf{C}_{12} \\ \mathbf{0} & -\lambda\mathbf{I} \end{vmatrix} = 0,$$

which reduces to

$$|-\lambda\mathbf{I}||\mathbf{C}_{11} - \lambda\mathbf{I}| = 0.$$

But $\mathbf{C}$ is symmetric, and hence $\mathbf{C}_{11}$ is symmetric; λ is either zero or a nonzero characteristic root of $\mathbf{C}_{11}$ and hence is real. If $\mathbf{B}$ is non-negative, the proof is the same. To prove (2) we note that λ is a root of $|\mathbf{A} - \lambda\mathbf{B}| = 0$ if and only if it is also a root of $|\mathbf{B}^{-1}||\mathbf{A} - \lambda\mathbf{B}| = 0$. But $|\mathbf{B}^{-1}||\mathbf{A} - \lambda\mathbf{B}| = |\mathbf{B}^{-1}\mathbf{A} - \lambda\mathbf{I}|$, and by (1) the values of λ that satisfy $|\mathbf{B}^{-1}\mathbf{A} - \lambda\mathbf{I}| = 0$ are real. ▮

Note that part (1) of the theorem may not be true if neither $\mathbf{A}$ nor $\mathbf{B}$ is non-negative; and part (2) may not be true if $\mathbf{B}$ is not positive definite.

Example 12.2.5. If we define $\mathbf{A}$ and $\mathbf{B}$ by

$$\mathbf{A} = \begin{bmatrix} 2 & 4 \\ 4 & 3 \end{bmatrix}, \qquad \mathbf{B} = \begin{bmatrix} 1 & -2 \\ -2 & 6 \end{bmatrix},$$

then we note that $\mathbf{B}$ is positive definite. We obtain $\mathbf{AB}$ to be

$$\mathbf{AB} = \begin{bmatrix} -6 & 20 \\ -2 & 10 \end{bmatrix} \quad \text{and} \quad |\mathbf{AB} - \lambda\mathbf{I}| = \begin{vmatrix} -6-\lambda & 20 \\ -2 & 10-\lambda \end{vmatrix}.$$

The characteristic equation is $\lambda^2 - 4\lambda - 20 = 0$, and the characteristic roots are $\lambda_1 = 2 - 2\sqrt{6}$; $\lambda_2 = 2 + 2\sqrt{6}$, which are real (notice that $\mathbf{AB}$ is not symmetric). To find the values of λ that satisfy $|\mathbf{A} - \lambda\,\mathbf{B}| = 0$, we compute

$$|\mathbf{A} - \lambda\mathbf{B}| = \begin{vmatrix} 2-\lambda & 4+2\lambda \\ 4+2\lambda & 3-6\lambda \end{vmatrix},$$

and $|\mathbf{A} - \lambda\mathbf{B}| = 0$ gives us the polynomial equation

$$2\lambda^2 - 31\lambda - 10 = 0,$$

and the roots are clearly real.

When quadratic forms are used in statistics, it is often desirable to be able to diagonalize matrices by orthogonal transformations (see Theorem 1.8.8), since, of course, this reduces quadratic forms to sums of squares. For example, it may be easier to study Eq. (12.1.5) if the matrices $\mathbf{A}_1$ and $\mathbf{A}_2$ can be transformed to diagonal matrices. Suppose that there is an orthogonal matrix $\mathbf{P}$ such that $\mathbf{P}'\mathbf{A}_1\mathbf{P} = \mathbf{D}_1$ and $\mathbf{P}'\mathbf{A}_2\mathbf{P} = \mathbf{D}_2$ where $\mathbf{D}_1$ and $\mathbf{D}_2$ are diagonal matrices. If we make the transformation from the vector $\mathbf{y}$ to a vector $\mathbf{x}$ by $\mathbf{x} = \mathbf{P}'\mathbf{y}$, then Eq. (12.1.5) can be written as

$$\mathbf{x}'\mathbf{P}'\mathbf{P}\mathbf{x} = \mathbf{x}'\mathbf{P}'\mathbf{A}_1\mathbf{P}\mathbf{x} + \mathbf{x}'\mathbf{P}'\mathbf{A}_2\mathbf{P}\mathbf{x},$$

which simplifies to

$$\mathbf{x}'\mathbf{x} = \mathbf{x}'\mathbf{D}_1\mathbf{x} + \mathbf{x}'\mathbf{D}_2\mathbf{x}.$$

Since $\mathbf{D}_1$ and $\mathbf{D}_2$ are diagonal matrices, the quadratic forms are sums of squares in the x_i. Orthogonal transformations are important for at least two reasons: (1) sums of squares transform into sums of squares (in the above, $\mathbf{y}'\mathbf{y} = \mathbf{x}'\mathbf{x}$); (2) if $\mathbf{y}$ is a multivariate normal vector with mean vector $\mathbf{0}$ and covariance matrix $\sigma^2\mathbf{I}$, then $\mathbf{x}$ has the same distribution. For these and other reasons not stated here, it is important to know conditions under which it is possible to diagonalize two (or more) matrices by the same orthogonal transformation. This is the subject of the next theorem and its corollaries.

Theorem 12.2.12

Let $\mathbf{A}$ and $\mathbf{B}$ be $k \times k$ symmetric matrices. A necessary and sufficient condition that an orthogonal matrix $\mathbf{P}$ exists such that $\mathbf{P}'\mathbf{A}\mathbf{P}$ and $\mathbf{P}'\mathbf{B}\mathbf{P}$ are each diagonal is that $\mathbf{AB} = \mathbf{BA}$ (or $\mathbf{AB}$ is a symmetric matrix).

Proof: First we shall prove the sufficiency part of the theorem. Let $\mathbf{R}$ be an orthogonal matrix such that $\mathbf{R}'\mathbf{A}\mathbf{R} = \mathbf{D}$ where $\mathbf{D}$ is a diagonal matrix with characteristic roots on the diagonal. In fact we can write $\mathbf{D}$ as

$$\mathbf{D} = \begin{bmatrix} \lambda_1\mathbf{I}_1 & \mathbf{0} & \cdots & \mathbf{0} \\ \mathbf{0} & \lambda_2\mathbf{I}_2 & \cdots & \mathbf{0} \\ \vdots & \vdots & & \vdots \\ \mathbf{0} & \mathbf{0} & \cdots & \lambda_m\mathbf{I}_m \end{bmatrix},$$

where $\lambda_1, \lambda_2, \ldots, \lambda_m$ are the distinct characteristic roots of $\mathbf{A}$ and $\mathbf{I}_i$ is the $k_i \times k_i$ identity matrix and $\sum_{i=1}^{m} k_i = k$. Let $\mathbf{C} = \mathbf{R}'\mathbf{B}\mathbf{R}$. Since we assume that $\mathbf{AB} = \mathbf{BA}$, this implies

$$\mathbf{R}'\mathbf{A}\mathbf{R}\mathbf{R}'\mathbf{B}\mathbf{R} = \mathbf{R}'\mathbf{B}\mathbf{R}\mathbf{R}'\mathbf{A}\mathbf{R} \quad \text{or} \quad \mathbf{DC} = \mathbf{CD}.$$

We shall partition $\mathbf{C}$, so that $\mathbf{DC} = \mathbf{CD}$ is

$$\begin{bmatrix} \lambda_1\mathbf{I}_1 & \mathbf{0} & \cdots & \mathbf{0} \\ \mathbf{0} & \lambda_2\mathbf{I}_2 & \cdots & \mathbf{0} \\ \vdots & \vdots & & \vdots \\ \mathbf{0} & \mathbf{0} & \cdots & \lambda_m\mathbf{I}_m \end{bmatrix} \begin{bmatrix} \mathbf{C}_{11} & \mathbf{C}_{12} & \cdots & \mathbf{C}_{1m} \\ \mathbf{C}_{21} & \mathbf{C}_{22} & \cdots & \mathbf{C}_{2m} \\ \vdots & \vdots & & \vdots \\ \mathbf{C}_{m1} & \mathbf{C}_{m2} & \cdots & \mathbf{C}_{mm} \end{bmatrix}$$

$$= \begin{bmatrix} \mathbf{C}_{11} & \mathbf{C}_{12} & \cdots & \mathbf{C}_{1m} \\ \mathbf{C}_{21} & \mathbf{C}_{22} & \cdots & \mathbf{C}_{2m} \\ \vdots & \vdots & & \vdots \\ \mathbf{C}_{m1} & \mathbf{C}_{m2} & \cdots & \mathbf{C}_{mm} \end{bmatrix} \begin{bmatrix} \lambda_1\mathbf{I}_1 & \mathbf{0} & \cdots & \mathbf{0} \\ \mathbf{0} & \lambda_2\mathbf{I}_2 & \cdots & \mathbf{0} \\ \vdots & \vdots & & \vdots \\ \mathbf{0} & \mathbf{0} & \cdots & \lambda_m\mathbf{I}_m \end{bmatrix}.$$

This implies $\mathbf{C}_{ij} = \mathbf{0}$ if $i \neq j$, since $\lambda_i \neq \lambda_j$ if $i \neq j$. Let $\mathbf{Q}_i$ be an orthogonal matrix such that $\mathbf{Q}_i'\mathbf{C}_{ii}\mathbf{Q}_i = \mathbf{D}_i$ for $i = 1, 2, \ldots, m$ where $\mathbf{D}_i$ is a diagonal matrix. Define $\mathbf{Q}$ by

$$\mathbf{Q} = \begin{bmatrix} \mathbf{Q}_1 & \mathbf{0} & \cdots & \mathbf{0} \\ \mathbf{0} & \mathbf{Q}_2 & \cdots & \mathbf{0} \\ \vdots & \vdots & & \vdots \\ \mathbf{0} & \mathbf{0} & \cdots & \mathbf{Q}_m \end{bmatrix}.$$

Clearly $\mathbf{Q}$ is an orthogonal matrix and $\mathbf{Q}'\mathbf{C}\mathbf{Q} = \mathbf{D}^*$ where $\mathbf{D}^*$ is a diagonal matrix, but $\mathbf{Q}'\mathbf{D}\mathbf{Q} = \mathbf{D}$. Hence

$$\mathbf{Q}'\mathbf{R}'\mathbf{A}\mathbf{R}\mathbf{Q} = \mathbf{D} \quad \text{and} \quad \mathbf{Q}'\mathbf{R}'\mathbf{B}\mathbf{R}\mathbf{Q} = \mathbf{D}^*,$$

but since $\mathbf{R}$ and $\mathbf{Q}$ are orthogonal matrices, we set $\mathbf{RQ} = \mathbf{P}$ and $\mathbf{P}$ is an orthogonal matrix such that $\mathbf{P}'\mathbf{A}\mathbf{P}$ and $\mathbf{P}'\mathbf{B}\mathbf{P}$ are diagonal matrices. To prove the necessary part of the theorem, assume there exists an orthogonal matrix such that $\mathbf{P}'\mathbf{A}\mathbf{P} = \mathbf{D}_1$ and $\mathbf{P}'\mathbf{B}\mathbf{P} = \mathbf{D}_2$, where $\mathbf{D}_1$ and $\mathbf{D}_2$ are diagonal matrices. But $\mathbf{D}_1\mathbf{D}_2 = \mathbf{D}_2\mathbf{D}_1$, and this implies

$$\mathbf{P}'\mathbf{A}\mathbf{P}\mathbf{P}'\mathbf{B}\mathbf{P} = \mathbf{P}'\mathbf{B}\mathbf{P}\mathbf{P}'\mathbf{A}\mathbf{P},$$

which in turn implies $\mathbf{AB} = \mathbf{BA}$, and the theorem is proved. ∎

That the theorem can be extended to more than two matrices is stated in the corollary below, and the proof is similar to the proof for the theorem.

Corollary 12.2.12.1

Let $\mathbf{A}_1, \mathbf{A}_2, \ldots, \mathbf{A}_n$ *be symmetric* $k \times k$ *matrices. A necessary and sufficient condition that there exists an orthogonal matrix* $\mathbf{P}$ *such that* $\mathbf{P}'\mathbf{A}_i\mathbf{P}$ *is a diagonal matrix for each* $i = 1, 2, \ldots, n$ *is* $\mathbf{A}_i\mathbf{A}_j = \mathbf{A}_j\mathbf{A}_i$ (or $\mathbf{A}_i\mathbf{A}_j$ *is symmetric) for all* i *and* j.

Corollary 12.2.12.2

In Corollary 12.2.12.1, *if* $\mathbf{A}_i\mathbf{A}_j = \mathbf{D}_{ij}$ *for each and every* $i \neq j = 1, 2, \ldots, n$, *where* $\mathbf{D}_{ij}$ *is a diagonal matrix, then there exists an orthogonal matrix* $\mathbf{P}$ *such that* $\mathbf{P}'\mathbf{A}_i\mathbf{P}$ *is a diagonal matrix for each* $i = 1, 2, \ldots, n$.

In particular, note that if $\mathbf{A}_i\mathbf{A}_j = \mathbf{0}$ for each $i \neq j$, then the hypothesis of Corollary 12.2.12.1 is satisfied.

When the conditions of Theorem 12.2.12 are not satisfied (when $\mathbf{AB} \neq \mathbf{BA}$), there is no orthogonal matrix $\mathbf{P}$ such that $\mathbf{P}'\mathbf{AP}$ and $\mathbf{P}'\mathbf{BP}$ are each diagonal. However, it may be of interest to determine the conditions under which there exists a nonsingular matrix $\mathbf{Q}$, not necessarily orthogonal, such that $\mathbf{Q}'\mathbf{AQ}$ and $\mathbf{Q}'\mathbf{BQ}$ are each diagonal. This is the subject of the next theorem. The proof for the first part is straightforward and is left for the reader. The proof for the second part can be found in [N–2].

Theorem 12.2.13

Let $\mathbf{A}$ *and* $\mathbf{B}$ *be* $k \times k$ *symmetric matrices.*

(1) *If* $\mathbf{A}$ *is positive definite, there exists a nonsingular matrix* $\mathbf{Q}$ *such that* $\mathbf{Q}'\mathbf{AQ} = \mathbf{I}$ *and* $\mathbf{Q}'\mathbf{BQ} = \mathbf{D}$, *where* $\mathbf{D}$ *is a diagonal matrix and the diagonal elements are the roots* λ *of the polynomial equation* $|\mathbf{B} - \lambda\mathbf{A}| = 0$.

(2) *If* $\mathbf{A}$ *and* $\mathbf{B}$ *are both non-negative (neither has to be positive definite), there exists a nonsingular matrix* $\mathbf{Q}$ *such that* $\mathbf{Q}'\mathbf{AQ}$ *and* $\mathbf{Q}'\mathbf{BQ}$ *are each diagonal.*

The next theorem contains a number of miscellaneous results on non-negative matrices.

Theorem 12.2.14

Let $\mathbf{A}$, $\mathbf{B}$, *and* $\mathbf{C}$ *be symmetric* $k \times k$ *matrices.*

(1) *There exists a scalar* t *such that* $\mathbf{A} + t\mathbf{I}$ *is positive definite.*

(2) *If* $\mathbf{A}$ *and* $\mathbf{B}$ *are nonsingular and* $\mathbf{A} - \mathbf{B}$ *is positive definite, then* $\mathbf{B}^{-1} - \mathbf{A}^{-1}$ *is positive definite.*
(3) *If* $\mathbf{A}$ *is positive definite, there exists a positive scalar* t *such that* $\mathbf{A} + t\mathbf{B}$ *is positive definite.*
(4) *If* $\mathbf{A} - \mathbf{B}$ *is non-negative and* $\mathbf{B} - \mathbf{C}$ *is non-negative, then* $\mathbf{A} - \mathbf{C}$ *is non-negative.*
(5) *If* $\mathbf{A}$ *and* $\mathbf{B}$ *are nonsingular, such that* $\mathbf{x}'\mathbf{A}\mathbf{x} > \mathbf{x}'\mathbf{B}\mathbf{x}$ *for each and every vector* $\mathbf{x} \neq \mathbf{0}$, *then* $\mathbf{x}'\mathbf{A}^{-1}\mathbf{x} < \mathbf{x}'\mathbf{B}^{-1}\mathbf{x}$ *for each and every vector* $\mathbf{x} \neq \mathbf{0}$.
(6) *If* $\mathbf{A}$ *is positive definite, then* $|\mathbf{A}| \leq a_{11}a_{22} \ldots a_{kk}$.
(7) *If* $\mathbf{A}$ *is positive definite,* $\mathbf{B}$ *is non-negative, and* $\mathbf{A} - \mathbf{B}$ *is non-negative, then* $|\mathbf{A} - \mathbf{B}| \leq |\mathbf{A}|$.
(8) *Let* $\mathbf{A}$ *be positive definite. For any* $k \times 1$ *vectors* $\mathbf{x}$ *and* $\mathbf{y}$, *the following inequality holds:*

$$(\mathbf{x}'\mathbf{y})^2 \leq (\mathbf{x}'\mathbf{A}\mathbf{x})(\mathbf{y}'\mathbf{A}^{-1}\mathbf{y});$$

and the equality holds if and only if there is a scalar a *such that* $\mathbf{A}\mathbf{x} = a\mathbf{y}$.
(9) *If* $\mathbf{A}$ *is a positive definite* $k \times k$ *matrix and* $\mathbf{B}$ *is a non-negative* $k \times k$ *matrix, then*

$$\lambda_1 \leq \frac{\mathbf{x}'\mathbf{B}\mathbf{x}}{\mathbf{x}'\mathbf{A}\mathbf{x}} \leq \lambda_k$$

for each and every vector $\mathbf{x} \neq \mathbf{0}$ *where* $\lambda_1 \leq \lambda_2 \leq \cdots \leq \lambda_k$ *are the roots of* $|\mathbf{B} - \lambda\mathbf{A}| = 0$.

Proof:

(1) Let $\mathbf{B} = \mathbf{A} + t\mathbf{I}$ and let $\mathbf{P}$ be an orthogonal matrix such that $\mathbf{P}'\mathbf{A}\mathbf{P} = \mathbf{D}$, where $\mathbf{D}$ is diagonal, and let $t > \max_i |d_{ii}|$. We obtain

$$\mathbf{P}'\mathbf{B}\mathbf{P} = \mathbf{D} + t\mathbf{I},$$

and the characteristic roots of $\mathbf{B}$ are $d_{ii} + t$. Clearly $d_{ii} + t > 0$ for $i = 1, 2, \ldots, k$, and thus the characteristic roots of $\mathbf{B}$ are positive, and hence $\mathbf{B}$ is positive definite.
(2) Let $\mathbf{Q}$ be a nonsingular matrix such that

$$\mathbf{Q}'(\mathbf{A} - \mathbf{B})\mathbf{Q} = \mathbf{I}, \quad \text{or} \quad \mathbf{Q}'\mathbf{A}\mathbf{Q} - \mathbf{Q}'\mathbf{B}\mathbf{Q} = \mathbf{I}.$$

Let $\mathbf{P}$ be an orthogonal matrix such that $\mathbf{P}'(\mathbf{Q}'\mathbf{A}\mathbf{Q})\mathbf{P} = \mathbf{D}_1$ where $\mathbf{D}_1$ is diagonal. But since

$$\mathbf{P}'(\mathbf{Q}'\mathbf{B}\mathbf{Q})\mathbf{P} = \mathbf{P}'\mathbf{Q}'\mathbf{A}\mathbf{Q}\mathbf{P} - \mathbf{I} = \mathbf{D}_1 - \mathbf{I},$$

it is clear that $\mathbf{P}'(\mathbf{Q}'\mathbf{B}\mathbf{Q})\mathbf{P}$ is also a diagonal matrix. Denote it by $\mathbf{D}_2$ and we have

$$\mathbf{D}_1 - \mathbf{D}_2 = \mathbf{I},$$

from which we obtain $d_{ii}^{(1)} - d_{ii}^{(2)} = 1$ or $d_{ii}^{(1)} > d_{ii}^{(2)}$ for $i = 1, 2, \ldots, k$. From this we get

$$\frac{1}{d_{ii}^{(2)}} > \frac{1}{d_{ii}^{(1)}}$$

for $i = 1, 2, \ldots, k$, since $d_{ii}^{(1)} \neq 0$ and $d_{ii}^{(2)} \neq 0$. Hence

$$\mathbf{D}_2^{-1} - \mathbf{D}_1^{-1} = \mathbf{D}_3$$

where $d_{ii}^{(3)} > 0$ for $i = 1, 2, \ldots, k$, and this implies that $\mathbf{D}_3$ (and hence $\mathbf{D}_2^{-1} - \mathbf{D}_1^{-1}$) is positive definite. But

$$\begin{aligned}\mathbf{D}_2^{-1} - \mathbf{D}_1^{-1} &= (\mathbf{P}'\mathbf{Q}'\mathbf{B}\mathbf{Q}\mathbf{P})^{-1} - (\mathbf{P}'\mathbf{Q}'\mathbf{A}\mathbf{Q}\mathbf{P})^{-1} \\ &= \mathbf{P}'\mathbf{Q}^{-1}\mathbf{B}^{-1}(\mathbf{Q}^{-1})'\mathbf{P} - \mathbf{P}'\mathbf{Q}^{-1}\mathbf{A}^{-1}(\mathbf{Q}^{-1})'\mathbf{P}.\end{aligned}$$

But if $\mathbf{D}_2^{-1} - \mathbf{D}_1^{-1}$ is positive definite, it follows that $\mathbf{C}'(\mathbf{D}_2^{-1} - \mathbf{D}_1^{-1})\mathbf{C}$ is also positive definite if we set $\mathbf{C} = \mathbf{P}'\mathbf{Q}'$. We obtain

$$\mathbf{C}'(\mathbf{D}_2^{-1} - \mathbf{D}_1^{-1})\mathbf{C} = \mathbf{B}^{-1} - \mathbf{A}^{-1}$$

and (2) is proved.

(3) If $\mathbf{B} = \mathbf{0}$, then any positive number t will work. Assume $\mathbf{B} \neq \mathbf{0}$ and let $\mathbf{Q}$ be a nonsingular matrix such that $\mathbf{Q}'\mathbf{A}\mathbf{Q} = \mathbf{I}$ and $\mathbf{Q}'\mathbf{B}\mathbf{Q} = \mathbf{D}$ where $\mathbf{D}$ is a diagonal matrix. Since $\mathbf{B} \neq \mathbf{0}$, there is at least one diagonal element of $\mathbf{D}$ that is nonzero. Let

$$0 < t < \min_{d_{ii} \neq 0} \left| \frac{1}{d_{ii}} \right|,$$

and hence $1 + td_{ii} > 0$ for all $i = 1, 2, \ldots, k$, and $\mathbf{I} + t\mathbf{D}$ is a positive definite matrix. Hence $(\mathbf{Q}^{-1})'[\mathbf{I} + t\mathbf{D}]\mathbf{Q}^{-1}$ is also positive definite; but this matrix is equal to $\mathbf{A} + t\mathbf{B}$, and (3) is proved.

(4) This result follows from the fact that the sum of two non-negative matrices is a non-negative matrix.

(5) If $\mathbf{x}'\mathbf{A}\mathbf{x} > \mathbf{x}'\mathbf{B}\mathbf{x}$ for each and every vector $\mathbf{x} \neq \mathbf{0}$, this implies that $\mathbf{x}'(\mathbf{A} - \mathbf{B})\mathbf{x} > 0$ for each and every vector $\mathbf{x} \neq \mathbf{0}$, which in turn implies that $\mathbf{A} - \mathbf{B}$ is positive definite. Therefore, the hypothesis of part (2) of this theorem is satisfied and hence $\mathbf{B}^{-1} - \mathbf{A}^{-1}$ is positive definite, and therefore $\mathbf{x}'(\mathbf{B}^{-1} - \mathbf{A}^{-1})\mathbf{x} > 0$ for each and every vector $\mathbf{x} \neq \mathbf{0}$. The result follows.

(6) We shall use induction to prove part (6). If $\mathbf{A}$ is a 1×1 matrix, the result is obviously true. Assume that it is true for every $(k-1) \times (k-1)$ positive definite matrix $\mathbf{A}$ (that is, assume $|\mathbf{A}| \leq a_{11}a_{22} \ldots a_{k-1,k-1}$). Let $\mathbf{A}^*$ be the $k \times k$ positive definite matrix

$$\mathbf{A}^* = \begin{bmatrix} \mathbf{A} & \mathbf{a} \\ \mathbf{a}' & a_{kk} \end{bmatrix}.$$

By Theorem 8.2.1 we have

$$|\mathbf{A}^*| = |\mathbf{A}||a_{kk} - \mathbf{a}'\mathbf{A}^{-1}\mathbf{a}| = |\mathbf{A}|(a_{kk} - \mathbf{a}'\mathbf{A}^{-1}\mathbf{a}) \leq |\mathbf{A}|a_{kk},$$

since $\mathbf{A}^{-1}$ is positive definite, and hence $\mathbf{a}'\mathbf{A}^{-1}\mathbf{a} \geq 0$. But by hypothesis, $|\mathbf{A}| \leq a_{11}a_{22} \ldots a_{k-1,k-1}$, and hence $|\mathbf{A}^*| \leq a_{11}a_{22} \ldots a_{kk}$, and the result follows by induction.

(7) Let $\mathbf{P}$ be a nonsingular matrix such that $\mathbf{P}'\mathbf{A}\mathbf{P} = \mathbf{I}$ and $\mathbf{P}'\mathbf{B}\mathbf{P} = \mathbf{D}$, where $\mathbf{D}$ is diagonal. Then

$$\mathbf{P}'(\mathbf{A} - \mathbf{B})\mathbf{P} = \mathbf{I} - \mathbf{D};$$

but since $\mathbf{A} - \mathbf{B}$ is non-negative, it follows that $\mathbf{I} - \mathbf{D}$ is non-negative, and hence $1 - d_{ii} \geq 0$ for $i = 1, 2, \ldots, k$. Also, since $\mathbf{B}$ is non-negative, so is $\mathbf{D}$, and hence $d_{ii} \geq 0$ for $i = 1, 2, \ldots, k$. Therefore, we obtain $1 \geq d_{ii} \geq 0$ and $0 \leq 1 - d_{ii} \leq 1$. But

$$|\mathbf{I} - \mathbf{D}| = \prod_{i=1}^{k}(1 - d_{ii}) \leq \prod_{i=1}^{k} 1 = 1 = |\mathbf{I}|.$$

This implies

$$|\mathbf{P}'(\mathbf{A} - \mathbf{B})\mathbf{P}| \leq |\mathbf{P}'\mathbf{A}\mathbf{P}| \quad \text{or} \quad |\mathbf{P}'||\mathbf{A} - \mathbf{B}||\mathbf{P}| \leq |\mathbf{P}'||\mathbf{A}||\mathbf{P}|,$$

which in turn implies $|\mathbf{A} - \mathbf{B}| \leq |\mathbf{A}|$.

(8) For any scalar t and for any $k \times 1$ vectors $\mathbf{u}$ and $\mathbf{v}$, we obtain

$$(\mathbf{u} + t\mathbf{v})'(\mathbf{u} + t\mathbf{v}) \geq 0.$$

Simplifying, we obtain

$$\mathbf{v}'\mathbf{v}t^2 + 2t\mathbf{u}'\mathbf{v} + \mathbf{u}'\mathbf{u} \geq 0.$$

This implies that the discriminant in the quadratic formula cannot be positive. That is,

$$(\mathbf{u}'\mathbf{v})^2 - (\mathbf{v}'\mathbf{v})(\mathbf{u}'\mathbf{u}) \leq 0,$$

or $(\mathbf{u}'\mathbf{v})^2 \leq (\mathbf{u}'\mathbf{u})(\mathbf{v}'\mathbf{v})$, and the equality holds if and only if $\mathbf{u} + t\mathbf{v} = \mathbf{0}$ or, in other words, if and only if there is a scalar a such that $\mathbf{u} = a\mathbf{v}$. This is the famous Cauchy inequality. Since $\mathbf{A}$ is positive definite, there exists a nonsingular matrix $\mathbf{Q}$ such that $\mathbf{A} = \mathbf{Q}'\mathbf{Q}$. Let $\mathbf{u} = \mathbf{Qx}$ and $\mathbf{v} = \mathbf{Q}'^{-1}\mathbf{y}$. Substitute into the inequality and the result follows.

(9) The proof is left for the reader. ∎

Theorem 12.2.15

If $\mathbf{A} = [a_{ij}]$ is a positive definite matrix, then $\mathbf{A}^{(k)}$ is also positive definite for each positive integer k, where the ij-th element of $\mathbf{A}^{(k)}$ is a_{ij}^k; that is, $\mathbf{A}^{(k)} = [a_{ij}^k]$.

Proof: This theorem is a direct result of Theorem 12.2.8. ∎

Often it is important to be able to determine whether a given symmetric matrix is positive definite. The conditions in Theorem 12.2.2 are useful in a theoretical sense, but they may not be too valuable in some situations. The following theorem can sometimes be used.

Theorem 12.2.16

If $\mathbf{A}$ is a symmetric $k \times k$ matrix and if

$$a_{ii} > \sum_{\substack{j=1 \\ j \neq i}}^{k} |a_{ij}| \quad \textit{for} \quad i = 1, 2, \ldots, k,$$

then $\mathbf{A}$ is a positive definite matrix.

Proof: The proof of this theorem is a direct result of Theorem 8.11.7. ∎

Even though Theorem 12.2.16 may not be applicable to a given matrix **A**, it may be used on a matrix **P′AP** for some nonsingular matrix **P**, since **A** is positive definite if and only if **P′AP** is positive definite. For example, let

$$\mathbf{A} = \begin{bmatrix} 1 & 1 & 0 \\ 1 & 4 & 1 \\ 0 & 1 & 6 \end{bmatrix}.$$

Clearly Theorem 12.2.16 does not apply. Now obtain **B** where $\mathbf{B} = \mathbf{D}'\mathbf{A}\mathbf{D}$ and where

$$\mathbf{D} = \begin{bmatrix} 2 & 0 & 0 \\ 0 & 1 & 0 \\ 0 & 0 & 1 \end{bmatrix}.$$

Then

$$\mathbf{B} = \begin{bmatrix} 4 & 2 & 0 \\ 2 & 4 & 1 \\ 0 & 1 & 6 \end{bmatrix},$$

and we use Theorem 12.2.16 to prove that **B** is positive definite, but since **D** is nonsingular, it follows that **A** is also positive definite.

Theorem 12.2.17

If **A** *is any* $k \times k$ *matrix, then*

$$|\mathbf{A}| \le \prod_{i=1}^{k} \left[\sum_{j=1}^{k} a_{ij}^2 \right]^{1/2}.$$

Proof: If **A** is singular, the result is obvious. Assume $|\mathbf{A}| \neq 0$. Let $\mathbf{B} = \mathbf{A}'\mathbf{A}$ and thus **B** is positive definite. Also

$$|\mathbf{B}| = |\mathbf{A}'\mathbf{A}| = |\mathbf{A}'||\mathbf{A}| = |\mathbf{A}|^2.$$

By part (6) of Theorem 12.2.14, we obtain

$$|\mathbf{B}| \le \prod_{i=1}^{k} b_{ii}.$$

But $b_{ii} = \sum_{j=1}^{k} a_{ij}^2$, so we get

$$|\mathbf{B}| \le \prod_{i=1}^{k} \left[\sum_{j=1}^{k} a_{ij}^2 \right];$$

and since $|\mathbf{A}| = \sqrt{|\mathbf{B}|}$, the result follows. ▮

We shall give some simple illustrations of Theorems 12.2.14 through 12.2.17.

Example 12.2.6. Verify that the symmetric matrices **A** and **B** are positive definite.

$$\mathbf{A} = \begin{bmatrix} 3 & -1 & 1 \\ -1 & 4 & 0 \\ 1 & 0 & 2 \end{bmatrix}; \qquad \mathbf{B} = \begin{bmatrix} 2 & -1 & 0 \\ -1 & 3 & 1 \\ 0 & 1 & 2 \end{bmatrix}.$$

We can use (3b) of Theorem 12.2.2 to verify that **A** and **B** are positive definite. However, we can also use Theorem 12.2.16, which is much easier to apply.

Example 12.2.7. Verify that the symmetric matrix **C** is positive definite.

$$\mathbf{C} = \begin{bmatrix} 2 & -1 & -1 \\ -1 & 1 & 1 \\ -1 & 1 & 4 \end{bmatrix}.$$

Since Theorem 12.2.16 does not apply, we can use (3b) of Theorem 12.2.2.

Example 12.2.8. It is easy to show that if each element is squared in each of the matrices **A**, **B**, and **C** in the examples above, then the matrices are still positive definite. This illustrates Theorem 12.2.15.

Example 12.2.9. Define **A** and **B** by

$$\mathbf{A} = \begin{bmatrix} 4 & 3 \\ 3 & 4 \end{bmatrix}; \qquad \mathbf{B} = \begin{bmatrix} 3 & 2 \\ 2 & 2 \end{bmatrix}.$$

Note that $\mathbf{A} - \mathbf{B} = \mathbf{C} = \begin{bmatrix} 1 & 1 \\ 1 & 2 \end{bmatrix}$,

C is positive definite, and **A** and **B** are nonsingular. By part (2) of Theorem 12.2.14, $\mathbf{B}^{-1} - \mathbf{A}^{-1}$ is also positive definite. We obtain

$$\mathbf{B}^{-1} - \mathbf{A}^{-1} = \frac{1}{2}\begin{bmatrix} 2 & -2 \\ -2 & 3 \end{bmatrix} - \frac{1}{7}\begin{bmatrix} 4 & -3 \\ -3 & 4 \end{bmatrix} = \frac{1}{14}\begin{bmatrix} 6 & -8 \\ -8 & 13 \end{bmatrix},$$

which is positive definite.

The conditions of part (7) of Theorem 12.2.14 are also satisfied and clearly $|\mathbf{A} - \mathbf{B}| = 1$ and $|\mathbf{A}| = 7$ and $|\mathbf{A} - \mathbf{B}| < |\mathbf{A}|$.

Next we demonstrate part (6) since clearly $|\mathbf{A}| \le a_{11}a_{22}$ gives the result $7 < 16$.

We can illustrate Theorem 12.2.17, since

$$\prod_{i=1}^{2}\left[\sum_{j=1}^{2} a_{ij}^2\right]^{1/2} = [(25)(25)]^{1/2} = 25 \quad \text{and} \quad 7 < 25.$$

The proofs for the next two theorems are left for the reader to work out in the problems.

Theorem 12.2.18

Let $\{\mathbf{A}_1, \mathbf{A}_2, \ldots, \mathbf{A}_k\}$ be a collection of $n \times n$ positive definite matrices and let $\{a_1, a_2, \ldots, a_k\}$ be a set of positive scalars. The matrix $\mathbf{B}$ is also positive definite where $\mathbf{B} = \sum_{i=1}^{k} a_i \mathbf{A}_i$.

Corollary 12.2.18

If each matrix $\mathbf{A}_i$ in Theorem 12.2.18 is non-negative and each scalar a_i is non-negative, then $\mathbf{C}$ is non-negative where $\mathbf{C} = \sum_{i=1}^{k} a_i \mathbf{A}_i$.

Theorem 12.2.19

Let $\mathbf{C}$ be any $n \times n$ matrix. Then $\mathbf{x}'\mathbf{C}\mathbf{x} = 0$ for all $\mathbf{x} \in E_n$ if and only if $\mathbf{C} = -\mathbf{C}'$, that is, if and only if $\mathbf{C}$ is a skew-symmetric matrix.

Corollary 12.2.19

If $\mathbf{C} = \mathbf{C}'$, then $\mathbf{x}'\mathbf{C}\mathbf{x} = 0$ for all $\mathbf{x} \in E_n$ if and only if $\mathbf{C} = \mathbf{0}$.

In Theorem 8.2.1 we stated a number of results that related submatrices in a partitioned matrix $\mathbf{B}$ where

$$\mathbf{B} = \begin{bmatrix} \mathbf{B}_{11} & \mathbf{B}_{12} \\ \mathbf{B}_{21} & \mathbf{B}_{22} \end{bmatrix} \tag{12.2.4}$$

and where the size of $\mathbf{B}_{ij}$ is $n_i \times n_j$. For various results in the theorem, the hypothesis included the fact that $|\mathbf{B}_{11}| \ne 0$, $|\mathbf{B}_{22}| \ne 0$, and $|\mathbf{B}| \ne 0$. We now state some results on partitioned matrices when these hypotheses may not apply; in particular, we shall discuss partitioned non-negative matrices. These results are especially useful in statistics when discussing covariance matrices of multivariate normal distributions.

Theorem 12.2.20

If a non-negative $k \times k$ matrix $\mathbf{B}$ is partitioned as in Eq. (12.2.4), *then there exist matrices $\mathbf{F}$ and $\mathbf{G}$ such that*

$$\mathbf{B}_{21} = \mathbf{F}\mathbf{B}_{11}; \qquad \mathbf{B}_{12} = \mathbf{G}\mathbf{B}_{22};$$
$$\mathbf{B}_{12} = \mathbf{B}_{11}\mathbf{F}'; \qquad \mathbf{B}_{21} = \mathbf{B}_{22}\mathbf{G}'.$$

Proof: Since $\mathbf{B}$ is non-negative, there exists a $k \times k$ matrix $\mathbf{K}$ such that $\mathbf{K}'\mathbf{K} = \mathbf{B}$. If we partition $\mathbf{K}$ such that $\mathbf{K} = [\mathbf{K}_1, \mathbf{K}_2]$ where $\mathbf{K}_1$ has size $k \times k_1$, we obtain

$$\mathbf{B} = \mathbf{K}'\mathbf{K} = \begin{bmatrix} \mathbf{K}_1' \\ \mathbf{K}_2' \end{bmatrix} [\mathbf{K}_1, \mathbf{K}_2] = \begin{bmatrix} \mathbf{K}_1'\mathbf{K}_1 & \mathbf{K}_1'\mathbf{K}_2 \\ \mathbf{K}_2'\mathbf{K}_1 & \mathbf{K}_2'\mathbf{K}_2 \end{bmatrix} = \begin{bmatrix} \mathbf{B}_{11} & \mathbf{B}_{12} \\ \mathbf{B}_{21} & \mathbf{B}_{22} \end{bmatrix}$$

and $\mathbf{K}_i'\mathbf{K}_j = \mathbf{B}_{ij}$ for $i = 1, 2$; $j = 1, 2$. Consider $\mathbf{K}_1'\mathbf{K}_1 = \mathbf{B}_{11}$ and multiply on the left by $\mathbf{K}_2'\mathbf{K}_1'^{-}$ and obtain $\mathbf{K}_2'\mathbf{K}_1 = \mathbf{K}_2'\mathbf{K}_1'^{-}\mathbf{B}_{11}$ or $\mathbf{B}_{21} = \mathbf{F}\mathbf{B}_{11}$ where $\mathbf{F} = \mathbf{K}_2'\mathbf{K}_1'^{-}$. If we let $\mathbf{G} = \mathbf{K}_1'\mathbf{K}_2'^{-}$, we see that $\mathbf{B}_{12} = \mathbf{G}\mathbf{B}_{22}$. The remaining results clearly follow and the theorem is proved. ∎

Corollary 12.2.20.1

Let $\mathbf{B}$ be a $k \times k$ non-negative matrix that is partitioned as in Eq. (12.2.4). *The columns of $\mathbf{B}_{21}$ are in the column space of $\mathbf{B}_{22}$, and the columns of $\mathbf{B}_{12}$ are in the column space of $\mathbf{B}_{11}$.*

If we combine this corollary with Corollary 6.6.9.2, we obtain the following result.

Corollary 12.2.20.2

Let the $k \times k$ non-negative matrix $\mathbf{B}$ be partitioned as in Eq. (12.2.4). *The matrix $\mathbf{B}_{12}\mathbf{B}_{22}^{c}\mathbf{B}_{21}$ is invariant for any c-inverse of $\mathbf{B}_{22}$. Also the matrix $\mathbf{B}_{21}\mathbf{B}_{11}^{c}\mathbf{B}_{12}$ is invariant for any c-inverse of $\mathbf{B}_{11}$.*

When $\mathbf{B}$ is a positive definite matrix and is partitioned as in Eq. (12.2.4), the matrix $\mathbf{B}_{11 \cdot 2}$ defined by

$$\mathbf{B}_{11 \cdot 2} = \mathbf{B}_{11} - \mathbf{B}_{12}\mathbf{B}_{22}^{-1}\mathbf{B}_{21}$$

is useful in finding the inverse of **B** by partitioned matrices (see Theorem 8.2.1), and this quantity also occurs frequently in the conditional distributions of multivariate normal distributions. The same can be said for $\mathbf{B}_{22\cdot 1}$, defined by

$$\mathbf{B}_{22\cdot 1} = \mathbf{B}_{22} - \mathbf{B}_{21}\mathbf{B}_{11}^{-1}\mathbf{B}_{12}.$$

We have proved in Theorem 8.2.1 that if **B** is positive definite, then $\mathbf{B}_{11\cdot 2}$ and $\mathbf{B}_{22\cdot 1}$ are also positive definite (also in that case we know that $\mathbf{B}_{11}^{-1}$ and $\mathbf{B}_{22}^{-1}$ exist). If we know only that **B** is non-negative instead of positive definite, then we cannot be sure that these results hold. However, similar results can be stated in this situation and this is the context of the next theorem. (Also see Prob. 101.)

Theorem 12.2.21

Let the $k \times k$ *non-negative matrix* **B** *be partitioned as in* Eq. (12.2.4). *The results below follow, and wherever a c-inverse occurs, we mean that the quantity is invariant for any c-inverse of that particular matrix.*

(1) $\mathbf{B}_{12}\mathbf{B}_{22}^{c}\mathbf{B}_{22} = \mathbf{B}_{12}\mathbf{B}_{22}\mathbf{B}_{22}^{-} = \mathbf{B}_{12}.$

(2) $\mathbf{B}_{22}^{-}\mathbf{B}_{22}\mathbf{B}_{21} = \mathbf{B}_{22}\mathbf{B}_{22}^{c}\mathbf{B}_{21} = \mathbf{B}_{21}.$

(3) $\mathbf{B}_{11}\mathbf{B}_{11}^{c}\mathbf{B}_{12} = \mathbf{B}_{11}^{-}\mathbf{B}_{11}\mathbf{B}_{12} = \mathbf{B}_{12}.$

(4) $\mathbf{B}_{21}\mathbf{B}_{11}^{c}\mathbf{B}_{11} = \mathbf{B}_{21}\mathbf{B}_{11}\mathbf{B}_{11}^{-} = \mathbf{B}_{21}.$

(5) $\mathbf{B}_{11} - \mathbf{B}_{12}\mathbf{B}_{22}^{c}\mathbf{B}_{21}$ *is non-negative.*

(6) $\mathbf{B}_{22} - \mathbf{B}_{21}\mathbf{B}_{11}^{c}\mathbf{B}_{12}$ *is non-negative.*

(7) $$\mathbf{B}_{11}\mathbf{B}_{11}^{c}(\mathbf{B}_{11} - \mathbf{B}_{12}\mathbf{B}_{22}^{c}\mathbf{B}_{21}) = (\mathbf{B}_{11} - \mathbf{B}_{12}\mathbf{B}_{22}^{c}\mathbf{B}_{21})\mathbf{B}_{11}^{c}\mathbf{B}_{11} = \mathbf{B}_{11} - \mathbf{B}_{12}\mathbf{B}_{22}^{c}\mathbf{B}_{21}.$$

(8) $$\mathbf{B}_{22}\mathbf{B}_{22}^{c}(\mathbf{B}_{22} - \mathbf{B}_{21}\mathbf{B}_{11}^{c}\mathbf{B}_{12}) = (\mathbf{B}_{22} - \mathbf{B}_{21}\mathbf{B}_{11}^{c}\mathbf{B}_{12})\mathbf{B}_{22}^{c}\mathbf{B}_{22} = \mathbf{B}_{22} - \mathbf{B}_{21}\mathbf{B}_{11}^{c}\mathbf{B}_{12}.$$

Proof: The results (1), (2), (3), (4) follow by applying Corollaries 6.6.9.2 and 12.2.20.2. The results (7) and (8) follow directly from (1), (2), (3), and (4). To prove (5), we note by Theorem 12.2.1 that **P'BP** is non-negative for any matrix **P**. Define **P** by

$$\mathbf{P} = \begin{bmatrix} \mathbf{I} & \mathbf{0} \\ -\mathbf{B}_{22}^{c}\mathbf{B}_{21} & \mathbf{I} \end{bmatrix},$$

and we obtain

$$\mathbf{P}'\mathbf{B}\mathbf{P} = \begin{bmatrix} \mathbf{B}_{11} - \mathbf{B}_{12}\mathbf{B}_{22}^{c}\mathbf{B}_{21} & \mathbf{0} \\ \mathbf{0} & \mathbf{B}_{22} \end{bmatrix}.$$

Hence $\mathbf{B}_{11} - \mathbf{B}_{12}\mathbf{B}_{22}^{c}\mathbf{B}_{21}$ is non-negative. The result (6) can be proved by a similar procedure, and the theorem is proved. ▮

In Sec. 8.6 (see also Theorem 8.9.7), a number of theorems were stated on the subject of factoring a given matrix $\mathbf{A}$ into the product of two matrices of a certain form. The next theorem states another factorization that sometimes is useful in statistics.

Theorem 12.2.22

Let $\mathbf{A}$ be a $k \times k$ matrix. There exists an orthogonal matrix $\mathbf{R}$ and a non-negative matrix $\mathbf{B}$ such that $\mathbf{A} = \mathbf{RB}$. Furthermore, if $\mathbf{A}$ is nonsingular, then $\mathbf{B}$ is positive definite.

Proof: By Theorem 8.9.7, if $\mathbf{A}$ is a $k \times k$ matrix, there exist orthogonal matrices $\mathbf{P}$ and $\mathbf{Q}$ such that $\mathbf{PAQ} = \mathbf{D}$ where $\mathbf{D}$ is diagonal and $d_{ii} \geq 0$. Hence $\mathbf{D}$ and $\mathbf{QDQ}' = \mathbf{B}$ are non-negative. But

$$\mathbf{A} = \mathbf{P}'\mathbf{D}\mathbf{Q}' = \mathbf{P}'\mathbf{Q}'\mathbf{Q}\mathbf{D}\mathbf{Q}' = \mathbf{RB},$$

where $\mathbf{R} = \mathbf{P}'\mathbf{Q}'$ is the product of two orthogonal matrices, and is therefore orthogonal. If $\mathbf{A}$ is non-singular, then $d_{ii} > 0$ for all i and $\mathbf{D}$ is positive definite, as is $\mathbf{QDQ}' = \mathbf{B}$. This concludes the proof of the theorem. ▮

12.3 Idempotent Matrices

In Sec. 12.1 we indicated that quadratic forms with idempotent matrices are used extensively in statistical theory. In fact, if a random $n \times 1$ vector $\mathbf{y}$ has a multivariate normal density with covariance matrix equal to $\mathbf{I}$, then the quadratic form $\mathbf{y}'\mathbf{A}_i\mathbf{y}$ in Eq. (12.1.1) has a noncentral chi-square density if and only if $\mathbf{A}_i$ is an idempotent matrix; this is an important result in the analysis of variance.

Definition 12.3.1

Idempotent Matrices. *Let an $n \times n$ matrix* $\mathbf{B}$ *be such that*

(1) $\mathbf{B} = \mathbf{B}'$ and

(2) $\mathbf{B} = \mathbf{B}^2$.

Then $\mathbf{B}$ *is defined to be an idempotent matrix if* (2) *is satisfied and a symmetric idempotent matrix if* (1) *and* (2) *are both satisfied.*

Theorem 12.3.1

If $\mathbf{B}$ *is an $n \times n$ idempotent matrix of rank n, then* $\mathbf{B} = \mathbf{I}$. *If* $\mathbf{B}$ *is a symmetric idempotent matrix of rank less than n, then* $\mathbf{B}$ *is a positive semidefinite matrix.*

Proof: If $\mathbf{B}$ has rank n, then $\mathbf{B}^{-1}$ exists and we can multiply $\mathbf{B}^2 = \mathbf{B}$ on the left by $\mathbf{B}^{-1}$ and obtain $\mathbf{B} = \mathbf{I}$. If $\mathbf{B}$ has rank less than n, then we can use part (1a) of Theorem 12.2.2. ▮

Theorem 12.3.2

Let $\mathbf{B}$ *be any $n \times n$ matrix of rank p.*

(1) *If* $\mathbf{B}$ *is idempotent, then* $\mathbf{B}$ *has p nonzero characteristic roots and they are each equal to* $+1$.

(2) *If* $\mathbf{B}$ *is symmetric, then a necessary and sufficient condition that* $\mathbf{B}$ *is idempotent is that there are p nonzero characteristic roots of* $\mathbf{B}$ *and each is* $+1$.

Proof: First we shall prove part (1). By Eq. (3.2.1), if λ is a characteristic root of a matrix $\mathbf{B}$, then for some nonzero vector $\mathbf{x}$ we have

$$\mathbf{Bx} = \lambda\mathbf{x}.$$

Multiply on the left by $\mathbf{B}$ and obtain

$$\mathbf{B}(\mathbf{Bx}) = \mathbf{B}^2\mathbf{x} = \lambda(\mathbf{Bx}) = \lambda(\lambda\mathbf{x}) = \lambda^2\mathbf{x},$$

but since $\mathbf{B}$ is idempotent, $\mathbf{B} = \mathbf{B}^2$, and we get

$$\lambda^2\mathbf{x} = \mathbf{B}^2\mathbf{x} = \mathbf{Bx} = \lambda\mathbf{x}$$

or

$$\lambda(\lambda - 1)\mathbf{x} = \mathbf{0}.$$

So, since $\mathbf{x}$ is nonzero, we must have either $\lambda = 1$ or $\lambda = 0$. Thus the characteristic roots of an idempotent matrix must be equal to either zero or one. But since the rank was assumed to be p and, by Theorem 9.1.5, rank $(\mathbf{B}) = \text{tr}\ (\mathbf{B}) = \sum \lambda_i = p$, there are exactly p nonzero roots and they are each equal to $+1$.

To prove part (2), let $\mathbf{P}$ be an orthogonal matrix such that $\mathbf{P}'\mathbf{B}\mathbf{P} = \mathbf{D}$ where $\mathbf{D}$ is a diagonal matrix with the characteristic roots of $\mathbf{B}$ on the diagonal. Clearly $\mathbf{D} = \mathbf{D}^2$ if and only if the nonzero diagonal elements are equal to $+1$; and $\mathbf{D} = \mathbf{D}^2$ if and only if $\mathbf{B} = \mathbf{B}^2$. ▌

Theorem 12.3.3

Let $\mathbf{A}$ be an $n \times n$ symmetric matrix such that $\mathbf{A}^t = \mathbf{A}^{t+1}$ for some positive integer t; then $\mathbf{A}$ is an idempotent matrix.

Proof: Let $\mathbf{P}$ be an orthogonal matrix such that $\mathbf{P}'\mathbf{A}\mathbf{P} = \mathbf{D}$ where $\mathbf{D}$ is a diagonal matrix with the characteristic roots of $\mathbf{A}$ displayed on the diagonal. From the fact that $\mathbf{A}^t = \mathbf{A}^{t+1}$, we obtain $\mathbf{D}^t = \mathbf{D}^{t+1}$, and hence each element on the diagonal of $\mathbf{D}$ is either unity or zero. Thus $\mathbf{D}^2 = \mathbf{D}$ and $\mathbf{P}'\mathbf{A}\mathbf{P}\mathbf{P}'\mathbf{A}\mathbf{P} = \mathbf{P}'\mathbf{A}\mathbf{P}$, which implies that $\mathbf{A}^2 = \mathbf{A}$. ▌

Theorem 12.3.4

If $\mathbf{B}$ is an $n \times n$ symmetric idempotent matrix and if the i-th diagonal element is equal to either zero or unity, then each off-diagonal element in the i-th row and the i-th column is zero.

Proof: This follows, since in an $n \times n$ symmetric idempotent matrix the i-th diagonal element is equal to the sum of squares of the elements in the i-th row (column). ▌

Theorem 12.3.5

Let $\mathbf{A}$ be an $n \times n$ (symmetric) idempotent matrix; then

(1) *$\mathbf{A}'$ is a (symmetric) idempotent matrix,*
(2) *$\mathbf{P}'\mathbf{A}\mathbf{P}$ is a (symmetric) idempotent matrix if $\mathbf{P}$ is orthogonal,*
(3) *$\mathbf{P}\mathbf{A}\mathbf{P}^{-1}$ is an idempotent matrix where $\mathbf{P}$ is nonsingular,*
(4) *$\mathbf{I} - \mathbf{A}$ is a (symmetric) idempotent matrix,*
(5) *if $\mathbf{A}$ is also a normal matrix (that is, if $\mathbf{A}'\mathbf{A} = \mathbf{A}\mathbf{A}'$), then $\mathbf{A}'\mathbf{A}$ and $\mathbf{A}\mathbf{A}'$ are symmetric idempotent matrices,*
(6) *$\mathbf{A}^n$ is a (symmetric) idempotent matrix, where n is any positive integer.*

Proof: In each case the proof is obtained by multiplication. ▌

Example 12.3.1. The following are idempotent matrices:

$$\mathbf{A}_1 = \begin{bmatrix} \frac{1}{2} & \frac{1}{2} \\ \frac{1}{2} & \frac{1}{2} \end{bmatrix}, \qquad \mathbf{A}_2 = \begin{bmatrix} \frac{1}{2} & -\frac{1}{2} \\ -\frac{1}{2} & \frac{1}{2} \end{bmatrix} = \mathbf{I} - \mathbf{A}_1,$$

$$\mathbf{A}_3 = \begin{bmatrix} 1 & 0 \\ 1 & 0 \end{bmatrix}, \qquad \mathbf{A}_4 = \begin{bmatrix} 0 & 0 \\ -1 & 1 \end{bmatrix} = \mathbf{I} - \mathbf{A}_3,$$

$$\mathbf{A}_5 = \begin{bmatrix} 1 & 0 & 0 \\ 0 & \frac{2}{3} & \frac{\sqrt{2}}{3} \\ 0 & \frac{\sqrt{2}}{3} & \frac{1}{3} \end{bmatrix}, \qquad \mathbf{A}_6 = \begin{bmatrix} 0 & 0 & 0 \\ 0 & \frac{1}{3} & -\frac{\sqrt{2}}{3} \\ 0 & -\frac{\sqrt{2}}{3} & \frac{2}{3} \end{bmatrix} = \mathbf{I} - \mathbf{A}_5,$$

$$\mathbf{A}_7 = \begin{bmatrix} 1 & 1 \\ 0 & 0 \end{bmatrix} = \mathbf{A}_3', \qquad \mathbf{A}_8 = \begin{bmatrix} \frac{6 + 8\sqrt{2}}{3} & -\frac{2 + 3\sqrt{2}}{3} \\ \frac{10 + 21\sqrt{2}}{3} & -\frac{3 + 8\sqrt{2}}{3} \end{bmatrix}.$$

Note: $\mathbf{A}_8 = \mathbf{PAP}^{-1}$, where

$$\mathbf{P} = \begin{bmatrix} 1 & 2 \\ 2 & 5 \end{bmatrix} \quad \text{and} \quad \mathbf{A} = \mathbf{A}^2 = \begin{bmatrix} \frac{2}{3} & \frac{\sqrt{2}}{3} \\ \frac{\sqrt{2}}{3} & \frac{1}{3} \end{bmatrix}.$$

The next three theorems can be used extensively in the analysis of variance.

Theorem 12.3.6

Let $\mathbf{A}_0 = \sum_{i=1}^{k} \mathbf{A}_i$ *where each* $\mathbf{A}_i$ *is an* $n \times n$ *symmetric matrix. Any two of the conditions* (1), (2), (3) *below imply the remaining condition.*

(1) $\mathbf{A}_0 = \mathbf{A}_0^2$;

(2) $\mathbf{A}_i = \mathbf{A}_i^2, \quad i = 1, 2, \ldots, k;$

(3) $\mathbf{A}_i \mathbf{A}_j = \mathbf{0}, \quad i \neq j; \quad i = 1, 2, \ldots, k; \quad j = 1, 2, \ldots, k.$

Proof: Assume first that (1) and (2) are true. We have $\mathbf{A}_0^2 = \mathbf{A}_0$, which gives

$$\mathbf{A}_0^2 = \left(\sum_{i=1}^{k} \mathbf{A}_i\right)^2 = \sum_{i=1}^{k} \mathbf{A}_i^2 + \sum_{j=1}^{k} \sum_{\substack{i=1 \\ i \neq j}}^{k} \mathbf{A}_i \mathbf{A}_j = \sum_{i=1}^{k} \mathbf{A}_i + \sum_{j=1}^{k} \sum_{\substack{i=1 \\ i \neq j}}^{k} \mathbf{A}_i \mathbf{A}_j = \mathbf{A}_0 = \sum_{i=1}^{k} \mathbf{A}_i .$$

So we obtain

$$\sum_{j=1}^{k} \sum_{\substack{i=1 \\ i \neq j}}^{k} \mathbf{A}_i \mathbf{A}_j = \mathbf{0},$$

and from this we get

$$\operatorname{tr}\left(\sum_{j=1}^{k} \sum_{\substack{i=1 \\ i \neq j}}^{k} \mathbf{A}_i \mathbf{A}_j\right) = 0.$$

Since each $\mathbf{A}_i$ is symmetric and idempotent, each $\mathbf{A}_i$ is non-negative; so from part (4) of Theorem 12.2.4, it follows that $\mathbf{A}_i \mathbf{A}_j = \mathbf{0}$ if $i \neq j$; so (1) and (2) imply (3).

Next assume (1) and (3). We obtain

$$\mathbf{A}_0 = \mathbf{A}_0^2 = \left(\sum_{i=1}^{k} \mathbf{A}_i\right)^2 = \sum_{i=1}^{k} \mathbf{A}_i^2 \quad \text{or} \quad \sum_{i=1}^{k} \mathbf{A}_i = \sum_{i=1}^{k} \mathbf{A}_i^2 .$$

Multiplying both sides of this equation by $\mathbf{A}_q$ for $q \neq 0$, we get

$$\mathbf{A}_q^2 = \mathbf{A}_q^3,$$

since $\mathbf{A}_q \mathbf{A}_i = \mathbf{0}$ if $i \neq q$. By Theorem 12.3.3 it follows that $\mathbf{A}_q$ is idempotent. Repeat for $q = 1, 2, \ldots, k$; thus (1) and (3) imply (2).

Next assume (2) and (3). We get

$$\mathbf{A}_0^2 = \left(\sum_{i=1}^{k} \mathbf{A}_i\right)^2 = \sum_{i=1}^{k} \mathbf{A}_i^2 + \sum_{i=1}^{k} \sum_{\substack{j=1 \\ i \neq j}}^{k} \mathbf{A}_i \mathbf{A}_j = \sum_{i=1}^{k} \mathbf{A}_i = \mathbf{A}_0 ,$$

so $\mathbf{A}_0$ is idempotent, and the proof of the theorem is complete. If $\mathbf{A}_0 = \mathbf{I}$ in Theorem 12.3.6, then condition (2) implies (3) and vice versa. ∎

A variation of Theorem 12.3.6 is given below.

Theorem 12.3.7

Let $\mathbf{A}_i$, $i = 1, 2, \ldots, k$, *be* $n \times n$ *symmetric matrices of rank* n_i *such that* $\sum_{i=1}^{k} \mathbf{A}_i = \mathbf{I}$. *If* $\sum_{i=1}^{k} n_i = n$, *then*

(1) $\mathbf{A}_i \mathbf{A}_j = \mathbf{0}, \quad i \neq j = 1, 2, \ldots, k,$ *and*

(2) $\mathbf{A}_i = \mathbf{A}_i^2, \quad i = 1, 2, \ldots, k.$

This theorem states that if the sum of the ranks of the $\mathbf{A}_i$ is equal to the rank of the sum of the $\mathbf{A}_i$, then the $\mathbf{A}_i$ are disjoint and idempotent. The proof is left for the reader.

Theorem 12.3.8

Let $\mathbf{A}_i$, $i = 1, \ldots, k$, *be* $n \times n$ *symmetric idempotent matrices of rank* n_i, *and let* $\mathbf{A}_{k+1}$ *be an* $n \times n$ *non-negative matrix. Further suppose* $\mathbf{I} = \sum_{i=1}^{k+1} \mathbf{A}_i$. *Then* $\mathbf{A}_{k+1}$ *is symmetric idempotent of rank* $n - \sum_{i=1}^{k} n_i$, *and* $\mathbf{A}_i \mathbf{A}_j = \mathbf{0}$ *for all* $i \neq j = 1, 2, \ldots, k + 1$.

Proof: This theorem follows directly from Theorem 12.3.6. ▮

Theorem 12.3.9

Let $\mathbf{A}_0 = \sum_{i=1}^{k} \mathbf{A}_i$ *where each* $\mathbf{A}_i$ *is an* $n \times n$ *symmetric matrix for* $i = 1, 2, \ldots, k$ *and let* $\mathbf{A}_0 = \mathbf{A}_0^2$. *Then condition* (1) *below implies that*

(a) $\mathbf{A}_i = \mathbf{A}_i^2$ *for* $i = 1, 2, \ldots, k$

and

(b) $\mathbf{A}_i \mathbf{A}_j = \mathbf{0}$ *for all* $i \neq j$; $\quad i = 1, 2, \ldots, k; \quad j = 1, 2, \ldots, k.$

(1) $\mathbf{A}_i$ *is non-negative for* $i = 1, 2, \ldots, k$; *and*

$$\operatorname{tr}(\mathbf{A}_0) \leq \operatorname{tr}\left(\sum_{i=1}^{k} \mathbf{A}_i^2\right).$$

Proof: By Theorem 12.3.6, either (a) or (b) implies the other, since we have assumed that $\mathbf{A}_0 = \mathbf{A}_0^2$. Therefore, in the proof we shall show that condition

(1) implies (b) and hence also implies (a). By the hypothesis of the theorem, we get

$$\mathbf{A}_0 = \mathbf{A}_0^2 = \left(\sum_{i=1}^{k} \mathbf{A}_i\right)^2 = \sum_{i=1}^{k} \mathbf{A}_i^2 + \sum_{j=1}^{k} \sum_{\substack{i=1 \\ i \neq j}}^{k} \mathbf{A}_i \mathbf{A}_j$$

and

$$\mathrm{tr}\,(\mathbf{A}_0) = \mathrm{tr}\,(\sum_{i=1}^{k} \mathbf{A}_i^2) + \mathrm{tr}\left[\sum_{j=1}^{k} \sum_{\substack{i=1 \\ i \neq j}}^{k} \mathbf{A}_i \mathbf{A}_j\right].$$

But since $\mathrm{tr}\,(\mathbf{A}_0) \leq \mathrm{tr}\,(\sum \mathbf{A}_i^2)$, we get

$$\sum_{j=1}^{k} \sum_{\substack{i=1 \\ i \neq j}}^{k} \mathrm{tr}\,(\mathbf{A}_i \mathbf{A}_j) \leq 0;$$

but each $\mathbf{A}_i$ is assumed to be non-negative, so, by (3) and (4) of Theorem 12.2.4, it follows that $\mathbf{A}_i \mathbf{A}_j = \mathbf{0}$ for each $i = 1, 2, \ldots, k$; $j = 1, 2, \ldots, k$; $i \neq j$. ∎

Example 12.3.2. For a brief illustration of these theorems, consider an $n \times m$ matrix $\mathbf{X}$ of rank $m \leq n$. We partition $\mathbf{X}$ as follows:

$$\mathbf{X} = [\mathbf{X}_1, \mathbf{X}_2],$$

where $\mathbf{X}_1$ has size $n \times m_1$ and $\mathbf{X}_2$ has size $n \times m_2$. We define matrices $\mathbf{A}_1$ and $\mathbf{A}_2$ as follows:

$$\mathbf{A}_1 = \mathbf{I} - \mathbf{X}(\mathbf{X}'\mathbf{X})^{-1}\mathbf{X}', \qquad \mathbf{A}_2 = \mathbf{X}_1(\mathbf{X}_1'\mathbf{X}_1)^{-1}\mathbf{X}_1',$$

$$\mathbf{A}_3 = \mathbf{X}(\mathbf{X}'\mathbf{X})^{-1}\mathbf{X}' - \mathbf{X}_1(\mathbf{X}_1'\mathbf{X}_1)^{-1}\mathbf{X}_1'.$$

We get the matrix equation

$$\mathbf{I} = \mathbf{A}_1 + \mathbf{A}_2 + \mathbf{A}_3$$

and the equation in quadratic forms,

$$\mathbf{y}'\mathbf{y} = \mathbf{y}'\mathbf{A}_1\mathbf{y} + \mathbf{y}'\mathbf{A}_2\mathbf{y} + \mathbf{y}'\mathbf{A}_3\mathbf{y},$$

for a random vector $\mathbf{y}$.

As explained in Sec. 12.1, it is important in analysis of variance theory to determine if the $\mathbf{A}_i$ are idempotent and disjoint. It is generally known that each $\mathbf{A}_i$ is non-negative because of the method of computing each quadratic form. We shall use Theorem 12.3.8. Since $\mathbf{A}_0 = \mathbf{I}$, it is idempotent, and it is easy to verify that $\mathbf{A}_1$ and $\mathbf{A}_2$ are idempotent. Assume that we know by the method of constructing $\mathbf{A}_3$ that it is non-negative; the result of the theorem states that $\mathbf{A}_3$ is also idempotent and $\mathbf{A}_i\mathbf{A}_j = \mathbf{0}$ for all $i \neq j$. Also

$$\text{rank}\,(\mathbf{I}) = n; \quad \text{rank}\,(\mathbf{A}_1) = \text{tr}\,(\mathbf{A}_1) = n - m; \quad \text{rank}\,(\mathbf{A}_2) = \text{tr}\,(\mathbf{A}_2) = m_1;$$

hence by the result of the theorem, the rank of $\mathbf{A}_3$ is $n - (n - m) - m_1 = m - m_1$. We could have used Theorem 12.3.6 or 12.3.7 instead of 12.3.8.

Example 12.3.3. For another illustration, consider Eq. (12.1.5),

$$\sum_{i=1}^{n} y_i^2 = n\bar{y}^2 + \sum_{i=1}^{n} (y_i - \bar{y})^2,$$

or, equivalently,

$$\mathbf{y}'\mathbf{y} = \mathbf{y}'\mathbf{A}_1\mathbf{y} + \mathbf{y}'\mathbf{A}_2\mathbf{y}.$$

Clearly each quadratic form is non-negative. We obtain the matrix equation

$$\mathbf{I} = \mathbf{A}_1 + \mathbf{A}_2,$$

and clearly $\mathbf{A}_1 = (1/n)\mathbf{J}$ is symmetric idempotent. The rank of $\mathbf{A}_1$ is equal to

$$\text{tr}\,(\mathbf{A}_1) = \text{tr}\left(\frac{1}{n}\mathbf{J}\right) = \frac{1}{n}\,\text{tr}\,(\mathbf{J}) = 1.$$

The hypothesis of Theorem 12.3.8 is satisfied; hence $\mathbf{A}_2$ is idempotent of rank $n - 1$ and $\mathbf{A}_1\mathbf{A}_2 = \mathbf{0}$. Earlier in this chapter we showed that Theorem 12.2.6 can be used in this example to obtain the result that $\mathbf{A}_1\mathbf{A}_2 = \mathbf{0}$.

The remainder of this section is devoted to some miscellaneous theorems on idempotent matrices.

Theorem 12.3.10

Let $\mathbf{A}_i$, $i = 1, 2, \ldots, k$, be symmetric idempotent matrices such that

$$\mathbf{A}_i\mathbf{A}_j = \mathbf{0},\ i \neq j, \quad \text{and} \quad \sum_{i=1}^{k} \mathbf{A}_i = \mathbf{I},$$

and let α_i be real numbers such that $\alpha_i > 0$, $i = 1, 2, \ldots, k$. Then **V** *is a positive definite matrix, where*

$$\mathbf{V} = \sum_{i=1}^{k} \alpha_i \mathbf{A}_i .$$

Proof: Since $\mathbf{A}_i$ and $\mathbf{A}_j$ commute (that is, $\mathbf{A}_i\mathbf{A}_j = \mathbf{0}$ and $\mathbf{A}_j\mathbf{A}_i = \mathbf{0}$, so $\mathbf{A}_i\mathbf{A}_j = \mathbf{A}_j\mathbf{A}_i$ for all $i \neq j$), there exists an orthogonal matrix **P** such that $\mathbf{P}'\mathbf{A}_i\mathbf{P} = \mathbf{E}_i$ for each i, where $\mathbf{E}_i$ is a diagonal matrix with only zeros and ones on the diagonal. Thus $\mathbf{P}'\mathbf{V}\mathbf{P} = \sum_{i=1}^{k}\alpha_i\mathbf{E}_i$, and $\mathbf{P}'\mathbf{V}\mathbf{P}$ is a diagonal matrix with α_i on the diagonal. Hence the α_i are the characteristic roots of **V**, and, since $\alpha_i > 0$ for every i, **V** is positive definite. It might appear that $\mathbf{P}'\mathbf{V}\mathbf{P}$ could have some zero diagonal elements, but this is impossible, since $\sum_{i=1}^{k} \mathbf{A}_i = \mathbf{I}$, and hence

$$\mathbf{P}'\left(\sum_{i=1}^{k} \mathbf{A}_i\right)\mathbf{P} = \mathbf{P}'\mathbf{I}\mathbf{P}, \quad \text{or} \quad \sum_{i=1}^{k} \mathbf{E}_i = \mathbf{I};$$

so $\sum_{i=1}^{k}\alpha_i\mathbf{E}_i$ can have no zero elements on the diagonal. ∎

Theorem 12.3.11

Let $\mathbf{A}_i$, $i = 1, 2, \ldots, k$, be symmetric idempotent matrices such that $\mathbf{A}_i\mathbf{A}_j = \mathbf{0}$, $i \neq j$, and let α_i be scalars such that $\alpha_i > 0$, $i = 0, 1, \ldots, k$. Then **V** *is positive definite, where*

$$\mathbf{V} = \alpha_0\mathbf{I} + \sum_{i=1}^{k} \alpha_i \mathbf{A}_i .$$

The proof is very similar to the proof for Theorem 12.3.10 and is left for the reader.

Corollary 12.3.11

In Theorem 12.3.11,

$$\mathbf{V}^{-1} = \beta_0\mathbf{I} + \sum_{i=1}^{k} \beta_i \mathbf{A}_i ,$$

where

$$\beta_0 = \alpha_0^{-1}; \quad \beta_i = \frac{-\alpha_i}{\alpha_0(\alpha_0 + \alpha_i)}; \quad i = 1, 2, \ldots, k.$$

Proof: By simply multiplying it can be shown that $\mathbf{V}\mathbf{V}^{-1} = \mathbf{I}$. ∎

Example 12.3.4. To illustrate Theorem 12.3.10, let us define $\mathbf{A}_1$ and $\mathbf{A}_2$ by

$$\mathbf{A}_1 = \begin{bmatrix} \frac{1}{2} & -\frac{1}{2} \\ -\frac{1}{2} & \frac{1}{2} \end{bmatrix} \qquad \mathbf{A}_2 = \begin{bmatrix} \frac{1}{2} & \frac{1}{2} \\ \frac{1}{2} & \frac{1}{2} \end{bmatrix}.$$

Clearly $\mathbf{A}_1\mathbf{A}_2 = \mathbf{0}$, $\mathbf{A}_1 + \mathbf{A}_2 = \mathbf{I}$ and $\mathbf{A}_1$ and $\mathbf{A}_2$ are symmetric and idempotent. If we define $\mathbf{V}$ by

$$\mathbf{V} = 2\mathbf{A}_1 + 3\mathbf{A}_2 = \begin{bmatrix} \frac{5}{2} & \frac{1}{2} \\ \frac{1}{2} & \frac{5}{2} \end{bmatrix},$$

$\mathbf{V}$ is seen to be positive definite.

To illustrate Theorem 12.3.11 and Corollary 12.3.11, we shall use $\mathbf{A}_1$ and $\mathbf{A}_2$ as defined above and define $\mathbf{V}$ by

$$\mathbf{V} = 6\mathbf{I} + \mathbf{A}_1 + 4\mathbf{A}_2 = \begin{bmatrix} \frac{17}{2} & \frac{3}{2} \\ \frac{3}{2} & \frac{17}{2} \end{bmatrix}.$$

Clearly $\mathbf{V}$ is positive definite and $\mathbf{V}^{-1}$ can be written

$$\mathbf{V}^{-1} = \frac{1}{6}\mathbf{I} - \frac{1}{42}\mathbf{A}_1 - \frac{4}{60}\mathbf{A}_2 = \frac{1}{70}\begin{bmatrix} \frac{17}{2} & -\frac{3}{2} \\ -\frac{3}{2} & \frac{17}{2} \end{bmatrix}.$$

Theorem 12.3.12

Let $\mathbf{A}$ *be an* $n \times n$ *symmetric idempotent matrix; then*

$$\mathbf{B} = \mathbf{I} - 2\mathbf{A}$$

is a symmetric orthogonal matrix.

Proof: $\mathbf{BB}' = (\mathbf{I} - 2\mathbf{A})(\mathbf{I} - 2\mathbf{A})' = \mathbf{I}$, and $\mathbf{B} = \mathbf{B}'$. ▮

Theorem 12.3.13

Let $\mathbf{A}$ *be an* $n \times n$ *idempotent matrix, and let* $\mathbf{B}$ *be an* $n \times n$ *idempotent matrix; then the matrix* $\mathbf{A} \times \mathbf{B}$ *is idempotent.*

Proof: $(\mathbf{A} \times \mathbf{B})(\mathbf{A} \times \mathbf{B}) = \mathbf{A}^2 \times \mathbf{B}^2 = \mathbf{A} \times \mathbf{B}$. ▮

Theorem 12.3.14

For every $n \times n$ *matrix* $\mathbf{A}$, *there are nonsingular matrices* $\mathbf{P}$ *and* $\mathbf{Q}$ *of size* $n \times n$ *such that* $\mathbf{PAQ}$ *is a symmetric idempotent matrix.*

Proof: Select $\mathbf{P}$ and $\mathbf{Q}$ such that $\mathbf{PAQ} = \mathbf{R}_r$ where

$$\mathbf{R}_r = \begin{bmatrix} \mathbf{I}_r & \mathbf{0} \\ \mathbf{0} & \mathbf{0} \end{bmatrix}.$$

Clearly

$$\mathbf{R}_r\mathbf{R}_r = \mathbf{R}_r \quad \text{and} \quad \mathbf{R}_r' = \mathbf{R}_r. \quad ▮$$

Theorem 12.3.15

Let $\mathbf{P}$ *be an* $m \times n$ *matrix with* $n \geq m$ *such that* $\mathbf{PP}' = \mathbf{I}$, *that is, the rows are orthogonal and normal; then* $\mathbf{P}'\mathbf{P}$ *is an* $n \times n$ *symmetric idempotent matrix.*

Proof: Clearly $(\mathbf{P}'\mathbf{P})' = \mathbf{P}'\mathbf{P}$ and $(\mathbf{P}'\mathbf{P})(\mathbf{P}'\mathbf{P}) = \mathbf{P}'\mathbf{IP} = \mathbf{P}'\mathbf{P}$. ▮

Theorem 12.3.16

Let $\mathbf{A}$ *be any* $n \times n$ *symmetric matrix; then it can be written as a linear combination of symmetric, disjoint idempotent matrices, each of rank one; that is,* $\mathbf{A}$ *can be written as*

$$\mathbf{A} = \sum_{i=1}^{n} \lambda_i \mathbf{A}_i,$$

where $\mathbf{A}_i^2 = \mathbf{A}_i$, $\mathbf{A}_i' = \mathbf{A}_i$, $\mathbf{A}_i\mathbf{A}_j = \mathbf{0}$ *for all* $i \neq j$ *and* λ_i *are characteristic roots of* $\mathbf{A}$, *and* $\mathbf{A}_i$ *has rank* 1.

Proof: Since $\mathbf{A}$ is symmetric, there exists an orthogonal matrix $\mathbf{P}$, such that $\mathbf{P}'\mathbf{A}\mathbf{P} = \mathbf{D}$, where $d_{ii} = \lambda_i$ are the characteristic roots of $\mathbf{A}$. We write $\mathbf{A} = \mathbf{P}\mathbf{D}\mathbf{P}'$, and if we write $\mathbf{P} = [\mathbf{p}_1, \mathbf{p}_2, \ldots, \mathbf{p}_n]$, it follows that

$$\mathbf{P}\mathbf{D} = [\lambda_1\mathbf{p}_1, \lambda_2\mathbf{p}_2, \ldots, \lambda_n\mathbf{p}_n],$$

where $\mathbf{p}_i$ is the i-th column of $\mathbf{P}$. Thus,

$$\mathbf{A} = \sum_{i=1}^{n} \lambda_i \mathbf{p}_i \mathbf{p}_i' = \sum_{i=1}^{n} \lambda_i \mathbf{A}_i,$$

where $\mathbf{A}_i = \mathbf{p}_i\mathbf{p}_i'$. But $\mathbf{A}_i = \mathbf{A}_i'$ and $\mathbf{A}_i^2 = \mathbf{A}_i$. Also $\mathbf{A}_i\mathbf{A}_j = \mathbf{p}_i\mathbf{p}_i'\mathbf{p}_j\mathbf{p}_j' = \mathbf{0}$ if $i \neq j$, since $\mathbf{p}_i'\mathbf{p}_j = 0$ if $i \neq j$. Also rank $(\mathbf{A}_i) = 1$, since $\mathbf{p}_i$ is $n \times 1$ for each $i = 1, 2, \ldots, n$, and the proof is complete. ▮

Theorem 12.3.17

If $\mathbf{A}$ and $\mathbf{B}$ are $n \times n$ matrices, then $\mathbf{AB}$ and $\mathbf{BA}$ are idempotent matrices if either

(1) $\mathbf{ABA} = \mathbf{A}$, *or*

(2) $\mathbf{BAB} = \mathbf{B}$.

Proof: Multiply (1) on the left by $\mathbf{B}$ to show that $\mathbf{BA}$ is idempotent. The remaining part of the theorem is proved similarly. ▮

Theorem 12.3.18

If $\mathbf{A}$ and $\mathbf{B}$ are $n \times n$ idempotent matrices, then $\mathbf{AB}$ and $\mathbf{BA}$ are idempotent matrices if $\mathbf{AB} = \mathbf{BA}$.

Proof: The proof is obtained by multiplication. ▮

The proofs of the next three theorems are left for the reader.

Theorem 12.3.19

Let $\mathbf{A}$ be any $m \times n$ matrix. Then $\mathbf{A}'\mathbf{A}$ is an idempotent matrix if and only if $\mathbf{A}\mathbf{A}'$ is an idempotent matrix.

Theorem 12.3.20

Let $\mathbf{A}$ be any $m \times n$ matrix. Then $\mathbf{A}'\mathbf{A}$ is idempotent if and only if $\mathbf{A}'$ is a c-inverse of $\mathbf{A}$.

Theorem 12.3.21

If $\mathbf{A}$ *and* $\mathbf{B}$ *are* $n \times n$ *symmetric idempotent matrices, then* $\mathbf{A} - \mathbf{B}$ *is a symmetric idempotent matrix if and only if* $\mathbf{B}(\mathbf{A} - \mathbf{B}) = \mathbf{0}$.

Theorem 12.3.22

Let $\mathbf{A}$ *be any symmetric* $n \times n$ *matrix of rank* r. *Then* $\mathbf{A}$ *can be written as a linear combination of* r *disjoint, symmetric idempotent matrices each of rank* 1; *that is,*

$$\mathbf{A} = \sum_{i=1}^{r} d_i \mathbf{A}_i,$$

where $\mathbf{A}_i^2 = \mathbf{A}_i$, $\mathbf{A}_i = \mathbf{A}_i'$, $\mathbf{A}_i\mathbf{A}_j = \mathbf{0}$ *all* $i \neq j$.

Proof: There exists an orthogonal matrix $\mathbf{P}$ such that

$$\mathbf{P}'\mathbf{A}\mathbf{P} = \begin{bmatrix} \mathbf{D} & \mathbf{0} \\ \mathbf{0} & \mathbf{0} \end{bmatrix}$$

where $\mathbf{D}$ is an $r \times r$ matrix with the nonzero roots of $\mathbf{A}$ on the diagonal. Now

$$\mathbf{A} = \mathbf{P}\begin{bmatrix} \mathbf{D} & \mathbf{0} \\ \mathbf{0} & \mathbf{0} \end{bmatrix}\mathbf{P}' = [\mathbf{p}_1, \mathbf{p}_2, \ldots, \mathbf{p}_n]\begin{bmatrix} d_1 & 0 & \cdots & 0 \\ 0 & d_2 & \cdots & 0 \\ \vdots & \vdots & & \vdots \\ 0 & 0 & \cdots & 0 \end{bmatrix}\begin{bmatrix} \mathbf{p}_1' \\ \vdots \\ \mathbf{p}_n' \end{bmatrix}$$

$$= \sum_{i=1}^{r} d_i \mathbf{p}_i \mathbf{p}_i' = \sum_{i=1}^{r} d_i \mathbf{A}_i,$$

and clearly the $\mathbf{A}_i$ are disjoint, symmetric, and idempotent. ▌

Corollary 12.3.22

If $\mathbf{A}$ *is a symmetric idempotent matrix of rank* r, *then it can be written as the sum of* r *disjoint, symmetric idempotent matrices of rank* 1.

12.4 Tripotent Matrices

In this section we define tripotent matrices and state some theorems that are useful in statistical theory. The main reason these matrices are useful is that, if a matrix $\mathbf{C}$

is symmetric and tripotent, it can be written as the difference of two disjoint symmetric idempotent matrices. For example, if $\mathbf{y}$ is an $n \times 1$ random vector that has a normal density with covariance matrix $\mathbf{I}$, then $\mathbf{y}'\mathbf{C}\mathbf{y}$ is distributed as the difference of two noncentral chi-square random variables. This situation sometimes occurs in estimating variance components.

Definition 12.4.1

Tripotent Matrices. *Let* $\mathbf{C}$ *be an* $n \times n$ *matrix such that*

(1) $\mathbf{C} = \mathbf{C}'$,
(2) $\mathbf{C} = \mathbf{C}^3$.

If (2) *is satisfied,* $\mathbf{C}$ *is defined to be a tripotent matrix. If* (1) *and* (2) *are both satisfied,* $\mathbf{C}$ *is defined to be a symmetric tripotent matrix.*

The proof for some of the theorems given below are left for the reader.

Theorem 12.4.1

Let $\mathbf{C}$ *be an* $n \times n$ *(symmetric) tripotent matrix.*

(1) *If* $\mathbf{P}$ *is any* $n \times n$ *orthogonal matrix, then* $\mathbf{P}'\mathbf{C}\mathbf{P}$ *is an* $n \times n$ *(symmetric) tripotent matrix.*
(2) *If* $\mathbf{P}$ *is any* $n \times n$ *nonsingular matrix, then* $\mathbf{P}^{-1}\mathbf{C}\mathbf{P}$ *is an* $n \times n$ *tripotent matrix.*
(3) $\mathbf{C}^2$ *is an* $n \times n$ *(symmetric) idempotent matrix.*
(4) $-\mathbf{C}$ *is an* $n \times n$ *(symmetric) tripotent matrix.*
(5) *A matrix* $\mathbf{C}$ *is equal to its c-inverse if and only if* $\mathbf{C}$ *is tripotent.*

Theorem 12.4.2

Let $\mathbf{C}$ *be any* $n \times n$ *tripotent matrix; then the characteristic roots of* $\mathbf{C}$ *are equal to* -1, 0, *or* $+1$.

Proof: The characteristic roots of $\mathbf{C}$ are given by λ where λ satisfies

$$\mathbf{C}\mathbf{x} = \lambda\mathbf{x}$$

for some nonzero vector $\mathbf{x}$.

If we multiply this equation on the left by $\mathbf{C}^2$, we obtain

$$\mathbf{C}^3\mathbf{x} = \lambda\mathbf{C}^2\mathbf{x} = \lambda\mathbf{C}(\mathbf{C}\mathbf{x}) = \lambda^2\mathbf{C}\mathbf{x} = \lambda^3\mathbf{x}.$$

But, since $\mathbf{C} = \mathbf{C}^3$, we have

$$\lambda\mathbf{x} = \lambda^3\mathbf{x},$$

or

$$\lambda(1 - \lambda^2)\mathbf{x} = \mathbf{0};$$

and since $\mathbf{x} \neq \mathbf{0}$, we have the desired results. ∎

Corollary 12.4.2

Let $\mathbf{C}$ *be a symmetric* $n \times n$ *matrix. A necessary and sufficient condition that* $\mathbf{C}$ *be a symmetric tripotent matrix is that the characteristic roots of* $\mathbf{C}$ *consist only of the numbers* $+1$, -1, *and* 0.

Theorem 12.4.3

Let $\mathbf{C}$ *be any* $n \times n$ *symmetric matrix. A necessary and sufficient condition that* $\mathbf{C}$ *is symmetric tripotent is that there exist two symmetric idempotent, disjoint* $n \times n$ *matrices* $\mathbf{A}$ *and* $\mathbf{B}$ *such that* $\mathbf{C} = \mathbf{A} - \mathbf{B}$. *These two matrices are unique and*

$$\mathbf{A} = \tfrac{1}{2}(\mathbf{C}^2 + \mathbf{C}), \qquad \mathbf{B} = \tfrac{1}{2}(\mathbf{C}^2 - \mathbf{C}).$$

Proof: By straightforward operations it is easy to show that for matrices $\mathbf{A}$ and $\mathbf{B}$ that are symmetric, disjoint, and idempotent, $\mathbf{A} - \mathbf{B}$ is symmetric tripotent. Also, if $\mathbf{C}$ is symmetric tripotent, then the matrices $\mathbf{A}$ and $\mathbf{B}$ exhibited are idempotent, disjoint, and symmetric. To show uniqueness, we can use Theorem 12.2.7, since $\mathbf{A}$ and $\mathbf{B}$ are symmetric idempotent and, hence, nonnegative. ∎

Theorem 12.4.4

Let $\mathbf{C}$ *be any* $n \times n$ *tripotent matrix; then* rank $(\mathbf{C}) = \text{tr}\,(\mathbf{C}^2)$.

Proof: Let $\mathbf{P}$ and $\mathbf{Q}$ be $n \times n$ nonsingular matrices such that

$$\mathbf{PCQ} = \begin{bmatrix} \mathbf{I}_{n_1} & \mathbf{0} \\ \mathbf{0} & \mathbf{0} \end{bmatrix} = \mathbf{E} \quad \text{or} \quad \mathbf{C} = \mathbf{P}^{-1}\mathbf{E}\mathbf{Q}^{-1},$$

where n_1 is the rank of $\mathbf{C}$. Now $\mathbf{C} = \mathbf{C}^3$ gives

$$\mathbf{P}^{-1}\mathbf{E}\mathbf{Q}^{-1}\mathbf{P}^{-1}\mathbf{E}\mathbf{Q}^{-1}\mathbf{P}^{-1}\mathbf{E}\mathbf{Q}^{-1} = \mathbf{P}^{-1}\mathbf{E}\mathbf{Q}^{-1} \quad \text{or} \quad \mathbf{E}\mathbf{Q}^{-1}\mathbf{P}^{-1}\mathbf{E}\mathbf{Q}^{-1}\mathbf{P}^{-1}\mathbf{E} = \mathbf{E}.$$

If we consider the trace of each matrix, we obtain

$$\operatorname{tr}(\mathbf{E}\mathbf{Q}^{-1}\mathbf{P}^{-1}\mathbf{E}\mathbf{Q}^{-1}\mathbf{P}^{-1}\mathbf{E}) = \operatorname{tr}(\mathbf{E}) = n_1 = \operatorname{rank}(\mathbf{C}).$$

But

$$\operatorname{tr}(\mathbf{E}\mathbf{Q}^{-1}\mathbf{P}^{-1}\mathbf{E}\mathbf{Q}^{-1}\mathbf{P}^{-1}\mathbf{E}) = \operatorname{tr}(\mathbf{P}^{-1}\mathbf{E}\mathbf{Q}^{-1}\mathbf{P}^{-1}\mathbf{E}\mathbf{Q}^{-1}) = \operatorname{tr}(\mathbf{C}^2),$$

and the result follows. ▌

Theorem 12.4.5

Let $\mathbf{C}$ be any $n \times n$ tripotent matrix with n_1 characteristic roots equal to $+1$, n_2 characteristic roots equal to -1, and n_3 characteristic roots equal to 0. Then

(1) $\frac{1}{2}\operatorname{tr}(\mathbf{C}^2 + \mathbf{C}) = n_1$,
(2) $\frac{1}{2}\operatorname{tr}(\mathbf{C}^2 - \mathbf{C}) = n_2$,
(3) $\operatorname{tr}(\mathbf{I} - \mathbf{C}^2) = n_3$,
(4) $\operatorname{tr}(\mathbf{C}) = n_1 - n_2$.

Theorem 12.4.6

Let $\mathbf{A}$ and $\mathbf{B}$ be symmetric $n \times n$ matrices.

(1) *If $\mathbf{A}$ and $\mathbf{B}$ are idempotent and $\mathbf{AB} = \mathbf{BA}$, then $\mathbf{A} - \mathbf{B}$ is a symmetric tripotent matrix.*
(2) *If $\mathbf{A}$ and $\mathbf{B}$ are idempotent, then $\mathbf{A}$, $-\mathbf{A}$, $\mathbf{B}$, and $-\mathbf{B}$ are symmetric tripotent matrices.*
(3) *A necessary and sufficient condition that $\mathbf{A}$ (or $\mathbf{B}$) is a tripotent matrix is that $\mathbf{A}^2$ (or $\mathbf{B}^2$) is an idempotent matrix.*

Theorem 12.4.7

If $\mathbf{C}$ is an $n \times n$ nonsingular tripotent matrix, then

$$\mathbf{C}^{-1} = \mathbf{C}, \quad \mathbf{C}^2 = \mathbf{I}, \quad \textit{and} \quad (\mathbf{C} + \mathbf{I})(\mathbf{C} - \mathbf{I}) = \mathbf{0}.$$

Example 12.4.1. The matrix **C** defined below is an example of a 3 × 3 symmetric tripotent matrix.

$$\mathbf{C} = \begin{bmatrix} \frac{1}{3} & -\frac{2}{3} & -\frac{2}{3} \\ -\frac{2}{3} & \frac{1}{3} & -\frac{2}{3} \\ -\frac{2}{3} & -\frac{2}{3} & \frac{1}{3} \end{bmatrix}.$$

Note that $\mathbf{C}^2 = \mathbf{I}$ and hence $\mathbf{C} = \mathbf{C}^{-1}$. It is easy to verify that $-\mathbf{C}$ is also tripotent. We can use Theorem 12.4.3 to obtain

$$\mathbf{B} = \frac{1}{2}(\mathbf{C}^2 - \mathbf{C}) = \begin{bmatrix} \frac{1}{3} & \frac{1}{3} & \frac{1}{3} \\ \frac{1}{3} & \frac{1}{3} & \frac{1}{3} \\ \frac{1}{3} & \frac{1}{3} & \frac{1}{3} \end{bmatrix}, \qquad \mathbf{A} = \frac{1}{2}(\mathbf{C}^2 + \mathbf{C}) = \begin{bmatrix} \frac{2}{3} & -\frac{1}{3} & -\frac{1}{3} \\ -\frac{1}{3} & \frac{2}{3} & -\frac{1}{3} \\ -\frac{1}{3} & -\frac{1}{3} & \frac{2}{3} \end{bmatrix},$$

and we see that **A** and **B** are idempotent, $\mathbf{AB} = \mathbf{0}$, and $\mathbf{C} = \mathbf{A} - \mathbf{B}$. Notice that

$$\operatorname{tr}\left[\frac{1}{2}(\mathbf{C}^2 + \mathbf{C})\right] = 2; \quad \operatorname{tr}\left[\frac{1}{2}(\mathbf{C}^2 - \mathbf{C})\right] = 1; \quad \operatorname{tr}(\mathbf{I} - \mathbf{C}^2) = 0;$$

and hence **C** has characteristic roots $\lambda_1 = 1$, $\lambda_2 = 1$, $\lambda_3 = -1$.

12.5 Projections

Another way to view "projections" is by a transformation of vectors. For instance, if we want to project a given vector **x** into a vector space S_n, this is equivalent to "moving" the vector **x** to a vector **y** by a suitable transformation matrix **A**, obtaining $\mathbf{y} = \mathbf{Ax}$.

In previous discussions (Sec. 4.4), we showed that if a vector **x** in E_n is projected into a subspace S_n of E_n, the projected vector $\mathbf{y}$ is obtained by the transformation $\mathbf{y} = \mathbf{Ax}$, where $\mathbf{A} = \mathbf{B}(\mathbf{B}'\mathbf{B})^{-1}\mathbf{B}'$ and where the set of column vectors of the $n \times p$ matrix $\mathbf{B}$ is a basis for S_n. We note that $\mathbf{A}$ is symmetric and idempotent; also the rank of $\mathbf{A}$ is equal to the dimension of the subspace S_n. We shall state and prove some theorems about projection of vectors and then formally define the projection of one vector space onto another vector space. Recall that we defined the projection of a vector $\mathbf{x}$ into a vector space S_n to be a vector $\mathbf{y}$ that has the following two properties: (1) $\mathbf{y}$ is in S_n; (2) the vector $\mathbf{y} - \mathbf{x}$ is orthogonal to S_n. These two properties uniquely define the projection $\mathbf{y}$ when $\mathbf{x}$ and a basis set for S_n are given.

Theorem 12.5.1

Let $\mathbf{A}$ *be an* $n \times n$ *matrix that transforms the vector* $\mathbf{x}$ *to the vector* $\mathbf{y}$ (*that is,* $\mathbf{y} = \mathbf{Ax}$) *such that* $\mathbf{y}$ *is the projection of the vector* $\mathbf{x}$ *into the subspace* S_n *that is spanned by the columns of* $\mathbf{A}$. *Then there is a symmetric idempotent matrix* $\mathbf{C}$ *such that* $\mathbf{y} = \mathbf{Cx}$; *namely,* $\mathbf{C} = \mathbf{AA}^-$.

Proof: The fact that $\mathbf{y}$ is the projection of $\mathbf{x}$ into S_n implies that the vector $\mathbf{x} - \mathbf{y}$ is orthogonal to S_n, which implies that $\mathbf{x} - \mathbf{y}$ is orthogonal to every set of vectors that spans S_n, which in turn implies that $(\mathbf{x} - \mathbf{y})'\mathbf{A} = \mathbf{0}$, or $\mathbf{A}'\mathbf{x} = \mathbf{A}'\mathbf{y} = \mathbf{A}'\mathbf{Ax}$, which gives us

$$\mathbf{A}'\mathbf{x} = \mathbf{A}'\mathbf{Ax}.$$

If we multiply this equation on the left by $\mathbf{A}'^-$ and simplify, we get $\mathbf{AA}^-\mathbf{x} = \mathbf{Ax} = \mathbf{y}$, or $\mathbf{y} = \mathbf{AA}^-\mathbf{x}$; and $\mathbf{AA}^-$ is symmetric idempotent, so the theorem is proved. ▮

Note: There may exist a matrix $\mathbf{A}$ that transforms a vector $\mathbf{x}$ to a vector $\mathbf{y}$ by $\mathbf{y} = \mathbf{Ax}$ such that $\mathbf{y}$ is the projection of $\mathbf{x}$ into a subspace S_n, where $\mathbf{A}$ is neither symmetric nor idempotent.

Theorem 12.5.2

Suppose that $\mathbf{y}$ *is the projection of a vector* $\mathbf{x}$ *in* E_n *into a subspace* S_n *of* E_n. *If* $\mathbf{x}$ *is in* S_n, *then it is transformed into itself.*

Proof: Let the columns of the $n \times m$ matrix $\mathbf{B}$ be a basis set for S_n, and the fact that $\mathbf{x}$ and $\mathbf{y}$ are both in S_n implies that there exist vectors $\mathbf{g}$ and $\mathbf{h}$ in E_n, such that $\mathbf{x} = \mathbf{Bg}$ and $\mathbf{y} = \mathbf{Bh}$. Also, since $\mathbf{y}$ is the projection of $\mathbf{x}$ into S_n, we must have $(\mathbf{y} - \mathbf{x})'\mathbf{B} = \mathbf{0}$, which implies $\mathbf{h}'\mathbf{B}'\mathbf{B} = \mathbf{g}'\mathbf{B}'\mathbf{B}$. If we multiply on the right by $\mathbf{B}^-$, and simplify, we obtain the result $\mathbf{h}'\mathbf{B}' = \mathbf{g}'\mathbf{B}'$, or, in other words, $\mathbf{x} = \mathbf{y}$, and the theorem is proved. ▮

Rather than projecting a single vector $\mathbf{x}$ into a vector space S_n, we are often interested in projecting every vector in E_n into a vector subspace S_n. When we do this, we say "E_n is projected into S_n." We shall now extend the previous ideas to include the projection of one vector space into another vector space.

Suppose that every vector $\mathbf{x}$ in E_n is transformed by the $n \times n$ matrix $\mathbf{A}$. We denote the resulting set by V, and V is defined by

$$V = \{\mathbf{y} : \mathbf{y} = \mathbf{Ax}, \mathbf{x} \in E_n\}.$$

Clearly V is a vector space. We are interested here in determining the matrix $\mathbf{A}$ such that each vector $\mathbf{x}$ in E_n is "projected" into the vector space V. To define this projection, we extend the ideas used above when a single vector was projected into a subspace.

Definition 12.5.1

Projection of E_n ***into a Vector Space*** V. *Let every vector* $\mathbf{x}$ *in* E_n $(n \geq 1)$ *be transformed into the vector space* V *by the matrix* $\mathbf{A}$, *such that the following two conditions hold.*

(1) *If* $\mathbf{x}$ *is transformed to* $\mathbf{y}$, *then the vectors* $\mathbf{y}$ *and* $\mathbf{x} - \mathbf{y}$ *are orthogonal (perpendicular); that is, for each* $\mathbf{x}$ *in* E_n, $\mathbf{Ax}$ *and* $\mathbf{x} - \mathbf{Ax}$ *are orthogonal.*
(2) *If* $\mathbf{x}$ *is a vector that is already in* V, *then it is transformed onto itself; that is,* $\mathbf{x} = \mathbf{Ax}$ *for all* $\mathbf{x} \in V$.

The transformation $\mathbf{y} = \mathbf{Ax}$ *is defined to be a projection of* E_n *into* V *if and only if the two conditions are satisfied.*

Note: When $\mathbf{A}$ is a matrix such that the definition is satisfied, we sometimes call $\mathbf{A}$ a projection matrix. However, this should be distinguished from "the projection of a vector."

We now state and prove some theorems about a projection matrix.

Theorem 12.5.3

An $n \times n$ matrix $\mathbf{A}$ *is a projection matrix if and only if* $\mathbf{A}$ *is a symmetric idempotent $n \times n$ matrix.*

Proof: If $\mathbf{A}$ is a symmetric idempotent $n \times n$ matrix, then, clearly, (1) and (2) of Def. 12.5.1 are satisfied, and hence $\mathbf{A}$ is a projection. Next assume that (1) and (2) of Def. 12.5.1 are satisfied. Condition (1) implies

$$\mathbf{x}'\mathbf{A}'(\mathbf{I} - \mathbf{A})\mathbf{x} = 0$$

for all $\mathbf{x}$ in E_n. By Theorem 12.2.19, $\mathbf{A}'(\mathbf{I} - \mathbf{A})$ must be skew symmetric; that is, $\mathbf{A}'(\mathbf{I} - \mathbf{A}) = -(\mathbf{I} - \mathbf{A}')\mathbf{A}$, from which we obtain $\mathbf{A}' + \mathbf{A} = 2\mathbf{A}'\mathbf{A}$. Since $\mathbf{Ax}$ is in V for every $\mathbf{x}$ in E_n, condition (2) implies that

$$\mathbf{Ax} = \mathbf{A}(\mathbf{Ax})$$

for every vector $\mathbf{x}$ in E_n, which in turn implies $\mathbf{A} = \mathbf{A}^2$; that is, $\mathbf{A}$ is idempotent. If we combine this result with the equation $\mathbf{A}' + \mathbf{A} = 2\mathbf{A}'\mathbf{A}$, we obtain $\mathbf{A} = \mathbf{A}'$, and this completes the proof of the theorem. ▮

The next theorem relates least squares and projections.

Theorem 12.5.4

The vector $\mathbf{x}_0$ *is an* LSS *solution to* $\mathbf{Ax} - \mathbf{g} = \mathbf{e}(\mathbf{x})$ *if and only if the vector* $\mathbf{Ax}_0$ *is a projection of the vector* $\mathbf{g}$ *into the column space of* $\mathbf{A}$.

Proof: If $\mathbf{x}_0$ is an LSS solution, it can be written as

$$\mathbf{x}_0 = \mathbf{Bg} + (\mathbf{I} - \mathbf{BA})\mathbf{h}$$

for some $n \times 1$ vector $\mathbf{h}$, where $\mathbf{ABA} = \mathbf{A}$, $\mathbf{AB} = \mathbf{B}'\mathbf{A}'$. So $\mathbf{Ax}_0 = \mathbf{ABg}$. But from the definition of $\mathbf{B}$ we find that $\mathbf{AB}$ is a symmetric idempotent matrix, and hence $\mathbf{Ax}_0$ is a projection of $\mathbf{g}$. Also $\mathbf{ABg} = \mathbf{A}(\mathbf{Bg})$ is a vector in the subspace spanned by the columns of $\mathbf{A}$, and hence $\mathbf{Ax}_0$ is a projection of $\mathbf{g}$ into the column space of $\mathbf{A}$. Next assume that $\mathbf{Ax}_0$ is the projection of $\mathbf{g}$ into the column space of $\mathbf{A}$. It follows then that for all $\mathbf{x} \in E_n$,

$$(\mathbf{Ax} - \mathbf{g})'(\mathbf{Ax} - \mathbf{g}) \geq (\mathbf{Ax}_0 - \mathbf{g})'(\mathbf{Ax}_0 - \mathbf{g}),$$

and hence $\mathbf{x}_0$ is an LSS to $\mathbf{Ax} - \mathbf{g} = \mathbf{e}(\mathbf{x})$. ▮

12.6 Additional Theorems

In this section we state some miscellaneous theorems that are directly or indirectly related to the material in this chapter.

The first theorem illustrates a relationship between P-matrices and positive definite matrices.

Theorem 12.6.1

Let $\mathbf{A}$ *be an* $n \times n$ *matrix. If* $\mathbf{A} + \mathbf{A}'$ *is positive definite, then* $\mathbf{A}$ *(and* $\mathbf{A}'$ *also) is a P-matrix (that is, all principal minors are positive).*

Proof: Since $\mathbf{A} + \mathbf{A}'$ is positive definite, it follows that $\mathbf{x}'(\mathbf{A} + \mathbf{A}')\mathbf{x} > 0$ for all $\mathbf{x} \neq \mathbf{0}$. This gives us $0 < \mathbf{x}'(\mathbf{A} + \mathbf{A})')\mathbf{x} = \mathbf{x}'\mathbf{A}\mathbf{x} + \mathbf{x}'\mathbf{A}'\mathbf{x} = 2\mathbf{x}'\mathbf{A}\mathbf{x}$ and hence $\mathbf{x}'\mathbf{A}\mathbf{x} > 0$ for all $\mathbf{x} \neq \mathbf{0}$. Thus by (2) of Theorem 11.2.3, $\mathbf{A}$ is a P-matrix if we use $\mathbf{D} = \mathbf{I}$. ∎

Theorem 12.6.2

Let $\mathbf{V}$ *be an* $n \times n$ *non-negative matrix of* rank $p > 0$ *and let* $\mathbf{X}$ *be an* $n \times p$ *matrix such that rank* $(\mathbf{X}'\mathbf{V}\mathbf{X}) = p$. *Then* $\mathbf{V} = \mathbf{V}\mathbf{X}(\mathbf{X}'\mathbf{V}\mathbf{X})^{-1}\mathbf{X}'\mathbf{V}$.

The proof of this theorem will be asked for in the problems.

The next five theorems contain some additional results about idempotent and tripotent matrices.

Theorem 12.6.3

Let $\mathbf{A}$ *be an* $n \times n$ *singular matrix. Then* $\mathbf{A}$ *can be written as the product of* n *idempotent matrices.*

The proof of this theorem will be omitted. It can be found in [B–1].

Theorem 12.6.4

If $\mathbf{C}$ *is an* $n \times n$ *symmetric tripotent matrix, then* $\mathbf{I} - \mathbf{C}^2$ *is an* $n \times n$ *symmetric idempotent matrix.*

The proof of this theorem and the next will be asked for in the problems.

Theorem 12.6.5

Let **A** *be an* $n \times n$ *symmetric matrix. There exists a nonsingular matrix* **R** *such that* **R′AR** *is a symmetric tripotent matrix.*

Theorem 12.6.6

Let **A** *be an* $n \times n$ *symmetric matrix. Then* rank $(\mathbf{I} - \mathbf{A}) = $ rank $(\mathbf{I}) -$ rank $(\mathbf{A})$ *if and only if* **A** *is an idempotent matrix.*

The proof will be left for the reader.

Theorem 12.6.7

Let **A** *be an* $n \times n$ *matrix. Then* **A** *is tripotent if and only if* rank $(\mathbf{A}) =$ rank $(\mathbf{A} + \mathbf{A}^2) +$ rank $(\mathbf{A} - \mathbf{A}^2)$.

We leave the proof for the reader. The proof of this theorem and others about tripotent matrices can be found in [A–5] and the references there.

The next several theorems concern *c*-inverses and *g*-inverses of matrices.

Theorem 12.6.8

$(\mathbf{AB})^c = \mathbf{B}^c\mathbf{A}^c$ *for any c-inverses if and only if* $\mathbf{A}^c\mathbf{ABB}^c$ *is idempotent.*

Proof: Assume $(\mathbf{A}^c\mathbf{ABB}^c)(\mathbf{A}^c\mathbf{ABB}^c) = \mathbf{A}^c\mathbf{ABB}^c$; multiply on the left and right by **A** and **B**, respectively, to get

$$(\mathbf{AB})(\mathbf{B}^c\mathbf{A}^c)(\mathbf{AB}) = \mathbf{AB}$$

and hence $\mathbf{B}^c\mathbf{A}^c = (\mathbf{AB})^c$. Next assume $(\mathbf{AB})^c = \mathbf{B}^c\mathbf{A}^c$.

We get $\mathbf{ABB}^c\mathbf{A}^c\mathbf{AB} = \mathbf{AB}$. Multiply on the left and right by $\mathbf{A}^c$ and $\mathbf{B}^c$, respectively, to get $(\mathbf{A}^c\mathbf{ABB}^c)(\mathbf{A}^c\mathbf{ABB}^c) = \mathbf{A}^c\mathbf{ABB}^c$ and the theorem is proved. ∎

Corollary 12.6.8

$(\mathbf{AB})^c = \mathbf{B}^c\mathbf{A}^c$ *if* $\mathbf{A}^c\mathbf{ABB}^c = \mathbf{BB}^c\mathbf{A}^c\mathbf{A}$.

Theorem 12.6.9

Let **A** *and* **B** *be* $m \times n$ *and* $n \times m$ *matrices, respectively, such that* **AB** *is idempotent and* rank $(\mathbf{AB}) =$ rank $(\mathbf{B})$. *Then* **A** *is a c-inverse of* **B**.

Proof: Assume $\mathbf{AB}$ and $\mathbf{B}$ have rank r and hence rank $(\mathbf{A}) \geq r$. Since $\mathbf{AB}$ is idempotent, we get $\mathbf{ABAB} = \mathbf{AB}$, which gives $\mathbf{AB}(\mathbf{I} - \mathbf{AB}) = \mathbf{0}$. We write $\mathbf{AB}$ into the full rank factorization of $\mathbf{A}$ and $\mathbf{B}$ to get $\mathbf{A} = \mathbf{A}_L\mathbf{A}_R$ and $\mathbf{B} = \mathbf{B}_L\mathbf{B}_R$ and finally $\mathbf{AB}_L\mathbf{B}_R = \mathbf{CB}_R$, where $\mathbf{C} = \mathbf{AB}_L$ is $m \times r$ of rank r. Thus we multiply both sides of $\mathbf{0} = \mathbf{AB}(\mathbf{I} - \mathbf{AB})$ on the left by $\mathbf{B}_L\mathbf{C}^-$ and get $\mathbf{0} = \mathbf{B}_L\mathbf{C}^-\mathbf{AB}(\mathbf{I} - \mathbf{AB}) = \mathbf{B}_L\mathbf{B}_R(\mathbf{I} - \mathbf{AB})$ and finally $\mathbf{B}(\mathbf{I} - \mathbf{AB}) = \mathbf{0}$ or $\mathbf{B} = \mathbf{BAB}$. So $\mathbf{A}$ is a c-inverse of $\mathbf{B}$. ▮

Theorem 12.6.10

Let $\mathbf{A}$ *and* $\mathbf{B}$ *be* $m \times n$ *and* $n \times m$ *matrices, respectively. Then* $\mathbf{A}$ *is a c-inverse of* $\mathbf{B}$ *if and only if* rank $(\mathbf{I} - \mathbf{AB}) =$ rank $(\mathbf{I}) -$ rank $(\mathbf{B}) = m -$ rank $(\mathbf{B})$.

The proof will be left for the reader. It can be found in [R–3] along with other necessary and sufficient conditions for a matrix to be a c-inverse.

The next theorem is a generalization of (2) of Theorem 12.2.14.

Theorem 12.6.11

Let $\mathbf{A}$, $\mathbf{B}$, *and* $\mathbf{A} - \mathbf{B}$ *be non-negative* $n \times n$ *matrices. A necessary and sufficient condition for* $\mathbf{B}^- - \mathbf{A}^-$ *to be a non-negative matrix is that* rank $(\mathbf{A}) =$ rank $(\mathbf{B})$.

The proof can be found in [M–10] along with some related results and some applications.

Theorem 12.6.12

Let $\mathbf{A}$ *be an* $n \times n$ *matrix with real characteristic roots* $\lambda_1, \ldots, \lambda_n$. *Then* k *of the roots are equal to* 1 *and* $n - k$ *of the roots are equal to zero if and only if*

$$\operatorname{tr}(\mathbf{A}^2) = \operatorname{tr}(\mathbf{A}^3) = \operatorname{tr}(\mathbf{A}^4) = k.$$

Proof: Assume $\operatorname{tr}(\mathbf{A}^2) = \operatorname{tr}(\mathbf{A}^3) = \operatorname{tr}(\mathbf{A}^4) = k$. This implies $0 = \operatorname{tr}(\mathbf{A}^2) - 2\operatorname{tr}(\mathbf{A}^3) + \operatorname{tr}(\mathbf{A}^4) = \Sigma\lambda_i^2 - 2\Sigma\lambda_i^3 + \Sigma\lambda_i^4 = \Sigma\lambda_i^2(1 - \lambda_i)^2$, which implies $\lambda_i = 0$ or $\lambda_i = 1$ for $i = 1, 2, \ldots, n$. But $\operatorname{tr}(\mathbf{A}^2) = k$ implies exactly k of the λ_i equal 1 and the remaining roots are zero. Further, if k of the roots of $\mathbf{A}$ are equal to 1 and the remaining roots are equal to zero, clearly $\operatorname{tr}(\mathbf{A}^2) = \operatorname{tr}(\mathbf{A}^3) = \operatorname{tr}(\mathbf{A}^4) = k$, so the proof is complete. ▮

Corollary 12.6.12.1

If all roots of $\mathbf{A}$ *are non-negative, then* k *of them are equal to* 1 *and* $n - k$ *are equal to zero if and only if* $\text{tr}(\mathbf{A}) = \text{tr}(\mathbf{A}^2) = \text{tr}(\mathbf{A}^3) = k$.

Corollary 12.6.12.2

Let $\mathbf{A}$ *be an* $n \times n$ *symmetric matrix and let* $\mathbf{\Sigma}$ *be an* $n \times n$ *non-negative matrix. Then* $\mathbf{A\Sigma}$ *has* k *characteristic roots equal to* 1 *and* $n - k$ *roots equal to zero if and only if* $\text{tr}(\mathbf{A\Sigma})^2 = \text{tr}(\mathbf{A\Sigma})^3 = \text{tr}(\mathbf{A\Sigma})^4 = k$.

Corollary 12.6.12.3

Let $\mathbf{A}$ *and* $\mathbf{\Sigma}$ *be non-negative* $n \times n$ *matrices. Then* $\mathbf{A\Sigma}$ *has* k *roots equal to* 1 *and* $n - k$ *roots equal to zero if and only if* $\text{tr}(\mathbf{A\Sigma}) = \text{tr}(\mathbf{A\Sigma})^2 = \text{tr}(\mathbf{A\Sigma})^3 = k$.

Problems

1: Show that if $\mathbf{A}$ is a positive definite 2×2 matrix,
   (1) $a_{11} + a_{22} - 2a_{12} > 0$,
   (2) $a_{11} + a_{22} + 2a_{12} > 0$,
   (3) $a_{22}(a_{11} + a_{22} - 2a_{12}) > (a_{12} - a_{22})^2$.
2. Let $\mathbf{A}$ and $\mathbf{B}$ be positive definite 2×2 matrices. Show

$$a_{11}b_{11} - 2a_{12}b_{12} + a_{22}b_{22} > 0.$$

3. If $\mathbf{A}$ is a positive definite $k \times k$ matrix, show that $a_{tt}a_{ss} > a_{ts}^2$ for all $t \neq s = 1, 2, \ldots, k$.
4. For the matrices below, determine whether each is positive definite, positive semidefinite, or neither. Use 1(a) and 1(b) of Theorem 12.2.2.

$$\mathbf{A} = \begin{bmatrix} 1 & 2 & -1 \\ 2 & 4 & -2 \\ -1 & -2 & 8 \end{bmatrix}; \quad \mathbf{B} = \begin{bmatrix} 2 & 1 & 1 \\ 1 & 1 & -1 \\ 1 & -1 & 5 \end{bmatrix}; \quad \mathbf{C} = \begin{bmatrix} 1 & 2 & 3 \\ 2 & 5 & 2 \\ 3 & 2 & 24 \end{bmatrix}.$$

5. Repeat Prob. 4, using 2(a) and 2(b) of Theorem 12.2.2.

6. Show that the matrix **A** below is positive definite, and find a matrix **P** such that $\mathbf{P}'\mathbf{P} = \mathbf{A}$.

$$\mathbf{A} = \begin{bmatrix} 1 & 0 & -1 \\ 0 & 2 & 1 \\ -1 & 1 & 2 \end{bmatrix}.$$

7. Find two non-negative, disjoint matrices **A** and **B** such that $\mathbf{C} = \mathbf{A} - \mathbf{B}$ where **C** is defined below.

$$\mathbf{C} = \begin{bmatrix} 2 & 3 \\ 3 & 2 \end{bmatrix}.$$

8. Let **A** and $\mathbf{A} + \mathbf{I}$ be nonsingular $k \times k$ matrices. Show that $\mathbf{A}^{-1} + \mathbf{I}$ is nonsingular.
9. If **A** is an orthogonal $k \times k$ matrix and $\mathbf{A} + \mathbf{I}$ is a nonsingular matrix, show that

$$(\mathbf{A} + \mathbf{I})^{-1} + [(\mathbf{A} + \mathbf{I})^{-1}]' = \mathbf{I}.$$

10. If **A** is a symmetric orthogonal matrix and $\mathbf{A} + \mathbf{I}$ is nonsingular, show that $\mathbf{A} = \mathbf{I}$.
11. If **A** is an orthogonal matrix and $\mathbf{A} + \mathbf{I}$ is nonsingular, show that **B** is a skew-symmetric matrix, where

$$\mathbf{B} = 2(\mathbf{A} + \mathbf{I})^{-1} - \mathbf{I}.$$

12. If **A** is a $k \times k$ skew-symmetric matrix and $\mathbf{A} + \mathbf{I}$ is nonsingular, show that **B** is an orthogonal matrix, where

$$\mathbf{B} = 2(\mathbf{I} + \mathbf{A})^{-1} - \mathbf{I}.$$

13. For any square matrix **A**, show that $\mathbf{A} + \mathbf{I}$ and $\mathbf{A} - \mathbf{I}$ commute.
14. If **P** is an orthogonal matrix such that $\mathbf{P}'\mathbf{A}\mathbf{P}$ is a diagonal matrix where **A** is a symmetric nonsingular matrix, show that $\mathbf{P}'\mathbf{A}^{-1}\mathbf{P}$ is also a diagonal matrix.
15. If **A** and **B** are $k \times k$ nonsingular matrices that commute, show that $\mathbf{A}^{-1}$ and $\mathbf{B}^{-1}$ also commute.
16. If **A** is any square matrix such that $\mathbf{I} + \mathbf{A}$ and $\mathbf{I} - \mathbf{A}$ are nonsingular, show that

$$2(\mathbf{I} - \mathbf{A})^{-1} - \mathbf{I} \quad \text{and} \quad 2(\mathbf{I} + \mathbf{A})^{-1} - \mathbf{I}$$

are nonsingular.

17. In Prob. 16, show that the inverse of $2(\mathbf{I} - \mathbf{A})^{-1} - \mathbf{I}$ is $2(\mathbf{I} + \mathbf{A})^{-1} - \mathbf{I}$.
18. If $\mathbf{A}$ and $\mathbf{I} + \mathbf{A}$ are $k \times k$ nonsingular matrices, show that

$$(\mathbf{A} + \mathbf{I})^{-1} + (\mathbf{A}^{-1} + \mathbf{I})^{-1} = \mathbf{I}.$$

19. Let $\mathbf{C}$ be defined by $c_{ij} = a_{ij}b_{ij}$, $i, j = 1, 2, 3$, where $\mathbf{A}$ and $\mathbf{B}$ are positive definite matrices defined below. Show that $\mathbf{C}$ is positive definite, and hence illustrate Theorem 12.2.8.

$$\mathbf{A} = \begin{bmatrix} 3 & 1 & -1 \\ 1 & 2 & 2 \\ -1 & 2 & 4 \end{bmatrix}; \qquad \mathbf{B} = \begin{bmatrix} 1 & -1 & 1 \\ -1 & 4 & 1 \\ 1 & 1 & 6 \end{bmatrix}.$$

20. Show with an example that the converse of Theorem 12.2.9 is not true.
21. Show that Theorem 12.2.10 may not be true if $\mathbf{A}$ is not positive definite.
22. For the matrices $\mathbf{A}$ and $\mathbf{B}$ in Prob. 19, show that
    (1) the characteristic roots of $\mathbf{AB}$ are real,
    (2) the roots of $|\mathbf{A} - \lambda\mathbf{B}| = 0$ are real.
    This illustrates Theorem 12.2.11.
23. If $\mathbf{A}$, $\mathbf{B}$, and $\mathbf{AB}$ are symmetric $k \times k$ matrices and $\mathbf{AB}$ is nonsingular, show that there exists an orthogonal matrix $\mathbf{P}$ such that $\mathbf{P}'\mathbf{ABP}$, $\mathbf{P}'\mathbf{A}^{-1}\mathbf{BP}$, $\mathbf{P}'\mathbf{AB}^{-1}\mathbf{P}$, and $\mathbf{P}'\mathbf{A}^{-1}\mathbf{B}^{-1}\mathbf{P}$ are diagonal matrices.
24. Let $\mathbf{A}$ be a $k \times k$ positive definite matrix, $\mathbf{a}$ be a $k \times 1$ vector, and a be a scalar such that $a > \mathbf{a}'\mathbf{A}^{-1}\mathbf{a}$; show that $\mathbf{A}^*$ is a positive definite matrix where $\mathbf{A}^*$ is defined by

$$\mathbf{A}^* = \begin{bmatrix} \mathbf{A} & \mathbf{a} \\ \mathbf{a}' & a \end{bmatrix}.$$

25. Let $\mathbf{A}$ be a $k \times k$ nonsingular matrix, let $\mathbf{a}$ be a $k \times 1$ vector, and let a be a scalar such that $a > \mathbf{a}'\mathbf{A}^{-1}\mathbf{a} \geq 0$. Show that the matrix $\mathbf{B}$ is nonsingular where $\mathbf{B}$ is defined by

$$\mathbf{B} = \mathbf{A} - \frac{1}{a}\mathbf{aa}'.$$

26. If $\mathbf{A}$ and $\mathbf{B}$ are non-negative $k \times k$ matrices, show that it is not always true that $\mathbf{x}'\mathbf{ABx} \geq 0$.
27. If $\mathbf{A} - \mathbf{B}$ is non-negative, show that it is not always true that $\mathbf{A}^2 - \mathbf{B}^2$ is non-negative.

28. A $k \times k$ matrix $\mathbf{A} = [a_{ij}]$ is defined to be *skew cross-symmetric* if $a_{ij} = -a_{k+1-i,\, k+1-j}$ for all $i = 1, \ldots, k$; $j = 1, 2, \ldots, k$. Show that the matrices **A** and **B** below are skew cross-symmetric.

$$\mathbf{A} = \begin{bmatrix} 0 & 1 & -3 \\ 0 & 0 & 0 \\ 3 & -1 & 0 \end{bmatrix}; \mathbf{B} = \begin{bmatrix} 1 & 1 & -3 \\ 2 & 0 & -2 \\ 3 & -1 & -1 \end{bmatrix}.$$

29. In Prob. 28, show that **AB** is a C-matrix (see Def. 8.15.3), and that $\mathbf{A} + \mathbf{B}$, $\mathbf{A} - \mathbf{B}$, and $\mathbf{A}'$ are skew cross-symmetric matrices.
30. Generalize the results of Prob. 29 to $k \times k$ skew cross-symmetric matrices.
31. Prove Theorem 12.2.8.
32. For the matrix **B** below, find a scalar t such that $\mathbf{B} + t\mathbf{I}$ is positive definite. See (1) of Theorem 12.2.14.

$$\mathbf{B} = \begin{bmatrix} -3 & 2 & 0 \\ 2 & 1 & 1 \\ 0 & 1 & -2 \end{bmatrix}.$$

33. Find a positive scalar t such that $\mathbf{A} + t\mathbf{B}$ is positive definite where **B** is defined in Prob. 32 and **A** is defined below. See (3) of Theorem 12.2.14.

$$\mathbf{A} = \begin{bmatrix} 3 & 1 & 0 \\ 1 & 1 & -1 \\ 0 & -1 & 4 \end{bmatrix}.$$

34. Show that the diagonal elements of a symmetric idempotent matrix **B** satisfy $b_{ii} \le 1$.
35. If $\mathbf{A} = \mathbf{A}'$ and $\mathbf{A}^n = \mathbf{A}^{n+2m-1}$, where m and n are any positive integers, show that **A** is symmetric idempotent.
36. Let $\mathbf{P} = \begin{bmatrix} \mathbf{P}_1 \\ \mathbf{P}_2 \end{bmatrix}$, where **P** is orthogonal. Show that $\mathbf{P}_1'\mathbf{P}_1$ is idempotent.
37. Show that $\mathbf{aa}'/\sum a_i^2$ is symmetric idempotent where $\mathbf{a} = [a_i]$ is an $n \times 1$ nonzero vector.
38. Let **1** be the $n \times 1$ unity vector and **I** the $k \times k$ identity matrix. Show that **A** is a symmetric idempotent $nk \times nk$ matrix where **A** is defined by $\mathbf{A} = (1/n)(\mathbf{1} \times \mathbf{I}) \times (\mathbf{1} \times \mathbf{I})'$.
39. Find an orthogonal matrix **P** such that $\mathbf{P}'(1/n)\mathbf{J}\mathbf{P} = \mathbf{D}$, where **D** is diagonal.
40. If $\mathbf{A}^2 = \alpha\mathbf{A}$, where α is a scalar, $\alpha \ne 0$, find the scalar β such that $\beta\mathbf{A}$ is idempotent.

41. Let $\mathbf{A}$ be an $n \times n$ symmetric idempotent matrix; show that, for every constant $\beta \neq -1$, $\mathbf{I} + \beta\mathbf{A}$ is nonsingular and find its inverse. If $\beta > -1$ show that $\mathbf{I} + \beta\mathbf{A}$ is positive definite.
42. If $\mathbf{A}$ is an $n \times n$ symmetric idempotent matrix, show that

$$\sum_{i=1}^{n} \sum_{j=1}^{n} a_{ij}^2 = \text{rank}\,(\mathbf{A}).$$

43. Let $\mathbf{A}$ be a symmetric $n \times n$ matrix of rank $n - 1$ such that $\mathbf{1}'\mathbf{A} = \mathbf{0}$; that is, every column of $\mathbf{A}$ adds to zero. Show that $\mathbf{B} = \mathbf{A} + (1/n)\mathbf{1}\mathbf{1}'$ is nonsingular and the inverse is $\mathbf{A}^- + (1/n)\mathbf{J}$.
44. Show that $\mathbf{A}^k + \mathbf{J}$ is nonsingular where $\mathbf{A}$ is defined in Prob. 43 and k is any positive integer.
45. Show that $\mathbf{C}$ is nonsingular where $\mathbf{C}$ is defined by

$$\mathbf{C} = \begin{bmatrix} \mathbf{A} & \mathbf{1} \\ \mathbf{1}' & 0 \end{bmatrix}$$

and $\mathbf{A}$ is defined in Prob. 43.
46. In Prob. 45 let $\mathbf{B} = \mathbf{C}^{-1}$ and partition $\mathbf{B}$ as

$$\mathbf{B} = \begin{bmatrix} \mathbf{B}_{11} & \mathbf{b}_{12} \\ \mathbf{b}_{21} & b \end{bmatrix},$$

where $\mathbf{B}_{11}$ is an $n \times n$ submatrix. Show that
(1) $b = 0$,
(2) $\mathbf{b}_{12} = (1/n)\mathbf{1}$,
(3) $\mathbf{1}'\mathbf{B}_{11} = \mathbf{0}$,
(4) $\mathbf{A}\mathbf{B}_{11}$ and $\mathbf{B}_{11}\mathbf{A}$ are idempotent.
47. Show that Theorem 12.3.4 is not necessarily true if $\mathbf{B}$ is an idempotent matrix, but not symmetric.
48. Prove Theorem 12.3.7.
49. Prove Theorem 12.3.8.
50. Exhibit a 3×3 matrix $\mathbf{A}$ of rank 2 that has two roots equal to $+1$ and one root equal to zero such that $\mathbf{A}$ is not idempotent. This illustrates the fact that (1) of Theorem 12.3.2 is not a sufficient condition for $\mathbf{A}$ to be idempotent.
51. Let $\mathbf{A}$ and $\mathbf{V}$ be $n \times n$ matrices and let $\mathbf{V}$ be nonsingular. If $\mathbf{AV}$ is idempotent, show that $\mathbf{VA}$ is also idempotent.
52. Let $\mathbf{A}$ and $\mathbf{V}$ be symmetric $n \times n$ matrices and let $\mathbf{V}$ be nonsingular. If $\mathbf{AV}$ is a symmetric idempotent matrix, show that $\mathbf{VA}$ is also a symmetric idempotent matrix.

53. Let **A** be a symmetric idempotent matrix. Prove that

$$-\frac{1}{2} \le a_{ij} \le \frac{1}{2} \quad \text{for all} \quad i \ne j.$$

54. Let **A** be a symmetric idempotent matrix. Prove that the sum of squares of the off-diagonal element of any row (or column) is less than or equal to 1/4.
55. If **A** is any $k \times k$ matrix such that $\mathbf{A}^n = \mathbf{A}^{n+1}$ for some positive integer n, show that $\text{tr}(\mathbf{A}) = \text{tr}(\mathbf{A}^2) = \text{tr}(\mathbf{A}^3) = \cdots = \text{tr}(\mathbf{A}^{n+1})$.
56. Let **V** be an $n \times n$ symmetric positive definite matrix such that $\mathbf{V} = \mathbf{P}'\mathbf{P}$, where **P** has size $n \times n$, and let **A** be a symmetric $n \times n$ matrix. Show that $\mathbf{PAP}'$ is a symmetric idempotent matrix if and only if **AV** is an idempotent matrix.
57. If **A** and **B** are symmetric $n \times n$ matrices and **AB** is idempotent (not necessarily symmetric), show that **BA** is also idempotent.
58. Let **A** be any $n \times n$ idempotent matrix and let t be any real number. Show that $(\mathbf{I} - \mathbf{A})(\mathbf{I} - t\mathbf{A})$ is an idempotent matrix.
59. Let **B** be a $k \times k$ matrix and **A** be a $k \times n$ matrix. Show that $\mathbf{A}'\mathbf{BA}$ is an idempotent matrix if $\mathbf{BAA}'$ is an idempotent matrix.
60. Show that the matrix **C** below is tripotent and find its characteristic roots.

$$\mathbf{C} = \frac{1}{12}\begin{bmatrix} 1 & 1 & -5 & 3 \\ 1 & 1 & -5 & 3 \\ -5 & -5 & 7 & 3 \\ 3 & 3 & 3 & -9 \end{bmatrix}.$$

61. In Prob. 60, show that $\mathbf{C}^2$ is a symmetric idempotent matrix.
62. In Prob. 60, find two matrices **A** and **B** that are symmetric, idempotent, and disjoint such that $\mathbf{C} = \mathbf{A} - \mathbf{B}$. See Theorem 12.4.3.
63. Use the result of Theorem 12.4.4 for the matrix **C** in Prob. 60 to show that $\text{rank}(\mathbf{C}) = 2$.
64. Let **P** be a $k \times k$ orthogonal matrix and let $\mathbf{P}_1$ and $\mathbf{P}_2$ be respectively k_1 and k_2 distinct rows from **P** such that $\mathbf{P}_1\mathbf{P}_2' = \mathbf{0}$. Show that **A** is a symmetric tripotent matrix where

$$\mathbf{A} = \mathbf{P}_1'\mathbf{P}_1 - \mathbf{P}_2'\mathbf{P}_2.$$

65. Let $\mathbf{C}_1$ and $\mathbf{C}_2$ be (symmetric) disjoint tripotent matrices. Show that $\mathbf{C}_1 + \mathbf{C}_2$ and $\mathbf{C}_1 - \mathbf{C}_2$ are (symmetric) tripotent matrices.

66. If $\mathbf{A}$ and $\mathbf{B}$ are positive definite $k \times k$ matrices, show that

$$|\mathbf{A}+\mathbf{B}|^{1/k} \geq |\mathbf{A}|^{1/k} + |\mathbf{B}|^{1/k},$$

where k is any positive integer.

67. Show that the rank of an $n \times n$ upper triangular idempotent matrix is equal to the number of nonzero diagonal elements.

68. Let $\mathbf{A}$ be a non-negative matrix. Show that $\mathbf{B}$ is positive definite for each and every a such that $0 < a \leq 1$ where

$$\mathbf{B} = a\mathbf{I} + (1-a)\mathbf{A}.$$

69. For any symmetric matrix $\mathbf{C}$, show that there exist two non-negative matrices $\mathbf{A}$ and $\mathbf{B}$ such that for each and every positive integer m

$$\mathbf{C}^m = \mathbf{A}^m + (-\mathbf{B})^m.$$

70. Let $\mathbf{C}_1$ and $\mathbf{C}_2$ be $k \times k$ symmetric disjoint matrices such that $\mathbf{C}_1 + \mathbf{C}_2$ is tripotent. Show that $\mathbf{C}_1$ and $\mathbf{C}_2$ are tripotent.

71. Let $\mathbf{A}$ be a $k \times k$ symmetric matrix and define δ_i by

$$\delta_1 = a_{11},\ \delta_2 = \begin{vmatrix} a_{11} & a_{12} \\ a_{21} & a_{22} \end{vmatrix}, \ldots, \delta_k = |\mathbf{A}|.$$

Show that $\delta_i \geq 0$ for $i = 1, 2, \ldots, k$ is not a sufficient condition for $\mathbf{A}$ to be non-negative.

72. Let $\mathbf{A}$ be a non-negative $k \times k$ matrix and define δ_i as in Prob. 71. Show that if $\delta_i = 0$ for $i = t$, then $\delta_i = 0$ for all $i > t$.

73. Prove Theorem 12.3.3 by using $\mathbf{A}^-$, the g-inverse of $\mathbf{A}$.

74. Let $\mathbf{A}$ be a $k \times n$ matrix. Show that $\mathbf{I} - 2\mathbf{A}\mathbf{A}^-$ is an orthogonal matrix. Show that $\mathbf{I} - 2\mathbf{A}\mathbf{A}^L$ is orthogonal for any L-inverse of $\mathbf{A}$.

75. Let $\mathbf{a}$ be a $k \times 1$ vector such that $\mathbf{a}'\mathbf{a} = 1$. Show that $\mathbf{I} - 2\mathbf{a}\mathbf{a}'$ is an orthogonal matrix.

76. If $\mathbf{P}$ is a symmetric orthogonal matrix, show that the characteristic roots of $\mathbf{P}$ are either $+1$ or -1. If $\mathbf{P} \neq \pm\mathbf{I}$, show that $\mathbf{P}$ has at least one root of each.

77. If $\mathbf{P}$ is an orthogonal symmetric matrix and $\mathbf{P} \neq \pm\mathbf{I}$, show that $\mathbf{I} + \mathbf{P}$ and $\mathbf{I} - \mathbf{P}$ are both singular.

78. Show that $\mathbf{C}$ is an idempotent matrix if and only if there exist two matrices $\mathbf{A}$ and $\mathbf{B}$, each symmetric idempotent, such that $\mathbf{C} = (\mathbf{AB})^-$. Show that $\mathbf{C} = \mathbf{BCA}$.

79. Let $\mathbf{C}$ be any tripotent matrix. Show that rank $(\mathbf{C})$ = rank $(\mathbf{C}^2)$.
80. If $\mathbf{A}$ is an $n \times n$ matrix and $\mathbf{A}^2$ has real, non-negative characteristic roots, show that $\mathbf{A}$ also has real roots.
81. Let $\mathbf{A}_{11}$ be any $n \times n$ symmetric idempotent matrix and let $\mathbf{B}$ be an $(m+n) \times (m+n)$ matrix defined by

$$\mathbf{B} = \begin{bmatrix} \mathbf{A}_{11} & \mathbf{A}_{12} \\ \mathbf{A}_{21} & \mathbf{A}_{22} \end{bmatrix}.$$

Show that $\mathbf{B}$ is a symmetric idempotent matrix if and only if $\mathbf{A}_{22}$ is a symmetric idempotent matrix and $\mathbf{A}_{12} = \mathbf{0}$.
82. Let $\mathbf{A}$ be a symmetric matrix. Show that if a c-inverse of $\mathbf{A}$ exists that is non-negative, then $\mathbf{A}$ must be non-negative.
83. Let $\mathbf{A}$ be any $n \times n$ non-negative matrix such that $\mathbf{A} = \mathbf{C}'\mathbf{C}$ where $\mathbf{C}$ has size $n \times n$ and let $\mathbf{B}$ be any c-inverse of $\mathbf{A}$. Show that $(\mathbf{CB})'(\mathbf{CB})$ is also a c-inverse of $\mathbf{A}$.
84. Show that an $n \times n$ matrix is idempotent if and only if its transpose is idempotent.
85. Show that $\mathbf{A}^-\mathbf{A} - \mathbf{A}'\mathbf{BA}$ is an idempotent matrix if $\mathbf{A}'\mathbf{BA}$ is an idempotent matrix.
86. If $\mathbf{A}$ is a symmetric idempotent matrix, show that $\mathbf{A}^-\mathbf{A} - \mathbf{A}$ and $\mathbf{AA}^- - \mathbf{A}$ are also symmetric idempotent matrices.
87. If $\mathbf{A}$ and $\mathbf{B}$ are each $n \times n$ symmetric idempotent matrices, show that tr $(\mathbf{AB}) \leq$ tr $(\mathbf{A})$.
88. In Prob. 87, show that tr $(\mathbf{AB})$ = tr $(\mathbf{A})$ if and only if $\mathbf{AB} = \mathbf{A}$.
89. If $\mathbf{C}_1$ and $\mathbf{C}_2$ are $n \times n$ symmetric matrices such that $\mathbf{C}_1\mathbf{C}_2 = \mathbf{0}$ and $\mathbf{C}_1 + \mathbf{C}_2$ is tripotent, show that $\mathbf{C}_1$ and $\mathbf{C}_2$ are each symmetric tripotent matrices.
90. If an $n \times n$ matrix $\mathbf{T}$ is an upper triangular, idempotent matrix with the first k diagonal elements equal to unity and the remaining diagonal elements equal to zero and $\mathbf{T}$ is partitioned so that

$$\mathbf{T} = \begin{bmatrix} \mathbf{T}_{11} & \mathbf{T}_{12} \\ \mathbf{0} & \mathbf{T}_{22} \end{bmatrix}$$

where $\mathbf{T}_{11}$ is a $k \times k$ matrix, show that $\mathbf{T}_{11} = \mathbf{I}$, $\mathbf{T}_{22} = \mathbf{0}$, and $\mathbf{T}_{12}$ is arbitrary.
91. If $\mathbf{A}$ and $\mathbf{B}$ are $n \times n$ non-negative matrices, show that

$$\sum_{j=1}^{n} \sum_{i=1}^{n} a_{ij} b_{ij} \geq 0.$$

92. In Prob. 91, show that if the equal sign holds, then $\mathbf{AB} = \mathbf{BA} = \mathbf{0}$.

93. If $\mathbf{A}$ is a positive definite $n \times n$ matrix, show for any positive integer m that $\mathbf{B}$ is positive definite where $b_{ij} = a_{ij}^m$.
94. If $\mathbf{A}$ is an $n \times n$ nonsingular matrix, $\mathbf{B}$ is an $m \times m$ matrix, and $\mathbf{C}$ is an $n \times m$ matrix, show that

$$|\mathbf{I} - \mathbf{ACBC}'| = |\mathbf{I} - \mathbf{C}'\mathbf{ACB}|.$$

95. If $\mathbf{A}$ is an $m \times n$ matrix and $\mathbf{B}$ is an $n \times m$ matrix such that $\mathbf{I} + \mathbf{BA}$ is nonsingular, show that $\mathbf{I} + \mathbf{AB}$ is also nonsingular and

$$(\mathbf{I} + \mathbf{AB})^{-1} = \mathbf{I} - \mathbf{A}(\mathbf{I} - \mathbf{BA})^{-1}\mathbf{B}.$$

96. If $\mathbf{A}$ is an $n \times n$ non-negative matrix and

$$\mathbf{A} = \begin{bmatrix} \mathbf{A}_{11} & \mathbf{A}_{12} \\ \mathbf{A}_{21} & \mathbf{A}_{22} \end{bmatrix}$$

where $\mathbf{A}_{11}$ is an $n_1 \times n_1$ matrix, show that $|\mathbf{A}| \le |\mathbf{A}_{11}||\mathbf{A}_{22}|$.
97. Let $\mathbf{A}$ be an $n \times n$ nonsingular matrix, $\mathbf{B}$ be an $n \times n$ symmetric matrix, and $\mathbf{AB}$ be an idempotent matrix. Show that $\mathbf{A}'\mathbf{B}$ is also an idempotent matrix.
98. If $\mathbf{A}$ is an $n \times n$ positive definite matrix, show that there exists a positive definite matrix $\mathbf{B}$ such that $\mathbf{A} = \mathbf{B}^2$.
99. Work Prob. 98 if the words positive definite are replaced by non-negative.
100. If $\mathbf{V}$ is an $n \times n$ positive definite matrix and $\mathbf{A}$ is a symmetric $n \times n$ matrix show that there exists a nonsingular matrix $\mathbf{R}$ such that $\mathbf{RAVR}^{-1} = \mathbf{D}$, where $\mathbf{D}$ is a diagonal matrix.
101. Let $\mathbf{X}$ be an $n \times p$ matrix such that $\mathbf{X} = [\mathbf{X}_1, \mathbf{X}_2]$, where $\mathbf{X}_1$ is $n \times p_1$ and $\mathbf{X}_2$ is $n \times p_2$ and $p_1 + p_2 = p$. Show that a c-inverse of $\mathbf{X}'\mathbf{X}$ is given by

$$(\mathbf{X}'\mathbf{X})^c = \begin{bmatrix} \mathbf{X}_1'\mathbf{X}_1 & \mathbf{X}_1'\mathbf{X}_2 \\ \mathbf{X}_2'\mathbf{X}_1 & \mathbf{X}_2'\mathbf{X}_2 \end{bmatrix}^c$$

$$= \begin{bmatrix} (\mathbf{X}_1'\mathbf{X}_1)^c + (\mathbf{X}_1'\mathbf{X}_1)^c\mathbf{X}_1'\mathbf{X}_2\mathbf{A}^c\mathbf{X}_2'\mathbf{X}_1(\mathbf{X}_1'\mathbf{X}_1)^c & -(\mathbf{X}_1'\mathbf{X}_1)^c(\mathbf{X}_1'\mathbf{X}_2)\mathbf{A}^c \\ -\mathbf{A}^c\mathbf{X}_2'\mathbf{X}_1(\mathbf{X}_1'\mathbf{X}_1)^c & \mathbf{A}^c \end{bmatrix},$$

where $\mathbf{A} = \mathbf{X}_2'[\mathbf{I} - \mathbf{X}_1(\mathbf{X}_1'\mathbf{X}_1)^c\mathbf{X}_1']\mathbf{X}_2$, where $\mathbf{A}^c$ is any c-inverse of $\mathbf{A}$ and where $(\mathbf{X}_1\mathbf{X}_1')^c$ is any c-inverse of $\mathbf{X}_1'\mathbf{X}_1$. (Note the similarity with Theorem 8.2.1.)
102. Let $\mathbf{V}$ be an $n \times n$ positive definite matrix. For each positive integer q, show that there exists an $n \times n$ unique positive definite matrix $\mathbf{B}$ such that $\mathbf{B}^q = \mathbf{V}$.

103. Consider the following Toeplitz matrix, which occurs in time series.

$$\mathbf{R} = \begin{bmatrix} 1 & \rho_1 & \rho_2 & \cdots & \rho_{n-1} \\ \rho_1 & 1 & \rho_1 & \cdots & \rho_{n-2} \\ \rho_2 & \rho_1 & 1 & \cdots & \rho_{n-3} \\ \vdots & & \vdots & & \vdots \\ \rho_{n-1} & \rho_{n-2} & \rho_{n-3} & \cdots & 1 \end{bmatrix}.$$

Is $\mathbf{R}$ positive definite for all ρ_i that satisfy $0 \le \rho_i < 1$?

104. If $\mathbf{A}$ and $\mathbf{B}$ are $n \times n$ symmetric matrices and if $\mathbf{A} + \mathbf{B}$ (or $\mathbf{A} - \mathbf{B}$) is positive definite, show that there exists a nonsingular matrix $\mathbf{R}$ such that $\mathbf{R}'\mathbf{A}\mathbf{R}$ and $\mathbf{R}'\mathbf{B}\mathbf{R}$ are each diagonal.

105. Show that if $\mathbf{A} = \mathbf{A}'$, then $\mathbf{A}$ is a tripotent matrix if and only if $\mathbf{A} = \mathbf{A}^-$.

106. Let $\mathbf{C}$ be an $n \times n$ matrix with characteristic roots -1, 0, and $+1$. Show that $\mathbf{C}$ is not necessarily tripotent.

107. Let $\mathbf{A}$ be an $n \times n$ matrix with characteristic roots 0 and $+1$. Show that $\mathbf{A}$ is not necessarily idempotent.

108. Prove Corollary 12.6.12.1.

109. Prove Corollary 12.6.12.2.

110. Let $\mathbf{A} = \mathbf{BC}$, where $\mathbf{A}$ is $n \times n$ of rank p and $\mathbf{C}$ is $p \times n$ of rank p. Show that $\mathbf{A}$ is idempotent if and only if $\mathbf{CB} = \mathbf{I}$.

111. Prove Theorem 12.3.21.

112. Prove Theorem 12.6.2.

113. Prove Theorem 12.6.4.

114. Prove Theorem 12.6.5.

115. If $\mathbf{A}$ and $\mathbf{B}$ are positive definite $n \times n$ matrices and $\mathbf{A} - \mathbf{B}$ is a non-negative $n \times n$ matrix, show that $\det(\mathbf{A}) \ge \det(\mathbf{B})$.

116. In Prob. 115, show that the roots of the polynomial $|\mathbf{A} - \lambda\mathbf{B}|$ are greater than or equal to 1.

117. Exhibit a 3×3 matrix with real characteristic roots such that $\mathrm{tr}(\mathbf{A}^2) = \mathrm{tr}(\mathbf{A}^3) = \mathrm{tr}(\mathbf{A}^4) = k$ and $\mathbf{A}$ is not idempotent.

References and Additional Readings

[A–1] Abramowitz, M., and I. Stegun. *Handbook of Mathematical Functions*. Washington, D.C.: National Bureau of Standards, Appl. Math. Ser. 55, 1970.

[A–2] Aitken, A. C. "On the Statistical Independence of Quadratic Forms in Normal Variates." *Biometrika, 37* (1950), 93–96.

[A–3] Anderson, T. W. *An Introduction to Multivariate Statistical Analysis*. New York: Wiley, 1958.

[A–4] Anderson, T. W. *The Statistical Analysis of Time Series*. New York: Wiley, 1971.

[A–5] Anderson, T. W., and George P. H. Styan. "Cochran's Theorem, Rank Additivity, and Tripotent Matrices." Tech. Report No. 43, Aug. 1980, Dept. of Statistics, Stanford Univ.

[B–1] Ballantine, C. S. "Products of Idempotent Matrices." *Linear Algebra and Its Applications, 19* (1978), 81–86.

[B–2] Bellman, Richard. *Introduction to Matrix Analysis*. New York: McGraw-Hill, 1960.

[B–3] Ben-Israel, A., and A. Charnes. "Contributions to the Theory of Generalized Inverses." *J. Soc. Indust. Appl. Math., 11* (1963), 667–699.

[B–4] Ben-Israel, A., and S. J. Wersan. "An Elimination Method for Computing the Generalized Inverse of an Arbitrary Complex Matrix." *J. of ACM, 10* (1963), 532–537.

[B–5] Ben-Israel, A. "An Iterative Method of Computing the Generalized Inverse of an Arbitrary Matrix." *Math. Comp., 10* (1965), 452–455.

[B–6] Ben-Israel, A., and T. N. E. Greville. *Generalized Inverses: Theory and Applications*. New York: Wiley, 1974.

[B–7] Berman, A., and R. J. Plemmons. "Eight Types of Matrix Monotonicity." *Linear Algebra and Its Applications, 13* (1976), 115–123.

[B–8] Berman, Abraham, and Robert J. Plemmons. *Nonnegative Matrices in the Mathematical Sciences*. New York: Academic Press, 1979.

[B–9] Bjerhammar, A. "Rectangular Reciprocal Matrices with Special Reference to Geodetic Calculations." *Bull. Geodesique, 52* (1951), 188–220.

[B–10] Boot, J. C. G. "The Computation of the Generalized Inverse of Singular or Rectangular Matrices." *Am. Math. Monthly, 70* (1963), 302–303.

[B–11] Bose, R. C. Analysis of Variance. Unpublished lecture notes, Univ. of North Carolina, Chapel Hill, 1959.

[B–12] Boullion, Thomas L., and Patrick L. Odell. *Generalized Inverse Matrices*. New York: Wiley, 1971.

[B–13] Bush, K. A., and I. Olkin. ''Extrema of Quadratic Forms with Applications to Statistics.'' *Biometrika, 46* (1959), 484–486.

[C–1] Campbell, S. L., and C. D. Meyer, Jr. *Generalized Inverses of Linear Transformations*. London: Pitman, 1979.

[C–2] Carpenter, O. ''Note on the Extension of Craig's Theorem to Non-Central Variates.'' *Ann. Math. Statist., 21* (1950), 455.

[C–3] Chipman, John S. ''On Least Squares with Insufficient Observations.'' *J. Am. Statist. Assoc., 59* (1964), 1078–1111.

[C–4] Chipman, J. S., and M. M. Rao. ''The Treatment of Linear Restrictions in Regression Analysis.'' *Econometrica, 32* (1964), 198–209.

[C–5] Cline, Randall E. ''Representations for the Generalized Inverse of a Partitioned Matrix.'' *J. Soc. Indust. Appl. Math., 12* (1964), 588–600.

[C–6] Cline, Randall E. ''Note on the Generalized Inverse of the Product of Matrices.'' *SIAM Rev., 6* (1964), 57–58.

[C–7] Cline, R. E. ''Representations for the Generalized Inverses for Sums of Matrices.'' *J. of SIAM, Num. An.*, Ser. B, *2* (1965), 99–114.

[C–8] Cochran, W. G. ''The Distribution of Quadratic Forms in a Normal System.'' *Proc. Camb. Philos. Soc., 30* (1934), 178.

[C–9] Cord, Marian S., and Richard J. Sylvester. ''The Property of Cross-Symmetry.'' *J. Soc. Indust. Appl. Math., 10* (1962), 632–637.

[C–10] Cornish, E. A. ''An Application of the Kronecker Product of Matrices in Multiple Regression.'' *Biometrics, 13* (1957), 19–27.

[C–11] Craig, A. T. ''Note on the Independence of Certain Quadratic Forms.'' *Ann. Math. Statist., 14* (1943), 195–197.

[D–1] Davis, Philip J. *Circulant Matrices*. New York: Wiley, 1979.

[D–2] Decell, Henry P., Jr. ''An Alternate Form of the Generalized Inverse of an Arbitrary Complex Matrix.'' *SIAM Rev., 7* (1965), 356–358.

[D–3] Decell, Henry P., Jr. ''An Application of the Cayley-Hamilton Theorem to Generalized Matrix Inversion.'' *SIAM Rev., 7* (1965), 526–528.

[D–4] Deemer, Walter L., and Ingram Olkin. ''The Jacobians of Certain Matrix Transformations Useful in Multivariate Analysis.'' *Biometrika, 38* (1951), 345–367.

[D–5] Drazin, M. P. ''On Diagonal and Normal Matrices.'' *Quart. J. Math., 2* (1951), 189–198.

[D–6] Dwyer, Paul S., and M. S. MacPhail. ''Symbolic Matrix Derivatives.'' *Ann. Math. Statist., 19* (1948), 517–534.

[D–7] Dwyer, Paul S. ''Some Applications of Matrix Derivatives in Multivariate Analysis.'' *J. Am. Statist. Assoc., 62* (1967), 607–625.

[E–1] Erdelyi, Ivan. ''On Partial Isometries in Finite-Dimensional Euclidean Spaces.'' *J. Soc. Indust. Appl. Math., 14* (1966), 453–467.

[F–1] Faddeev, D. K., and V. N. Faddeeva. *Computational Methods of Linear Algebra.* San Francisco: Freeman, 1963.

[F–2] Ferrar, W. L. *Algebra, A Textbook of Determinants, Matrices, and Algebraic Forms.* London: Oxford Univ. Press, 1941.

[F–3] Fiedler, M., and V. Ptak. ''On Matrices with Non-Positive Off-Diagonal Elements and Positive Principal Minors.'' *Czech. Math. J., 12* (1962), 382–400.

[F–4] Fiedler, M., and V. Ptak. ''Some Generalizations of Positive Definiteness and Monotonicity.'' *Numer. Math., 9* (1966), 163–172.

[F–5] Frame, J. S. ''Matrix Operations and Generalized Inverses.'' *IEEE Spectrum* (1964), 209–220.

[F–6] Frazer, R. A., W. J. Duncan, and A. R. Collar. *Elementary Matrices.* London: Cambridge Univ. Press, 1963.

[G–1] Gantmacher, F. R. *The Theory of Matrices,* vols. 1 and 2. New York: Chelsea, 1959.

[G–2] Good, I. J. ''The Inverse of a Centrosymmetric Matrix.'' *Technometrics, 12* (1970), 925–928.

[G–3] Gray, Robert M. ''Toeplitz and Circulant Matrices: II.'' Tech. Report No. 6504-1, Apr. 1977, Stanford University.

[G–4] Graybill, F. A., and G. Marsaglia. ''Idempotent Matrices and Quadratic Forms in the General Linear Hypothesis.'' *Ann. Math. Statist., 28* (1957), 678–686.

[G–5] Graybill, Franklin A. *An Introduction to Linear Statistical Models,* vol. 1. New York: McGraw-Hill, 1961.

[G–6] Graybill, F. A., C. D. Meyer, and R. J. Painter. ''Note on the Computation of the Generalized Inverse of a Matrix.'' *SIAM Rev., 8* (1966), 522–524.

[G–7] Graybill, Franklin A. *Theory and Application of the Linear Model.* North Scituate, Mass.: Duxbury Press, 1976.

[G–8] Greenberg, B. G., and A. E. Sarhan. ''Matrix Inversion, Its Interest and Application in Analysis of Data.'' *J. Am. Statist. Assoc., 54* (1959), 755–766.

[G–9] Grenander, Ulf, and Murray Rosenblatt. *Statistical Analysis of Stationary Time Series.* New York: Wiley, 1957.

[G–10] Grenander, U., and G. Szegö. *Toeplitz Forms and Their Applications*. Berkeley: Univ. of California Press, 1958.

[G–11] Greville, T. N. E. "The Pseudo Inverse of a Rectangular or Singular Matrix and Its Application to the Solution of Systems of Linear Equations." *SIAM Rev., 1* (1959), 38–43.

[G–12] Greville, T. N. E. "Some Applications of the Pseudoinverse of a Matrix." *SIAM Rev., 2* (1960), 15–22.

[G–13] Greville, T. N. E. "Note on Fitting of Functions of Several Independent Variables." *J. Soc. Indust. Appl. Math., 9* (1961), 109–115.

[H–1] Hawkins, D., and H. A. Simon. "Note: Some Conditions of Macroeconomic Stability." *Econometrica, 17* (1949), 245–248.

[H–2] Hedayat, A., and W. D. Wallis. "Hadamard Matrices and Their Applications." *Ann. Statist., 6* (1978), 1184–1238.

[H–3] Hogg, R. V., and A. T. Craig. "On the Decomposition of Certain χ^2 Variables." *Ann. Math. Statist., 29* (1959), 608–610.

[H–4] Householder, A. S. *The Theory of Matrices in Numerical Analysis*. Waltham, Mass.: Ginn/Blaisdell, 1964.

[H–5] Hsu, P. L. "On Symmetric, Orthogonal, and Skew-Symmetric Matrices." *Proc. Math. Soc. Edinburgh,* Ser 2. (1948), 34–44.

[I–1] Ingham, A. E. "An Integral which Occurs in Statistics." *Proc. Camb. Philos. Soc., 29* (1933), 271–276.

[J–1] John, Peter W. M. "Pseudoinverses in the Analysis of Variance." *Ann. Math. Statist., 35* (1964), 895–896.

[J–2] John, P. W. M. *Statistical Design and Analysis of Experiments*. New York: Macmillan, 1971.

[K–1] Kurkjian, B., and M. Zelen. "Applications of the Calculus of Factorial Arrangements I: Block and Direct Product Designs." *Biometrika, 50* (1963), 63–73.

[L–1] Lancaster, H. O. "Traces and Cumulants of Quadratic Forms in Normal Variables." *J. Roy. Statist. Soc.* (B), *16* (1954), 247–254.

[L–2] Lewis, T. O., T. L. Boullion, and P. L. Odell. "A Bibliography on Generalized Matrix Inverses." *Proceedings of a Symposium on the Theory and Application of Generalized Inverses of Matrices,* Math. Ser. No. 4, 1968, Texas Technological College, Lubbock.

[L–3] Luther, Norman Y. "Decomposition of Symmetric Matrices and Distributions of Quadratic Forms." *Ann. Math. Statist., 36,* 683–690.

[M–1] MacDuffee, C. C. *The Theory of Matrices*. New York: Chelsea, 1946.

[M–2] Magnus, Jan R., and H. Neudecker. "The Commutation Matrix: Some Properties and Applications." *Ann. Statist., 7* (1979), 381–394.

[M–3] Marcus, M., and H. Minc. *A Survey of Matrix Theory and Matrix Inequalities.* Boston: Allyn & Bacon, 1964.

[M–4] Marcus, Marvin, and Henryk Minc. *Introduction to Linear Algebra.* New York: Macmillan, 1965.

[M–5] Marsaglia, George. "Conditional Means and Covariances of Normal Variables with Singular Covariance Matrix." *J. Am. Statist. Assoc., 59* (1964), 1203–1204.

[M–6] Meek, D. S. "A New Class of Matrices with Positive Inverses." *Linear Algebra and Its Applications, 15* (1976), 253–260.

[M–7] Meiman, Louise Bradt. "Circulants." Unpublished Masters Thesis, 1967, Colorado State Univ.

[M–8] Mentz, Raul Pedro. "On the Inverse of Some Covariance Matrices of Toeplitz Type." *J. of SIAM, 31* (Nov. 1976), 426–437.

[M–9] Meyer, C. D., Jr., and M. W. Stadelmaier. "Singular M-Matrices and Inverse Positivity." *Linear Algebra and Its Applications, 22* (1978), 139–156.

[M–10] Milliken, George A., and Fikri Akdeniz. "A Theorem on the Difference of the Generalized Inverses of Two Nonnegative Matrices." *Commun. Statist.—Theor. Meth.,* A6, *1* (1977), 73–79.

[M–11] Moore, E. H. "On the Reciprocal of the General Algebraic Matrix." Abstract, *Bull. Am. Math. Soc., 26* (1920), 394–395.

[M–12] Moore, E. H. "General Analysis, Part 1." *Mem. Am. Philos. Soc., 1* (1935), 197–209.

[M–13] Mustafi, Chandan K. "The Inverse of a Certain Matrix, with an Application," *Ann. Math. Statist., 38* (1967), 1289–1292.

[N–1] Neudecker, H. "Some Theorems on Matrix Differentiation with Special Reference to Kronecker Matrix Products." *Am. Statist. Assoc. J., 64* (Sept. 1969), 953–963.

[N–2] Newcomb, Robert W. "On the Simultaneous Diagonalization of Two Semidefinite Matrices." *Quart. Appl. Math., 19* (1960), 144–146.

[O–1] Olkin, Ingram. Note on "The Jacobians of Certain Matrix Transformations Useful in Multivariate Analysis." *Biometrika, 40* (1953), 43–46.

[O–2] Olkin, Ingram. "A Class of Integral Identities with Matrix Argument." *Duke Math. J., 26* (1959), 207–214.

[P–1] Penrose, R. "A Generalized Inverse for Matrices." *Proc. Camb. Philos. Soc., 51* (1955), 406–413.

[P–2] Penrose, R. "On Best Approximate Solutions of Linear Matrix Equations." *Proc. Camb. Philos. Soc., 52* (1956), 17–19.

[P–3] Perron, O. "Grundlagen für eine Theorie des Jacobischen Kettenbruchalgorithmus." *Math. Ann., 64* (1907), 1–76.

[P–4] Poole, George, and Thomas Boullion. "A Survey of M-Matrices." *SIAM Rev.*, *16* (Oct. 1974), 419–427.

[P–5] Pringle, R. M., and A. A. Rayner. *Generalized Inverse Matrices with Applications to Statistics*. London: Griffin, 1971.

[P–6] Pyle, L. D. "Generalized Inverse Computations Using the Gradient Projection Method." *J. of ACM, 11* (1964), 422–428.

[R–1] Raktoe, B. L., A. Hedayat, and W. T. Federer. *Factorial Designs*. New York: Wiley, 1981.

[R–2] Rao, C. R. "A Note on a Generalized Inverse of a Matrix with Applications to Problems in Mathematical Statistics." *J. Roy. Statist. Soc.*, Ser. B., *24* (1962), 152–158.

[R–3] Rao, C. R., and S. K. Mitra. *Generalized Inverse of Matrices and Its Applications*. New York: Wiley, 1971.

[R–4] Rao, C. Radhakrishna. *Linear Statistical Inference and Its Applications,* 2nd ed. New York: Wiley, 1973.

[R–5] Ray, W. D. "The Inverse of a Finite Toeplitz Matrix." *Technometrics, 12* (Feb. 1970), 153–156.

[R–6] Rayner, A. A., and R. M. Pringle. "A Note on Generalized Inverses in the Linear Hypothesis Not of Full Rank." *Ann. Math. Statist., 38* (1967), 271–273.

[R–7] Rehnqvist, Lars. "Inversion of Certain Symmetric Band Matrices." *BIT, 12* (1972), 90–98.

[R–8] Rohde, Charles A. "Generalized Inverses of Partitioned Matrices." *J. Soc. Indust. Appl. Math., 13* (1965), 1033–1035.

[R–9] Rosen, J. B. "Minimum and Basic Solutions to Singular Linear Systems." *J. Soc. Indust. Appl. Math., 12* (1964), 156–162.

[R–10] Roy, S. N. *Some Aspects of Multivariate Analysis*. New York: Wiley, 1957.

[R–11] Roy, S. N., and A. E. Sarhan. "On Inverting a Class of Patterned Matrices." *Biometrika, 43* (1956), 227–231.

[R–12] Roy, S. N., B. G. Greenberg, and A. E. Sarhan. "Evaluation of Determinants, Characteristic Equations and Their Roots for a Class of Patterned Matrices." *J. Roy. Statist. Soc.*, Ser. B, *22* (1960), 348–359.

[S–1] Scroggs, James E., and Patrick L. Odell. "An Alternate Definition of a Pseudoinverse of a Matrix." *J. SIAM, 14* (1966), 796–810.

[S–2] Searle, S. R. "Additional Results Concerning Estimable Functions and Generalized Inverse Matrices." *J. Roy. Statist. Soc.*, Ser. B, *27* (1965), 486–490.

[S–3] Searle, S. R. *Linear Models*. New York: Wiley, 1971.

[S–4] Seber, G. A. F. *Linear Regression Analysis*. New York: Wiley, 1977.

[S–5] Shah, B. V. "On a Generalization of the Kronecker Product Designs." *Ann. Math. Statist., 30* (1959), 48–54.

[S–6] Shanbhag, D. N. "On the Distribution of a Quadratic Form." *Biometrika, 57* (1970), 222.

[S–7] Smith, Ronald L. "Moore-Penrose Inverses of Block Circulant and Block *k*-Circulant Matrices." *Linear Algebra and Its Applications, 16* (1977), 237–245.

[S–8] Stein, F. Max. *Introduction to Matrices and Determinants*. Belmont, Calif.: Wadsworth, 1967.

[V–1] Varga, Richard S. *Matrix Iterative Analysis*. Englewood Cliffs, N.J.: Prentice-Hall, 1962.

[V–2] Varga, Richard S. "On Recurring Theorems on Diagonal Dominance." *Linear Algebra and Its Applications, 13* (1976), 1–9.

[W–1] Widom, H. "Toeplitz Matrices." In *Studies in Modern Analysis*. Washington, D.C.: Math. Assoc. of America, 1965.

[Z–1] Zohar, S. "Toeplitz Matrix Inversion: The Algorithm of W. F. Trench." *J. Assoc. Comput. Mach., 16* (1969), 592–601.

Index